Essentials of Bridge Engineering

Sixth Edition

Essentials of Bridge Engineering

Sixth Edition

David Johnson Victor
Former Professor of Civil Engineering
Indian Institute of Technology Madras, Chennai

Oxford & IBH Publishing Co. Pvt. Ltd.
New Delhi
(*A Unit of* **CBS Publishers & Distributors** Pvt Ltd)

CBS Publishers & Distributors Pvt Ltd

New Delhi • Bengaluru • Chennai • Kochi • Kolkata • Mumbai
Hyderabad • Jharkhand • Nagpur • Patna • Pune • Uttarakhand

Essentials of
Bridge Engineering
Sixth Edition

ISBN-13: 978-81-204-1717-5
ISBN-10: 81-204-1717-8

Reprint: 2017, 2019, 2021

OXFORD & IBH
New Delhi
(A Unit of CBS Publishers & Distributors Pvt Ltd)

Published by Satish Kumar Jain and produced by Varun Jain for
CBS Publishers & Distributors Pvt Ltd
4819/XI Prahlad Street, 24 Ansari Road, Daryaganj, New Delhi 110 002, India.
Ph: 011-23289259, 23266861, 23266867 Website: www.cbspd.com
Fax: 011-23243014 e-mail: delhi@cbspd.com; cbspubs@airtelmail.in.

Corporate Office: 204 FIE, Industrial Area, Patparganj, Delhi 110 092
Ph: 011-4934 4934 Fax: 011-4934 4935 e-mail: publishing@cbspd.com; publicity@cbspd.com

Branches

- **Bengaluru:** Seema House 2975, 17th Cross, K.R. Road, Banasankari 2nd Stage, Bengaluru 560 070, Karnataka
  Ph: +91-80-26771678/79 Fax: +91-80-26771680 e-mail: bangalore@cbspd.com
- **Chennai:** 7, Subbaraya Street, Shenoy Nagar, Chennai 600 030, Tamil Nadu
  Ph: +91-44-26680620, 26681266 Fax: +91-44-42032115 e-mail: chennai@cbspd.com
- **Kochi:** 42/1325, 1326, Power House Road, Opp KSEB, Kochi 682 018, Kerala, India
  Ph: +91-484-4059061-65,67 Fax: +91-484-4059065 e-mail: kochi@cbspd.com
- **Kolkata:** 6/B, Ground Floor, Rameswar Shaw Road, Kolkata-700014 (West Bengal), India
  Ph: +91-33-2289-1126, 2289-1127, 2289-1128 e-mail: kolkata@cbspd.com
- **Mumbai:** 83-C, Dr E Moses Road, Worli, Mumbai-400018, Maharashtra
  Ph: +91-22-24902340/41 Fax: +91-22-24902342 e-mail: mumbai@cbspd.com

Representatives

- Hyderabad 0-9885175004
- Jharkhand 0-9811541605
- Nagpur 0-9421945513
- Patna 0-9334159340
- Pune 0-9623451994
- Uttarakhand 0-9716462459

Printed at Chaman Enterprises, Daryaganj, New Delhi, India

To My Daughters

Esther Malini and Nalini Prema

Foreword to the First Edition

The importance of transportation for the prosperity of any country cannot be overemphasized; and its realisation by the Planning Commission of our country is reflected in the increased allocation of funds for transportation projects in the national Five Year Plans. A major portion of the expenditure on roads and railways will be on the construction of major and minor bridges and urban traffic structures like grade-separated intersections. It is, therefore, evident that we have a heavy bridge building programme ahead of us.

The subject of bridge engineering is being given greater emphasis in the new curricula for engineering studies at the first degree and the postgraduate levels. Considerable difficulty has, however, been experienced in the past in finding suitable textbooks which emphasize and describe the current professional practice obtaining in the country. Dr. Victor's book seems to be well designed to bridge this gap and meet the long felt need.

This book discusses the essential details of all the components of a complete bridge, and is perhaps unique in this respect. The author has introduced a new notation for detailing reinforcement and nomograms for the design of slab bridges. He has incorporated in the text many useful ideas from his past experience. The illustrative examples should be of much value for a beginner to grasp the application of the code regulations to actual design situations. The book is developed in such a manner as to prepare a student to fit into a bridge design or construction organisation without much difficulty.

Dr. Victor has been active in the area of bridge engineering by way of teaching, research, publications and organisation of special short courses and technical reports. He deserves support for this book in its attempt to present a professional approach to the subject with accent on modern trends in bridge design and construction in this country.

P. SIVALINGAM
Director of Technical Education
Government of Tamil Nadu

Chennai
February 1973

Preface to the Sixth Edition

It is very encouraging to note that the book has been adopted as a text book for the subject in most engineering colleges and technical institutions in India. As a senior citizen with varied experience during his successful career, the author derives satisfaction at the opportunity to make a humble contribution to his profession through the publication of this book first in 1973 and since then through the second to the fifth editions to guide numerous engineering students in this country.

The sixth edition now presented contains additional information on many aspects of bridge engineering. The text is brought up-to-date with reference to revisions of relevant standard codes of practice and in conformity with the advances in field practice.

As in the earlier editions, this book emphasizes the essential concepts of the entire field of bridge engineering. It presents an overview of the principles and practice of bridge design, construction and maintenance in a simple and concise manner, as the book is primarily addressed to students in engineering colleges and engineers in the early part of their career. With a view to encourage the serious reader to enhance his knowledge of the subject, a number of actual applications having special features have been mentioned along with reference to the source of literature.

Though the current trend in design offices is towards the use of computer programs for design and drawing, this text presents design calculations in a manner suitable for hand calculations, because the author firmly believes that prospective bridge engineers should first understand the essential concepts thoroughly and should develop a feel for the approximate shape and size of the various components before attempting designs using available software.

A book of this nature has to necessarily refer to published material from numerous sources, including those which are specifically mentioned at the end of each chapter. The author gratefully records his indebtedness to the sources in the technical literature and specially to the authors of the above references.

The author is also grateful to the many colleagues in his profession, practicing engineers and recent graduates, who communicated to him encouraging comments regarding the text and helpful suggestions for its improvement. Sincere thanks are due to the publishers for their effective cooperation.

Chennai
March 2007

D. Johnson Victor

Preface to the First Edition

Bridge engineering is one of the most fascinating fields in Civil Engineering, calling for expertise in many areas, ranging from surveying to statistics, runoff calculations to rubble masonry, steel to structural concrete, and materials to modern methods of construction. A successful Bridge Engineer has to have an appreciation of aesthetics and economics, besides ability in analysis and dexterity in design. The materials and procedures involved in the construction of any sizeable bridge are quite varied. For instance, a prestressed concrete road bridge would require a proprietary system of prestressing, high-grade concrete and high-tensile steel for girders, normal reinforced concrete for deck slab, stone masonry for substructure, piling or caissons for foundations, neoprene for bearings, bituminous mastic and copper sheet for expansion joints, aluminium tubing for road signs and lighting posts, steel and wooden shuttering, different construction machinery, etc. It is, therefore, evident that a textbook of the present size cannot attempt to cover exhaustively the entire field of bridge engineering, but can only highlight the essentials of the subject.

This book is primarily intended as a textbook for a two-semester course in bridge engineering at the university level in India. Reference has been made to the relevant codes of practice, and the treatment reflects the professional practice as obtaining currently in this country. To the knowledge of the author, no parallel textbook with this type of coverage is presently available; and it was to fill this gap that the author ventured to write this book. It is hoped that the teacher would find the book convenient for classroom use, while the serious student would use it as a beginner's text. It would also serve as a reference material for the practicing engineer.

This book has been compiled from the lecture notes of the author prepared for teaching a one-semester course in Bridge Engineering at the Master of Technology level at the Indian Institute of Technology Madras, Chennai. The contents have also been used in a short term course organised by the author at the above Institute in January 1972 for the benefit of practicing engineers. The author has benefited by the comments of the students and participants in the above courses.

The treatment of the subject as given in the book and some of the ideas expressed are original with the author. However, the bulk of the basic information has been derived from many sources including those specially mentioned at the end of each chapter. The author gratefully acknowledges his indebtedness to these earlier writers.

The author records his gratitude to Mr. P. Sivalingam, Director of Technical Education, Government of Tamil Nadu, Chennai for kindly writing the Foreword for this book and for his encouragement. The author also acknowledges his indebtedness to Professor Dr. P.C. Varghese, Head of the Department of Civil Engineering, Indian Institute of Technology Madras, Chennai for helpful guidance and valuable suggestions.

Sincere thanks are due to the author's wife, Leena, for her patience and understanding in putting up with the inconveniences when the author was engaged with the preparation of this text.

The author would appreciate suggestions from the readers for improvement of the text.

Chennai
February 1973

D.J. Victor

Contents

Chapter 1

Introduction

1.1 Definition and Basic Forms

A bridge is a structure providing passage over an obstacle without closing the way beneath. The required passage may be for a road, a railway, pedestrians, a canal or a pipeline. The obstacle to be crossed may be a river, a road, railway or a valley.

There are six basic forms of bridge structures: beam bridges, truss bridges, arch bridges, cantilever bridges, suspension bridges and cable stayed bridges.

The beam bridge carries vertical loads by flexure. The truss bridge of simple span behaves like a beam because it carries vertical loads by bending. The top chords are in compression, and the bottom chords are in tension, while the vertical and diagonal members are either in tension or compression depending on their orientation. Loads are carried primarily in compression by the arch bridge, with the reactions at the supports (springing) being both vertical and horizontal forces. A cantilever bridge generally consists of three spans, of which the outer spans, known as anchor spans, are anchored down to the shore, and these cantilever over the channel. A suspended span is rested at the ends of the two cantilevers, and acts as a simply supported beam or truss. The cantilevers carry their loads by tension in the upper chords and compression in the lower chords. A suspension bridge carries vertical loads from the deck through curved cables in tension. These loads are transferred to the ground through towers and through anchorages. In a cable stayed bridge, the vertical loads on the deck are carried by the nearly straight inclined cables which are in tension. The towers transfer the cable forces to the foundation through vertical compression. The tensile forces in the stay cables induce horizontal compression in the deck. The above description presents the basic concepts of the different bridge forms in a simplified manner.

1.2 Components of a Bridge

The main parts of a bridge structure are:

(a) Decking, consisting of deck slab, girders, trusses, etc.;
(b) Bearings for the decking;
(c) Abutments and piers;
(d) Foundations for the abutments and the piers;
(e) River training works, like revetment for slopes for embankment at abutments, and aprons at river bed level;
(f) Approaches to the bridge to connect the bridge proper to the roads on either side; and
(g) Handrails, parapets and guard stones.

Some of the components of a typical bridge are shown in Fig. 1.1. The components

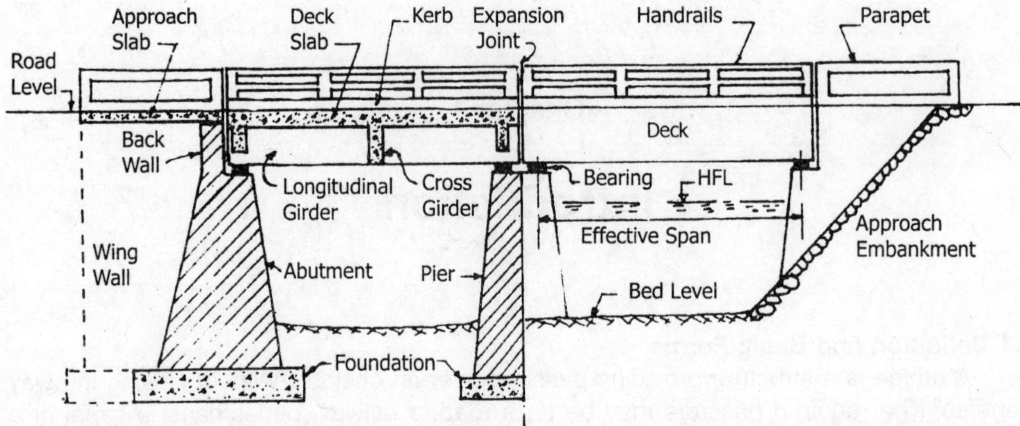

Figure 1.1 Components of a Typical Bridge.

above the level of bearings are grouped as superstructure, while the parts below the bearing level are classed as substructure. The portion below the bed level of a river bridge is called the foundation. The components below the bearing and above the foundation are often referred as substructure.

1.3 Classification

Bridges may be classified in many ways, as below:

(a) According to function as aqueduct (canal over a river), viaduct (road or railway over a valley), pedestrian, highway, railway, road-cum-rail or a pipeline bridge.

(b) According to the material of construction of superstructure as timber, masonry, iron, steel, reinforced concrete, prestressed concrete, composite or aluminium bridge.

(c) According to the form or type of superstructure as slab, beam, truss, arch, cable stayed or suspension bridge.

(d) According to the inter-span relations as simple, continuous or cantilever bridge.

(e) According to the position of the bridge floor relative to the superstructure as deck, through, half-through or suspended bridge.

(f) According to the method of connections of the different parts of the superstructure, particularly for steel construction, as pin-connected, riveted or welded bridge.

(g) According to the road level relative to the highest flood level of the river below, particularly for a highway bridge, as high-level or submersible bridge.

(h) According to the method of clearance for navigation as high-level, movable- bascule, movable-swing, movable-lift or transporter bridge.

(i) According to the length of bridge (total length between the inner faces of dirtwalls) as culvert (less than 6 m), minor bridge (6 to 60 m), major bridge (above 60 m) or a long span bridge when the main span of the major bridge is above 120 m.

(j) According to degree of redundancy as determinate or indeterminate bridge.

(k) According to the anticipated type of service and duration of use as permanent, temporary, military (pontoon, Bailey) bridge.

These classifications are not mutually exclusive. Any one type may overlap with others. For example, a multi-span highway bridge may consist of steel trusses of the through type and may be a high-level bridge over a river.

1.4 Importance of Bridges

Thomas B. Macaulay once said: "Of all inventions, the alphabet and the printing press alone excepted, those inventions which abridge distance have done the most for the civilization of our species". Since ancient times, bridges have been the most visible testimony to the contribution of engineers. Bridges have always figured prominently in human history. Cities have sprung up at a bridgehead or where at first a river could be forded at any time of the year; examples are London, Oxford, Cambridge, and Innsbruck. Some bridges embody the spirit and character of a people or a place, as the Brooklyn bridge for New York City, the Golden Gate Bridge for San Fransisco, the Tower bridge for London, the Harbour bridge for Sydney and the Howrah bridge for Kolkata. Bridges add beauty to cities, e.g., the bridges across the river Seine in Paris and the bridges across the river Thames in London. They enhance the vitality of the cities and aid the social, cultural and economic improvements of the areas around them.

Great battles have been fought for cities and their bridges. The mobility of an army at war is often affected by the availability or otherwise of bridges to cross rivers. That is why military training puts special emphasis on learning how to destroy bridges during combat and while retreating and how to build new ones quickly while advancing.

A bridge is an important element in a transportation system, as its capacity governs the capacity of the system, its failure or defective performance will result in serious disruption of traffic flow, and also the cost per km of bridge structure is many times that for the road or rail track on either side of the bridge. It is prudent, therefore, to devote special attention in design to ensure adequate strength and durability, consistent with safety and cost.

1.5 Historical Development

1.5.1 Early Bridges

The history of development of bridge construction is closely linked with the history of human civilization[1-4]. The efficiency and sophistication of design and the ingenious construction procedures kept pace with the advances in science, materials and technology. Fig. 1.2 outlines the development of the various forms. Nature fashioned the first bridges. The tree fallen accidentally across a stream was the earliest example of a beam type bridge. Similarly, the natural rock arch formed by erosion of the loose soil below and the creepers hanging from tree to tree allowing monkeys to cross from one bank to the other were the earliest forebears of the arch and the suspension bridges, respectively. The primitive man imitated nature and learned to build beam and suspension bridges. The ancient who felled a tree deliberately so that it fell across a stream and afforded him a crossing was the first bridge builder. Since the primitive man was a wanderer in search of food and shelter from the elements, the first structures he built were bridges.

Around 4000 BC, men were settling down to community life and were giving more thought for permanence of bridges. The lake dwellers in Switzerland pioneered the timber trestle construction which led later to timber bridges. At this time, the prototype of the

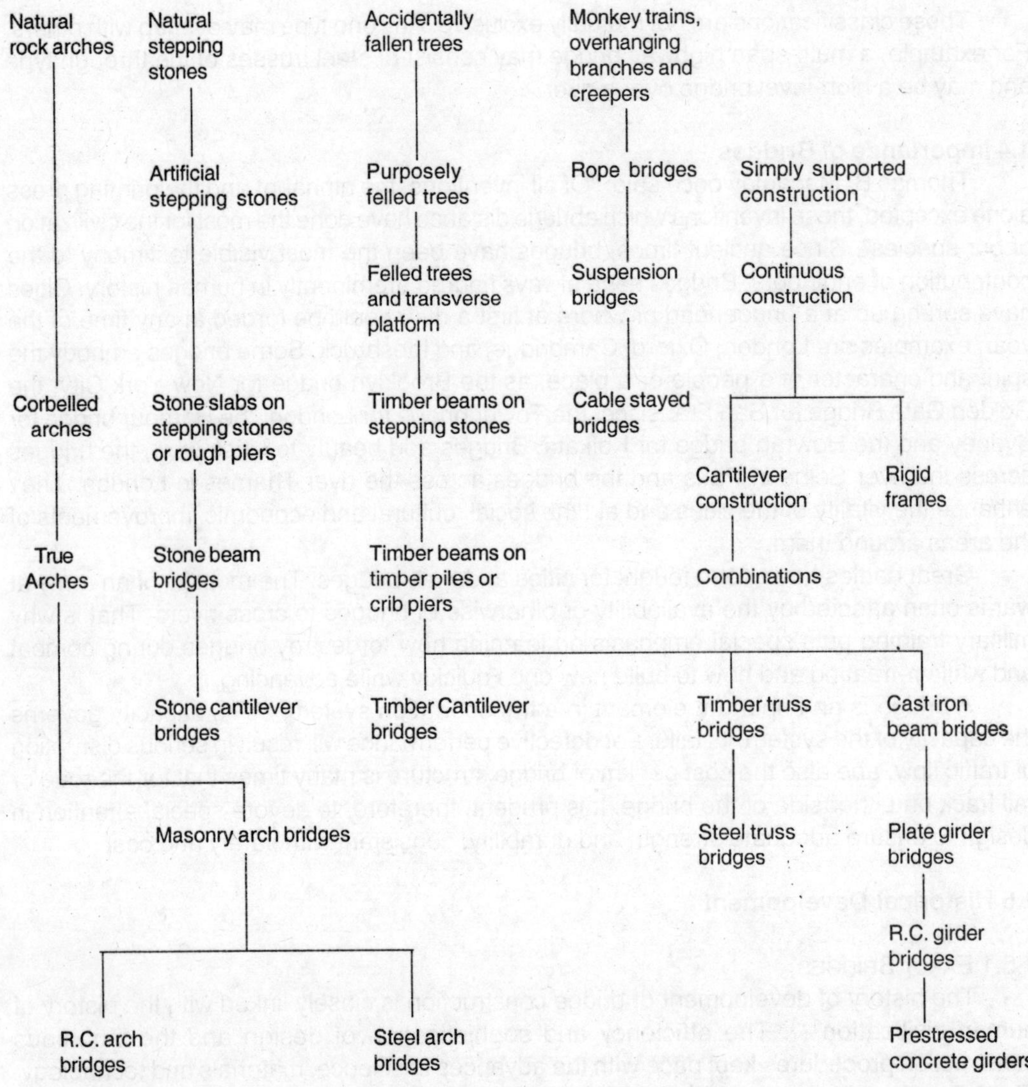

Figure 1.2 Development of Bridges.

modern suspension bridge was developed in India and China. Two parallel ropes were stretched between two banks and a level pedestrian platform was hung. India was also the birthplace of the cantilever bridge. Planks of wood, which were anchored at the two banks with heavy stones, were corbelled out progressively towards the midstream until the gap could be spanned by a single plank. During this time, the Mesopotamians developed the true arch bridge, the arch rib being made up of bricks or stones on end. The oldest bridge still standing is a pedestrian stone slab bridge across the Meles River in Smyrna, Turkey, which according to legend was used by the ancient Greek epic poet Homer and is at least 2800 years old.

1.5.2 Stone Masonry Bridges

The Romans were the pioneers in the art of bridge building. They introduced four significant developments: the discovery and use of natural cement; the development of the cofferdam; widespread use of semi-circular masonry arch; and the concept of public works. Between 200 BC and 260 AD, the Romans built many magnificent stone arch bridges. The arches were semi-circular using massive piers, so that, if one span got damaged in war, the others would remain standing. One famous bridge of this period is the Ponte Milvio across the Tiber river in Rome built in 109 BC and still in use[5]. Another magnificent example of Roman bridge building is the aqueduct, the Pont du Gard, built at Nimes, France, in 14 AD, consisting of three tiers of semi-circular arches composed of large rectangular stones assembled without mortar. The bridge expresses simplicity with repetition, symmetry and pleasant proportions. Yet another fine example of Roman bridge building is the Alcantara bridge at Toledo in Spain built by Caius Julius Lacer for Emperor Trajan in 98 AD. The arches have 29 m spans, with voussoirs accurately shaped without need for mortar for the joints. It is a plain, unadorned structure, but noble in its proportions and majestic in its simplicity. There is an inscription in Latin over the central tower: "I have left a bridge that shall remain for eternity". This epitomises the typical passionate desire of the Roman for eternal fame. Bridge construction was considered so important in ancient Rome that the Roman emperors adopted the title 'Pontifex Maximus', meaning 'Chief Bridge builder'.

The Chinese were building stone arch bridges since 250 B.C. The Zhaozhou Bridge (also known as the Anji bridge) built in 605 AD is a notable long-lived vehicular bridge today[6]. Situated about 350 km south-west of Beijing, it is a stone arch bridge of a single span of 37.6 m and rise of 7.2 m, with a roadway width of 9 m. The main arch ring is 1.03 m thick. Two small arches of 3.8 m and 2.85 m clear span penetrate each of the spandrels, and these serve to drain flood waters and also to reduce the bridge weight. The secret of its longevity was that the voussoirs were dressed exactly to match and there was no mortar joint. The form employed in this bridge was rarely seen in Europe prior to the late-16th century. In 1991, the Zhaozhou bridge was named among the world's cultural relics.

In the middle ages after the fall of Rome, the bridge building activity in Europe was mainly taken up by the religious orders. Medieval bridges adopted the pointed arch. With the pointed arch, the tendency to sag at the crown is less dangerous and the horizontal thrust at the springing is less. The Pont d'Avignon with 20 arch spans of about 34 m each built by St. Benezet over the Rhone river in 1188 and the Old London bridge across Thames river with 19 pointed arch spans of varied lengths built by Peter Colechurch in 1209 were examples of this period. Segmental curves were first used in Europe in the Ponte Vecchio Bridge, built at Florence, Italy in 1345. The medieval bridges were loaded with decorative and defensive towers, chapels, statues, shops and dwellings.

The dawn of Renaissance witnessed advances in theory, technical skill and mechanical appliances. In his treatise 'Four Books of Architecture' published in 1520, Palladio proposed the use of truss systems in bridge design. Galileo Galilei wrote the first book on the science of structural analysis in 1638 entitled Dialoghi delle Nouve Scienze (Dialogues on the New Science)[3]. Bridges were regarded as civic works of art and the bridge builder was recognised as a leader in progress and a creator of monuments. Stone masonry segmental arches were predominantly used in bridges. The Rialto Bridge in Venice, Italy, built in 1591 with a single segmental arch span of 27 m and a rise of 6 m, is typical of this period. Another magnificent example of this period is the Khaju bridge at Isfahan in Iran (1667) with eighteen pointed arches, carrying a 26 m wide roadway. The Bridge combined architecture and engineering in perfect harmony. It served also as a dam and included cool rooms for travelers.

The eighteenth century ushered in the age of reason. The first treatise on bridge engineering was published in 1716 by Henri Gautier, a French engineer. In 1716, the Corps des Ingenieurs de Ponts et Chaussees was founded for the scientific advancement of bridge construction; and in 1747, the Ecole de Ponts et Chaussees at Paris, the first bridge engineering school in the world, was founded, with Jean Perronet, 'Father of modern bridge building', as the first Director. Perronet perfected the masonry arch and introduced slender piers, his best work being the Pont de la Concorda at Paris built in 1791. John Rennie in England designed the New London bridge across Thames river using segmental masonry arches and the same was completed by his son in 1831. Another fine stone arch bridge is the Grosvenor bridge built by Harrison in 1832 with a span of 61 m. Many masonry arch bridges have been built in India adopting small spans, mainly for the railways, e.g. Thane railway bridge (1853).

1.5.3 Timber Bridges

Timber bridges served as an economic alternative during the years prior to the twentieth century, and were built adopting several structural types. The 400 m long timber trestle bridge across the Rhine built by Julius Caesar in 55 BC within a time span of ten days was a masterpiece of bold design and efficient site organization. The wooden arches of Japan (e.g. Kintaiko bridge, 1673) are examples of the primitive ingenuity and craft technology. Timber bridges have been used in Kashmir as cantilever bridges. Covered timber bridges, covered for protection of the timber from the weather, were popular during the late eighteenth century. Hans Grubenmann built a covered timber truss bridge with spans of 51 m and 58 m over the Rhine at Schaffhausen in 1755. Another notable timber bridge was the 'Colossus' bridge over the Schuylkill river at Fairmount, Pennsylvania, with an arch span of 104 m. This fine bridge, built in 1812 by Louis Wernwag, was destroyed by fire in 1838.

1.5.4 Iron Bridges

The Industrial Revolution ushered in the use of iron in bridges in place of stone and timber. The first iron bridge was built at Coalbrookdale in 1779 over the Severn in England by Abraham Darby and John Wilkinson. It consisted of five semicircular arch ribs in cast iron, joined together side by side to form a single arch span of 30 m. The construction details of the iron bridge followed the spirit of timber and masonry construction practice.

Wrought iron replaced cast iron in bridge construction during the period 1800-90. Wrought iron was ductile, malleable and strong in tension. In 1808, James Finley in Pennsylvania patented a design for a suspension bridge with wrought iron chain cables and level floor. Forty bridges of this design were built in USA within the next eight years. Wrought iron chains were used for a suspension bridge built by Thomas Telford across the Menai Straits in Wales in 1826 with a record-breaking span of 177 m. The Menai Straits bridge was the world's first iron suspension bridge for vehicles and also the world's first iron suspension bridge over sea water. Isambard Kingdom Brunel designed the Clifton suspension bridge in Bristol, UK in 1850 with a main span of 183 m using wrought iron chains, and the work was completed in 1864.

George Stephenson built the first iron railway bridge, the Gauntlet Viaduct, in 1823 on the Stockton-Darlington railway. The most famous of the early iron railway bridges is the

Britannia tubular bridge built by Robert Stephenson in 1850 across the Menai Strait. The bridge consisted of twin wrought iron tubes, continuous over four spans of 70, 140, 140 and 70 m. The design for this bridge was evolved using theory and model tests, and was the fore-runner for the present-day practice for the design of suspension bridges. The bridge continued in service till it was damaged by fire in 1970. Many truss bridges of the form Howe, Pratt, Whipple, Bolman, Fink, and Warren were built on the railways during this period. A notable example of the use of wrought iron is the Royal Albert bridge at Saltash, UK built in 1859 by Brunel.

A number of wrought iron railway bridges failed causing serious loss of lives. The most remembered failures are those of the Ashtabula, Ohio Howe truss bridge in 1876 killing 65 people and the Firth of Tay lattice girder bridge in Scotland in 1879 resulting in the death of 75 persons. The Ashtabula bridge collapsed due to derailment in a snow storm. The Tay bridge disaster was attributed to poor quality castings, unsuitable joints and inadequate wind resistance. These failures highlighted the inadequacy of wrought iron for bridges and led to a new era in bridge building, an era of specialization, research, careful detailing, thorough inspection and a more durable and stronger material - steel.

In 1884, Gustave Eiffel built the Garabit Viaduct (railway bridge) over the Truyere river near St. Flour, France. The 165 m main span of the bridge consisted of a crescent shaped two-hinged arch of wrought iron and represented the world's longest arch span when completed. The wide lateral spacing of the arch in plan towards the hinged supports increased the lateral stiffness of the structure, besides enhancing the visual impact of the crescent form. The arch gets narrower but deeper as it rises from the hinges. As mentioned by Billington[7], "the form is handsome in pure profile (its two-dimensional aspect) but in addition it provides visual surprise and delight from different perspectives (its three-dimensional aspect)".

1.5.5 Steel Arch and Truss Bridges

Though steel is said to have been known in China by 200 BC and in India by 500 BC, its widespread use materialized only in the latter half of nineteenth century after the discovery of the Bessemer process in 1856. Steel was first used extensively in the Eads Bridge at St. Louis, Missouri, built in 1874 as a steel arch bridge of three spans of 153, 158 and 153 m. The Eads Bridge, designed by James Buchanan Eads, was also the first bridge to use pneumatic caissons in USA, the first bridge to make extensive use of cantilever method of erection, and the first to specify and test for elastic limit and ultimate strength for steel. The first all-steel bridge was built at Glasgow, South Dakota in 1878. Steel was also used in the cables and spans of Brooklyn bridge during 1869-83. The Forth Railway bridge built by Baker and Fowler in 1889 featured steel plates of 51 MPa tensile strength. The bridge has two cantilevered spans of 513 m each. The Alexandre III bridge over Seine in Paris, France was built in 1900 with a main span of 107 m, adopting shallow arch ribs made up of moulded steel segments.

Inspired by the success of the Eads bridge, many fine steel arch bridges were built. Notable among these are the Hell Gate bridge at New York with a span of 297 m built in 1917, the Bayonne bridge (1931) with 504 m span and the Sydney Harbour bridge at Sydney, Australia, with a span of 503 m built in 1932. The deck type arch span is aesthetically the most pleasing. The Henry Hudson bridge built in 1936 with a span of 244 m and the Rainbow bridge at Niagara Falls built in 1941 with a span of 290 m are outstanding examples of beautiful steel arch bridges. The world's longest steel arch bridge has been the New River Gorge bridge in West Virginia built in 1977 using weathering steel with a span of 519 m. Currently, the Lupu bridge in Shanghai, built in 2004 with a span of 550 m, is the world's longest steel arch span bridge. Rising like a beautiful rainbow across the Huangpu river, the Lupu bridge adopted a combination of construction techniques of the arch bridge, the cable stayed bridge and the suspension bridge.

Consequent on the introduction of steel, the earlier truss forms yielded place to more efficient forms such as the Baltimore, Parker, Pennsylvania and K-truss types. In India, many major bridges were built with steel decks in the late nineteenth century and early twentieth century to carry the railway tracks across the major rivers. An example of such a bridge is the Upper Sone Railway bridge built in 1899 with an overall length of 3.1 km consisting of 93 spans of 30.5 m each. Truss bridges have been used economically in the span range of 100 to 200 m mainly for railway bridges. The world's longest simple steel truss bridge span is the suspended span of J.J. Barry bridge across Delaware river in USA built in 1973 with a span of 251 m.

1.5.6 CANTILEVER BRIDGES

The world's first modern cantilever bridge was built in 1867 by Heinrich Gerber across the river Main at Hassfurt, Germany, with a main span of 129 m. The world's most famous cantilever bridge is the Firth of Forth bridge in Scotland, built in 1889 with two main spans of 521 m. This bridge was designed by John Fowler and Benjamin Baker soon after the failure of the Tay bridge. So the design wind pressure was adopted conservatively as 2.68 kN/m^2, which was about 5.5 times the estimated wind pressure that toppled the Tay bridge. The bridge featured steel plates of specified minimum ultimate tensile strength of 463 MPa (30 tons/in^2). The designers had to devise innovative methods to present the concept of the cantilever bridge to the financiers. The maintenance of this bridge, particularly the needed routine painting to prevent rusting of the steel members, is a continuous and costly task.

The world's longest span cantilever bridge was built in 1917 at Quebec, over the St. Lawrence river, with a main span of 549 m. The first attempt to construct this bridge ended in failure due to miscalculation of the dead load and buckling of the web plates of the compression chord members near the south pier. The design was then revised and the structure was rebuilt. The Howrah bridge over the Hooghly river at Kolkata, built in 1943 with a main span of 457 m, has elegant aesthetics and possesses pleasing proportions among the suspended span, cantilever arms and the anchor spans. It was a notable achievement at the time of its construction. The Osaka Port bridge (also known as the Minato bridge), completed in 1974 with a main span of 510 m, applied advanced techniques of welding and was built with precision and without any accident. In view of the larger weight and labour involved compared with a cable stayed bridge of the same clear span, the cantilever bridge is not very popular at present.

1.5.7 BOX GIRDER BRIDGES

Developments in welding technology and precision gas cutting techniques in the post second world war period facilitated the economical fabrication of monolithic structural steel box girders characterized by the use of thin stiffened plates and the closed form of cross section. The Cologne Deutz bridge built in 1952 was the world's first slender steel box girder bridge. Box girder bridges have exceptional torsional rigidity resulting in better transverse load distribution. Hence box girder steel bridges were constructed in many parts of the world. However, three major failures occurred under construction during 1970-71: at Milford Haven, UK and Melbourne, Australia in 1970 and at Koblenz, Germany in 1971. These collapses were attributable more to inefficient detailing of steelwork than to incorrect design of the effective section. After a number of studies, stringent requirements for design and workmanship have been prescribed. The Rio-Niteroi bridge in Brazil (1976) with 200-300-200 m spans is a record–span steel box girder bridge.

1.5.8 Reinforced Concrete Bridges

Concrete was used in 1840 for a 12 m span bridge across the Garoyne Canal at Grisoles in France. The first reinforced concrete bridge was built by Adair in 1871 as a 15 m span bridge across the Waveney at Homersfield, England. Soon after, a 6 m arch was built in 1889 at Golden Gate Park in San Francisco and a girder bridge was built in 1893 by Hennebique as an approach to a mill at Don, France. The adaptability of reinforced concrete to any architectural form and the increased efficiency in concrete construction resulted in its widespread use in bridge building. Robert Maillart in Switzerland produced fine arch bridges in reinforced concrete, utilizing the integrated structural action of thin arch slabs with monolithically cast stiffening beams, e.g. Salginatobel bridge and Schwandbach bridge (1930 and 1933, respectively). These bridges by Maillart are classical examples of structural art, exemplifying the design criteria: efficient use of materials, economy in cost and enhanced aesthetic expression. Another elegant arch bridge is the Bixby Creek bridge in California built in 1932 with a main span of 109.7 m and deck at 79.2 m above the stream bed. The longest span concrete arch bridge is the Wanxiang bridge in China built in 1996 with a span of 420 m and a rise-to-span ratio of 1:5.

The use of reinforced concrete for road bridges has become popular in India since the beginning of the twentieth century[8]. The bridge types adopted include: (a) simply supported slabs; (b) simply supported T-beam spans; (c) balanced cantilever with suspended spans; (d) arch and bow string girder; and (e) continuous or framed structure. Solid slab simply supported bridges were common in the 1920s. T-beam bridges have been used widely in the span range of 10 m to 25 m. Elegant arch bridges were built during 1920 to 1950. The Dum Dum bridge at Kolkata, built in 1926 with two arches of 24 m each, is the first major reinforced concrete arch bridge in India. The Coronation bridge (1941) across Teesta river in Bengal with a main arch span of 81.7 m and rise of 39.6 m is a particularly elegant structure, which has also proved to be durable. The Third Godavari Railway bridge built in 1996 with 28 spans of 97.5 m is a recent example of elegant concrete bowstring girder bridges.

1.5.9 Prestessed Concrete Bridges

The application of prestressing to concrete was pioneered by Eugene Freyssinet in France. He demonstrated the concept of artificial introduction of beneficial stresses in concrete members by the use of high strength steel tendons. His early works of Le Veurdre bridge (67.5-72-62.5 m spans) in 1911, Plougastel bridge (3 spans of 180 m) in 1930 and the Marne bridge at Esbly (74 m span) in 1939 represented path breaking efforts in concrete development. The first prestressed concrete (PSC) bridge in USA was the Walnut Lane bridge (22.6 + 48.8 + 22.6 m spans) in Philadelphia, PA built in 1949, while the first PSC bridge in Canada was the Mosquitto Creek bridge in Vancouver, BC, constructed in 1952. The Bendorf bridge over the Rhine in Germany with a main span of 208 m built in 1965 by Finsterwalder using free cantilever method with cast-in-situ concrete and short high tensile bar tendons marked a breakthrough in prestressed concrete bridge construction. Jean Muller designed the Choisy bridge in France in 1965 as the first PSC bridge using match cast precast segments and epoxy joints[9]. The Confederation bridge in Canada built in 1997 is a spectacular achievement in prestressed concrete. The bridge has a total length of 12.9 km. The main bridge has 44 spans of 250 m, involving extensive use of precast components, high-performance concrete and innovative construction techniques. The share of prestressed concrete bridges in the total of all new bridges has grown steadily since 1950 : e.g. the share in USA in 1994 is reported to be over 50% of all bridges built [10].

The construction in 1948 of three prestressed concrete railway bridges with spans ranging from 12.8 m to 19.2 m on the Assam Rail Link near Siliguri heralded the use of prestressing in bridge construction in India. The first prestressed concrete highway bridge built in India was the Palar bridge near Chingleput, built in 1954 with 23 spans of 27 m each. Since then, many prestressed concrete bridges have been successfully built in this country using innovative designs and construction techniques. The adoption of segmental cantilever construction facilitated the construction of girder bridges of longer spans and slender sections. Notable examples include the Barak bridge (1961) with a clear span of 122 m and the Ganga bridge at Patna (1982) with 45 spans of 121 m each and two end spans of 63.5 m. The incremental launching method has been used successfully in 1994 in the case of the Panvel Viaduct (spans 9 x 40 m + 2 x 30 m) on the Konkan railway and later for the Yamuna bridge on the Delhi Metro.

1.5.10 Cable Stayed Bridges

The cable stayed bridge is specially suited in the span range of 200 to 900 m and thus provides a transition between the continuous box girder bridge and the suspension bridge. It was developed in Germany during the post-World War II years in an effort to save steel which was then in short supply. The Stromsund bridge in Sweden, built in 1957 with a main span of 183 m, and the Dusseldorf North bridge built in 1958 with a span of 260 m are early examples. Since then many cable stayed bridges have been built all over the world, chiefly because they are economical and also because they are aesthetically attractive. Another well-known bridge in this category is the Maracaibo Lake bridge in Venezuela designed by Ricardo Morandi of Italy and built in 1963. The Sunshine Skyway bridge (1987) designed by Eugene Figg and Jean Muller over Tampa Bay in Florida, has a main span of 360 m with prestressed concrete deck and single-plane cables. The Dames Point bridge at Jacksonville, Florida, built in 1987 with a span of 390 m is the longest cable stayed bridge in USA. Designed by Howard Needles and Finsterwalder, the bridge features H-shaped R.C. towers and two-plane cables supporting R.C. deck girders. Currently, the Tatara bridge in Japan (1999) with a span of 890 m is the longest cable stayed bridge in the world. The Millau viaduct, completed in 2005, with six spans of 350 m and two spans of 240 m, supported on towers up to 235 m height is a unique cable stayed bridge.

India's first cable stayed vehicular bridge is the Akkar bridge in Sikkim completed in 1988 with two spans of 76.2 m each. The Second Hooghly bridge (Vidyasagar Setu), completed in 1992, with a central span of 457.2 m and two side spans of 182.9 m each, is a notable engineering achievement in India.

Table 1.1. The Progress in Bridge Building as Recorded in Successive Record Span Lengths.

Year	Bridge	Location	Type	Main Span Metres
B.C. 219	Martorell	Spain	Stone arch	37
A.D. 14	† Nera river	Lucca, Italy	Stone arch	43
104	† Trajan's	Danube river	Timber arch	52
1377	† Trezzo	Italy	Stone arch	77
1758	† Wettingen	Switzerland	Timber arch	119
1816	† Schuylkill falls	Philadelphia, Pa.	Suspension	124
1820	† Union (Tweed)	Berwick, England	Chain	137
1826	Menai strait	Wales	Chain	177
1834	† Fribourg	Switzerland	Suspension	265
1849	Wheeling	Ohio river	Suspension	308
1851	† Lewiston	Niagara river	Suspension	318
1867	Cincinnati	Ohio river	Suspension	322
1869	† Clifton	Niagara falls	Suspension	386
1883	Brooklyn	New York City	Suspension	486
1889	Forth	Scotland	Cantilever	521
1917	Quebec	Canada	Cantilever	549
1929	Ambassador	Detroit, Mich.	Suspension	564
1931	George Washington	New York City	Suspension	1067
1937	Golden Gate	San Francisco, Calif.	Suspension	1280
1964	Verrazano Narrows	New York City	Suspension	1298
1981	Humber	Humber, U.K.	Suspension	1410
1998	Akashi Kaikyo	Japan	Suspension	1991

† Not standing

1.5.11 Suspension Bridges

The concept of suspension bridges has originated from India and China. A few short bridges exist in the Himalayan region. However, the modern suspension bridges are mostly in USA, Europe, Japan and China. The suspension bridge has come to stay as the type best suited for very long spans. Thomas Telford built the Menai Straits eyebar suspension bridge with 177 m main span in 1826. The Wheeling bridge across the Ohio river in USA built by Charles Ellet in 1849 with a span of 308 m was the first long span wrought iron wire cable suspension bridge in the world. The Brooklyn bridge with a main span of 486 m , designed by John Roebling and completed in 1886 by Washington Roebling and Emily Warren-Roebling, was the first steel wire cable suspension bridge in the world. Other suspension bridges followed raising the record span. Othmar Amman designed the George Washington bridge in 1931. This was the first modern suspension bridge with a main span of 1067 m, nearly double the previous record. The Golden Gate bridge at San Francisco was completed in 1937 by Joseph Strauss with a record span of 1280 m. In 1940, the Tacoma Narrows bridge at Puget Sound, Washington, was opened to traffic as a beautiful bridge with a span of 853 m and stiffening plate girders only 2.4 m deep. The bridge collapsed the same year during a 68 km/h gale due to aerodynamic instability. This failure forcefully brought home the need for consideration of aerodynamic effects on suspension bridges and triggered many theoretical studies on the problem.

Table 1.2 The World's Longest Spans for Various Bridge Types.

Type	Bridge	Location	Date	Main Span Metres
Cable suspension	Akashi Kaikyo	Kobe-Naruto, Japan	1998	1991
Cable stayed - Steel	Tatara	Japan	1999	890
Cable stayed - Concrete Girder	Skarsund	Norway	1991	530
Cantilever	Quebec	Canada	1917	549
Steel arch	Lupu	Shanghai, China	2004	550
Concrete arch	Wanxiang	Yangzi river, China	1996	420
Continuous truss	Astoria	Oregon, USA	1966	376
Eyebar suspension	Florianopolis	Brazil	1926	340
Continuous steel girder	Niteroi	Brazil	1974	300
Simple truss	J.J. Barry	Delaware river, USA	1973	251
Prestressed concrete girder	Stolmasundet	Austevoll, Norway	1998	301
Concrete girder	Villeneure	France	1939	78
Tubular girder	Britannia	Menai strait	1850	140
Simple girder	Harlem river	New York city	1951	101
Masonry arch	Wuchao river	Hunan, China	1990	120
Vertical lift	Arthur Kill	Elizabeth, N.J.	1959	170
Wichert truss	Homestead	Pittsburg, Pa.	1937	163
Swing span	El- Ferdan	Suez, Egypt	2001	340
Bascule	Sault Ste. Marie	Michigan	1914	102
Single-leaf bascule	16th street	Chicago, Illinois	1919	79

The Severn bridge (1966) with a span of 988 m became a land mark in bridge construction. Its deck consisted of an all-welded steel stiffened box girder, streamlined and tapered at the edges. The shape was obtained after extensive wind-tunnel tests. This concept has since become the salient feature of European bridge design, and has been adopted for the Bosporus bridge (1973) at Istanbul and the Humber Estuary bridge at Humber, England (1981). The Humber bridge with a main span of 1410 m was the longest span bridge till 1998. Currently, the Akashi-Kaikyo bridge in Japan, completed in 1998, is the longest span bridge with a main span of 1991 m and side spans of 960 m. This great bridge is indicative of the irresistible urge of man to continually thrust forward. According to D.B. Steinman, bridge spans as large as 3000 m are practically feasible and will be built in the future.

1.5.12 Conclusion

The above discussion shows that the early bridges prior to the Industrial Revolution used designs in stone and timber. Better understanding of the properties of materials and advances in analysis led to applications first in cast iron, then in wrought iron and structural steel, and still later in reinforced concrete and prestressed concrete. Evolution of bridges also followed the industrial advancement. Thus innovations in bridge design were initiated mainly in UK in the early 19th century, while advances were registered in USA and Switzerland from the late 19th century to mid 20th century. Significant strides were made in Germany and Japan in subsequent decades.

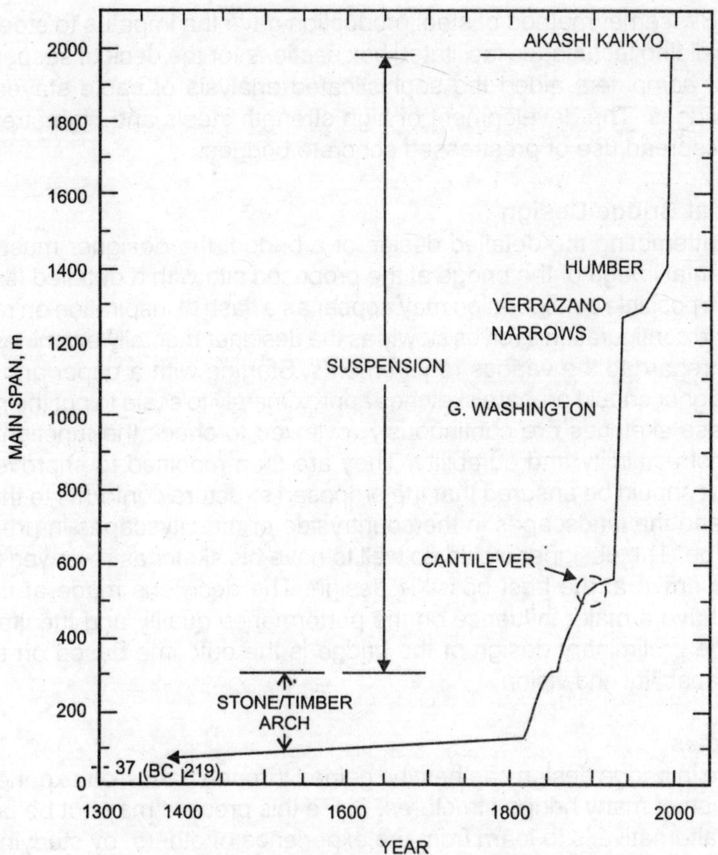

Figure 1.3 Growth in Bridge Spans.

The progress in bridge construction as recorded by successive record span lengths can be seen from Table 1.1 and Fig. 1.3. The world's longest spans for various bridge types are listed in Table 1.2. Data on the world's notable bridges of different categories are included in Tables C.1 to C.10 in Appendix C. In terms of total length, the longest bridge in the world is the Pontchartrain bridge, New Orleans, USA with a total length of 38.6 km built in 1956. A chronology of selected developments of bridges is listed in Appendix D to facilitate easy reference.

The design of long span bridges always poses a challenge to the ingenuity and to the perseverance of the designer. Every long span bridge brings in new problems in design concepts and new construction details. The designer chooses his solution with full freedom of decision, but with full responsibility for the success of the work. In fact, long span bridges are creative works, and serve as landmarks in the art of bridge building.

The above review serves to illustrate how bridge engineering has developed into a specialized subject drawing upon the advances in materials, theoretical analysis, construction techniques, computer applications and concepts in many branches of engineering. For example, the Bessemer method of steel production gave the impetus to steel bridges. The advances in welding techniques facilitated box sections for the deck of suspension bridges. The advent of computers aided the sophisticated analysis of cable stayed bridges and suspension bridges. The development of high strength steels and high strength concrete led to the widespread use of prestressed concrete bridges.

1.6 Conceptual Bridge Design

Prior to attempting the detailed design of a bridge, the designer must first visualize and form a mental image of the bridge at the proposed site with a detailed fitment of all the components. A possible configuration may appear as a flash of inspiration on rare occasions. But normally the configuration evolves slowly as the designer mentally examines the imagined structure with regard to the various requirements. Starting with a paper and a soft-leaded pencil, the designer should prepare sketches approximately to scale to put the mental images on paper. These sketches are continuously reviewed to check the function, applicability, economy, constructability and durability. They are then modified to improve on the likely performance. It should be ensured that the proposed structure conforms to the surrounding environment and the landscapes in the countryside or the cityscapes in urban centres as the case may be. The designer would do well to have his sketches reviewed by colleagues with a view to arrive at the best possible design. The decisions made at the conceptual design stage have a major influence on the performance quality and the life-cycle cost of the bridge. The preliminary design of the bridge is the outcome based on the designer's experience, capability and vision.

1.7 Case studies

Expertise in bridge design can best be gained through hands-on experience in design and construction of many bridge structures. Since this process may not be possible for all, the next best alternative is to learn from the experience of others, by studying in detail the evolution of bridge engineering through the available reports on the design provisions, construction features and performance data of different bridges, besides visiting sites of bridges under construction and in service. The aim of the case study is to examine the reasons for major decisions with regard to the selection of site, the choice of the structural form, the materials and construction techniques adopted, and the performance of the bridge. The engineer should try to get a thorough understanding of the design concepts of the structural system through a fundamental enquiry. Detailed studies of various examples of bridges with regard to their positive and negative aspects will help to create a data bank in the mind, which will lead to the ability to select and rate one's own design solutions. Rapid strides have been registered in the recent past in bridge design, detailing, construction technology and materials used for bridge construction. In order to facilitate identification of suitable case studies, the names of a number of important bridges with significant features as well as the names of a few famous bridge engineers have been mentioned at various points of discussion in the text along with references to the published literature. The author would urge the serious reader to develop a feel for the subject through a critical study of these references.

1.8 Standard Specifications

Bridge design is a complex problem, calling for creativity and practicability, while satisfying the basic requirements of safety and economy. Standard specifications and codes of practice have been evolved by the concerned government agencies and professional institutions, based on years of observation, research and development. The primary purpose of these codes is to ensure satisfactory design, detailing and construction of the structures, to achieve adequate safety, quality and durability. Conformance to the standards also affords reasonable protection against legal liability arising out of failures due to no fault of the designer. Since the public roads and railways in India are owned and controlled by government, the bridges built on them should follow specifications laid down by the respective authorities. All highway bridges in India have to be built in accordance with the Indian Roads Congress (IRC) Codes, besides Specifications prescribed by the Ministry of Road Transport & Highways, Government of India (MORTH). Similarly, the design of railway bridges should conform to Indian Railway Standard (IRS) Codes including the Bridge Rules and the Specifications laid down by the Research, Design & Standards Organisation (RDSO) of the Indian Railways. Wherever applicable, the specifications of the Bureau of Indian Standards should also be complied with. Some important specifications from these codes are indicated and discussed in this text. Unless mentioned otherwise, reference to 'the Code' in this text refers to the relevant provision in the IRC Bridge Code. The specifications in these codes undergo revision frequently. The bridge engineer is expected to be conversant with the different clauses in the latest revisions of the above codes.

1.9 References

1. Steinman, D.B., and Watson, S.R., 'Bridges and their builders', Dover Publications, New York, 1957, 401 pp.
2. Shirley-Smith, H., 'World's great bridges', English Language Book Society, London, 1964, 250 pp.
3. Bennet, D., 'The history and aesthetic development of bridges', In Ryall, M.J., et al,(Eds.), 'The Manual of Bridge Engineering', Thomas Telford, London, 2002, pp. 1-41.
4. DeLony, E., 'Context for World Heritage Bridges', International Council on Monuments and Sites, Paris, France, 1996.
5. Melarano, M., 'Preliminary design of bridges for architects and engineers', Marcell Bekkar, New York, 1998, 534 pp.
6. Li, G., and Xiao, R., 'Bridge design practice in China', In Chen, W-F, and Duan, L., (Eds.), 'Bridge Engineering Handbook', CRC Press, Boca Raton, Florida, USA, 2000, pp. 63-1 to 37.
7. Billington, D.P., 'The Tower and the Bridge', Basic Books Inc., New York, 1983, 306 pp.
8. Subba Rao, T.N., 'Trends in the construction of concrete bridges', Gammon Bulletin, Special Issue, March 1985, 23 pp.
9. Muller, J.M., 'Design practice in Europe', In Chen, W-F, and Duan, L., (Eds.), 'Bridge Engineering Handbook', CRC Press, Boca Raton, Florida, USA, 2000, pp. 64-1 to 43.
10. Committee on Concrete Bridges, 'Concrete Bridges', http://gulliver.trb.org/publications/millenium/00019.pdf.

Chapter 2

Investigation for Bridges

2.1 Need for Investigation

Before a bridge can be built at a particular site, it is essential to consider many factors, such as the need for a bridge, the present and future traffic, stream characteristics, subsoil conditions, alternative sites, aesthetics and cost.

The aim of the investigation is to select a suitable site at which a bridge can be built economically, at the same time satisfying the demands of traffic, the stream, safety and aesthetics. The investigation for a major bridge project should cover studies on technical feasibility and economic considerations and should result in an investigation report. The success of the final design will depend on the thoroughness of the information furnished by the officer in charge of the investigation.

2.2 Selection of Bridge Site

The choice of the right site is a crucial decision in the planning and designing of a bridge. It may not be possible always to have a wide choice of sites for a bridge. This is particularly so in case of bridges in urban areas and flyovers. For river bridges in rural areas, usually a wider choice may be available.

The characteristics of an ideal site for a bridge across a river are:

(a) A straight reach of the river;
(b) Steady river flow without serious whirls and cross currents;
(c) A narrow channel with firm banks;
(d) Suitable high banks above high flood level on each side;
(e) Rock or other hard inerodible strata close to the river bed level;
(f) Economical approaches which should not be very high or long or liable to flank attacks of the river during floods; the approaches should be free from obstacles such as hills, frequent drainage crossings, sacred places, graveyards, or built up areas or troublesome land acquisition;
(g) Proximity to a direct alignment of the road to be connected;
(h) Absence of sharp curves in the approaches;
(i) Absence of expensive river training works; and
(j) Avoidance of excessive underwater construction.

For selecting a suitable site for a major bridge, the investigating engineer should make a reconnaissance survey for about one km on the upstream side and one km on the downstream side of the proposed bridge site and should journey along the road for about one km on either side of the road from the bridge site in order to form a general impression of the landscape and to decide on the type of structure best suited to the site. Care should be taken to investigate a number of probable alternative sites and then decide on the site which is likely to serve the needs of the bridge at the least cost. A brief description of the reasons for selection of a particular site should be furnished in the investigation report along with salient details of alternative sites investigated and rejected.

When the river to be crossed is a meandering river, the bridge should be located at a nodal point, i.e. the location where the river regime is constant serving as a fulcrum about which the river channels swing laterally.

To the extent possible, it is desirable to align the bridge at right angles to the river, i.e., to provide a square crossing, which facilitates minimum span length, deck area and pier lengths, with accompanying economies. Further, a square crossing involves simpler designs and detailing. Sometimes, a skew crossing which is inclined to the centre line of the river at an angle different from a right angle has to be provided in order to avoid costly land acquisition or sharp curves on the approaches. A skew bridge usually poses more difficulties in design, construction and maintenance. The location of the bridge in relation to the alignment of the approaches may be decided as below[1]: (a) For bridges of total length less than 60 m, the alignment of the approaches will govern; (b) For bridges of total length between 60 m and 300 m, both the proper alignment of the approaches and the requirements of a good bridge site should be considered together in ascertaining the appropriate site; and (c) For major bridges over 300 m length, the requirements of a good bridge site will govern the alignment.

2.3 Preliminary Data to be Collected

The engineer in charge of the investigation for a major bridge should collect the following information:

(a) Name of the stream, road and the identification mark allotted to the crossing and location in km to centre of crossing;
(b) Location of the nearest GTS (Great Trigonometric Survey) bench mark with its reduced level;
(c) Present and anticipated future volume and nature of traffic on the road at the bridge site;
(d) Hydraulic data pertaining to the river, including the highest flood level (HFL), ordinary flood level (OFL) and low water level (LWL), size, shape, slope and nature of the catchment, possibility of subsequent changes in the catchment like afforestation, deforrestation, and urban development intensity and frequency of rainfall in the catchment, and probability of large trees or rolling debris floating down the stream;
(e) Soil profile along the probable bridge sites over the length of the bridge and approaches;
(f) Navigational requirements, if any, for the stream;
(g) Need for large scale river training works;
(h) Liability of the site to earthquake disturbances;
(i) Availability, quality and location of the nearest quarries for stones for masonry and for concrete aggregates;
(j) Nearest place of availability of cement, steel and timber;
(k) Means of transport for materials;
(l) Availability of unskilled and skilled labour for different trades required for construction;

(m) Facilities required for housing labour during construction;
(n) Important details of the bridges, if any, crossing the same river within a reasonable distance of the proposed bridge;
(o) Availability of electric power; and
(o) Details of any utilities and services to be provided for (e.g., telephone cables, power cables, water supply pipes) along with relevant information on size and arrangement.

Some of these items are discussed in detail in the later sections.

2.4 Preliminary Drawings

The following drawings should be prepared at the time of investigation:

(i) *An index map* drawn to a suitable scale (usually 1: 50000) showing the proposed location of the bridge, the alternative sites investigated and rejected, the existing communications, the general topography of the area, and the important towns, etc., in the vicinity. A typical index map for Godavari second bridge is shown in Fig. 2.1.

(ii) *A contour survey plan* of the stream showing all topographical features for a sufficient distance on either side of the site to permit a clear indication of the features that would influence the location and design of the bridge and its approaches. All sites worth considering should be indicated on the plan. The plan may be drawn to a suitable scale, such as 1 :

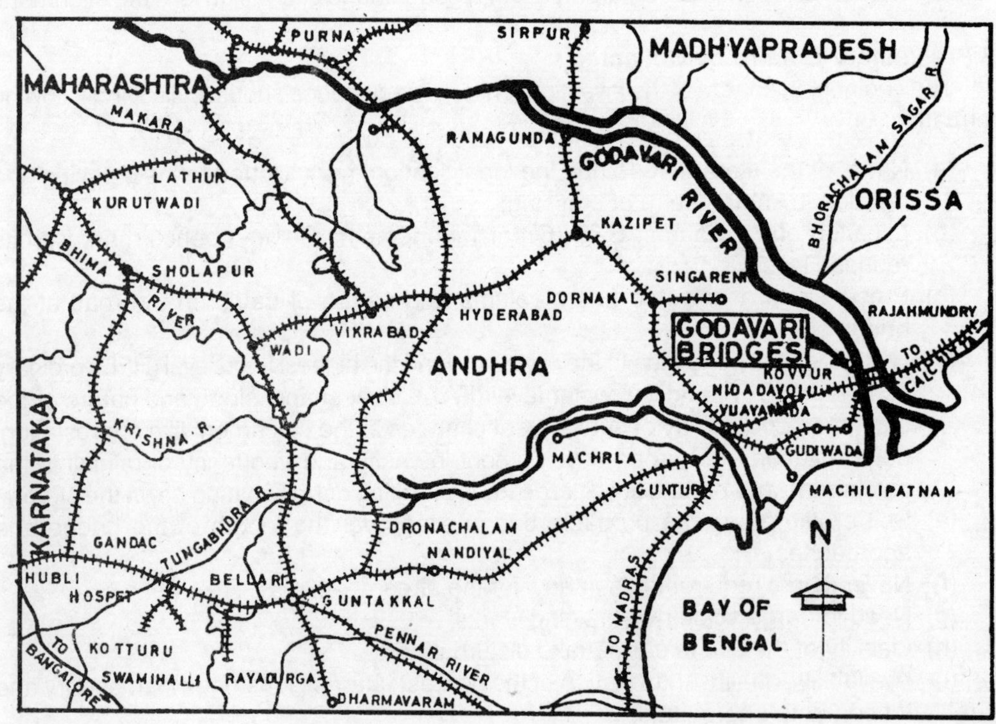

Figure 2.1 Index Map for Godavari Second Bridge.

1000 to 1 : 5000, and should cover distances on either side of the proposed bridge site of 100 m, 300 m and 1500 m for catchment areas of 3 km^2, 15 km^2 and over 15 km^2, respectively.

(iii) *A site plan* to a suitable scale (preferably 1 : 1000) showing the details of the selected site and the details of the stream to a distance of 100 to 200 metres upstream and downstream of the selected site. The plan should include the following details:

(a) Name of the stream, road and nearest distance marker;
(b) Approximate outlines of the banks and channels at high water level and low water level;
(c) Direction of flow;
(d) The alignment of existing approaches and the proposed crossing with its approaches;
(e) The angle and direction of skew, if the proposed alignment is on a skew;
(f) The name of the nearest inhabited identifiable locality at either end of the crossing on the roads leading to the site;
(g) Location and reduced level of the bench mark used as datum, connected to GTS bench mark wherever available;
(h) Location of the longitudinal section and cross-sections of road and stream taken within the area of the plan;
(i) The locations of trial pits and borings with their identification numbers:
(j) The locations of all nullahs, buildings, wells, outcrops of rocks and other possible obstructions to a road alignment.

In case of a crossing over an obstruction other than a stream, similar applicable details will have to be furnished. A typical site plan for Godavari second bridge is shown in Fig. 2.2.

(iv) *A cross-section* of the river at the proposed bridge site to a scale of about 1 : 1000 horizontally and about 1 : 100 vertically. The cross-section should include the following information:

(a) The name of the stream, road, and chainage;
(b) The river bed line with banks up to a level above the highest flood level;
(c) The nature of surface soil in bed, bank and approaches;
(d) The low water level, ordinary flood level and the highest flood level;
(e) If the stream is tidal, lowest and highest tide levels and mean sea level.

(v) *A longitudinal section* of the stream showing the site of the bridge with the HFL, OFL, LWL and the bed levels at suitably spaced intervals along the approximate centre line of the deep water channel. Suitable scale may be used for horizontal distance, but the vertical scale should not be less than 1 : 1000.

(vi) *Additional cross-sections* of the stream at suitable distance both upstream and downstream of the proposed bridge site. These should indicate the distance from the bridge site and also show the flood levels and low water level. The cross-sections should extend to adequate horizontal distances on either side so as to show the banks a little beyond HFL. Approximate distances, upstream and downstream of the selected bridge site, at which these cross-sections should be taken are 150 m, 300 m and 400 m for catchment areas of less than 2.5 km^2 , 2.5 to 10.0 km^2 and over 10.0 km^2 , respectively.

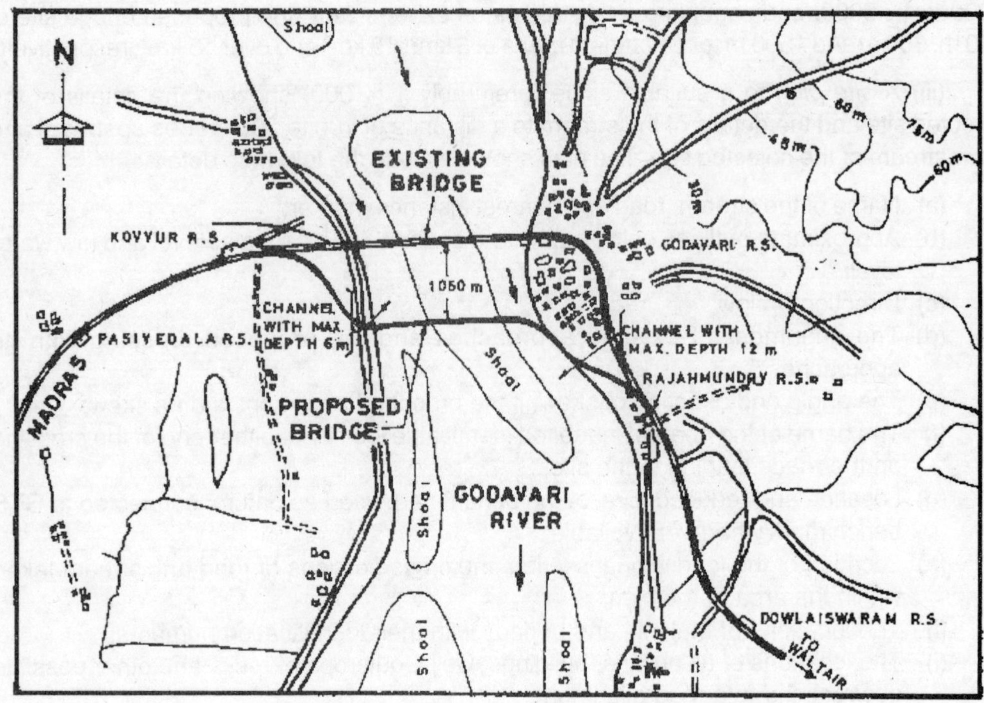

Figure 2.2 Site plan for Godavari Second Bridge.

(vii) *Catchment area map* for the river at the proposed bridge site is to be prepared by tracking the line of the ridge of the watershed from the topographical maps of Survey of India drawn to a scale of 1 : 50000 or 1 : 25000 if available. The included area may be computed by placing a transparent squared paper on the map and counting the number of squares.

(viii) *Soil profile* should be determined by subsoil exploration along each of the probable bridge alignments. Wherever possible, borings should be taken at the possible locations of the abutments and piers. The drawing should show the bed and banks as well as the classification and levels of the layers of the subsoil. The levels of the rock or other hard soil suitable for resting foundations should be clearly marked.

Every drawing should contain the identification particulars of the crossing. North point and directions of flow of the river should be indicated in the index map, contour survey plan and the site plan.

2.5 Determination of Design Discharge

2.5.1 Method

The maximum discharge which a bridge across a natural stream is to be designed to pass can be estimated by the following methods:

(a) By using one of the empirical formulae applicable to the region;

(b) By using a rational method involving the rainfall and other characteristics for the area;
(c) By the area-velocity method, using the hydraulic characteristics of the stream such as cross-sectional area, and the slope of the stream;
(d) By unit hydrograph method;
(e) From any available records of the flood discharges observed at the bridge site or at any other site in the vicinity.

It is desirable to estimate the flood discharge by at least two of the above methods, and the maximum discharge determined by judgement by the engineer responsible for the design. These methods are briefly explained here.

2.5.2 Empirical Formulae

Empirical formulae for flood discharge from a catchment have been proposed of the form:

$$Q = CA^n \quad \text{... (2.1)}$$

where Q = maximum flood discharge in m³ per second
A = catchment area in square kilometres
C = constant depending on the nature of the catchment and location
n = constant.

A popular empirical formula of the above type is Ryve's formula given by Equation (2.2).

$$Q = CA^{2/3} \quad \text{... (2.2)}$$

The value of C is taken as 6.8 for flat tracts within 25 km of the coast, 8.5 for areas between 25 and 160 km of the coast and 10 for limited areas near the hills.

These empirical formulae are oversimplified and much depends on the assumption of the correct value for C. A reliable value for C for any particular region can only be derived by a careful statistical analysis of a large volume of observed flood and catchment data. Even such a value will not be valid for any other region. Hence the reliability of an empirical formula of this nature is extremely limited[2].

2.5.3 Rational Method

A rational formula for flood discharge should take into account the intensity, distribution and duration of rainfall as well as the area, shape, slope, permeability and initial wetness of the catchment (drainage basin). The area of the catchment is a major contributing factor for the runoff. The shape of the catchment affects the peak discharge, long and narrow basins yielding less than pear shaped basins. Steep slopes result in shorter time of concentration than flatter slopes.

Many complicated formulae are available in treatises on hydrology. A typical rational formula[3] is:

$$Q = A.I_o.\lambda \quad \text{... (2.3)}$$

where Q = maximum flood discharge in m³ per second

A = catchment area in km²
I_o = peak intensity of rainfall in mm per hour
λ = a function depending on the characteristics of the catchment in producing the peak run-off

$$= \frac{0.56\,Pf}{t_c + 1}$$

t_c = concentration time in hours

$$= \left(0.87 \times \frac{L^3}{H}\right)^{0.385}$$

L = distance from the critical point to the bridge site in kilometres
H = difference in elevation between the critical point and the bridge site in metres
P = coefficient of run-off for the catchment characteristics, from Table 2.1
f = a factor to correct for the variation of intensity of rainfall I_o over the area of the catchment, from Table 2.2.

Table 2.1 Value of P in Rational Formula.

Surface	P
Steep bare rock, and also city pavements	0.90
Rock, steep but with thick vegetation	0.80
Plateaus, lightly covered	0.70
Clayey soils, stiff and bare	0.60
Clayey soils, lightly covered	0.50
Loam, lightly cultivated	0.40
Loam, largely cultivated	0.30
Sandy soil, light growth	0.20
Sandy soil, heavy brush	0.10

In Equation (2.3), I_o measures the role of clouds in the region and λ represents the role of the catchment in producing the peak runoff. The values of A, L and H can be obtained from Survey of India topographical maps. I_o is to be obtained from the Meteorology Department. Considerable judgment and experience are called for in assessing the value of P. Any error in the latter will diminish the reliability of the result of the laborious calculations involved in this method.

Table 2.2 Value of Factor *f* in Rational Formula.

Area km²	*f*	Area km²	*f*
0	1.000	80	0.760
10	0.950	90	0.745
20	0.900	100	0.730
30	0.875	150	0.675
40	0.845	200	0.645
50	0.820	300	0.625
60	0.800	400	0.620
70	0.775	2000	0.600

2.5.4 Area-Velocity Method

The area-velocity method based on the hydraulic characteristics of the stream is probably the most reliable among the methods for determining the flood discharge. The velocity obtaining in the stream under the flood conditions is calculated by Manning's or similar formula: Manning's formula is used here. The discharge Q is given by Equation (2.4).

$$Q = A.V \qquad \text{... (2.4)}$$

where Q = discharge in m^3 per second

A = wetted area in m^2

V = velocity of flow in metres per second

$= \frac{1}{n} . R^{0.67} . S^{0.5}$

n = coefficient of roughness, from Table 2.3

S = slope of stream

R = hydraulic mean depth in metres

$= \frac{\text{wetted area in } m^2}{\text{wetted perimeter in metres}}$

Since the cross-section of the stream is usually plotted with different scales for horizontal and vertical distances, the wetted perimeter cannot be scaled off directly, but has to be calculated. The wetted line is divided into a convenient number of parts and the partial length along the perimeter computed as hypotenuse of the right triangle with the horizontal and vertical lengths of the element as the two sides. The sum of such parts gives the wetted perimeter P. Similarly the wetted area A is calculated as the sum of partial areas of elements obtained as the product of the horizontal interval and the mean depth to bed below the flood level considered at the two ends of the element. The hydraulic mean radius can then be computed as A/P. Typical calculations for A, P and R are indicated in Example 2 in section 2.5.9.

Table 2.3 Roughness Coefficient *n* in Manning's Formula for Natural Streams.

	Surface	Perfect	Good	Fair	Bad
1.	Clean straight banks, no rifts or deep pools	0.025	0.028	0.030	0.033
2.	Same as (1), but some weeds and stones	0.030	0.033	0.035	0.040
3.	Winding, some pools and shoals, clean	0.035	0.040	0.045	0.050
4.	Same as (3), more effective slope and sections	0.040	0.045	0.050	0.055
5.	Same as (3), some weeds and stones	0.033	0.035	0.040	0.045
6.	Same as (4), stony section	0.045	0.050	0.055	0.060
7.	Sluggish river reaches, rather weedy	0.050	0.060	0.070	0.080
8.	Highly weedy reaches	0.075	0.100	0.125	0.150

The quantity S in Equation (2.4) denotes the slope of the stream and is a difficult quantity for evaluation. The normal practice is to compute the slope from the bed levels at two cross-sections over a long distance. This may lead to unreliable results, since it is difficult to take any particular level in a cross-section as "the bed level". The author has found it desirable to compute the slope of the stream from the low water levels or water levels at any one time at the proposed site and at one section each upstream and downstream of the proposed site.

The success or otherwise of the use of this method depends on the correct determination of the flood levels. Considerable judgement tempered with experience will be called for in order to correctly assess the evidence in this connection. If a railway track is near the bridge site, the maximum flood mark will be usually available from the markings on railway cross drainage works. In the case of new road formation in sparsely inhabited or undeveloped areas, the investigation engineer has to come to his conclusion on the maximum flood level based on his evaluation of the evidence from the elderly inhabitants of the area and the observation of the banks, deposit of debris on tree trunks, etc.

2.5.5 Unit Hydrograph Method

The runoff from a basin can also be estimated using the unit hydrograph method, originally formulated by Sherman[4]. A hydrograph is the graphical representation of discharge in a stream plotted against time due to a rain storm of specified intensity, duration and areal pattern. For any given drainage basin, the hydrographs of runoff due to two rain storms will be similar, their ordinates being proportional to the intensity of rainfall. A unit hydrograph is defined as the runoff hydrograph representing a unit depth (1 mm) of direct run off as a result of rainfall excess occuring uniformly over the basin and at a uniform rate for a specified duration (e.g. 6 hours or 12 hours). The area under a unit hydrograph represents the volume of rainfall excess due to a rain of 1 mm over the entire basin.

The peak discharge is computed as below. The storm hydrograph for the basin for a particular rainfall excess is plotted from documented data of runoff rates in m^3/hour and time in hours. The base flow is separated from the direct runoff. The volume of direct runoff is computed from the area under the storm hydrograph. This volume divided by the area of basin gives the direct runoff in terms of depth of flow d (expressed in mm) over the basin. The ordinates of the unit hydrograph are obtained by dividing the corresponding ordinates of the storm hydrograph by d. Direct runoff for any given storm can then be calculated by multiplying the maximum ordinate of the unit hydrograph by the depth of runoff over the area. (The depth of runoff corresponding to the given storm can be computed in proportion to the depth relating to the storm hydrograph already plotted for deriving the unit hydrograph.) The maximum runoff rate is obtained by adding the base flow to the maximum direct runoff rate.

The unit hydrograph method assumes that the storm occurs uniformly over the entire basin and that the intensity of the rainfall is constant for the duration of the storm. While these assumptions may be reasonable for small basins, they are not normally satisfactory for large catchments of over 5000 km^2 .

2.5.6 Estimation From Flood Marks

If flood marks can be observed on an existing bridge structure near the proposed site, the flood discharge passed by the structure can be estimated reasonably well, by applying an appropriate formula available in treatises on hydraulics. It is possible by inspection to ascertain the flood levels soon after a flood. Sometimes, these flood marks can be identified even years after a flood, but it is desirable to locate these as soon after the flood as possible.

2.5.7 Design Discharge

The design discharge may be taken as the maximum value obtained from at least two of the methods mentioned. If the value so obtained exceeds the next high value by more than 50%, then the maximum design discharge is limited to 1.5 times the lower estimate. Freak discharges of high intensity due to the failure of a dam or tank constructed upstream of the bridge site need not be catered for. From consideration of economy, it is not desirable to aim to provide for the passage of the very extraordinary flood that may ever happen at a particular site. It may be adequate to design for a flood occurring once in 50 years and to ensure that rarer floods be passed without excessive damage to the structure.

2.5.8 EXAMPLE 1

Problem

Determine the design discharge at a bridge site after computing the maximum discharge by (i) Empirical Method,(ii) Rational Method, and (iii) Area-Velocity Method, for the following data:

Catchment area	=	160 km²
Distance of site from coast	=	12 km
Distance of critical point to bridge site	=	16 km
Difference in elevation between the critical point and the bridge site	=	96 m
Peak intensity of rainfall	=	60 mm/h
Surface of catchment is loam, largely cultivated.		
Cross sectional area of stream at MFL at bridge site	=	120 m²
Wetted perimeter of stream at MFL at bridge site	=	90 m
Stream condition - Clean straight banks, Fair condition.		
Slope of stream	=	1/500.

Solution

(a) *By Empirical Method*

According to Ryve's formula, maximum flood discharge (Q) is obtained as

$$Q = C.A^{2/3}$$

Here $C = 6.8$, $A = 160\ \text{km}^2$

Substituting, $Q = 6.8 \times (160)^{2/3} = 201\ \text{m}^3/\text{s}$.

(b) *By Rational Method*

Using the notations in Section 2.5.3,

$A = 160\ \text{km}^2$, $I_o = 60$ mm/h,

$P = 0.3$, $f = 0.669$,

$L = 16$ km, $H = 96$ m

Substituting the above values,

$$t_c = \left[\frac{0.87 \times 16^3}{96}\right]^{0.385} = 4.04 \text{ hours}$$

$$\lambda = \frac{0.56 \times 0.3 \times 0.669}{4.04 + 1} = 0.0223$$

$$Q = 160 \times 60 \times 0.0223 = 214 \text{ m}^3/\text{s}$$

(c) *By Area-Velocity Method*

Using the notations in Section 2.5.4,

A = 120 m² , n = 0.03,

P = 85 m, s = 1/500 = 0.002

R = 120/90 = 1.33

$$V = \frac{1}{0.03} \times (1.33)^{0.67} \times (0.002)^{0.5} = 1.80 \text{ m/s}$$

$$Q = 120 \times 1.80 = 216 \text{ m}^3/\text{s}$$

(d) *Design Discharge*

Since the values obtained from the three methods are reasonably close, the maximum of the three values is adopted for design.

Design discharge = 216 m³ /s

2.5.9 EXAMPLE 2

Problem

Compute the wetted area, the wetted perimeter and the hydraulic mean radius at a cross section of a stream whose cross sectional details are as indicated in the first three columns of Table 2.4. HFL = 97.96 m.

Solution

The computation is performed as in Table 2.4.

Wetted area A = 607.98 m²

Wetted perimeter P = 123.69 m

Hydraulic mean radius = A/P = 4.92 m

2.6 Linear Waterway

When the water course to be crossed is an artificial channel for irrigation or navigation, or when the banks are well defined for natural streams, the linear waterway should be the full width of the channel or stream.

For large alluvial stream with undefined banks, the required effective linear waterway may be determined using Lacey's formula given in Equation (2.5).

$$W = C\sqrt{Q} \qquad \text{... (2.5)}$$

where W = the effective linear waterway in metres

Q = the designed maximum discharge in m³ /s

C = a constant, usually taken as 4.8 for regime channels, but may vary from 4.5 to 6.3 according to local conditions.

The effective linear waterway is the total width of the waterway of the bridge minus the mean submerged width of the piers and their foundation down to the mean scour level.

Table 2.4 Stream Cross Section Details for Example 2

HFL = 97.96 m

Chainage	Distance	Bed level	(HFL – BL)	Area	Difference in BL	Wetted Perimeter
	x	(BL)			y	$\sqrt{(x^2+y^2)}$
m	m	m	m	m²	m	m
60 L	0	97.96	0	0	0	0
55 L	5	95.70	2.26	5.65	2.26	5.49
50 L	5	94.60	3.36	14.05	1.10	5.12
40 L	10	91.80	6.16	47.60	2.80	10.38
20 L	20	91.40	6.56	127.20	0.40	20.00
PLS	20	90.90	7.06	136.20	0.50	20.01
20 R	20	91.10	6.86	139.20	0.20	20.00
40 R	20	95.16	2.80	96.60	4.06	20.41
50 R	20	95.80	2.16	49.60	0.64	20.01
55 R	5	96.08	1.88	10.10	0.28	1.40
62 R	7	97.96	0	6.58	1.88	7.25
Total				607.98		123.69

Note: PLS is the point on longitudinal section of stream
60 L means 60 m to left of PLS
62 R means 62 m to right of PLS

It is not desirable to reduce the linear waterway below that for regime condition. If a reduction is effected, special attention should be given to afflux and velocity of water under the bridge. With reduced waterway, velocity would increase and greater scour depths would be involved, requiring deeper foundations. Thus, any possible saving from a smaller linear waterway will be offset by the extra expenditure on deeper foundations and protective works. In view of the deficiencies of the assumptions made in the computations for design discharge and for the effective waterway by Lacey's formula, it is often prudent to adopt the full natural width for the linear waterway, taking care not to succomb to the trap of overconfidence in apparently precise methods of calculation.

Afflux is the heading up of water over the flood level caused by constriction of waterway at a bridge site. It is measured by the difference in levels of the water surfaces upstream and downstream of the bridge. Afflux can be computed from Equation (2.6).

$$x = \frac{V^2}{2g}\left[\frac{L^2}{c^2 L_1^2} - 1\right] \quad \text{... (2.6)}$$

where x = afflux
V = velocity of normal flow in the stream
g = acceleration due to gravity

L = width of stream at HFL
L_1 = linear waterway under the bridge
c = coefficient of discharge through the bridge, taken as 0.7 for sharp entry and 0.9 for bell mouthed entry.

The afflux should be kept minimum and limited to 300 mm. Afflux causes increase in velocity on the downstream side, leading to greater scour and requiring deeper foundations. The road formation level and the top level of guide bunds are dependent on the maximum water level on the upstream side including afflux.

The increased velocity under the bridge should be kept below the allowable safe velocity for the bed material. Typical values of safe velocities are as below:

Loose clay or fine sand	up to 0.5 m/s
Coarse sand	0.5 to 1.0 m/s
Fine gravel, sandy or stiff clay	1.0 to 1.5 m/s
Coarse gravel, rocky soil	1.5 to 2.5 m/s
Boulders, rock	2.5 to 5.0 m/s

2.7 Economical Span

Considering only the variable items, for a given linear waterway, the total cost of the superstructure increases and the total cost of substructure decreases with increase in the span length. The most economical span length is that for which the cost of superstructure equals the cost of substructure. This condition may be derived as below:

Let A = cost of approaches
B = cost of two abutments, including foundations
L = total linear waterway
s = length of one span
n = number of spans
P = cost of one pier, including foundation
C = total cost of bridge.

Assuming that the cost of superstructure of one span is proportional to the square of the span length, total cost of superstructure equals $n.ks^2$, where k is a constant.

The cost of railings, flooring, etc., is proportional to the total length of the bridge and can be taken as K'L.

$$C = A + B + (n-1)\,P + nks^2 + K'L$$

For minimum cost, dC/ds should be zero.

Substituting n = L/s and differentiating, and equating the result of differentiation to zero, we get $P = ks^2$

Therefore, for an economical span, the cost of superstructure of one span is equal to the cost of substructure of the same span.

The economical span (s_e) can then be computed from Equation (2.7).

$$s_e = \sqrt{P/k} \qquad \text{... (2.7)}$$

P and k are to be evaluated as average over a range of possible span lengths.

2.8 Location of Piers and Abutments

Piers and abutments should be so located as to make the best use of the foundation conditions available. Normally, the span lengths for a river bridge would be influenced by the hydraulic considerations, subsoil profile affecting the foundation requirements, height of piers, floating debris expected during floods, availability of handling machinery and skilled labour for construction and cost considerations. As far as possible, the most economical span as above may be adopted. If navigational or aesthetic requirements are to be considered, the spans may be suitably modified. As a rule, the number of spans should be kept low, as piers obstruct water flow. If piers are necessary, an odd number of spans is to be preferred. Placing a pier at the deepest portion of an active channel may be avoided by suitably adjusting the number and length of the span.

For small bridges with open foundations and solid masonry piers and abutments, the economical span is approximately 1.5 times the total height of the pier or abutments, while that for masonry arch bridges it is about 2.0 times the height of the keystone above the foundation. For major bridges with more elaborate foundations, the question has to be examined in greater detail.

The alignment of piers and abutments should be, as far as possible, parallel to the mean direction of flow in the stream. If any temporary variation in the direction and velocity of the stream current is anticipated, suitable protective works should be provided to protect the substructure against the harmful effects on the stability of the bridge structure.

2.9 Vertical Clearance Above HFL

For high level bridges, a vertical clearance should be allowed between the highest flood level (HFL) and the lowest point of the superstructure. This is required to allow for any possible error in the estimation of the HFL and the design discharge. It also allows floating debris to pass under the bridge without damaging the structure. It is, therefore, logical to provide higher values of clearance for greater discharges as in Table 2.5.

For arched bridges, the clearance below the crown of the intrados of the arch should not be less than one-tenth of the maximum depth of the water plus one-third of the rise of the arch intrados. For structures provided with metallic bearings, the clearance between the base of the bearings and the highest flood level taking afflux into account is not to be less than 500 mm. For irrigation channels, the vertical clearance may be relaxed at the discretion of the engineer-in-charge.

The difference between vertical clearance and freeboard is sometimes not clearly understood. While vertical clearance is the difference in level between HFL and the lowest point of the superstructure, freeboard is associated with the approaches and guide bunds. The freeboard at any point is the difference between the highest flood level after allowing for afflux, if any, and the formation level of road embankment on the approaches or top level of guide bunds at that point. For high level bridges, the freeboard should not be less than 1750 mm.

2.10 Subsoil Exploration

The determination of a reasonably accurate soil profile at each of the proposed bridge sites is essential for correctly deciding the location and type of foundation. The subsoil investigation should aim to provide adequate data to the designer at a cost consistent with the value of the information. Availability of correct and reliable data would enable the construction contractors to estimate their costs correctly and to plan their work intelligently,

Table 2.5 Minimum Vertical Clearance.

Discharge m^3 per second	Minimum vertical clearance mm
Below 0.3	150
0.3 to 3.0	450
3.1 to 30.0	600
31 to 300	900
301 to 3000	1200
Over 3000	1500

resulting in a better job at a lower cost. Defects in bridge structures attributable to serious errors in subsoil exploration cannot be easily rectified later. Guidelines for subsoil exploration are available in IRC Bridge Code - Section VII[5].

The sub-surface investigations for bridges can be carried out in two stages: preliminary and detailed. The aim of the preliminary investigation is to get a general idea about the nature of soil strata. The methods used may include the study of existing geological information, besides field investigations with sounding rods, auger borings, wash borings, test pits and possibly geophysical methods.

Sounding rods consist of solid bars of tool steel of l5 to 20 mm diameter. The bottom end of the first length (usually about 3 m) is pointed. The top end is threaded so that additional lengths of about 2 m can be coupled. The sounding rod is first churned into the soil by hand to about 2 m and then driven with a 50 kN hammer. Additional sections are added as necessary. The resistance to driving is to be carefully interpreted to give a rough idea of the type of soil met with. This method is suitable only as a preliminary investigation to determine the location of rock at shallow depths.

Earth auger and post-hole auger are used singly for shallow depths and in conjunction with a 60 mm pipe casing and additional coupled rods for larger depths up to about 15 to 20 m. Samples are taken out in the augers and examined. When samples will not stick to the auger, wash borings are used. Here, a water jet is forced down a wash pipe (drill rod itself, if hollow). The water rises through the annular space between the casing and the wash pipe, carrying fragments of soil which is collected and examined.

After a study of the data collected in the preliminary investigation, the bridge site, the type of structure with span layout, and the location and type of foundations are tentatively determined. The detailed exploration should cover the entire length of the bridge and should extend at either end of the bridge for a distance of about twice the depth of foundation of the abutment. The depth of investigation should generally extend to a depth about twice the width of foundation below the level of the foundation, and should preferably be extended into firm soil or rock.

Borings have to be taken at suitable intervals, including the probable locations of abutments and piers. The data required are: (i) nature of soil deposit, (ii) depths and thicknesses of soil strata, (iii) location of ground water table, (iv) depth to rock bed, and (v) engineering properties of soils and rock. Standard penetration test (SPT) values expressed as number of blows per 300 mm penetration are also determined at different depths below the surface. Core drills are used to take out samples of rock to examine the thickness of rock layer, and to ensure that the layer reached is not a boulder but rock of adequate thickness.

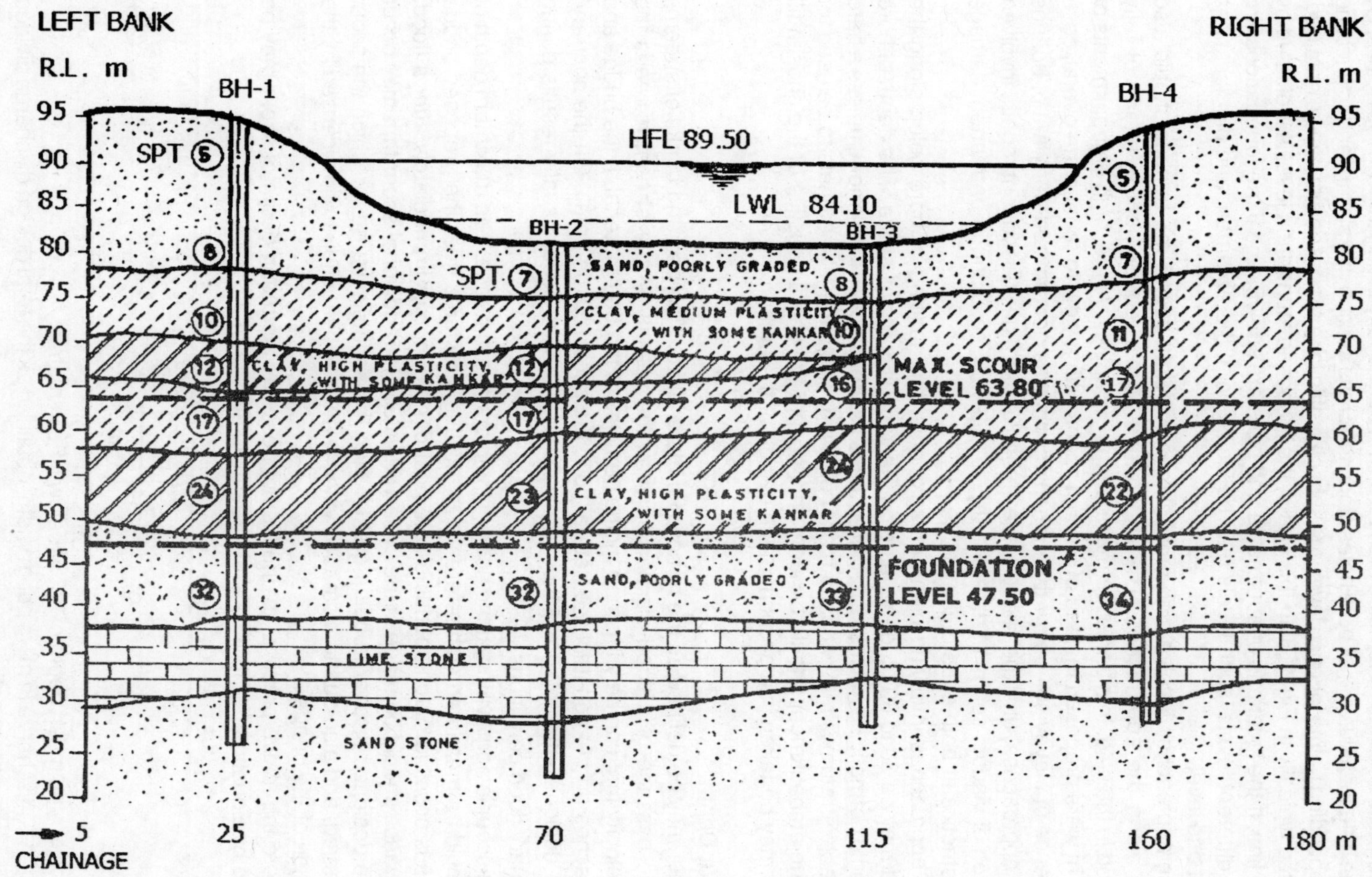

Figure 2.3 Typical Subsoil Profile.

Boring charts for the bore holes are first drawn individually listing the various details. By plotting the data for a number of bore holes along the cross section of a stream and connecting the corresponding points for each layer, the subsoil profile for a proposed site can be obtained, as indicated typically in Fig. 2.3. In this figure, the proposed foundation level is indicated by a dashed line. If the locations of abutments and piers are known, and if the foundations are rested at different levels, such levels may be indicated at the abutment and pier locations instead of one level.

If the approaches involve high embankments (height over 6 m), borings should also be taken along the approaches. The importance of this requirement is often not fully understood in the field. In a particular major bridge of cantilever type, high embankments of about 12 m were proposed instead of viaduct spans with a view to reduce the overall cost. The presence of clayey soil below the embankments was not realized (for want of borings along the approaches) until the failure of the embankment. The subsequent repairs involved introduction of viaduct spans of different structural layout, resulting in permanent damage to the aesthetics at both ends of the bridge.

For approaches involving high embankments, the spacing of bore holes along the alignment may be 30 to 80 m, depending on the variability of soil strata. The depth of bore holes should normally be not less than three times the height of embankment, unless a hard impenetrable stratum is met with at a higher elevation. Also the fill materials to be used for forming the embankment should be identified, so that the characteristics of the chosen fill materials may be taken into account in the design of the embankment.

2.11 Scour Depth

Scour may be defined as the removal of material from the bed and banks of streams during the passage of a flood discharge, when the velocity of the stream exceeds the limiting velocity that can be withstood by the particles of the bed material. If the bridge and its approaches do not constrict the natural flow, the scour will be small. On the contrary, when the designer attempts to reduce the waterway, severe scour usually results during extraordinary flood conditions.

The scour is aggravated at the nose of the piers and at bends. The maximum depth of scour should be measured with reference to existing structures near the proposed bridge site, if this is possible. Such soundings are best done during or immediately after a flood. Due allowance should be made in the observed values for additional scour that may occur due to the designed discharge being greater than the flood discharge for which the scour was observed, and also due to increased velocity due to obstruction to flow caused by the construction of the bridge.

Where the above practical method is not possible, the mean depth of scour may be computed by Equation (2.8) for natural streams in alluvial beds.

$$d_{sm} = 1.34\left(\frac{D_b^{\,2}}{K_{sf}}\right)^{0.33} \qquad \text{... (2.8)}$$

where d_{sm} = the mean depth of scour below HFL in metres

D_b = discharge in m^3/s per m width, obtained as the total design discharge divided by the effective linear waterway

K_{sf} = silt factor for a representative sample of the bed material, as in Table 2.6, taken as 1.76 times the square root of the particle size in mm (weighted mean diameter of particle determined as indicated in Appendix 2 of IRC:5-1998).

Table 2.6 Silt Factor.

Type of bed material	Size of particles mm	Silt factor
Very fine silt	0.08	0.50
Fine silt	0.12	0.60
Fine silt	0.16	0.70
Medium silt	0.23	0.85
Standard silt	0.32	1.00
Medium sand	0.50	1.25
Coarse sand	0.73	1.50
Heavy sand	1.29	2.00

In order to provide for an adequate margin of safety, the design discharge for the above calculation is increased by 30%, 25 to 20%, 20 to 10% and 10% for catchment areas of below 500 km², between 500 and 5000 km², between 5000 to 25000 km² and over 25000 km², respectively. When the effective linear waterway L is less than the regime width W, the value of d_{sm} computed from Equation (2.8) is to be increased by multiplying the same by the factor $(W/L)^{0.67}$.

The maximum depth of scour D below the HFL is to be taken as below:

(i) in a straight reach — $1.27\ d_{sm}$
(ii) at a moderate bend — $1.50\ d_{sm}$
(iii) at a severe bend — $1.75\ d_{sm}$
(iv) at a right angled bend — $2.00\ d_{sm}$
(v) at noses of piers — $2.00\ d_{sm}$
(vi) near abutments — $1.27\ d_{sm}$

The minimum depth of foundations below HFL is kept at 1.33 D for erodible strata. If the river is of a flashy nature and the bed does not submit readily to the scouring effects of the floods, the maximum depth of scour should be assessed by observations and not by the above calculations.

When a bridge is located close to the mouth of a river joining the sea, the possibility exists for the situation of the high tide opposing the flood discharge, resulting in heading up of the water level in the river. At the end of the high tide, the flood discharge may be relatively sudden, which may cause scour in excess of the values computed by Equation (2.8). Considerable engineering judgment is required in assessing the required depth of foundation in such cases.

2.12 Traffic Projection

The present traffic on the road at the proposed bridge site should be determined by traffic survey. With the construction of the bridge, the future traffic is likely to increase. The estimated traffic volume over the next 20 years should be assessed reasonably, taking into consideration the socio-economic conditions in the region and likely further development. This information is necessary to compute the benefits from the construction of the bridge.

2.13 Investigation Report

The engineer in charge of the investigation should prepare a detailed investigation report incorporating the various data collected, drawings prepared and calculations made as discussed in the earlier sections. The report should discuss the need for the bridge, the various sites considered, any special model studies conducted for river training and protection works and the criteria adopted for the design of the various components. An approximate cost estimate should be presented, along with comments on the economic feasibility indicating the value of the benefit-cost ratio or the internal rate of return. The report should also include any special features and precautions to be noted by the engineer during construction of the project.

2.14 Choice of Bridge Type

The choice of an appropriate type of bridge and planning of its basic features usually constitute a crucial decision to be taken by the bridge engineer. The designer must consider all the preliminary data made available to him from the detailed investigation before arriving at a solution. The entire completed structure should be the most suitable to carry the desired traffic, adequately strong to support the incident loads, economical in first cost and maintenance, and aesthetically pleasing.

Some of the factors influencing the choice of the bridge type and its basic features are the following:

1. The need to economize on the overall construction cost to the community by combining the railway and highway requirements may necessitate a road-cum-rail bridge in two tiers across a very wide river, e.g., Godavari second bridge.
2. Large navigational clearances required may dictate the use of particular types such as arches, cantilever bridges, cable stayed construction or suspension bridges, e.g., Howrah bridge.
3. Long and high approaches may be too costly at a plain coastal area for a railway line with low volume of traffic and it may be desirable to have a low level structure with a movable (bascule, swing or lift) span to cater to navigation, e.g., railway bridge at Pamban near Rameswaram.
4. A high level structure with uninterrupted traffic as on a National Highway and the need to reduce the number of piers may necessitate a cantilever bridge or a cable stayed bridge or a series of simply supported trusses, e.g., Zuari bridge, Ganga bridge at Patna.
5. The climatic and environmental conditions would preclude the use of some types and require some others. For example, the corrosive atmosphere has dictated the use of cantilever construction with precast segments for the prestressed concrete navigation span for the road bridge at Pamban near Rameswaram and has precluded the use of a cable stayed steel bridge.
6. Deck bridges are preferred to through bridges for highway traffic because of the better view of the surrounding scenery, e.g., Rainbow bridge near Niagara Falls.
7. The topographic and soil conditions at a site may limit the choice to a few general possibilities, e.g., a rocky ravine area is ideal for an arch bridge as in Salginatobel bridge.

8. Weak subsoil conditions may lead to the use of simply supported spans instead of continuous spans, e.g., bridges in areas subject to mining subsidence.
9. Shortage of funds may necessitate the adoption of a submersible bridge instead of a high level bridge on a road with low volume of traffic, and this in turn may result in reinforced concrete slab decking, e.g., Hosmatta bridge on Bisley Ghat Road in S. Kanara District.
10. The type of traffic may restrict the choice of bridge type. For railway traffic, steel trusses or steel cantilever types are preferable to suspension bridges.
11. The personal preferences or company specialization of the designer/construction firm may influence the type of bridge finally adopted, especially when competitive tenders are obtained for long span bridges with freedom to submit alternative designs. A firm specializing in prestressed concrete cantilever construction and another firm specializing in cable stayed steel bridges will offer different designs for the same bridge site, each design emphasizing the specialization of the concerned firm. Similarly, a builder possessing efficient piling rigs and another with expertise in well sinking will try to promote their approaches to the foundations for a multi-span bridge across a wide river.

The above discussion indicates that no hard-and-fast rules can be prescribed for the choice of bridge type for all cases. Each situation will call for individual study and decision. Good designs result from serious search for the best solution for the given situation, taking into account the technological possibilities of materials and construction methods, and based on a profound knowledge of structural behaviour besides aiming for an aesthetic structure. The author would urge every young engineer aspiring to become a successful bridge designer to develop the capability by making detailed study of the special features of existing bridges as well as failures, by making freehand sketches of various alternative designs, and by acquiring through practice an aesthetic and engineering sense to judge the merits of alternative designs.

2.15 Importance of Proper Investigation

The importance of careful investigation for a bridge cannot be over-emphasized. Expensive errors can be eliminated if the officer-in-charge of investigation acts diligently. It may be helpful to note that serious mistakes have actually occurred in the past due to defective investigation. Some examples listed by the Indian Roads Congress[6] are given below:

(i) Bridge sited immediately downstream of a junction of two big rivers. (Sites upstream should have been investigated.)

(ii) Bridge undermined after construction owing to regression of soft rock in the river bed. A series of water-falls immediately downstream were not taken into account.

(iii) The designs of two bridges were interchanged and each was built over the stream for which the other was intended. (Error occurred due to omission of reference to neighbouring villages, streams themselves and exact bridge sites of the bridges in their plans.)

(iv) Site involving deep foundations was selected, while a better investigation would have revealed the presence of a nearby downstream site with shallower foundations.

(v) Bridge much longer than necessary was built. (In this case, the investigation engineer failed to realize that the widespread of the river at the selected site was due to the presence of an old causeway and that the river admitted much narrower stream widths at about 0.8 km above and below the selected site.)

(vi) More money was spent on the approaches to a bridge than on the bridge itself. Had the bridge been built 70 metres upstream, the total cost would have been halved.

2.16 References

1. 'IRC: 5-1998 Standard specifications and code of practice for road bridges: Section I - General features of design', Indian Roads Congress, 1998, 38 pp.
2. Victor, D.J., 'The investigation, design and construction of submersible bridges', Journal of Indian Roads Congress, Vol. XXIV-1, Oct. 1959, pp. 181-213.
3. IRC:SP:13-2004, 'Guidelines for the design of small bridges and culverts', Indian Roads Congress, 2004, 106 pp.
4. Linsley, R.K., and Franzini, 'Water resources engineering", McGraw-Hill Kogakusha, Tokyo, 1979. 716 pp.
5. 'IRC: 78-2000 Standard specifications and code of practice for road bridges: Section VII - Foundations and substructure', Indian Roads Congress, 2000, 97 pp.
6. 'Bridging India's rivers, Vol. I', Indian Roads Congress, New Delhi, p. 7.

Chapter 3

Standard Specifications for Road Bridges

3.1 General

The Indian Roads Congress (IRC) has formulated Standard Specifications and Codes of Practice for Road Bridges with a view to establish a common procedure for the design and construction of road bridges in India. The specifications are collectively known as the Bridge Code. Prior to the formulation of the IRC Bridge Code, there was no uniform code for the whole country. Each State (or province) had its own rules about the standard loadings and stresses. As an example, the evolution of standard loadings for highway bridges in Madras State is given in Fig. 3.1[1]. Currently, we would follow the IRC Bridge Code. Some of the provisions of the Code are reviewed in this chapter.

3.2 Indian Roads Congress Bridge Code

The Indian Roads Congress (IRC) Bridge Code as available now consists of eight sections as below[2–10].

(1) Section I – General features of design
(2) Section II – Loads and stresses
(3) Section III – Cement concrete (plain and reinforced)
(4) Section IV – Brick, stone and block masonry
(5) Section V – Steel road bridges
(6) Section VI – Composite construction
(7) Section VII – Foundations and substructure
(8) Section IX – Bearings

Section I gives the specifications for the preliminary data to be collected, determination of design discharge, clearances, foundations, etc. Section II specifies the loadings for which the bridges have to be designed. The other sections give rules for guidance in the design of the bridge superstructure in masonry, reinforced concrete, steel and composite construction, foundations and bearings. General guidelines for the design of prestressed concrete bridges are given in a separate publication[11]. Some of the important clauses of the code are discussed in this chapter. The codal provisions undergo frequent changes, and hence the reader is advised to familiarize himself with the relevant code in its latest amended form.

3.3 Width of Carriageway

The width of carriageway required will depend on the intensity and volume of traffic anticipated to use the bridge. The width of carriageway is expressed in terms of traffic lanes, each lane meaning the width required to accommodate one train of Class A vehicles.

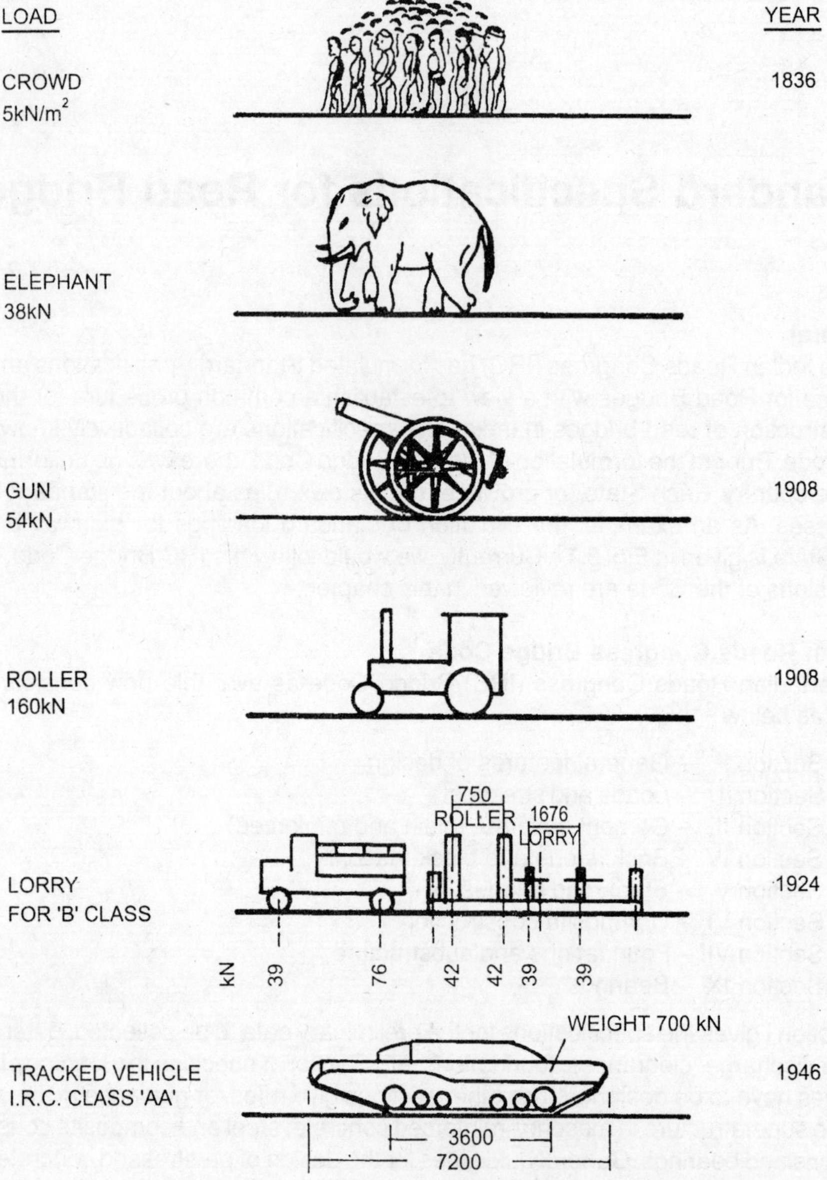

Figure 3.1 Evolution of Standard Loadings for Design of Bridges.

Except on minor village roads, all bridges must provide for at least two-lane width. The minimum width of carriageway is 4.25 m for a one-lane bridge and 7.5 m for a two-lane bridge. For every additional lane, a minimum of 3.5 m must be allowed. Bridges allowing traffic on both directions must have carriageways of two or four lanes or multiples of two lanes. Three-lane bridges should not be constructed, as these will be conducive to the occurrence of accidents. In the case of a wide bridge, it is desirable to provide a central

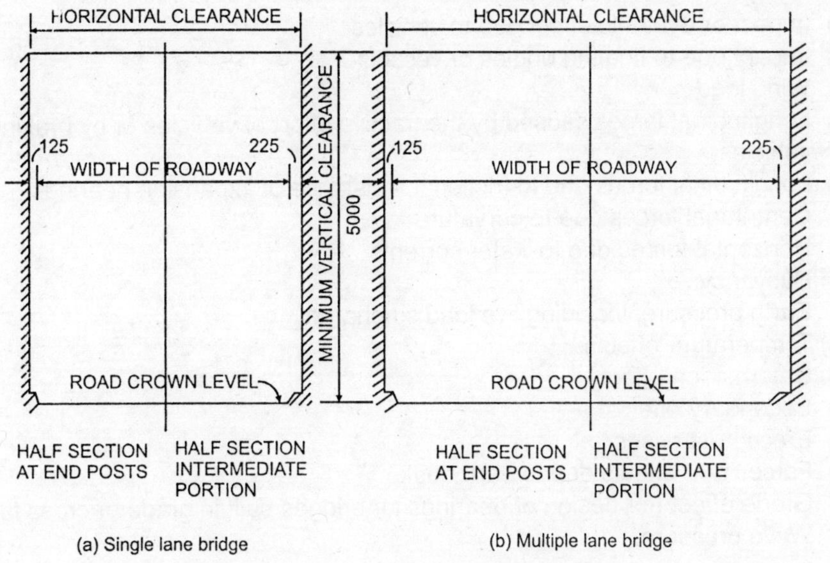

Figure 3.2 Clearance Diagram for Highway Traffic.

verge of at least 1.2 m width in order to separate the two opposing lines of traffic; in such a case, the individual carriageway on either side of the verge should provide for a minimum of two lanes of traffic. If the bridge is to carry a tramway or railway in addition, the width of the bridge should be increased suitably.

From consideration of safety and effective utilization of carriageway, it is desirable to provide footpath of at least 1.5 m width on either side of the carriageway for all bridges. In urban areas, it may be necessary also to provide for separate cycle tracks besides the carriageway.

3.4 Clearances

The horizontal and vertical clearances required for highway traffic are given in Fig. 3.2, wherein the maximum width and depth of a moving vehicle are assumed as 3300 mm and 4500 mm, respectively. The left half section of each diagram shows the main fixed structure between end posts of/on arch ribs, whereas the right half section shows the intermediate portions. For a bridge constructed on a horizontal curve with super-elevated surface, the minimum vertical clearance is to be measured from the super-elevated level of the roadway. The horizontal clearance should be increased on the inner side of the curve by an amount equal to 5 m multiplied by the super-elevation.

3.5 Loads to be Considered

While designing road bridges and culverts, the following loads, forces and stresses should be considered, where applicable:

(a) Dead load
(b) Live load
(c) Snow load

(d) Impact or dynamic effect due to vehicles
(e) Impact due to floating bodies or vessels
(f) Wind load
(g) Longitudinal forces caused by the tractive effort of vehicles or by braking of vehicles
(h) Longitudinal forces due to frictional resistance of expansion bearings
(i) Centrifugal forces due to curvature
(j) Horizontal forces due to water currents
(k) Buoyancy
(l) Earth pressure, including live load surcharge
(m) Temperature effects
(n) Deformation effects
(o) Secondary effects
(p) Erection stresses
(q) Forces and effects due to earthquake
(r) Grade effect (for design of bearings for bridges built in grade or cross fall)
(s) Wave pressure.

The basic philosophy governing the design of bridges is that a structure should be designed to sustain with a defined probability all actions likely to occur within its intended life span. In addition, the structure should maintain stability during unprecedented actions and should have adequate durability during its life span. Typical combinations of loads and forces to be considered in design and allowable increases in permissible stresses for certain combinations are given in the Code[3]. It will be necessary to ensure that when steel members are used, the maximum stress under any combination does not exceed the yield strength of the steel. Based on observations from recorded earthquakes, it is not considered probable that wind load and earthquake will occur simultaneously.

3.6 Dead Load

The dead load carried by a bridge member consists of its own weight and the portions of the weight of the superstructure and any fixed loads supported by the member. The dead load can be estimated fairly accurately during design and can be controlled during construction and service. As a guide in estimating the dead loads, the unit weights of materials may be assumed as given in Table 3.1.

3.7 IRC Standard Live Loads

Live loads are those caused by vehicles which pass over the bridge and are transient in nature. These loads cannot be estimated precisely, and the designer has very little control over them once the bridge is opened to traffic. However, hypothetical loadings which are reasonably realistic need to be evolved and specified to serve as design criteria.

There are four types of standard loadings for which road bridges are designed:

(a) IRC Class AA Loading: This loading consists of either a tracked vehicle of 700 kN or a wheeled vehicle of 400 kN with dimensions as shown in Fig. 3.3. The tracked vehicle simulates a combat tank used by the army. The ground contact length of the track is 3.6 m and the nose to tail length of the vehicle is 7.2 m. The nose to tail spacing between two successive vehicles shall not be less than 90 m. For two-lane bridges and culverts, one train

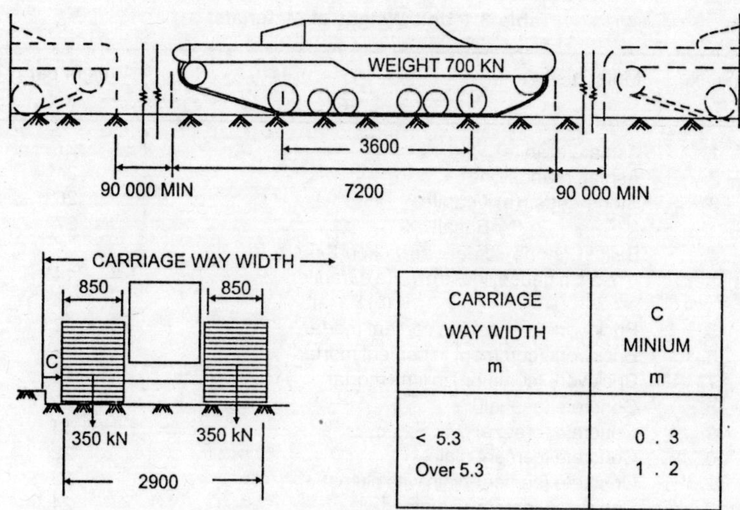

CARRIAGE WAY WIDTH m	C MINIUM m
< 5.3	0 . 3
Over 5.3	1 . 2

(a) TRACKED VEHICLE

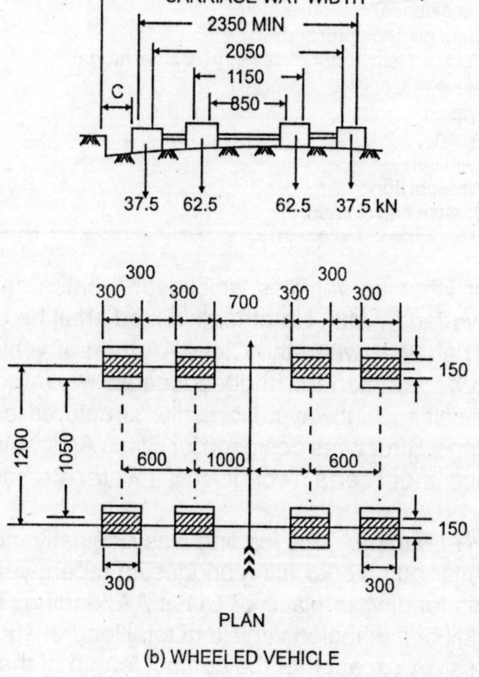

(b) WHEELED VEHICLE

Figure 3.3 IRC Class AA Loading.

Table 3.1 Unit Weight of Materials.

Sl. No.	Materials	Weight per m^3 kN
1.	Ashlar (granite)	27
2.	Ashlar (sandstone)	24
3.	Stone setts: (a) Granite	26
	(b) Basalt	27
4.	Ballast (stone screened, broken 2.5 cm to 7.5 cm gauge, loose): (a) Granite	14
	(b) Basalt	16
5.	Brickwork (pressed) in cement mortar	22
6.	Brickwork (common) in cement mortar	19
7.	Brickwork (common) in lime mortar	18
8.	Concrete (asphalt)	22
9.	Concrete (breeze)	14
10.	Concrete (cement-plain)	22
11.	Concrete (cement-plain with plums)	23
12.	Concrete (cement-reinforced)	24
13.	Concrete (cement-prestressed)	25
14.	Concrete (lime-brick aggregate)	19
15.	Concrete (lime-stone aggregate)	21
16.	Earth (compacted)	18
17.	Gravel	18
18.	Macadam (binder premix)	22
19.	Macadam (rolled)	26
20.	Sand (loose)	14
21.	Sand (wet compressed)	19
22.	Coursed rubble stone masonry (cement mortar)	26
23.	Stone masonry (lime mortar)	24
24.	Water	10
25.	Wood	8
26.	Cast iron	78
27.	Wrought iron	77
28.	Steel (rolled or cast)	78

of Class AA tracked or wheeled vehicles whichever creates severer conditions shall be considered for every two-lane width. No other live load shall be considered on any part of the above two-lane carriageway when the Class AA train of vehicles is on the bridge. The Class AA loading is to be adopted for bridges located within certain specified municipal localities and along specified highways. Normally, structures on National Highways are provided for these loadings. Structures designed for Class AA loading should also be checked for Class A loading, since under certain conditions, severer stresses may be obtained under Class A loading.

(b) IRC Class 70R Loading: This loading was originally included in the Appendix to the bridge code for use for rating of existing bridges. In recent years, there is an increasing tendency to specify this loading in place of Class AA loading. This loading consists of a tracked vehicle of 700 kN or a wheeled vehicle of total load of 1000 kN. The tracked vehicle is similar to that of Class AA except that the contact length of the track is 4.57 m, the nose to tail length of the vehicle is 7.92 m and the specified minimum spacing between successive vehicles is 30 m. The wheeled vehicle is 15.22 m long and has seven axles with loads totaling to 1000 kN. In addition, the effects on the bridge components due to a bogie loading

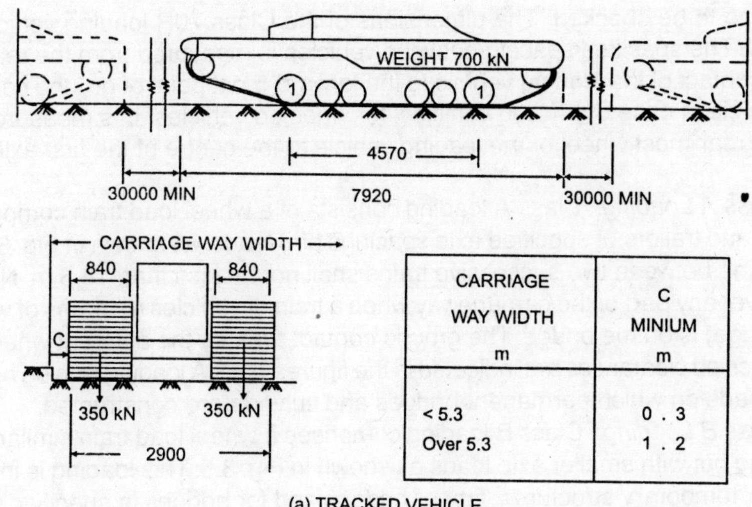

(a) TRACKED VEHICLE

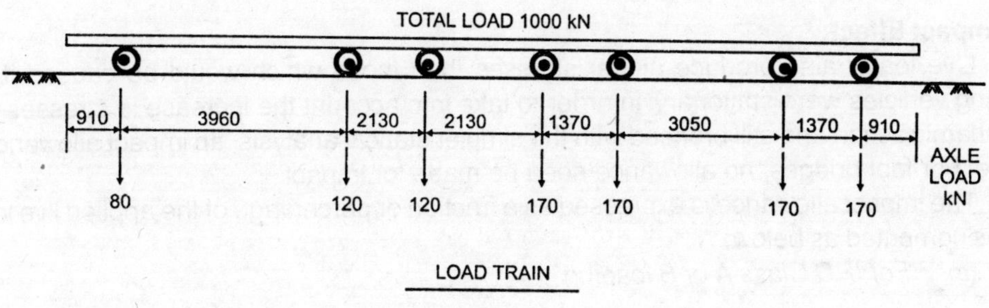

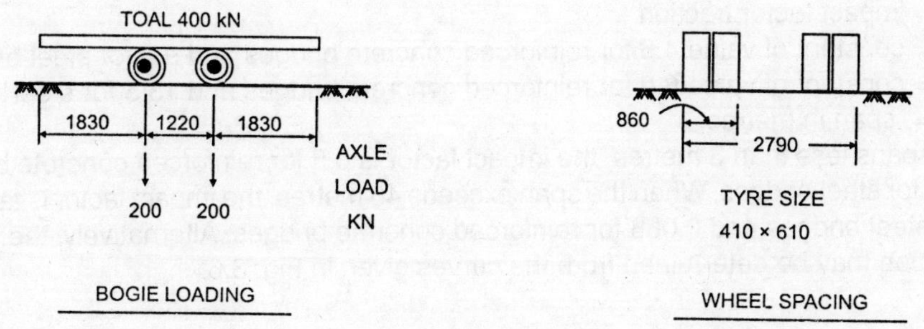

(b) WHEELED VEHICLE

Figure 3.4 IRC Class 70R Loading.

of 400 kN are also to be checked. The dimensions of the Class 70R loading vehicles are shown in Fig. 3.4. The specified spacing between vehicles is measured from the rear-most point of ground contact of the leading vehicle to the forward-most point of ground contact of the following vehicle in case of tracked vehicles; for wheeled vehicles, it is measured from the centre of the rear-most wheel of the leading vehicle to the centre of the first axle of the following vehicle.

(c) *IRC Class A Loading:* Class A loading consists of a wheel load train composed of a driving vehicle and trailers of specified axle spacings and loads, as shown in Fig. 3.5. The nose to tail spacing between two successive trains shall not be less than 18.5 m. No other live load shall cover any part of the carriageway when a train of vehicles (or trains of vehicles for multi-lane bridge) is on the bridge. The ground contact area for the different wheels and the minimum specified clearances are indicated in the figure. Class A loading is to be normally adopted on all roads on which permanent bridges and culverts are constructed.

(d) *IRC Class B Loading:* Class B loading comprises a wheel load train similar to that of Class A loading but with smaller axle loads as shown in Fig. 3.5. This loading is intended to be adopted for temporary structures, timber bridges and for bridges in specified areas.

The standard loads are to be arranged in such a manner as to produce the severest bending moment or shear at any section considered. The loading vehicles are to be aligned so as to travel parallel to the length of the bridge. Clause 207.4 of the Code specifies the combination of live load to be considered for single lane and multi-lane bridges.

3.8 Impact Effect

Live load trains produce higher stresses than those which would be caused if the loading vehicles were stationary. In order to take into account the increase in stresses due to dynamic action and still proceed with the simpler statical analysis, an impact allowance is made. For foot bridges, no allowance need be made for impact.

The impact allowance is expressed as a fraction or percentage of the applied live load, and is computed as below:

(*a*) *For IRC Class A or B loading*

$$I = \frac{A}{B+L} \qquad \text{... (3.1)}$$

where I = impact factor fraction

A = constant of value 4.5 for reinforced concrete bridges and 9.0 for steel bridges

B = constant of value 6.0 for reinforced concrete bridges and 13.5 for steel bridges

L = span in metres.

For spans less than 3 metres, the impact factor is 0.5 for reinforced concrete bridges and 0.545 for steel bridges. When the span exceeds 45 metres, the impact factor is taken as 0.154 for steel bridges and 0.088 for reinforced concrete bridges. Alternatively, the impact factor fraction may be determined from the curves given in Fig. 3.6.

(*b*) *For IRC Class AA or 70R loading*

(i) For spans less than 9 m

(a) For tracked vehicle .. 25% for spans up to 5 m linearly reducing to 10% for spans of 9 m.

(b) For wheeled vehicle .. 25%

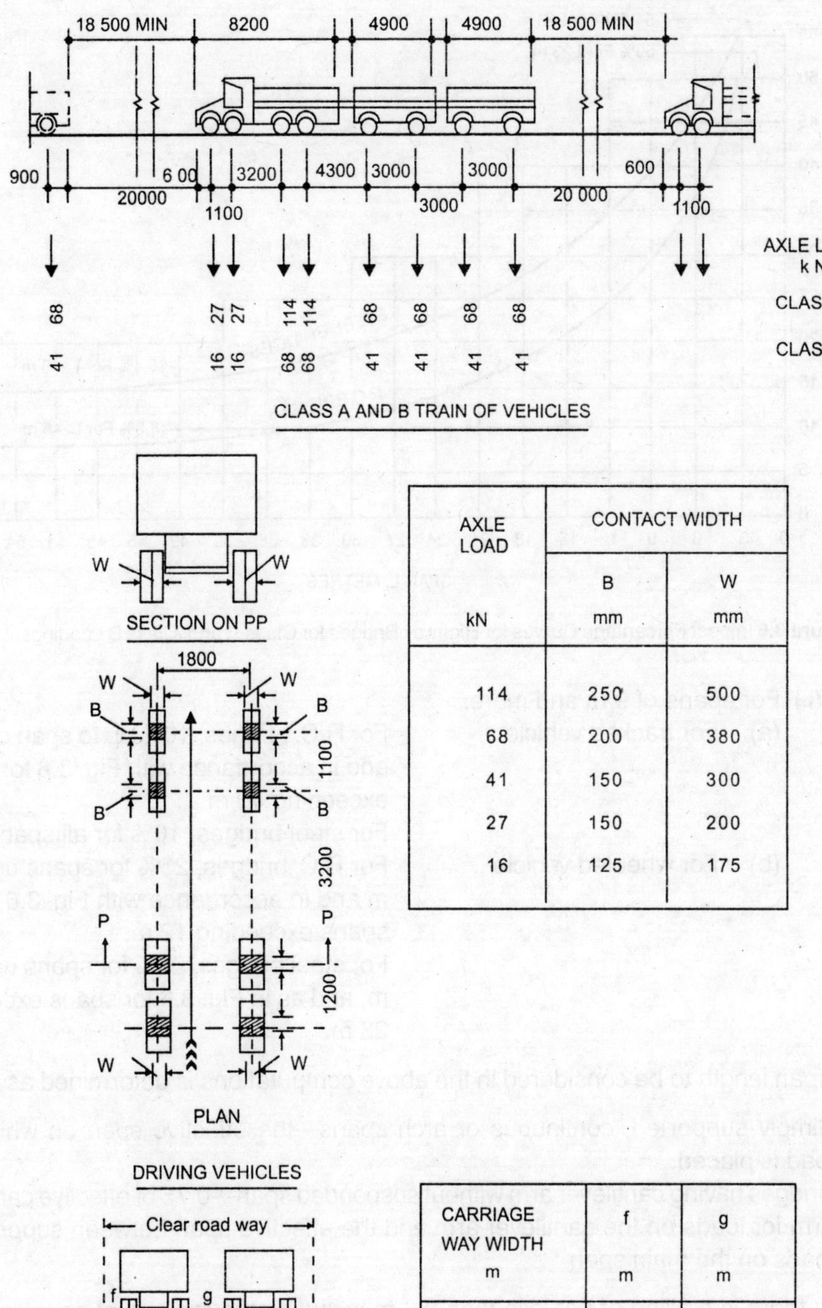

AXLE LOAD kN	CONTACT WIDTH	
	B mm	W mm
114	250	500
68	200	380
41	150	300
27	150	200
16	125	175

CARRIAGE WAY WIDTH m	f m	g m
5.5 to 7.5	0.15	0.4 to 1.2
Over 7.5	0.15	1.2

Figure 3.5 IRC Class A and B Loadings.

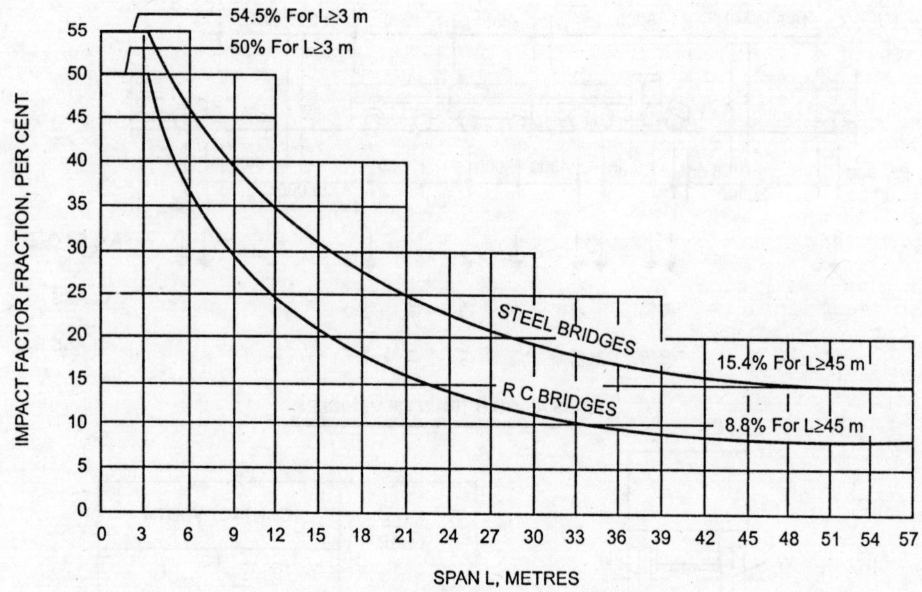

Figure 3.6 Impact Percentage Curves for Highway Bridges for Class A and Class B Loadings.

(ii) For spans of 9 m and more.

(a)	For tracked vehicle ..	For R.C. bridges, 10% up to span of 40 m and in accordance with Fig. 3.6 for spans exceeding 40 m, For steel bridges, 10% for all spans.
(b)	For wheeled vehicle ..	For R.C. bridges, 25% for spans up to 12 m and in accordance with Fig. 3.6 for spans exceeding 12 m. For steel bridges, 25% for spans up to 23 m, and as in Fig. 3.6 for spans exceeding 23 m.

The span length to be considered in the above computations is determined as below:

(i) Simply supported, continuous or arch spans—the effective span on which the load is placed.

(ii) Bridges having cantilever arm without suspended span—0.75 of effective cantilever arm for loads on the cantilever arm and the effective span between supports for loads on the main span.

When there is a filling of not less than 0.6 m including the road crust as in spandrel filled arches, the impact allowance may be taken as half that computed by the above procedure.

Full impact allowance should be made for design of bearings. But for computing the pressure at different levels of the substructure, a reduced impact allowance is made by multiplying the appropriate impact fraction by a factor as below:

(i) At the bottom of bed block 0.5
(ii) For the top 3 m of the sub-structure below the bed block 0.5 decreasing uniformly to zero
(iii) For portion of sub-structure more than 3 m below the bed block 0.0

3.9 Review of IRC Loadings

Thomas[12,13] has reported a comparative study of the IRC loadings with the loadings of seven other countries. He has shown that the IRC loading is the most severe for a single lane bridge, but is less severe than the French, German, Japanese and British loadings for a two-lane bridge. Further, the loadings are complicated in application to design, especially if Class 70R, Class AA and Class A loadings are to be considered in the design to determine the severest effects.

Very little information is available on the basis for the IRC loadings. While considerable refinement in the methods of analysis and design has been achieved, studies on the accuracy and adequacy of the assumptions of loadings have been neglected. The laborious computations involved in applying the IRC loadings to an actual design may create an impression that the design moments are being assessed precisely. In fact, the IRC loadings have little relation to the vehicles currently in use in the country. The Class AA tracked vehicle load of 700 kN is by no means an accurate representation of present military tanks, and a specified tail-to-nose distance of 90 m is not likely to be observed in the event of any emergency. Similarly, axle loads and spacings specified for wheel trains need not be exact. While trucks manufactured in our country could perhaps be controlled, imported vehicles may not satisfy these specifications. Thus the design moments and shears assessed from these hypothetical loadings after laborious computations can at best be only approximate. The value of refinement of knowledge and accuracy of prediction of the behaviour of structures under load is considerably diminished if it is not matched by corresponding precision of estimation of the loadings that cause that behaviour.

Even basic anomalies exist in the prescribed loadings. For example, the nose to tail spacing between two successive vehicles of Class AA tracked vehicle is 90 m while that for Class 70R is 30 m, though the vehicles are very similar in both cases. Further, the justification for the use in India of severer loadings than in advanced countries deserves serious consideration. In view of the above, the author strongly advocates the dropping of Class 70R loading and the development of simpler and more realistic loadings.

With a view to stimulate efforts towards development of simplified standard loadings, the author[14] proposed in 1968 equivalent simplified loadings applicable for slab bridges up to 7.6 m. The proposed loading consisted of a uniformly distributed load applied in conjunction with a knife edge load. The magnitudes were indicated for heavy loading, standard loading and light loading, corresponding to IRC Class AA, Class A and Class B loadings. Thomas[13] has subsequently evolved a new loading standard on similar lines but justified in greater detail and over a wider span range. However, till the IRC revise their standards, the current loadings as in Section 3.7 are to be adopted in designs.

The basis for IRC provisions regarding impact is not clear. The actual impact factor will depend on the bridge span, the surface characteristics of the bridge and the spring constant of the vehicle. Systematic studies are needed to derive realistic impact factor for conditions in our country. Field experiments in Britain by Mitchell[15] indicated that the impact effect need not be considered for the full live load but need only be applied to the heaviest axle or the pair of adjacent wheels causing the maximum moment or shear. Based on the above study, and practice in some other countries, Thomas[13] has advocated that the impact allowance be taken as 30 per cent and that this allowance be applied only on the heaviest axle or the pair of adjacent wheels, which produces the greatest bending moment or shear as the case may be. However, till the bridge code is revised, the rules given in Section 3.8 are to be followed in design.

Discussing the anamolies inherent in the present codes of different countries, Reddi[16] has pleaded for realism in loading standards and harmonisation of the provisions in the different national codes.

3.10 Application of Live Loads on Deck Slabs

Any rational method may be used for calculating the effect of concentrated loads on deck slabs. The disposition of the loading should be so arranged as to produce the maximum bending moment or shear for the deck slab.

A detailed description of the effect of concentrated loads on one-way and two-way slabs is given in Appendix A.

3.11 Wind Load

All bridge structures should be designed for the wind forces as below. Though the wind forces are dynamic in nature, the forces can be approximated as equivalent static loads. These forces are considered to act horizontally and in such a direction as to cause the maximum stresses in the member under consideration.

The intensity of wind force is taken from Table 3.2. These values are to be doubled for the Kathiawar Peninsula and the coastal regions of West Bengal and Orissa. It may be noted that the velocity and wind pressure increase non-linearly with the height above the ground or water level.

Table 3.2 Wind Velocities and Wind Pressures.

H m	V km/h	P kN/m^2	H m	V km/h	P kN/m^2
0	80	0.40	30	147	1.41
2	91	0.52	40	155	1.57
4	100	0.63	50	162	1.71
6	107	0.73	60	168	1.83
8	113	0.82	70	173	1.93
10	118	0.91	80	177	2.02
15	128	1.07	90	180	2.10
20	136	1.19	100	183	2.17
25	142	1.30	110	186	2.24

Note: H = average height in metres of the exposed surface above the mean retarding surface (ground or bed level or water level).
V = horizontal velocity of wind in kilometres per hour at height H
P = horizontal wind pressure kN/m^2 at height H.

The area on which the wind force is assumed to act is determined as below:

(i) For a deck structure:

The area of the structure as seen in elevation including the floor system and railing, less area of perforations in the railings or parapets.

(ii) For a through or half-through structure:

The area of the elevation of the windward truss, plus half the area of elevation above the deck level of all other trusses or girders.

The wind load acting on any exposed moving live load will be assumed to act at a height of 1.5 m above the roadway and to have a value 3 kN per linear metre. For the purpose of this calculation, the clear distance between the trailers of a train of vehicles should not be omitted.

When the wind velocity at deck level exceeds 130 km/h, no live load need be considered to be acting on the bridge.

The total assumed wind force shall not be less than 4.5 kN per linear metre in the plane of the loaded chord and 2.25 kN per linear metre in the plane of the unloaded chord on through or half-through truss, latticed or similar spans and not less 4.5 kN per linear metre on deck spans.

A wind pressure of 2.4 kN/m^2 on the unloaded structure shall be used if it produces greater stresses than those produced by the combined wind forces as stated above.

3.12 Longitudinal Forces

Longitudinal forces are caused in road bridges due to any one or more of the following:

(a) Tractive effort caused through acceleration of the driving wheels;
(b) Braking effect due to application of brakes to the wheels;
(c) Frictional resistance offered to the movement of free bearings due to change of temperature or any other cause.

Braking effect is invariably greater than the tractive effort. It is computed as follows:

(i) Single lane or two-lane bridge:

20% of the first train load plus 10% of the loads in succeeding trains or parts thereof on any one lane only. If the entire first train is not on the full span, the braking force is taken as 20% of the loads actually on the span. No impact allowance is included for this computation.

(ii) Multi-lane bridge:

As in (i) above for the first two lanes plus 5% of the loads on the lanes in excess of two.

The force due to braking effect shall be assumed to act along a line parallel to the roadways and 1.2 m above it. While transferring the force to the bearings, the change in the vertical reaction at the bearings should be taken into account.

The longitudinal force due to friction at a free bearing is to be limited to the sum of the dead and live load reactions at the bearing multiplied by the appropriate coefficient of friction assumed as below:

Steel roller bearings	0.03
Concrete roller bearings	0.05
Sliding bearings of hard copper alloy	0.15
Sliding bearings of steel on cast iron or steel	0.50
Sliding bearings of concrete over concrete with bitumen layer in between	0.60
Sliding bearings of teflon on stainless steel	0.05
Other types	Values to be decided by the engineer-in-charge based on available data.

The longitudinal force at a fixed bearing shall be taken as the algebraic sum of the longitudinal forces at the free bearings and the force due to the braking effect on the wheels.

For bridge structures without bearings such as arches, rigid frames, etc., the effects of braking force shall be calculated in accordance with standard methods of analysis of indeterminate structures.

The effects of longitudinal forces and all other horizontal forces should be calculated up to a level where the resultant passive earth resistance of the soil below the deepest scour level balances these forces.

3.13 Centrifugal Forces

Where a road bridge is located on a curve, the effects of centrifugal forces due to movement of vehicles should be taken into account.

The centrifugal force is given by Equation (3.2).

$$C = \frac{WV^2}{12.95\,R} \qquad \text{... (3.2)}$$

where C = centrifugal force in kN acting normally to the traffic (1) at the point of action of the wheel loads or (2) uniformly distributed over every metre length on which the uniformly distributed loads act

W = live load (1) in kN for wheel loads and (2) in kN/m for uniformly distributed live load

V = design speed in km per hour

R = radius of curvature in metres.

The centrifugal force is assumed to act at a height of 1200 mm above the level of the carriageway. The force is not increased for impact effect.

3.14 Horizontal Forces Due to Water Currents

Any part of a bridge structure which may be submerged in running water should be designed to sustain safely the horizontal pressure due to the force of the current.

On piers parallel to the direction of the water current, the intensity of pressure is given by Equation (3.3).

$$P = KW(V^2/2g) \qquad \text{... (3.3)}$$

where P = intensity of pressure in kN/m^2 due to the water current

W = unit weight of water in kN/m^3

V = velocity of current in m/s at the point where the pressure intensity is being calculated.
g = acceleration due to gravity in m/s^2
K = a constant depending on the shape of pier, taken as 1.50 for square ended piers, 0.66 for circular piers or for piers with semi-circular cutwaters, 0.5 to 0.9 for triangular cutwaters, and 1.25 for trestle type piers.

With the usual values of W and g, Equation (3.3) reduces to Equation (3.4).

$$P = 0.5\, KV^2 \qquad \text{... (3.4)}$$

The value of V^2 in Equation (3.3) is assumed to vary linearly from zero at the point of deepest scour to the square of the maximum velocity at the free surface of water. The maximum velocity at surface for the purpose of this clause is to be taken as $\sqrt{2}$ times the maximum mean velocity (v) of the current. In other words, the value of V^2 at the free surface is assumed to be $2v^2$.

When the current strikes the pier at an angle, the velocity is resolved into two components, parallel and normal to the pier. The pressures are then computed using the values of the components in Equation (3.3), with K assumed as 1.5 for all except circular piers.

To provide for the possible variation of the direction of the current from the direction assumed in the design, allowance should be made in the design of the piers for an extra variation in the current direction of 20 degrees.

3.15 Buoyancy Effect

Wherever submersion in water of a part or whole of a structure is possible, the forces due to buoyancy should be considered. In the case of submersible bridges, full buoyancy should be considered for the superstructure, piers and abutments.

For high level bridges, buoyancy forces due to submerged part of the substructure and foundations should be taken into account. For the piers submerged in water and for foundations in sand, full buoyancy is allowed. For other soils, a suitable proportion may be assumed.

3.16 Earth Pressure

Earth pressures computed according to a rational theory should be considered in the case of structures designed to retain earthfills[17,18]. Coulomb's theory is acceptable, subject to the modification that the centre of pressure exerted by the backfill, when considered dry, is located at an elevation of 0.42 of the height of the wall above the base, instead of 0.33 of that height. The minimum pressure for a retaining structure is that from a fluid weighing 4.8 kN/m^3.

The thrust P according to Coulomb's theory may be expressed by Equation (3.5) with the notations as indicated in Fig. 3.7.

$$P = 0.5\, wh^2 \left[\frac{\csc\theta \sin(\theta - \phi)}{\sqrt{\sin(\theta + z)} + \sqrt{\dfrac{\sin(\phi + z)\sin(\phi - \delta)}{\sin(\theta - \delta)}}} \right]^2 \qquad \text{... (3.5)}$$

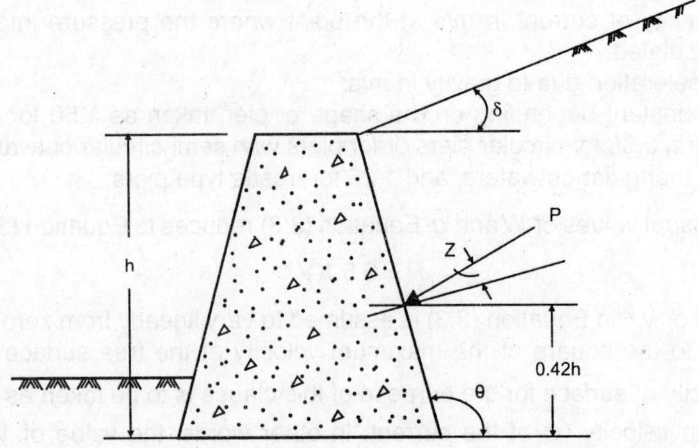

Figure 3.7 Notation for Equation (3.5).

where P = total active pressure, acting at a height of 0.42 h inclined at z to the normal to the wall on the earth side

w = unit weight of earthfill

h = height of wall

θ = angle subtended by the earthside wall with the horizontal on the earthside

ϕ = angle of internal friction of the earthfill

z = angle of friction of the earthside wall with the earth

δ = inclination of the earthfill surface with the horizontal.

If θ equals 90 degrees and z equals δ, the conditions conform to Rankine's theory, and Equation (3.5) reduces to the form of Equation (3.6).

$$P = 0.5\, wh^2.\cos\delta\,\frac{\cos\delta - \sqrt{\cos^2\delta - \cos^2\phi}}{\cos\delta + \sqrt{\cos^2\delta - \cos^2\phi}} \qquad ...(3.6)$$

Further, when the backfill is level, i.e. when δ equals zero, Equation (3.6) further reduces to the more familiar form given in Equation (3.7).

$$P = 0.5\, wh^2 \cdot \frac{1-\sin\phi}{1+\sin\phi} = 0.5\, wh^2.\tan^2\left(45 - \frac{\phi}{2}\right) \qquad ...(3.7)$$

For the design of abutments, the effect of concentrated live loads on the surface may be reckoned as equivalent surcharge and may be computed by any rational method of design, such as the one using Spangler's equation. As per Clause 217.1 of the Bridge Code, all abutments and return walls should be designed for a live load surcharge equivalent to 1.2 m height of earthfill.

For major bridges, it is necessary to provide an adequately designed reinforced concrete approach slab covering the entire width of the roadway with one end resting on the dirtwall of the abutment and extending for a length not less than 3.5 m into the approach. The approach slab may be of M30 grade concrete and reinforced with 12 mm diameter rods at 150 mm centres in each direction both at top and bottom. Thorough drainage of the backfilling material should be ensured by means of a filter medium to 600 mm thickness and weep holes at about 1 m spacing in both directions.

3.17 Temperature Effects

Daily and seasonal variations in temperature occur causing material to shorten with a fall in temperature and lengthen with a rise in temperature. These variations have two components: a uniform change over the entire bridge deck and a temperature gradient caused by the difference in temperatures at the top and the bottom of the deck.

Suitable provisions should be made for stresses or movements resulting from variations in temperature. The probable rise and fall in temperature shall be determined from meteorological records for the locality in which the bridge is located. In case of massive concrete members the time lag between air temperature and the interior temperature should be considered. The coefficient of expansion per degree centigrade shall be taken as 11.7 x 10^{-6} for steel and reinforced concrete structures and as 10.8 x 10^{-6} for plain concrete structures.

3.18 Deformation Stresses

Deformation stresses are considered for steel bridges only. A deformation stress is defined as the bending stress in any member of an open-web girder caused by the vertical deflection of the girder combined with the rigidity of the joints. No other stresses are included in this definition. The design, manufacture and erection of steel bridges should be so arranged as to keep the deformation stresses to a minimum. If detailed computations are not made to provide otherwise, the deformation stresses should be assumed to be not less than 16 per cent of the dead and live load stresses.

3.19 Secondary Stresses

In steel structures, secondary stresses are caused due to eccentricity of connections, floor beam loads applied at intermediate points in a panel, cross girders being connected away from panel points, lateral wind loads on the end posts of through girders, and movement of supports. Secondary stresses are brought into play in reinforced concrete structures due either to the movement of supports or to the deformations in the geometrical shape of the structure or its member, resulting from causes such as rigidity of end connection or loads applied at intermediate points of trusses or restrictive shrinkage of concrete floor beams. For reinforced concrete members, the shrinkage coefficient for design purposes may be taken as 0.0002. All bridges should be designed and constructed in such a manner that the secondary stresses are reduced to a minimum and these stresses should be allowed for in the design.

3.20 Erection Stresses

The stresses that are likely to be induced in members during erection should be considered in design. It is possible that the erection stresses may by different from those which the member will be subjected to in actual service.

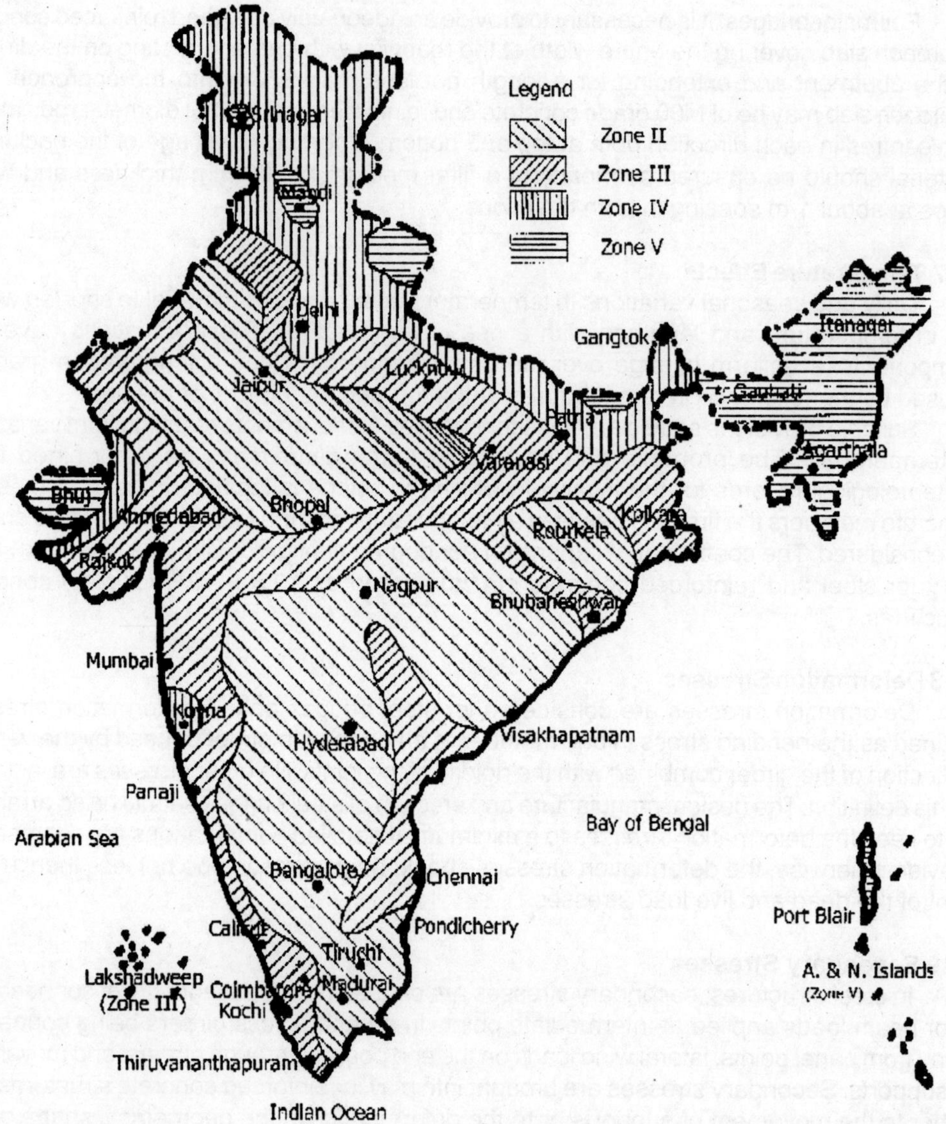

Figure 3.8 Seismic Zones in India.

3.21 Seismic Effects

3.21.1 Seismic Force

If a bridge is located in a region subject to earthquake, allowance should be made in the design for the seismic force, and earthquake resistant features should be incorporated in the structural details of design. The seismic force is considered a function of the dead weight of the structure, ground motion during earthquake, period of vibration of the bridge and the type of soil.

For the purpose of determining the earthquake force, the Indian codes [3, 19] (IRC:6-2000 with interim measures for seismic provisions - 2002 and IS:1893-2002) divide the country into four zones designated as Zones II to V as shown in Fig. 3.8, which figure is to be taken as a diagram and not as a 'map'. Zone I indicates areas not liable to earthquake damage, while Zone V comprises areas of known occurrences of earthquakes, the other zones being intermediate to the above two zones. The classification Zone I used in the earlier versions of the codes has now been dropped, meaning that no area in the country is considered free from earthquakes. More than 50% of the land area of the country lies in seismic zones III to V. Many major cities lie in seismically vulnerable areas. For example, Srinagar, Guwahati and Agatala are in Zone V; New Delhi, Chandigarh, Patna and Kolkata are located in Zone IV; Mumbai, Ahmedabad and Chennai lie in Zone III. The demarcation is subject to revision based on fresh observations.

All bridges in Zone V and Zone IV, and bridges of total length over 60 m in Zones III and II should be designed for seismic forces. Smaller bridges of length less than 60 m in Zones III and II need not be designed for seismic forces. The seismic force will have both horizontal and vertical components acting simultaneously. Masonry arch bridges above 10 m span should not be built in Zones IV and V.

The horizontal seismic force (F_{eq}) to be resisted may be computed from Equation (3.8).

$$F_{eq} = A_h (G + L_a) \qquad \text{... (3.8)}$$

where G = dead load above the section considered

L_a = appropriate live load under seismic condition, as specified in Clause 202.3

A_h = horizontal seismic coefficient

$$= \frac{(Z/2)(S_a/g)}{(R/I)}$$

Z = zone factor taken as 0.36, 0.24, 0.16 and 0.10 for Zones V, IV, III and II, respectively

I = importance factor, taken as 1.5 for important bridges, and as 1.0 for other bridges

T = fundamental period of the bridge member, in seconds, for horizontal vibrations

R = response reduction factor, taken as 2.5

S_a/g = average response acceleration coefficient for 5% damping depending on T

The earthquake response is aggravated in case of soft soils such as marine clay and silt. Sandy soils are susceptible to liquefaction. Hard and sound rock is the most desirable founding strata. The above aspect is important in higher seismic zones and for long and important bridges[20].

The value of S_a/g may be assumed as below:

For rocky and hard soil sites

$$S_a/g = \begin{cases} 2.50 & 0.0 \le T \le 0.40 \\ 1.00/T & 0.4 \le T \le 4.00 \end{cases}$$

For medium soil sites

$$S_a/g = \begin{cases} 2.50 & 0.0 \le T \le 0.55 \\ 1.36/T & 0.55 \le T \le 4.00 \end{cases}$$

For soft soil sites

$$S_a/g = \begin{cases} 2.50 & 0.0 \le T \le 0.67 \\ 1.67/T & 0.67 \le T \le 4.00 \end{cases}$$

The value of T may be calculated by any rational method.

In the absence of calculation of T for small bridges, the value of S_a/g may be taken as 2.5.

The above equation is applicable for spans less than 150 m. For longer spans, special studies may have to be undertaken. The value of horizontal seismic coefficient for portion below scour depth is taken as zero. The horizontal seismic force is taken to act through the centre of gravity of all the loads under consideration. The direction of the force should be such that the resultant stresses in the member under consideration are maximum. Based on past experience, it is not considered probable that seismic and wind forces act simultaneously, and hence these two need not be considered together. Similarly, the maximum seismic force and the highest flood need not be considered to occur simultaneously. The vertical seismic coefficient, taken as half the horizontal seismic coefficient, shall be considered for bridges in zones IV and V.

The seismic force due to live load can be ignored when acting in the direction of the traffic, but should be taken into account when acting in the direction perpendicular to traffic. The superstructure should be designed to resist horizontal and vertical seismic forces and should have a factor of safety of 1.5 against overturning in the transverse direction. In Zones IV and V, special precaution should be taken to secure the superstructure to the piers to prevent it from being dislodged off its bearings during an earthquake. For this purpose, seismic arresters may be provided and designed for 2 F_{eq} . Pier and abutment caps may be adequately dimensioned to withstand severe ground shaking.

3.21.2 Earthquake Magnitude and Earthquake Intensity

Earthquake is a natural process of readjustment of the earth's crust along a weak zone known as a fault. The earth's dynamic system causes a strain in the rock formation. When the accumulated strain exceeds the elastic limit of the rock, the rock slips along the faults and generates seismic waves at the focus. This sudden release of the stored energy causes an earthquake. The point on the surface of the earth directly over the focus is called the epicentre of the earthquake. The severity of an earthquake can be expressed in terms of both magnitude and intensity. It is necessary to distinguish between earthquake magnitude and earthquake intensity.

Earthquake magnitude is a measure of the energy released by the earthquake. It is denoted by a number on the Richter scale, which is based on the logarithm to the base of 10 of the ratio of the amplitude of an observed earthquake to the standard maximum trace obtained from a horizontal pendulum seismograph with a natural period of 0.8 second. Magnitude is thus represented by a single instrumented value. The earthquakes of the twentieth century have been observed to be below 8.6 on the Richter scale, though the

1897 Assam earthquake had a magnitude of 8.7. The 1920 earthquake in Ganzu, China measured 8.6, the 1950 Assam earthquake had a rating of 8.3, the 1995 Kobe, Japan earthquake measured 7.2, while the 2001 Bhuj (Gujarat) earthquake had a magnitude of 7.9 on the Richter scale.

Earthquake intensity indicates the extent of shaking experienced at a given location due to a particular earthquake. It is usually referred by a Roman numeral on the Modified Mercalli Intensity (MMI) scale, which is composed of 12 increasing levels of intensity ranging from imperceptible shaking to catastrophic destruction. Intensity of shaking at a location depends not only on the magnitude of the earthquake, but also on the distance of the site from the centre of earthquake and the geology of the area.

The four zones demarcated by IS:1893 as shown in Fig. 3.8 are based on the expected intensity of shaking in future earthquakes. The zones I, II, III, IV and V correspond to areas with potential for shaking intensities on MMI scale of V or less, VI, VII, VIII and IX or more, respectively[21]. The demarcation is not exact, as the 1993 Killari earthquake (magnitude 6.4 on the Richter scale) occurred in seismic zone I as per the then existing code.

3.21.3 Design Philosophy

The magnitude of a future earthquake and its intensity for any particular site are not amenable to precise prediction. If the structure were to respond elastically during an earthquake, the forces generated in the structure would be very large and it would be impractical to design the structure for the full force. For example, a structure located in zone IV (which corresponds to MMI of VIII) and having a natural period of 0.2 second may, if it were to remain elastic, sustain a maximum lateral force of about 75% of its self weight, corresponding to a peak ground acceleration of 0.75 g. In contrast, the IS and IRC codes provide for a design coefficient of about 0.05 g for the above structure. The difference is expected to be catered by the post-yield behaviour of the structure (ductility) and the normally available higher strength over the design strength (overstrength) in the structure[22]. In other words, the design seismic force for the structure can be taken as the expected maximum seismic force for fully elastic response, divided by a factor called the response reduction factor. The response reduction factor is defined as the product of the overstrength factor (usually about 2 to 3) and the ductility factor (about 4 to 5). The modern design philosophy is to permit a structure to deform inelastically to dissipate the energy while maintaining adequate strength to avoid collapse during a severe earthquake.

The seismic design codes of advanced countries, such as the AASHTO (American Association of State Highway and Transport Officials) and CALTRANS (California Department of Transportation) codes of USA and TNZ code (Bridge Manual of Transit New Zealand) provide detailed guidelines for the computation of the design seismic force[23–25]. According to the AASHTO code, the elastic forces in the members and connections generated under the maximum probable earthquake are first computed. Using appropriate reduction factors (detailed in the code), the design force in the member or connection is assessed. The CALTRANS code is similar to the AASHTO code with some difference in the method of assessing the reduction factors. The TNZ code approach involves the choice of an intended mode of structural behaviour during strong shaking, followed by design and detailing of members to ensure the selected behaviour. For a more elaborate review of the above three codes, the reader may see References 22 and 26.

The earthquake resistant design requires evaluation of seismic performance at two levels: (a) Functional evaluation, wherein it is aimed to achieve a structure which will withstand with minimal damage a moderate earthquake with a return period of about 60% of the bridge life; and (b) Safety evaluation, which tries to secure a structure that will survive with quickly repairable damage a maximum credible earthquake with a very low probability of occurring during the design life of the bridge. These new design criteria attempt to control the structural response of the bridge to earthquake ground motions.

3.21.4 Special Provisions

If the fundamental period of the bridge is beyond the range of normal earthquakes, the bridge would not be seriously affected by the earthquake induced forces. Most earthquakes are observed to have a period in the range 0.2 second to 0.8 second. Using base isolators such as elastomeric bearings, the natural period can be increased to a level beyond the above range. However, the reduced stiffness of the system causes increased displacements, which needs to be arrested by seismic restrainers in the longitudinal and transverse directions. An example of successful application of this technique is the New Surajbari bridge, which survived with minimum damage in the 2001 Bhuj earthquake, though located within 40 km of the epicentre [20].

The displacements along and transverse to bridge spans occurring during seismic shaking are to be controlled. Special vertical hold-down devices are needed at the piers and abutments, in situations where the vertical upward seismic reaction under design seismic conditions exceeds 50% of the dead load reaction. The hold-down device should be designed for a minimum force which is the greater of 10% of the dead load reaction or 1.2 times the net uplift force.

Horizontal linkage elements between adjacent spans and restraining devices are to be provided to prevent undesirable movements of adjacent superstructure units at supports. These elements are to be designed for acceleration coefficient times the weight of the lighter of the two adjoining spans. Transverse girder stops may be provided at the piers to prevent a transverse movement of the superstructure.

The seating width should be adequate to prevent any possibility of dislodging of the span from the support. Guidelines for determining the required width are to be evolved.

3.22 Barge/Ship Impact

Bridges located across navigable waters should provide for possible effects of barge/ship impact on piers in the navigable portion. Their design may necessitate special studies. The estimation of the force due to ship/barge collision is a complex problem. The force depends on the characteristics of the vessel, its velocity and the geometry of collision, besides characteristics of the bridge. Energy absorbing devices adjacent to the piers may be provided if large ships are expected. According to Clause 223.2 of the Bridge Code, the design impact force for collision may be at least 1000 kN acting at a height of 1 m above the HTL/HFL. Fenders and sacrificial caissons may be used when smaller barges negotiate the navigable waters.

3.23 Vehicle Collision Loads

Bridge piers built in the median or in the vicinity of the carriageway supporting the superstructure are likely to suffer damage due to collision of vehicles. In such cases, the piers and their foundations should be designed to withstand the collision loads. The collision load may be taken as 1000 kN when acting parallel to the carriageway and 500 kN when acting normal to the carriageway, the point of application being 0.75 m to 1.5 m above the carriageway. When the substructure is protected by a suitable fencing system, the collision loads may be taken as half the above, with the point of application being 1.5 m above the carriageway level.

3.24 References

1. 'Practice of evaluation of bridges in Madras State', Highways Research Station, Madras, Research Note No. 16, January 1966, 27 pp.
2. 'IRC: 5-1998 Standard specifications and code of practice for road bridges, Section I - General features of design (Seventh revision)', Indian Roads Congress, 1998, 38 pp.
3. 'IRC: 6-2000 Standard specifications and code of practice for road bridges, Section II - Loads and stresses', Indian Roads Congress, 2000, 55 pp; and ' Modified Clause for Interim Measures for Seismic Provisions, 2002', Indian Highways, January 2003, pp. 105-111.
4. 'IRC: 21-2000 Standard specifications and code of practice for road bridges, Section III - Cement concrete (plain and reinforced)', Indian Roads Congress, 2000, 80 pp.
5. 'IRC: 40-2002 Standard specifications and code of practice for road bridges, Section IV - Brick, stone and cement concrete block masonry', Indian Roads Congress, 2002, 40 pp.
6. 'IRC: 24-2001 Standard specifications and code of practice for road bridges, Section V - Steel road bridges', Indian Roads Congress, 2001, 157 pp.
7. 'IRC. 22-1986 Standard specifications and code of practice for road bridges, Section VI - Composite construction', Indian Roads Congress, 1991, 32 pp; and 'Errata No. 1 to IRC: 22-1986', Indian Highways, July 2006, pp. 63-64.
8. 'IRC: 78-2000 Standard specifications and code of practice for road bridges, Section VII - Foundations and substructure', Indian Roads Congress, 2000, 97 pp.
9. 'IRC: 83-1982 Standard specifications and code of practice for road bridges, Section IX - Bearings - Part I: Metallic bearings', Indian Roads Congress, 1982, 27 pp.
10. 'IRC. 83 (Part II) - 1987 Standard specifications and code of practice for road bridges. Section IX - Bearings - Part II: Elastomeric bearings', Indian Roads Congress, 1996, 29 pp.
11. 'IRC: 18-2000 Design criteria for prestressed concrete road bridges (post-tensioned concrete)", Indian Roads Congress, 61 pp; and Amendment No. 4, Indian Highways, May 2006, p. 72.
12. Thomas, P.K., 'A Comparative study of highway bridge loadings in different countries', Transport and Road Research Laboratory, UK, Supplementary Report 135 UC, 1975, 47 pp.
13. Thomas, P.K., 'The evolution of a new highway bridge loading standard for India', Indian Roads Congress Journal, Vol. 36-2, November 1975, pp. 147-185.
14. Victor, D.J., and Chettiar, C.G., 'Simplified loadings for slab bridges', Proceedings of National Seminar on Design and Construction of Roads and Bridges, Bombay, October, 1968, Vol. II, pp. 21-27.
15. Mitchell, R.G., 'Dynamic stresses in cast iron girder bridges', National Buildings Studies Research Paper No. 19, H.M. Stationery Office, 1954.
16. Reddi, S.A., 'Loading considerations in concrete bridges: A plea for realism', In Dhir, R.K., et al, (Eds.), 'Role of concrete bridges in sustainable development', Thomas Telford, London, 2003, pp. 183-194.
17. Taylor, D.W., 'Fundamentals of soil mechanics', Asia Publishing House, 1960, 700 pp.
18. Leonards, G.A., 'Foundation engineering', McGraw-Hill Book Co., New York, 1962, 1136 pp.
19. 'IS. 1893-2002 Indian standard criteria for earthquake resistant design of structures', Bureau of Indian Standards, 2002.
20. Reddi, S.A., et al, 'Seismic loading standards and performance of bridges', In Dhir, R.K., et al,

(Eds.), 'Role of concrete bridges in sustainable development', Thomas Telford, London, 2003, pp. 215-223.

21. Murty, C.V.R., and Jain, S.K., 'Seismic performance of bridges in India during past earthquakes', The Bridge and Structural Engineer, Journal of ING-IABSE, New Delhi, Vol. 27, No. 4, Dec. 1997, pp. 45-79.
22. Jain, S.K., and Murty, C.V.R., 'A state-of-the-art review on seismic design of bridges - Part I: Historical development and AASHTO code', Indian Concrete Journal, Vol. 72, Feb. 1998, pp. 79-86.
23. AASHTO 'Standard specifications for seismic design of highway bridges', American Association of State Highway and Transport Officials, Washington, D.C., USA, 1992.
24. CALTRANS 'Bridge design specifications', California Department of Transportation, Sacremento, California, CA, USA, 1991.
25. TNZ 'Bridge Manual : Design and evaluation', Transit New Zealand, Wellington, New Zealand, 1991.
26. Jain, S.K., and Murty, C.V.R., 'A state-of-the-art review on seismic design of bridges - Part II: CALTRANS, TNZ and Indian codes', Indian Concrete Journal, Vol. 72, March 1998, pp. 129-138.

Chapter 4

Standards for Railway Bridges

4.1 General

Indian Railways have been pioneers in the construction of bridges. Currently, there are about 116 000 bridges of all types and spans on the Indian Railways, making an average of two bridges per route km. Nearly 20% of the bridges are girder bridges, while arch bridges account for about 19%, followed by slab culverts at 25% and others at 19%.

Railway bridges in India are to be built to conform to the Indian Railway Standards (IRS) laid down by the Ministry of Railways, Government of India, as below [1-5]:

(a) The loads to be considered in design are given in IRS Bridge Rules.
(b) The details of design of steel bridges should conform to IRS Steel Bridge Code.
(c) The details of design of bridge members in plain, reinforced and prestressed concrete should be in accordance with IRS Concrete Bridge Code.
(d) Masonry and plain concrete arch bridges should be detailed so as to conform to IRS Arch Bridge Code.
(e) The substructure for bridges should be in accordance with IRS Bridge Substructure Code.

Railway tracks are classified according to the width of track (gauge) and according to the importance of the line.

There are three gauges used on Indian Railways with the width of track between the inner faces of rails as indicated below:

(a) Broad Gauge (BG)	1676	mm	(5'6")
(b) Metre Gauge (MG)	1000	mm	(3'3-3/8")
(c) Narrow Gauge (NG)	762	mm	(2'6")

The tracks are also classified according to the importance and traffic intensity as:

(i) Main line
(ii) Branch line.

Some of the important clauses from the Indian railway standards are discussed in this Chapter. For further details, the reader should refer to the latest versions of the codes mentioned above.

4.2 Loads to be Considered

The following items should be considered, wherever applicable, while computing stresses in the bridge members:

(a) Dead load
(b) Live load
(c) Dynamic effects
(d) Forces due to curvature or eccentricity of track
(e) Temperature effect
(f) Frictional resistance of expansion bearings
(g) Longitudinal forces
(h) Racking force
(i) Wind pressure effect
(j) Forces on parapets
(k) Forces and effects due to earthquake
(l) Erection forces and effects.

4.3 Dead Load

Dead load is the weight of the structure itself together with permanent loads carried on the structure.

4.4 Live Load

4.4.1 Standard Axle and Train Loads

Railway bridges should be designed for one of the following standards of loading:

(a)	Broad Gauge ...	(i)	Modified BG Loading –1987
		(ii)	Branch line
(b)	Metre Gauge ...	(i)	Modified MG Loading – 1988
		(ii)	Main Line
		(iii)	Branch Line
		(iv)	Standard C
(c)	Narrow Gauge ...	(i)	'H' Class
		(ii)	'A' Class Main Line
		(iii)	'B' Class Branch Line

In every case, the standard loading consists of two locomotives of specified axle loadings and spacings plus a train of specified loading per linear metre on both sides of the locomotives. Separate and detailed drawings for each of these standard loadings are given in IRS Bridge Rules.

The salient details of one type of loading each for Modified Broad Gauge loading and Modified Metre Gauge Loading are shown in Fig. 4.1. It should be noted that Branch Line standards are to be adopted only for "branch lines, which are obviously never likely to be other than branch lines". Since this latter condition is difficult to forecast, it would be prudent to generally design bridges for modified BG or MG standards. In view of the desirability of having a uniform gauge throughout the country, it is advantageous to design the railway bridges to Modified BG standard, even if the track is for MG to start with, so that later conversion will be easy.

IRS Bridge Rules also furnish tables giving equivalent uniformly distributed loads (EUDL) on each track and also the coefficient of dynamic augment (CDA) applicable to different

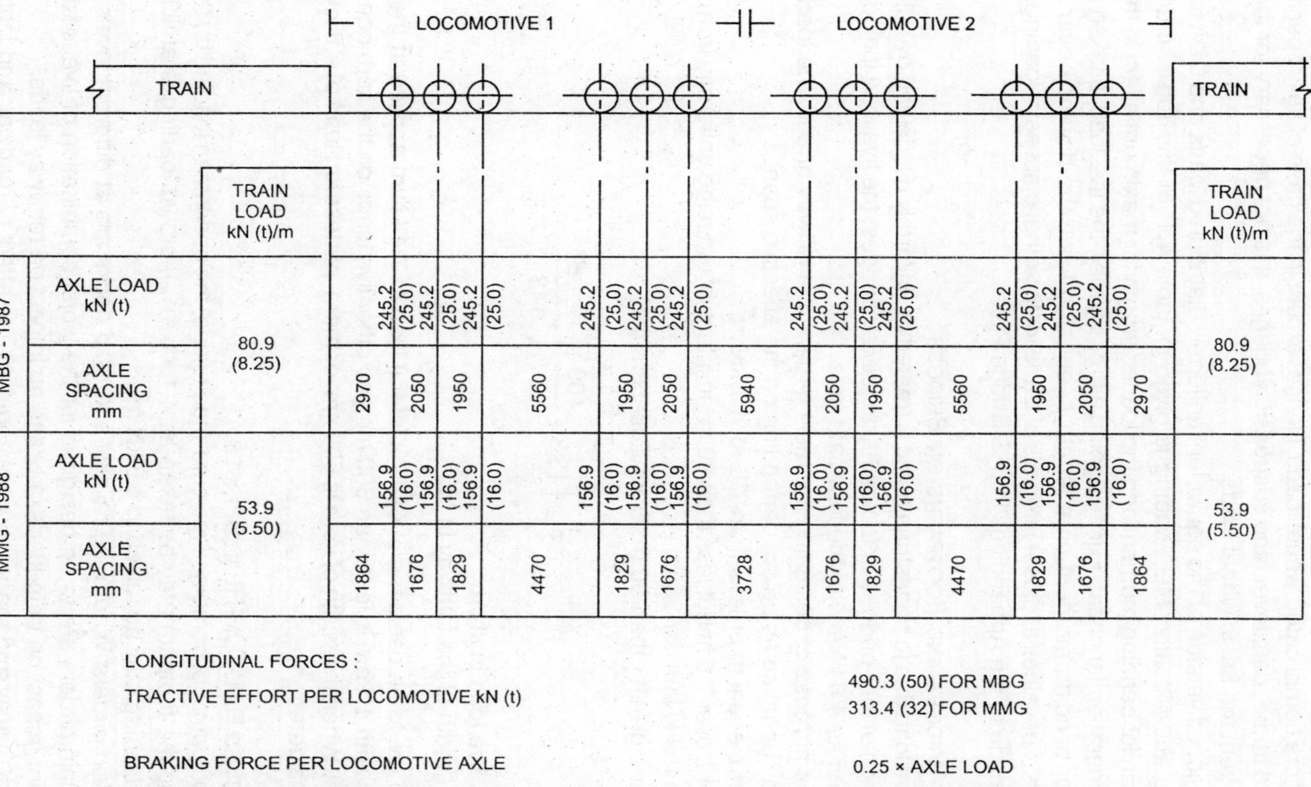

Figure 4.1 Typical Patterns of MBG-1987 and MMG-1988 Loadings

spans. These tables have been so derived as to yield, by the simpler procedure applicable for uniform loading, the design maximum moments and shears as would be obtained by the use of the more complicated wheel loads. Hence, except in the case of very special bridges, such as rigid frame, cantilever, and suspension bridges, it is not necessary for the designer to proceed from the basic wheel loads.

An abridged version of the equivalent uniformly distributed loads and CDA for Modified BG Loading and Modified MG Loading is given in Table 4.1. In this Table, L denotes the effective span for bending moment and the loaded length for maximum effect in the member under consideration for shear. For designing a cross girder, the live load is taken as half the total load for bending moment on a length L, equal to twice the distance over centres of cross girders. For values at closer intervals of L and for the other classes of loading, reference should be made to the Appendices in IRS Bridge Rules.

4.4.2 Foot-Bridges and Footpaths on Bridges

The live load due to pedestrian traffic is treated as uniformly distributed over the footway. For the design of foot-bridges or footpaths on railway bridges, the live load including impact is to be taken as 4.8 kN/m^2 of the foot-path area.

For the purpose of designing the main girders of a railway bridge, the loading on the footpath may be taken as below, depending on the effective span, L:

(a) For L equal to or less than 7.5 m: 4.1 kN/m^2

(b) For L over 7.5 m but less than 30 m: an intensity reducing uniformly from 4.1 kN/m^2 for 7.5 m to 2.9 kN/m^2 for 30 m.

(c) For L greater than 30 m: From Equation (4.1).

$$P = \left(13.3 + \frac{400}{L}\right)\left(\frac{17 - B}{143}\right) \qquad \dots (4.1)$$

where P = live load in kN/m^2

B = width of the footway in m.

Kerbs are to be designed to carry a lateral load of 7.4 kN/m applied at the top of the kerb. If the width of kerb is less than 600 mm, no other live loads on the kerb need be taken into account. When the width of kerb is 600 mm or more, pedestrian load of 4.8 kN/m should also be considered.

4.4.3 Combined Rail-Cum-Road Bridges

(a) *Footpath:* For combined rail-cum-road bridge, the pedestrian loading for the purpose of designing the footpath may be taken as 4.1 kN/m^2. If crowd loading is anticipated, the pedestrian loading may be raised to 4.8 kN/m^2.

(b) *Main girders*: Where the railway and road decks are at different levels or side by side, the main girders are to be designed for the worst combination of live loads, with full allowance for impact on train-loads only and no impact on roadway loads.

Where railway and road decks are common, the effect of roadway and footpath loads on main girders is to be provided by a minimum distributed load of 1.9 kN/m^2 over the whole area of roadways and footpaths not occupied by the train-load.

Table 4.1 EUDL, CDA and Longitudinal Loads for Modified BG Loading-1987 and Modified MG Loading - 1988.

L	CDA for Bending Moment and Shear	Modified BG Loading – 1987				Modified MG Loading – 1988			
		EUDL (Total) for B.M.	EUDL (Total) for Shear	Tractive Effort	Braking Force	EUDL (Total) for B.M.	EUDL (Total) for Shear	Tractive Effort	Braking Force
m		kN	kN	kN	kN	kN	kN	kN	kN
1	1.000	490	490	81	62	314	314	89	57
2	1.000	490	519	164	123	314	365	118	78
3	1.000	490	662	164	123	326	452	118	78
4	0.950	596	778	245	184	429	536	157	118
5	0.877	741	888	245	184	501	616	157	118
6	0.817	838	985	245	185	581	685	157	124
7	0.765	911	1068	327	221	644	755	176	135
8	0.721	981	1154	409	276	714	819	209	157
9	0.683	1040	1265	409	276	774	871	262	196
10	0.650	1101	1377	490	331	828	934	262	198
12	0.594	1377	1589	490	331	953	1061	314	235
15	0.531	1631	1801	490	368	1138	1252	353	253
20	0.458	1964	2168	735	496	1421	1532	471	353
25	0.408	2356	2586	735	565	1677	1833	523	401
30	0.372	2727	2997	981	662	1991	2144	628	486
40	0.324	3498	3815	981	816	2589	2748	628	594
50	0.293	4253	4630	981	978	3099	3269	628	702
60	0.271	5051	5442	981	1140	3625	3818	628	810
70	0.255	5831	6254	981	1301	4178	4372	628	918
80	0.243	6603	7065	981	1463	4727	4922	628	1026
90	0.233	7391	7876	981	1625	5274	5470	628	1134
100	0.225	8201	8686	981	1787	5822	6017	628	1242
110	0.219	9011	9496	981	1949	6365	6562	628	1349
120	0.213	9820	10306	981	2110	6908	7106	628	1457
130	0.209	10630	11115	981	2272	7451	7649	628	1565

(c) *Floor members*: Floor members, which carry roadway and railway loads simultaneously, should be designed for the maximum effect, including impact, which may be imposed by either class of load separately or together.

If the road and railway are both on the same alignment, the floor members should be designed for the maximum effect of either class of load.

4.4.4 DISTRIBUTION OF LIVE LOAD

All simply supported span bridges without ballasted floor and bridges of span over 8 m with ballasted deck may be designed for a uniformly distributed load equal to the appropriate EUDL as given in Table 4.1. For bridges upto 8 m span with ballasted deck, separate EUDL tables are given in IRS Bridge Rules.

The sleeper may be assumed to distribute the live load uniformly on the top of the ballast over the area of contact given below:

	Type I	*Type II*
		under each rail seat
Broad gauge	2745 x 254 mm	760 x 330 mm
Metre gauge	1830 x 230 mm	610 x 270 mm

The load under the sleeper may be assumed to be dispersed by the ballast at an angle not greater than half horizontal to one vertical, and all deck slabs should be designed for both types of sleepers.

The effective width of dispersion through the R.C. deck slab is to be limited to one-fourth span on each side of the loaded area in case of simply supported, fixed or continuous span. In case of cantilever slabs, the effective dispersion width is not to exceed one-fourth the loaded length on each side of the loaded area. Further the dispersion width should not be assumed to exceed the total width of decking for slabs spanning in the longitudinal direction, and minimum axle spacing in the case of slabs spanning in the transverse direction.

The distribution of wheel loads on steel troughing or steel or wooden beams spanning transversely to the track and supporting the rails directly shall be designed in accordance with the continuous elastic support theory.

4.5 Dynamic Effect

When a train moves over a bridge, an additional impact load is caused due to factors such as fast travel of the load, uneven track, rough joints, imperfectly balanced driving wheels and lateral sway. The increase in load due to dynamic effects should be considered by adding a load equivalent to a coefficient of dynamic augment (CDA) multiplied by the live load giving the maximum stress in the member under consideration. Values of CDA, applicable for speeds up to 160 km/h in BG and 100 km/h in MG, can be computed as indicated below:

(a) *For steel bridges on BG and MG*

(i) For single track spans

$$CDA = 0.15 + 8/(6 + L) < 1.0 \qquad \text{... (4.2)}$$

where L is loaded length of span in m for the position of the train giving the maximum stress in the member under consideration; 1.5 times the cross girder spacing in the case of stringers (rail bearers); and 2.5 times the cross girder spacing in the case of cross girders.

(ii) For main girders of double track spans with two girders

$$CDA = 0.72\,[0.15 + 8/(6 + L)] < 0.72 \qquad \text{... (4.3)}$$

(iii) For intermediate girders of multiple track spans

$$CDA = 0.60\,[0.15 + 8/(6 + L)] < 0.60 \qquad \text{... (4.4)}$$

(iv) For cross girder carrying two or more tracks

$$CDA = 0.72\,[0.15 + 8/(6 + L)] < 0.72 \qquad \text{... (4.5)}$$

(b) *Steel bridges on narrow gauge*

$$CDA = 91.5 / (91.5 + L) \qquad \text{... (4.6)}$$

(c) *Railway pipe culverts, arch bridges, concrete slabs and concrete girders for all gauges*

(i) When the depth of fill is less than 900 mm

$$CDA = 0.5\,(2 - d/0.9)\,[0.15 + 8/(6 + L)] \quad \text{... (4.7)}$$

where d = depth of fill, and

$$[0.15 + 8/(6 + L)] < 1.0$$

(ii) When the depth of fill is 900 mm

$$CDA = 0.5\,[0.15 + 8/(6 + L)] < 0.5 \quad \text{... (4.8)}$$

(iii) When the depth of fill exceeds 900 mm, the coefficient of dynamic augment should be uniformly decreased to zero within the next 3 m of the fill.

Fill is the distance from the underside of the sleeper to the crown of an arch or the top of a slab or pipe. The CDA values from Equations (4.7) and (4.8) are applicable to both single and multiple track bridges. On multiple track arch bridges of spans exceeding 15 m, two-thirds of the above impact shall be allowed.

(d) *Foot bridges*

No impact allowance need be made.

4.6 Forces Due to Curvature and Eccentricity of Track

Where a bridge is on a horizontal curve, allowance should be made for centrifugal action of the moving load, considering all tracks on the bridge to be occupied.

The horizontal force due to centrifugal force is computed from Equation (4.9).

$$C = WV^2 / (12.95\,R) \quad \text{... (4.9)}$$

Where C = horizontal force in kN per m run of span
W = equivalent distributed live load in kN per m run
V = maximum speed in km per hour
R = radius of the curve in m.

This force is assumed to act at a height of 1830 mm for Broad Gauge and 1450 mm for Metre Gauge above rail level. Design should consider both the horizontal force and the resulting overturning moment.

Also, the extra loads on one girder due to the additional reaction on one rail and to the lateral displacement of the track should be considered. These loads are calculated under the following two conditions:

(i) Live load running at the maximum speed
(ii) Live load standing with half normal dynamic augment.

4.7 Temperature Effect

Where any portion of the structure is not free to expand or contract under variation of temperature, allowance should be made for stresses arising from this condition. In computing these stresses, the coefficient of expansion may be assumed as 11.7×10^{-6} per degree centigrade for steel and reinforced concrete and as 10.8×10^{-6} per degree centigrade for plain concrete.

4.8 Frictional Resistance of Expansion Bearings

Frictional resistance of expansion bearings has to be taken into account, the coefficients of friction being assumed as below:

For roller bearings	..	0.03
For sliding bearings of steel on steel or cast iron		0.25
For sliding bearings of steel on ferroasbestos	..	0.20
For sliding bearings of steel on copper alloy	..	0.15
For sliding bearings of PTFE/Elastomeric type	..	0.10

For expansion or contraction of the structure, due to variation of temperature under dead load, the friction on the expansion bearing shall be considered as an additional load throughout the chord to which the bearing plates are attached.

4.9 Longitudinal Forces

Longitudinal loads are caused due to one or more of the following causes:

(a) the tractive effort of the driving wheels of the locomotives
(b) the braking force due to application of the brakes to all braked vehicles
(c) resistance to the movement of bearings due to change in temperature.

These forces are considered as acting horizontally through the knuckle pins, or through the girder seat where the girders have sliding bearings. For spans supported on sliding bearings, the horizontal loads shall be considered as being divided equally between the two ends. In case of spans with roller bearings at one end, the horizontal force shall be considered to act through the fixed end.

For railway bridges, the longitudinal force due to either the tractive effort or the braking force for the loaded length L can be obtained from Table 4.1.

The loaded length L is taken as below :

(a) The length of one span, when considering the effect of the longitudinal loads on
   (i) the girders,
   (ii) the stability of abutments,
   (iii) the stability of piers under the condition of span loaded, or when piers carry one fixed and one roller bearing.

(b) The length of two spans, when considering the stability of piers carrying fixed or sliding bearings for the condition of both spans loaded. In this latter case, the total longitudinal force is to be divided between the two spans in proportion to their lengths.

(c) For determining the value of tractive effort, L should not be taken to exceed 29 m for BG and 27 m for MG. Where the structure carries more than one track, the longitudinal loads shall be considered to act simultaneously on all tracks. The maximum effect on any girder with two tracks so occupied shall be allowed for, but with more than two tracks a suitable reduction may be made on the loads for the additional tracks.

4.10 Racking Forces

Lateral bracings of the loaded deck of railway spans should be designed to resist, in addition to the wind and centrifugal forces, a lateral load due to racking forces of 5.9 kN/m treated as a moving load. This lateral load need not be considered for computing the stresses in chords or flanges of main members.

For spans up to 20 m, it is not necessary to calculate wind stresses, provided bracings are designed for a lateral load of 8.8 kN/m treated as moving load in addition to the centrifugal load, if any.

4.11 Forces on Parapets

Railings or parapets should have a minimum height above the adjacent roadway or footway surface of 1 m less one half of the horizontal width of the top rail or top of the parapet. They are to be designed to resist a horizontal force and a vertical force each of 1.5 kN/m applied simultaneously at the top of the railing or parapet.

4.12 Wind Pressure Effect

The basic wind pressure is to be obtained from the meteorological records, if available or from IS: 875[6]. No live load on the bridge need be considered when the basic wind pressure at deck level exceeds the following limits:

Broad Gauge bridges	1.5 kN/m^2
Metre Gauge and Narrow Gauge bridges	1.0 kN/m^2
Foot-bridges	0.75 kN/m^2

The basic wind pressure is considered as applied to the exposed area as below:

(a) *For unloaded spans*

One and a half times the horizontal projected area of the spans for decks other than plate girders.

For plate girders, the area of the windward girder plus a fraction as below of the area of the leeward girder:

For spacing of leeward girder

less than half its depth	0
half depth to full depth	0.25
full depth to 1.5 depth	0.50
1.5 depth to 2.0 depth	1.00

(b) *For loaded spans*

The area as above for the unloaded portion, plus the area of the windward girder above and below the moving load plus the horizontal projected area of the moving load.

For railway bridges, the height of the moving load is the distance between the top of the highest stack for which the bridge is designed and the rail level, less 600 mm. In case of foot-bridges, the height of the moving load is to be taken as 2 m throughout the span.

The wind pressure effect is considered as a horizontal force acting in such a direction that the resultant stresses in the member under consideration are the maximum. The following effects of wind pressure are to be considered:

(i) Lateral effect on the top chords and wind bracing considered as a horizontal girder.
(ii) The same effect on the bottom chords.
(iii) The vertical loads on the main girders due to the overturning effect of the wind on the span and on the live load.
(iv) Bending and direct stresses in the members transmitting the wind load from the top to the bottom chords or *vice versa*.

4.13 Forces Due to Earthquake

For the purpose of determining seismic forces, the country is divided into five zones (IS: 1893[7]) as shown in Fig. 3.8. Both horizontal and vertical seismic forces have to be taken into account for design of bridge structures.

The horizontal seismic force to be resisted is computed from Equation (4.10).

$$F_h = W_m . \alpha_h \quad \text{... (4.10)}$$

where F_h = horizontal seismic force to be resisted

W_m = weight of mass under consideration ignoring reduction due to buoyancy

α_h = $\beta . I . \alpha_0$

α_0 = basic horizontal seismic coefficient, taken as 0.08, 0.05, 0.04, 0.02 and 0.01 for zones V, IV, III, II and I, respectively.

I = a coefficient depending on the importance of the structure, taken as 1.5 for major bridges of over 300 *m* linear waterway, and as 1.0 for other bridges.

β = a coefficient depending upon the soil foundation system, the value varying from 1.0 to 1.5 as given in the code.

The vertical seismic force to be resisted (F_v) is estimated from Equation (4.11).

$$F_v = W_m . \alpha_v \quad \text{... (4.11)}$$

where

$$\alpha_v = 0.5\, \alpha_h$$

The bridge as a whole and every part of it should be designed and constructed to resist the stresses produced by seismic effects. For horizontal acceleration, the stresses should be calculated as the effect of force applied horizontally at the centre of mass of the elements of the bridges into which it is conveniently divided for the purpose of design. The forces can come from any horizontal direction.

Seismic forces need not be considered for bridges in zones I to III and for bridges of spans less than 15 m.

4.14 Erection Forces

The forces which may act temporarily during erection should be considered. It is permissible to allow for stresses during erection different from those which the member will be subjected to during actual working.

4.15 Earth pressure on abutments

The horizontal earth pressure P acting on the abutment may be computed from Equation (4.12), the location of its application above any horizontal section considered being obtained from Equation (4.13).

$$P = 0.5\, wh(h + 2h_s)\left(\frac{1 - \sin\phi}{1 + \sin\phi}\right) \quad \text{... (4.12)}$$

$$y = \frac{1}{3} h \left(\frac{h + 3h_s}{h + 2h_s}\right) \quad \text{... (4.13)}$$

where w = weight of the fill in kN/m³

h = height of fill in m up to formation level above the horizontal section considered

h_s = equivalent height of surcharge in m due to dead and live loads
ϕ = angle of repose of the fill
y = height in m above section at which P acts.

The surcharge due to live loads may be assumed to extend up to the front face of the ballast wall, and equal to the loads placed at the formation level as in Table 4.2.

Table 4.2 Surcharge Load.

Loading Standard	Surcharge load kN/m	Width of uniform distribution at formation level m
Broad Gauge M.L.	134	3.0
Metre Gauge M.L.	98	2.1
Narrow Gauge	83	1.8

The dispersion of the surcharge load below the formation level may be taken at a slope of 1 horizontal to 2 vertical; and for the design of wing and return walls, it may be taken as 1 horizontal to 1 vertical.

In the case of return walls, the horizontal earth pressure is computed as sum of two parts P_1 due to earthfill and P_2 due to surcharge as below:

$$P_1 = 0.5\, wh^2 . \left(\frac{1-\sin\phi}{1+\sin\phi}\right) \quad \text{... (4.14)}$$

acting at a height $h/3$ above the section considered

$$P_2 = \frac{S.h'}{(B+2D)}\left(\frac{1-\sin\phi}{1+\sin\phi}\right) \quad \text{... (4.15)}$$

acting at height of $h'/2$ above the section considered

where P_1 = pressure due to earth fill
P_2 = pressure due to surcharge
B = Length of sleeper, 2.75 m for BG and 1.83 m for MG
S = surcharge load from Table 4.2
D = depth from bottom of sleeper to a point at which a 45° line from the end of the sleeper cuts a vertical through the rear toe at the section considered
h' = height above the section considered, up to a point where the dispersion line of surcharge load meets the vertical through the rear toe.

4.16 References

1. 'Bridge Rules—Rules specifying the loads for designing the super-structure and sub-structure of bridges and for assessment of the strength of existing bridges', Govt. of India, Ministry of Railways, 1988, with correction slips.
2. 'Indian railway standard code of practice for the design of steel or wrought iron bridges carrying rail, road or pedestrian traffic', Govt. of India, Ministry of Railways, 1962.

3. 'Indian railway standard code of practice for reinforced concrete construction', Govt. of India, Ministry of Railways, 1962.
4. 'Indian railway standard code of practice for the design and construction masonry and plain concrete arch bridges', Govt. of India, Ministry of Railways, 1965.
5. 'Indian railway standard code of practice for the design of substructures for bridges', Govt. of India, Ministry of Railways, 1964.
6. 'IS: 875-1964, Indian standard code of practice for structural safety of buildings: Loading standards', Seventh reprint, Bureau of Indian Standards.
7. 'IS: 1893-2002, Indian standard criteria for earthquake resistant design of structures', Bureau of Indian Standards, 2002.

Chapter 5

General Design Considerations

5.1 General

As noted by Andrea Palladio, Italian architect and engineer of the sixteenth century, "bridges should befit the spirit of the community by exhibiting commodiousness, firmness and delight." In other words, a good design for a bridge structure should satisfy the following requirements: (i) Function; (ii) Aesthetics; and (iii) Economy. The function of a bridge is to provide a passage to the intended traffic over the particular obstruction with utmost safety and convenience. The load carrying capacity of the structure should be adequately above the normal service loads so that the probability of failure during the specified lifetime of the bridge is below the limit specified. The safety and serviceability of the bridge are ensured by limiting the deflections and vibrations during its service life.

The bridge codes and specifications[1-6] prescribe minimum requirements to ensure the functional performance of a bridge. Some of the factors to be considered in the design of the bridge structure are included in Sections 5.2 to 5.15. The design requirements relating to the site conditions would cover considerations of topography, geology and hydrology. Functional requirements would involve specifications for road or track alignment, cross-section, clearances, constructional schedule and durability. Designs should have structural clarity and be appropriate to their location and environment. The designs for railway bridges differ from those for highway bridges of similar size due to the following factors: (a) Higher value for the ratio of live load to dead load and the impact factor; (b) Preference for simple span structures to facilitate ease in repairs and replacement; and (c) Critical importance of maintenance without interruption to traffic.

The aesthetic requirements of a bridge structure; though very important, are difficult to codify. The author would plead for greater attention to be paid to aesthetics of bridges than is devoted at present, so that the design of each major bridge incorporates the visual design as a major element. However, the designer should attempt integration of the structure and the visual form rather than treating them as separate parts. Contrary to the common perception among engineers, the additional cost to improve the aesthetics of an otherwise functionally efficient bridge is not likely to exceed five per cent. A brief description of aesthetics of bridge design is included in Section 5.16.

The need for the consideration of economy in bridge structures cannot be over-emphasized. The criteria for economy vary from place to place and no rigid rules can be specified. The bridge engineer should consider several preliminary designs using different span arrangements, structural materials and construction techniques. He should examine the features of the above designs for function, aesthetics, durability, buildability and economy, and narrow down the choice to two or three alternatives. Using his experience and judgment, and applying the principles of Value Engineering[7], he should then choose the preferred alternative, which, besides being economical in initial cost, also results in low costs on periodical maintenance during its design life and permits ease of access for inspection and repair. In other words, the chosen design should facilitate minimum life cycle cost. The useful life span of a bridge is often taken as 100 years or two human life spans.

Based on the predominant material used for the superstructure, bridges are broadly classified as follows:

(a) Structural concrete bridges, including reinforced concrete and prestressed concrete bridges;
(b) Steel bridges including structural steel bridges, cable stayed bridges and suspension bridges;
(c) Composite bridges, using a combination of steel and concrete components; and
(d) Other types covering masonry, timber, pontoon and movable bridges.

The above classification has a limited use for convenience in discussion. An actual bridge may comprise of varied combinations. A cable stayed bridge, for example, may consist of steel cables for stays, steel-concrete composite deck for superstructure, steel pylons and reinforced concrete caisson foundation.

Concrete bridges are the most frequently used types for short and medium spans, being adaptable to the numerous forms to result in innovative and aesthetically pleasing structures.

5.2 Structural Concrete

5.2.1 Components of Concrete

Cement concrete is produced by mixing cement, sand (fine aggregate), crushed stone (coarse aggregate) and water in suitable proportions. Approved admixtures may be added to enhance any desired property of the concrete.

Cement used for bridge construction is normally any one of the following: (a) Ordinary Portland Cement (OPC), 33 grade, conforming to IS:269; (b) OPC, 43 grade, conforming to IS: 8112; (c) OPC, 53 grade, conforming to IS:12269; and (d) Rapid Hardening Portland Cement in accordance with IS:8041. Cement conforming to IS:8112 and IS:12269 are nowadays used for structural concrete for bridges. The age of cement at the time of use should not be more than 90 days for reinforced concrete and not more than 60 days for prestressed concrete.

All coarse and fine aggregates shall conform to IS: 383. Coarse aggregates shall consist of clean, hard, strong, dense, non-porous and durable pieces of crushed stone, gravel or shingle. The maximum size of the coarse aggregate should be less than one-quarter of the minimum size of the member or 10 mm less than the minimum lateral clearance between individual reinforcements or 10 mm less than the minimum clear cover to any reinforcement. The preferred nominal size of coarse aggregate is 20 mm for reinforced concrete and prestressed concrete. For plain concrete, the preferred nominal size may vary from 20 mm to 40 mm. Fine aggregates shall consist of hard, strong, durable, clean pieces of natural sand, crushed stone or gravel to size passing 4.75 mm sieve. Grading of aggregates shall be such as to produce a dense concrete of the specified strength with adequate workability, to enable placement in position without segregation and without the use of excessive water content.

Water used for mixing and curing shall be clean and free from materials harmful to concrete or reinforcement. Potable water is generally considered satisfactory for mixing concrete. As per IRC:21, solids in water should not exceed the limits as below: organic 200 mg/lit, inorganic 3000 mg/*l*, sulphates 500 mg/*l*, chlorides 200 mg/*l* and suspended matter 2000 mg/*l*. The pH value shall not be less than 6. Curing of concrete by water prevents drying up of the intrinsic moisture inside the capillaries of the concrete and thus aids hydration of cement (to gain strength) and reduces shrinkage cracking.

Admixtures are available for increasing the workability of concrete (plasticizers) facilitating the reduction of water cement ratio and for retardation of setting of cement during hot weather concreting. Concrete properties such as durability, strength and service life can be enhanced by use of suitable mineral and chemical admixtures.

5.2.2 Concrete Grade

Structural concrete is designated with reference to the characteristic compressive strength of standard (150 mm size) cube specimens at 28 days. Characteristic compressive strength of a mix is that level of concrete strength below which not more than 5 per cent of the test results are expected to fall. Concrete used for bridge construction is designated as grades M15, M20, M25, M30, M35, M40, M45, M50, M55 and M60, where the numeral part indicates the specified characteristic compressive strength in MPa. Thus a concrete designated as M20 grade will have a specified characteristic compressive strength of 20 MPa at 28 days. The properties relating to the different grades of concrete as prescribed in IRC:21 are listed in Table 5.1.

Normally, plain concrete will be of grade M15 to M20, while reinforced concrete will use grade M20 to M35. Concrete of grade M40 to M60 is used for prestressed concrete superstructure. The minimum cement content for plain concrete is 250 kg/m³ for minor works and 360 kg/m³ for major bridges, while the corresponding values for reinforced concrete are 310 kg/m³ and 380 kg/m³, respectively. The cement content may be increased by about

Table 5.1 Concrete Grades and Minimum Cement Content.

Grade Designation	Specified Characteristic Strength MPa	Minimum Cement Content kg/m³	Current Margin MPa	Target Mean Strength MPa
M15	15	250	10	25
M20	20	310	10	30
M25	25	360	11	36
M30	30	380	12	42
M35	35	400	12	47
M40	40	400	12	52
M45	45	400	13	58
M50	50	400	13	63
M55	55	400	14	69
M60	60	400	14	74

15 per cent when the concrete is exposed to marine surroundings or to sulphate attack. In case of underwater concreting, the quantity of cement may be increased by 10 per cent. For prestressed concrete, the cement content should not be less than 400 kg/m³ nor more than 540 kg/m³.

5.2.3 Permissible Stresses

The design value of Modulus of Elasticity (E_c) and the basic permissible stresses for concrete of different grades as given in IRC: 21 are shown in Table 5.2. IRC:21 allows the use of a constant value of 10 for the modular ratio (E_s/E_c) for calculating stresses in a reinforced concrete section. Permissible shear stress in concrete τ_c is to be determined as specified in Clause 304.7 of the Bridge Code.

Table 5.2 Properties and Basic Permissible Stresses of Concrete.

Properties	Concrete Grade								
	M15	M20	M25	M30	M35	M40	M45	M55	M60
1. Modulus of Elasticity (E_c), GPa (Design value)	26.0	27.5	29.0	30.5	31.5	32.5	33.5	36.0	37.0
2. Permissible Direct Compressive Stress (σ_{co}), MPa	3.75	5.00	6.25	7.50	8.75	10.00	11.25	13.75	15.00
3. Permissible Flexural Compressive Stress (σ_{cb}), MPa	5.00	6.67	8.33	10.00	11.67	13.33	15.00	18.30	20.00
4. Permissible Tensile Stress in Plain Concrete, MPa	0.40	0.50	0.61	0.67	0.67	0.67	0.67	0.67	0.67

Note: A modular ratio (E_s/E_c) of 10 may be adopted for calculating stresses in a reinforced concrete section.

5.2.4 Ready Mixed Concrete

At many bridge sites, specially in urban areas, preparation of concrete at the construction site becomes difficult due to non-availability of adequate space for storage and handling of the constituent materials and for mixing operations. When the construction activities in a city are of such magnitude as to assure a sustained demand for large volume of concrete, it is desirable to establish ready mixed concrete (RMC) plants in the outskirts of the city and to transport the concrete in special transit mixer trucks to the construction site at the right time. Though the use of RMC is not yet widespread in many Indian cities, this development is inevitable in the near future.

A typical RMC plant has the following components: (i) Central batching plant with a capacity of 30 to 200 m^3 of concrete per hour; (ii) Transit mixer trucks to transport concrete to construction sites with the help of rotating type transit mixers of capacity about 6 m^3; and (iii) Concrete pumps and conveyors to deliver concrete at the work sites.

Use of ready mixed concrete has several advantages. Firstly, the quality of the concrete is assured with lower standard deviation for the compressive strength. The automatic batching plant can be of the state-of-the-art technology. All the operations are carried out under strictly controlled conditions, as there is usually a quality control laboratory attached to the plant. Concrete grade and cement type could be specified. Secondly, the construction contractor is relieved of the inconvenience of procuring different materials at the required time. Thirdly, stock piling of huge quantities of materials like aggregates and cement at the construction site is eliminated, resulting in cleaner and less polluted surroundings at the work site. Fourthly, the use of RMC facilitates speedy construction through continuous mechanical operations including placement of concrete by pumping. Fifthly, the concreting operations can be managed with much less labour force, which also results in avoidance of unauthorised hutment colonies around the work site.

5.3 Concrete Mix Design

5.3.1 Principles of Mix Design

The design of a concrete mix is essentially the technique of determining the most economic proportions of cement, water, fine aggregate, coarse aggregate and additives, if any, in order to yield a plastic mix which can be satisfactorily compacted in a mould or formwork and which when hardened after curing will produce concrete of the desired strength. Many of the desired properties of hardened concrete such as the tensile strength, bond strength, durability and impermeability are related to the compressive strength. Hence it is usual to specify concrete by the 28-day characteristic compressive strength determined by testing a few cube specimens of 150 mm size.

The 28-day cube compressive strength of concrete depends on many factors, such as, water/cement ratio, quality of cement, type of aggregates, workability of the mix, aggregate/cement ratio, batching, mixing, placing, compaction and curing. The water/cement ratio is the main controlling factor for concrete strength. If concrete is fully compacted, the compressive strength for a given set of materials at a given age varies inversely with the ratio of the weight of water in the concrete to the weight of cement. Empirical curves relating the 28-day compressive strength of concrete to water/cement ratio for different qualities of cement are available[8] as shown typically in Fig. 5.1.

The grading of aggregates influences the workability of concrete. Three gradings referred as zones A, B and C as suggested in IRC:44-1976 are given in Table 5.3. The available aggregates may be combined suitably to conform to one of the suggested gradings.

An important attribute of the concrete affecting its performance is durability evidenced by its resistance to deterioration and the adverse effects of the environment. A low permeability of concrete enhances durability. For chosen aggregates, the cement content should be just sufficient to provide adequate workability with a low water-cement ratio so that concrete can be thoroughly compacted resulting in a dense and impermeable concrete.

5.3.2 Design Steps

The design of a concrete mix involves the following steps:

1. The given data should include the characteristic concrete strength required in the field at 28 days (f_{ck}), the type of cement, the maximum size of the aggregate, grading of available aggregates, the degree of workability required and the expected degree of quality control.

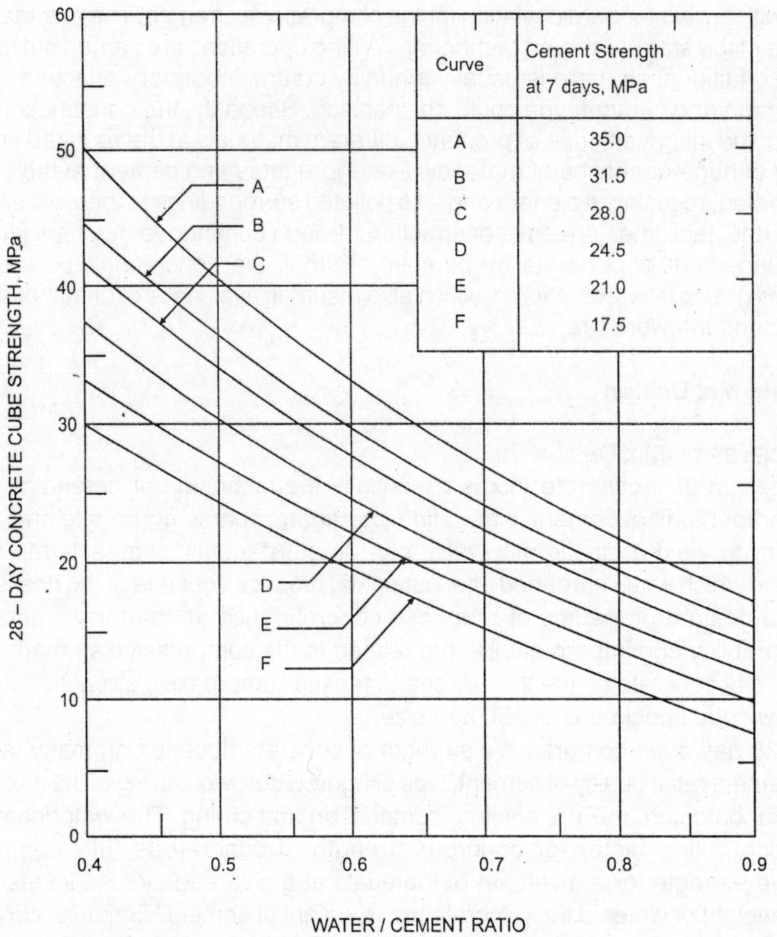

Figure 5.1 Typical Relationship between Water/Cement Ratio and Concrete Strength.

2. The desired proportion of the aggregates is determined by trying various combinations of the available aggregates or by a graphical procedure, to correspond to one of the three grading zones indicated in Table 5.3, the zone B being preferable.
3. The concrete mix should be so designed as to attain in the preliminary tests at 28 days a target strength determined as (f_{ck} + 1.64 *s*), where *s* is the standard deviation. The coefficient 1.64 corresponds to an acceptance level of 5 percent low values. The standard deviation may be assumed as given in IS: 10262 if the same is not determined for the site. Typical values of standard deviation for different grades of concrete and different degrees of control are indicated in Table 5.4. The target strength should normally be not less than 1.33 times the characteristic strength required. IRC: 21-2000 denotes the required increase over the characteristic strength to determine the target strength as 'current margin', and gives typical values of the current margin and the target strength for different grades of concrete. These values are included in Table 5.1.

4. The water/cement ratio corresponding to the target strength is obtained from available data as in Fig. 5.1.

Table 5.3 Suggested Coarse Aggregate Gradings.

Maximum size of Aggregate (mm)	Zone	Per cent passing sieve sizes (mm)			
		40	20	10	4.75
20	A	100	100	21-32	0
	B	100	100	31-40	0
	C	100	100	40-52	0
40	A	100	34-40	16-18	0
	B	100	40-45	18-20	0
	C	100	45-53	20-25	0

5. The water and sand contents per cubic metre of concrete are estimated in the first instance. For a particular water/cement ratio of 0.50, slump of 25 mm and a fineness modulus of 2.60 for sand, the values of water and sand contents for different maximum sizes of aggregate are given in Table 5.5. For each 0.05 increase or decrease in water/cement ratio, the sand content is increased or decreased by 1 per cent. For each 0.1 increase or decrease in fineness modulus of sand, the sand content is increased or decreased by 0.5 per cent. The water content is increased or decreased by 4 per cent of every 25 mm increase or decrease in slump.

Table 5.4 Typical Values of Standard Deviation.

Grade of Concrete	Standard Deviation, MPa		
	Very Good Control	Good Control	Fair Control
M10	2.0	2.3	3.3
M15	2.5	3.5	4.5
M20	3.6	4.6	5.6
M25	4.3	5.3	6.3
M30	5.0	6.0	7.0
M35	5.3	6.3	7.3
M40	5.6	6.6	7.6
M45	6.0	7.0	8.0
M50	6.4	7.4	8.4
M55	6.7	7.7	8.7
M60	6.8	7.8	8.8

6. The cement content is determined from the water content and water/cement ratio. The cement content should not be less than 250 kg/m^3 for plain concrete or 310 kg/m^3 for reinforced concrete for bridge components. The minimum cement content for prestressed concrete members is 400 kg/m^3. The maximum cement content for concrete is 540 kg/m^3.

Table 5.5 Water and Sand Content.

Maximum size of aggregate mm	Water content per cubic metre of concrete with saturated surface dry aggregates kg	Proportion of sand in total aggregate by absolute volume per cent	Entrapped air per cent
80	172.0	28.0	0.3
40	175.0	33.5	1.0
25	177.5	38.0	1.5
20	178.0	40.0	2.0

7. The sand content and the coarse aggregate content per m³ of concrete may be calculated separately from Equations (5.1) and (5.2)

$$V = 0.001\left[W + \frac{C}{S_c} + \frac{1}{P}\cdot\frac{S}{S_s}\right] \quad \text{... (5.1)}$$

$$V = 0.001\left[W + \frac{C}{S_c} + \frac{1}{(1-P)}\cdot\frac{A}{S_a}\right] \quad \text{... (5.2)}$$

where V = absolute volume of the wet mix
= gross volume (1.0 m³) minus volume of entrapped air (from Table 5.5)
W = water content in kg (= litre) per m³ of concrete
C = cement content in kg per m³ of concrete
S = fine aggregate content in kg per m³ of concrete
A = total aggregate content in kg per m³ of concrete
P = proportion (in decimal fraction) of sand in total aggregate determined on absolute volume basis
S_c, S_s, S_a = specific gravity of cement, sand and coarse aggregate, respectively.

8. With the above data, the mix proportions can be worked out. The actual mix proportions to be adopted in the field will be determined by allowing for the moisture available in the aggregates.

Concrete mix design arrived at as above can only be taken as a guide and not as a final design, because of the many assumptions made in the use of empirical data and the variability of material properties. Hence the design at any major construction site should be checked by field trials before adoption on a large scale.

5.3.3 Illustrative Example

Problem: To design a suitable concrete mix for a major bridge project for the following data.

Concrete grade : M20

Cement : Ordinary portland cement with specific gravity 3.15, having 7-day compressive strength with standard sand of 21 MPa

Coarse aggregate	:	Angular aggregate 20 mm size with grading corresponding to zone B of Table 5.3 and specific gravity 2.65.
Fine aggregate	:	Sand with fineness modulus of 2.48 and specific gravity 2.65.
Water absorption	:	0.4% for coarse aggregate and 0.6% for fine aggregate.
Free (surface) moisture	:	nil for coarse aggregate and 2% for fine aggregate
Degree of quality control	:	Good.

Solution

Characteristic strength = 20 MPa
From Table 5.4, standard deviation = 4.6 MPa
Target strength at 28 days = 20 + 1.64 × 4.6 = 27.5 MPa
Using curve E of Fig. 5.1, w/c = 0.48
From Table 5.5, for w/c = 0.5, and F.M. of sand = 2.60,
Water content per m^3 of concrete = 178 kg
Sand content = 40% of total aggregate
Adjustment needed for sand content due to w/c ratio = (0.48 – 0.50)/0.05 = – 0.4%
Adjustment needed for sand content due to F.M. = [(2.48 – 2.60)/0.1]0.5 = – 0.6%
Net change in sand content = – 1.0%
Required sand content = 39.0%
Cement content = 178/0.48 = 371 kg

From Table 5.5, entrapped air is 2%
Applying Equation (5.1),

$$0.98 = 0.001\left[178 + \frac{371}{3.15} + \frac{1}{0.39}\cdot\frac{S}{2.65}\right]$$

$$S = 707 \text{ kg}$$

Applying Equation (5.2),

$$0.98 = 0.001\left[178 + \frac{371}{3.15} + \frac{1}{0.61}\cdot\frac{A}{2.65}\right]$$

$$A = 1106 \text{ kg}$$

The nominal mix proportions are:

Water	Cement	Sand	Coarse Aggregate	Total Aggregate
178 kg	371 kg	707 kg	1106 kg	1813 kg
0.48	1	1.91	2.98	4.89

Aggregate / cement ratio = 4.89 : 1

The designed mix is 0.48 : 1 : 1.91 : 2.98 (by weight).

The actual quantities required in the field per bag of cement are then computed, allowing for free water in the aggregates.

Cement = 50 kg
Fine aggregate = 50 × 1.91 (1 + 0.02) = 97.4 kg
Coarse aggregate = 50 × 2.98 (1 – 0.004) = 148.4 kg
Water = 50 × 0.48 – 50 × 1.91 × 0.02 + 50 × 2.98 × 0.004 = 22.7 kg or 22.7 *l*

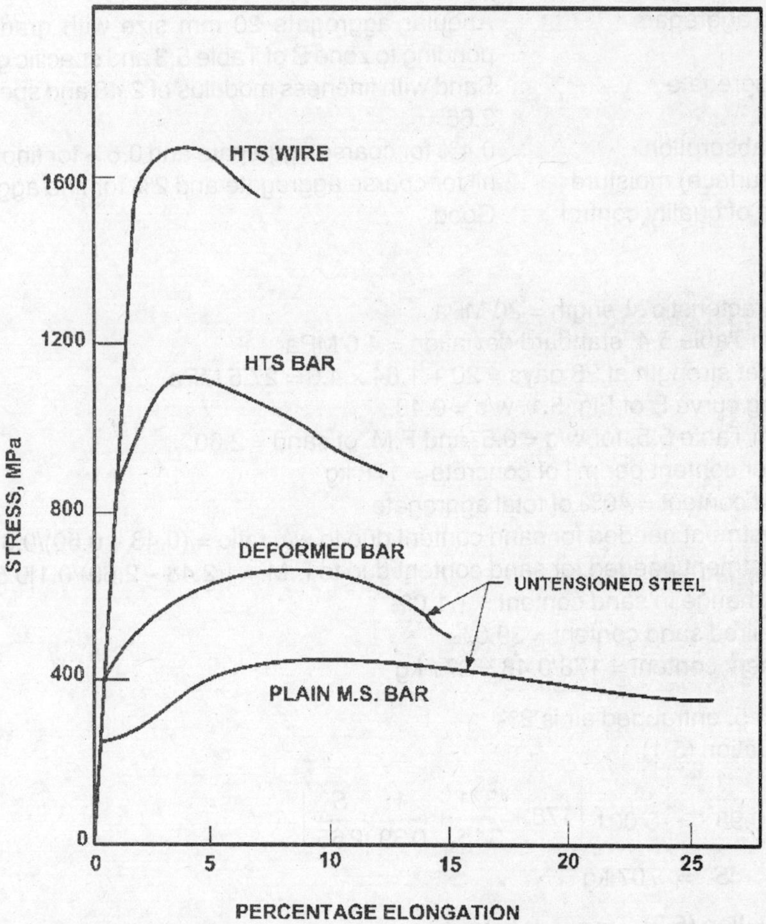

Figure 5.2 Typical Stress-Strain Diagrams for Steel.

5.4 Reinforcement

5.4.1 Bar Grades

The reinforcement in reinforced concrete members and the untensioned reinforcement in prestressed concrete members consist of steel bars conforming to one of the following types: (a) Mild steel bars according to IS: 432 (Part I)[9] with characteristic yield strength f_y of 240 MPa, denoted by bar grade Fe 240; (b) High yield strength deformed bar according to IS:1786[10] with yield strength f_y (determined as 0.2 percent proof stress) of 415 MPa, denoted by bar grade Fe 415; or High yield strength deformed bar according to IS:1786 with characteristic strength f_y of 500 MPa, denoted by bar grade Fe 500. Typical stress-strain diagrams for steel bars are shown in Fig. 5.2. The modulus of elasticity of steel E_s may be assumed as 200 GPa. The bars should be clean and free from loose mill scale and loose rust. Re-rolled bars are not to be used in bridge work.

With a view to ensure adequate ductility, IS:1786 requires a minimum elongation of 14.5 % for Fe 415 and 12.0 % for Fe 500. The minimum value of the ratio of ultimate tensile strength (UTS) to yield strength (YS), which is a measure of the toughness, is specified as 1.10 for Fe 415 and 1.08 for Fe 500. Resistance to severe earthquake effects demands adequate ductility and toughness in steel reinforcement. Currently, efforts are being made by the steel manufacturers to increase the strength, ductility and toughness through addition of small quantities of alloys such as Titanium and Vanadium to the steel prior to rolling.

Plain mild steel bars conforming to IS: 432 have been used in earlier construction. High yield strength deformed bars according to IS: 1786 are currently in use. With the latter bars, the yield strength and hence the permissible stress in steel are higher than the corresponding values for plain mild steel bars. They also give better crack control than plain bars and result in economies due to shorter lap lengths required. In recent years, the corrosion resistance of the reinforcing bars has been enhanced by coating the bars with epoxy after sand blasting to remove the mill scale. This is particularly important for bridges located in coastal areas.

5.4.2 Permissible Stresses

Permissible stresses in reinforcement as given in IRC:21 are shown in Table 5.6. Where a group of bars is used, the stress as above may be applied to the centre of gravity of the group, subject to the limitation that the stress in the extreme fibre of any bar should not exceed the permissible stress by more than 10 per cent. The modulus of elasticity of steel may be assumed as 2×10^5 MPa.

5.4.3 Bar Sizes and Spacings

The maximum size of a bar to be used in a bridge superstructure is limited to 40 mm; but preferably the size should be limited to 32 mm. Beyond this size, bending of bars will be difficult. In slabs, the size of main bars should be limited to 25 mm diameter, or one-tenth of the depth of the slab. The minimum size of a longitudinal bar in a column is 12 mm. The minimum size of any reinforcing bar including distributor or stirrup is 8 mm. The bar size is to be decided from the point of view of flexural as well as bond strength. The maximum diameter of shear reinforcement in beam webs, including cranked bars if any, should be limited to one-eighth the thickness of the web.

Table 5.6 Permissible Stresses in Reinforcing Bars.

Type of stress	Permissible stress, MPa		
	Bar grade Fe 240	Bar grade Fe 415	Bar grade Fe 500
Tension in flexure or combined bending	125	200	240
Tension in shear	125	200	200
Direct compression	115	170	205
Tension in helical reinforcement	95	95	95

The horizontal distance between two parallel reinforcing bars should be not less than the diameter of the bar (larger bar if unequal bars are used) or 10 mm more than the nominal size of the coarse aggregate used in the concrete. When immersion vibrators are used, sufficient space should be left between groups of bars to enable the vibrator to be inserted. The minimum vertical distance between two horizontal main reinforcing bars should be 12 mm or the maximum size of the coarse aggregate or the maximum diameter of the bar, whichever is greater. The pitch of bars of main tensile reinforcement in slabs should not exceed 300 mm or twice the effective depth whichever is smaller, and should also satisfy the crack control criteria indicated in section 5.4.4. Tables E.1 to E.3 in the Appendix give the perimeter and area of groups of round bars in tabular form.

5.4.4 Requirements For Crack Control

With a view to ensure control of cracking in concrete at the tensile face of reinforced concrete components, IRC:21 has prescribed that the diameter and spacing of reinforcing bars shall not exceed the limits as below:

	Diameter, mm	Spacing, mm
Slabs	25	150
Beams	32	150
Columns	32	300

If the above criteria could not be followed, the designs are to be checked for design surface crack widths as per procedure indicated in Appendix 1 of IRC: 21, and the crack width under sustained load (taken as dead load plus 50% of the live load) should be limited to 0. 2 mm for severe conditions of exposure and 0.3 mm for moderate conditions of exposure.

5.4.5 Cover

Inadequate cover to the reinforcement has often been the cause for durability problems in concrete structures, as such a deficiency leads to reduced resistance to both carbonation and chloride ion induced corrosion of the reinforcement. The workmanship at the construction site should be of good quality to ensure the realisation of the correct cover. The cover is measured as the thickness from the outer surface of the concrete to the nearest surface of the reinforcement, sometimes referred as 'clear cover'. The minimum clear cover to any reinforcement bar should be 40 mm. The minimum cover is to be increased to 50 mm when concrete members are exposed to severe conditions of exposure, such as marine environment. For the condition of alternate wetting and drying and in case of foundations, the minimum cover is to be increased to 75 mm. The above cover may be reduced by 5 mm in the case of factory made precast products with high level of quality assurance. Specially made polymer cover blocks are recommended to be used for ensuring proper cover to reinforcement.

5.5 Elastic Design Coefficients

Currently, only elastic design is used in bridge design. This is because the effects of dynamic loads on the performance characteristics of structures proportioned according to ultimate strength design methods are not yet fully understood.

In the design of reinforced concrete one-way slab or singly reinforced beam, we have to compute the design coefficients m, n, j and R for any given set of values of s_{cb} and s_{st}

Table 5.7 Typical Values of Elastic Design Coefficients.

Concrete Grade	Bar Grade	s_{cb} MPa	s_{st} MPa	n	j	R
M15	Fe 415	5.0	200	0.20	0.93	0.47
M20	Fe 415	6.7	200	0.25	0.92	0.77
M25	Fe 415	8.3	200	0.29	0.90	1.10
M30	Fe 415	10.0	200	0.33	0.89	1.47
M35	Fe 415	11.5	200	0.36	0.88	1.84
M40	Fe 415	11.5	200	0.36	0.88	1.84

Note: The value of the modular ratio m is adopted as 10, as permitted in IRC: 21.

Here s_{cb} and s_{st} are the permissible stresses in concrete and steel, m is the modular ratio, n is the ratio of the depth from the extreme compression fibre to the neutral axis to the effective depth, and j is the lever arm factor. R is the coefficient of resistance, which when multiplied by bd^2 gives the moment of resistance. The relationships are well known. The modular ratio is taken as 10 as per IRC: 21. Typical values of the elastic design coefficients for normally used concrete and steel grades are indicated in Table 5.7.

5.6 Curtailment of Bars

To prevent large changes in the moment of resistance, the locations of bar curtailment should be staggered. When bars are curtailed in the tension zone, certain precautions are necessary. It has been found that if the bars are cut exactly at the theoretical point of cutoff without any offset, premature shear cracks appear and lead to loss of strength of as much as thirty-three per cent[11]. Where bars are curtailed, they should be extended at least 12 diameters beyond the theoretical cutoff point and should preferably be bent up and anchored in the compression zone. In simply supported spans, at least one-third of the total bars required to resist the bending moment should be continued to the support in case of beams, and 50% of the bars in case of slabs. For a span continuous over a support, at least 25% of the steel required to resist the maximum positive bending moment should be carried over the support.

5.7 Bond, Anchorage and Splice

5.7.1 Anchorage Length

The length required for developing full strength of a bar with straight ends in tension at anchorage to keep the average bond stresses within allowable limits is known as the basic permissible value of anchorage length. This is denoted by L_o and is expressed as a multiple (n) of the diameter of the bar (ϕ): i.e., $L_o = n\phi$.

The value of n varies with the concrete grade, the bar grade and the bonding zone. The bonding zone is termed as unfavourable, if the horizontal bar or bar bent less than 45° is located such that more than 300 mm of fresh concrete is cast below the bar. Bonding zone is considered favourable if the bar is not located in the unfavourable zone. The values of n for different conditions are listed in Table 5.8.

The design anchorage length L_d for bars in tension is derived as a modification of L_o, to account for the provision of hooks and steel provided in excess of requirement.

Table 5.8 Basic Permissible Anchorage Length as Multiples of Bar Diameter.

Bonding Zone	Bar Grade	Value of n		[n = L_o/ϕ]		
		M20	M25	M30	M35	M40
I (favourable)	Fe 500	66	56	48	42	42
	Fe 415	55	46	40	35	35
	Fe 240	65	60	55	50	50
II (unfavourable)	Fe 500 Fe 415 Fe 240	1.4 times the values given for zone I				

$$L_d = \alpha . L_o \quad \ldots (5.3)$$

where α = 1 for bars with straight ends
= 0.7 for bars with end hooks
= A_{sd}/A_s when the steel area provided A_s is in excess of design requirement A_{sd} at full permissible stress

Also L_d should not be less than $L_{a,min}$, where $L_{a,min}$ is the greater of 12ϕ or 300 mm for bars with straight ends, or the greater of 6 ϕ or 150 mm for bars with end hooks. For bars in compression, the anchorage length is computed as for bars with straight ends, disregarding the end hooks.

5.7.2 Splices

When it is necessary to extend a bar beyond the available length, the additional length can be provided by splicing. Splices of reinforcement can be formed by: (a) lap of bars; (b) welded joints; or (c) joints with mechanical devices.

When lap splices are used, the area of bars spliced at any section should not exceed 50% of the total reinforcement area for Fe 500 and Fe 415 bars or 25% for Fe 240 bars. The splices should be staggered longitudinally at least by 1.3 times the required lap length. The lap length of bars in tension with straight ends or with end hooks should be K.L_d, where K is 1.4 for 25% splices and 1.8 for 50% splices. Additional stirrups should be provided along the lap length in beams. Lap splices are usually not allowed for bar sizes larger than 36 mm.

Welded joints may be provided when Fe 240 grade bars are used, and the full area may be considered effective. Welding of Fe 500 and Fe 415 bars is generally discouraged. However, welding of Fe 500 and Fe 415 bars may be permitted when the steel is proved to be weldable by chemical analysis. This is done by calculating the carbon equivalent (C.E.) from the chemical composition using the formula (5.4) and showing it to be 0.4 or less.

$$\text{C.E.} = C + \frac{Mn}{6} + \frac{Cr + Mo + V}{5} + \frac{Ni + Cu}{15} \quad \ldots (5.4)$$

where C, Mn, Cr, etc. represent the percentage of the element concerned in the chemical composition of the steel. At such welded joints, only 80 % of the bar area is considered effective. Butt welding is adopted for bar diameter more than 20 mm. The weld is built up so that the throat diameter is greater than the diameter of the bars joined, the increase in diameter being 10 to 25 % of the bar diameter. Lap weld with longitudinal beads may be used for bars of diameter less than 20 mm.

Bars may be joined by approved mechanical devices (couplers). The effectiveness of such devices is to be determined by static and fatigue tests.

5.7.3 Detailing of Reinforcement

Detailing of reinforcement should be performed taking due note of the construction sequence and construction joints. The aim in detailing is to provide bars where needed with minimum redundant steel and without restricting the fabrication of the formwork.The arrangement should be simple, limiting the variety of sizes and using straight bars to the extent possible. The detailing should consider the bar bending and fixing tolerances.

Splices of reinforcements should be carefully staggered. When using large diameter bars in T-beam ribs, care should be taken to ensure adequate clearance between rods and below the reinforcement.

The ends of main reinforcement bars are either hooked or bent at right angles to the length of the bar to provide adequate anchorage. According to IS:456, the anchorage value of a bend may be taken as 4 times the diameter of the bar (ϕ) for each 45° bend, subject to a maximum of 16ϕ. For a standard 90° bend, the anchorage value is 8ϕ, which includes a minimum extension of 4ϕ.

In the case of a 180° bend, called the standard U-type hook, the anchorage value is 16ϕ including a minimum extension of 4ϕ. The minimum turning radius for deformed bars is 4ϕ.

For stirrups and ties in columns, the necessary anchorage length is provided by: minimum extension of 8ϕ for 90° bend, or a minimum extension of 6ϕ for 135° bend, or a minimum extension of 4ϕ for 180° bend.

5.8 Shear in Reinforced Concrete Beams and Slabs

5.8.1 Shear Stress

The design shear stress τ at any cross section in a reinforced concrete beam or slab of uniform depth is calculated from Equation (5.5).

$$\tau = \frac{V}{bd} \qquad \text{... (5.5)}$$

where V = design shear across the section
d = effective depth of the section
b = breadth of the rectangular beam or slab, or the breadth of the web in case of flanged beam.

The design shear stress calculated from Equation (5.5) is limited to the maximum permissible shear stress τ_{max} in beams and 0.5 τ_{max} in slabs. The value of τ_{max} in N/mm^2 varies from 1.8 to 2.5 for concrete of grades M20 to M40.

5.8.2 Members without Shear Reinforcement

The permissible shear stress τ_c in concrete in beams and slabs without shear reinforcement is tabulated in IRC:21.

For slab section having $\tau > \tau_c$, shear reinforcement to cater to full shear should be provided. Beams in bridges will normally require shear reinforcement.

5.8.3 Members with Shear Reinforcement

Shear reinforcement is provided by: (a) Stirrups perpendicular to the longitudinal reinforcement; or (b) a combination of stirrups and bars bent up at 45° ($\alpha = 45°$), where not more than 50% of the shear shall be carried by bent up bars.

$$\text{Shear resisted by stirrups } V_s = s_s . A_{st} \quad \dots (5.6)$$

where s denotes the spacing of stirrups. When bent up bars are provided, the shear carried by bent up bars (V_b) is first computed, and the balance of shear ($V - V_b$) is designed to be resisted by the stirrups.

$$\text{Shear resisted by bent up bars } V_b = s_s . A_{st} \text{ Sin } \alpha \quad \dots (5.7)$$

The area of stirrups so computed should not be less than 0.0015 bs for $f_y \leq 415$ MPa. The maximum spacing of stirrups is limited to one-half the depth of the beam subject to a maximum of 300 mm. Stirrups shall pass around or otherwise be secured to the appropriate longitudinal tensile reinforcement.

5.9 Effective Flange Width for T-beams and L-beams

For the purpose of computing the moment of inertia or for computing the flexural stresses, the effective width of compression flange of beams with solid webs is taken as the smaller of the actual flange width or web width plus 0.2 L_o for T-beam or web width plus 0.1 L_o for *L*-beam, where L_o is the distance between points of zero moments (taken as 0.7 times the effective span for continuous beams). In order to avoid stress concentration, the junction of the web with the slab is splayed with the provision of a fillet. The obtuse angle at the junction should be at least 110°.

5.10 Minimum Reinforcement in Slabs and Beams

In slabs, the reinforcement in either direction should be not less than 0.12 per cent of the total cross sectional area when Fe 415 grade bars are used, and not less than 0.15 per cent of the total cross sectional area when Fe 240 grade bars are used.

For beams, the minimum tension reinforcement is 0.2 per cent of *bd* when using Fe 415 grade bars or 0.3 per cent of *bd* when using Fe 240 grade bars, where *b* is the breadth and *d* is the effective depth. For T- or L-beams, *b* is taken as the average width of concrete below the upper flange.

5.11 Concreting Operations

All the ingredients of concrete, i.e. cement, fine and coarse aggregates, water and admixtures if any, are weighed as per design and loaded into the power driven batch mixer. The concrete should be mixed well until it is of even colour and uniform consistency. The mixed concrete should be transported from the place of mixing to the location of final deposit as rapidly as practicable with due care to avoid segregation or loss of ingredients. Concrete may be transported by transit mixers or by pumping. The concrete should be placed in its final position before commencement of setting in such a manner as to avoid segregation.

Concrete should be thoroughly compacted during the placement and should be worked around the reinforcement, around embedded fixtures and into the corners of the formwork. Internal vibrators (and external vibrators where desired) are used to aid compaction. The prescribed cover to reinforcement is ensured by the use of manufactured chairs. The concrete should be properly cured, by keeping it wet for a minimum period of 14 days.

5.12 Prestressed Concrete

5.12.1 General

The application of the concept of prestressing to structural concrete members has opened up a wide spectrum of bridge types and has enlarged the span range possible with concrete. Prestressing may be defined as the application of a predetermined force to a structural member in such a manner that the combined internal stresses in the member resulting from this force and any possible condition of external loading will be counteracted to a desired degree. The prestress is usually imparted to concrete by straining the prestressing tendon relative to the concrete, thereby causing compressive stresses in concrete due to tension in the tensioned steel. The aim of the application of prestress is to avoid cracking of concrete due to flexural or principal tensile stresses under service loads.

Though the concept of prestressing originated with Jackson in 1886 and Doehring in 1888, the widespread use as in modern practice is mainly due to the pioneering efforts of E. Freyssinet since about 1926. A few basic requirements of prestressed concrete design as per IRC:18 and IS:1343[12] are indicated in this section and the application of prestressed concrete to bridges is discussed in Chapter 8. For a discussion of the basic theory of prestressed concrete and for hardware details of proprietary systems, reference may be made to standard text books.

There are three types of members in stressed concrete according to IS: 1343 defining the degree of prestress: (a) Type 1 – Full prestress; (b) Type 2 – Limited prestress; and (c) Type 3 – Partial prestress. For Type 1 members, tensile stress is not permitted in the section under any loading condition during construction or in service. Full prestressing is adopted in structures subjected to dynamic loads and structures exposed to corrosive atmosphere. Limited prestressing permits low values of tensile stresses below the cracking stress such that the entire section is effective. This may be adopted for structures which occassionally carry the maximum load. Under partial prestress, the crack width and strain are limited, and stresses under working load are computed using a cracked section neglecting concrete in tension. As a conservative measure, the present design practice in this country permits only full prestress for design of highway bridges.

5.12.2 Special Features

The special features of prestressed concrete as compared to normal reinforced concrete are listed below:

(a) The internal forces for a reinforced concrete member are functions of the applied load only, whereas such forces in a prestressed concrete member are affected to a large extent by the tension in the prestressing tendon. For an unloaded reinforced concrete member, the total tension (T) and the total compression (C) at a section are individually equal to zero. With increase in load and hence moment, the values of *C* and *T* increase. The lever arm is constant within elastic range and increases marginally towards the ultimate stage due to nonlinearity of the concrete stress block. On the other hand, the net compression *C* in a prestressed concrete member is nearly constant and is guided by the initial prestressing force. The lever arm varies with variation in the moment due to change in load.

(b) The elimination of cracking enhances the durability and service life of the structure.

(c) The dead load is reduced due to the use of materials of higher strength and due to the full section being effective. For smaller spans and under certain circumstances, it is possible to arrange the profile of the prestressing tendons in such a manner that a part of the dead load is counteracted.

(d) The quantity of reinforcement necessary for a bridge structure is reduced very considerably due to the use of high tensile steel. However, the ratio of unit cost of high tensile steel to that of untensioned reinforcement is so high at the present time that the overall economy resulting from the use of prestressed concrete is often difficult to predict except for long span bridges.

(e) The technique of construction subjects the materials to the severest condition obtainable in service during the construction period. Hence the materials, particularly the reinforcement, are automatically proof-tested during construction.

(f) The shear capacity of concrete is increased due to prestress, enabling the use of thinner webs. The resulting saving in dead weight is particularly significant in long span bridges.

(g) The elimination of cracking under service loads also leads to better resistance to fatigue.

(h) The structure can be precast as a number of segments of sizes convenient for the available handling facilities. These segments can be combined by prestress to make up the full structure. The precast elements can be made under factory conditions to a high degree of precision and a more efficient turnover of formwork and allied equipment.

(i) It is essential to take into account the procedure of construction even in the design of the bridge.

(j) The slender dimensions possible with prestressed concrete enhance the appearance of the finished structure and hence make it specially suitable for urban highway structures.

5.12.3 Concrete

The concrete used in prestressed concrete bridge work should be high strength high performance concrete designated by the specified characteristic compressive strength (f_{ck}) of 150 mm cubes at 28 days expressed in MPa. The characteristic compressive strength of concrete is defined as the strength of concrete, below which not more than 5 per cent of the test results are expected to fall. The minimum specified characteristic compressive strength is 40 MPa (grade M40) for prestressed concrete bridges. The concrete mix shall be designed in accordance with IS: 10262 to have a target mean strength defined as $f_{ck} + 1.64\ s$, where *s* is the standard deviation. The target mean strength of the concrete mixes of grade M40, M45 and M50 shall not be less than 52 MPa, 58 MPa and 63 MPa, respectively. The minimum cement content should be 400 kg/m^3 for all exposure conditions. Also the quantity of cement in the concrete mix shall not be more than 540 kg/m^3. The minimum cement content is specified to ensure adequate strength and the maximum value is specified to reduce shrinkage effects. The age of cement should not be more than 60 days for prestressed concrete works. The maximum water/cement ratio is to be restricted to 0.45 for severe exposure and to 0.50 for normal exposure.

The modulus of elasticity of concrete (E_c) may be computed from Equation (5.8).

$$E_c = 5000\sqrt{f_{ck}} \text{ MPa} \quad \dots (5.8)$$

When the value of the modulus of elasticity at any particular age of concrete is desired, the same may be computed by substituting the concrete strength at that age in the above equation.

5.12.4 Prestressing Steel

The high tensile steel (HTS) used for prestressing may be any one of the following: (a) Plain hard drawn wire conforming to IS: 1785; (b) Cold drawn indented wire as per IS: 6003; (c) High tensile steel bar to IS: 2090; (d) Uncoated stress relieved strand conforming to IS: 6006; and (e) Uncoated stress relieved low relaxation steel conforming to IS: 14268. Hard drawn wires used for bridges are generally of 5, 7 or 8 mm diameter having an ultimate strength of 1600 to 1400 MPa. The wires may be indented to enhance the bond characteristics. Alloy steel bars used for prestressing are hot rolled bars with sizes 9.5 mm to 32 mm with an ultimate tensile strength of about 1000 MPa. Strands, manufactured from hard drawn wires of small size, are available generally with a strand diameter of 13 mm and 15 mm. A typical strand is composed of seven wires, of which six wires are twisted on a pitch of about 12 to 15 diameters around a slightly larger central wire. The nominal area of a 13 mm strand is 100 mm^2, while that of a 15 mm strand is 150 mm^2. It would be desirable to schedule the arrival of HTS used for prestressing such that long storage at site and exposure to corrosive environment are avoided. The HTS should preferably be stored at site in godowns with humidity controlled up to 60 per cent.

Typical stress-strain diagrams for high tensile steel (HTS) and mild steel (MS) are included in Fig. 5.2. In case of HTS and deformed bars, the proof stress is taken to be equal to the stress corresponding to a permanent strain of 0.2 per cent. The proof stress of HTS bars is about 0.8 f_p, whereas that of wire or strand is about 0.85 f_p, where f_p is the ultimate tensile strength. The modulus of elasticity of steel for purposes of design may be taken nominally as 2.1 x 10^5 MPa for wires, 2.0 x 10^5 MPa for bars and 1.95 x 10^5 MPa for strands.

The strain possible in a MS tendon, if it were to be used for prestressing, would be in the order of 0.0006 to 0.0009. Such a level of strain would be lost within a short time due to creep, shrinkage and relaxation of steel, leaving no significant prestress. With HTS, the strain in steel at the time of transfer could be in the range of 0.004 to 0.005, permitting a substantial residual prestress after losses. Hence it is necessary to use HTS as the material for prestressing tendons.

5.12.5 Prestressing Systems

Various proprietary systems of prestressing are available. Prestressing tendons may consist of HTS wires, bars or strands. Individual wires of diameter 5 mm and 7 mm are sometimes used in pretensioned beams. The Freyssinet system with wire cables employs a wedge based anchorage system. This will involve a slip of about 3 mm. HTS bars of 9.5 to 32 mm diameter are used as tendons in certain systems. The bars are placed in the ducts cast between two anchor blocks, pulled from one end by a stressing jack and anchored by a nut assembly as in Dywidag system. This provides a positive anchorage with no slip.

Recent applications invariably use 7-wire strands of 13 or 15 mm diameter. The commonly used tendons have 7, 12 or 19 strands to suit the standard anchor blocks available, similar to Freyssinet system for wires. While stressing, the jack is placed over the tendon, gripping each strand and pulling till the required force is generated (checked with the elongation measured). Wedges are then placed around each strand and seated into the anchor block. On the release of the jack, the wedges grip the strand and transfer the force to the concrete through the anchor block.

5.12.6 Permissible Stresses in Concrete

(a) *Temporary stresses*

Stage prestressing is permissible; but it should be ensured that concrete should have attained a minimum strength of 20 MPa before any prestressing is applied.

The compressive stress produced due to loading during construction such as transport, handling, launching and erection should not exceed 0.5 f_{cj} or 20 MPa, where f_{cj} is the concrete strength at that time, subject to a maximum value of 30 MPa. At full transfer, the cube strength of concrete shall not be less than 0.8 f_{ck}. Temporary compressive stress in the extreme fibre of concrete shall not exceed 0.5 f_{cj} or 30 MPa. The temporary tensile stresses in the extreme fibres of concrete shall not exceed 0.1 of the permissible temporary compressive stress of the concrete.

(b) *Stress during service*

The compressive stress in concrete under service loads should not exceed one-third of the characteristic strength. No tensile stress shall be permitted in the concrete during service. If precast segmental elements are joined by prestressing, the stresses in the extreme fibres of concrete during service should always be compressive and the minimum compressive stress in an extreme fibre should not be less than five per cent of the maximum permanent compressive stress that may be developed in the same section.

(c) *Bearing stress behind anchorages*

The maximum allowable bearing stress, immediately behind the anchorage of a prestressing tendon in an adequately reinforced end block may be computed from Equation (5.9).

$$f_b = 0.48\, f_{cj} \sqrt{A_2 / A_1} \text{ limited to } 0.8\, f_{cj} \qquad \dots (5.9)$$

where f_b = the permissible compressive contact stress in concrete

A_1 = the bearing area of the anchorage converted in shape to a square of equivalent area

A_2 = the maximum area of the square that can be inscribed within the member without overlapping the corresponding area of adjacent anchorages, and concentric with the bearing area A.

The above value of f_b is permissible only if there is a projection of concrete of at least 50 mm or one fourth the side of the square relating to the bearing area A_1. When anchorages are embedded in concrete, the bearing stress may be investigated after accounting for the surface friction between the anchorage and the concrete. If adequate hoop reinforcement is provided at the anchorage, the value of f_b calculated from Equation (5.9) may be increased suitably. The pressure operating on the anchorage shall be taken before allowing for losses due to creep and shrinkage of concrete, but after allowing for losses due to elastic shortening, relaxation of steel and seating of anchorage.

5.12.7 Permissible Stresses in Prestressing Steel

Maximum jack pressure shall not exceed 90 per cent of the 0.1% proof stress or 80 percent of the ultimate tensile strength (UTS), whichever is lower.

5.12.8 Losses in Prestress

A part of the applied prestress is lost after a brief period due to elastic shortening, creep and shrinkage of concrete, relaxation of steel, friction, and seating of anchorages. The losses due to friction, seating of anchorages (slip) and elastic shortening are referred as instantaneous losses, while the losses due to shrinkage and creep of concrete and relaxation of steel are known as time dependant losses. The method of computation of these losses is indicated below:

(a) *Elastic shortening*

In a pretensioned beam, the concrete at the level of the tendon profile will undergo shortening due to the prestressing force and would result in reduction of the tensile strain in the tendons. The loss of prestress would be the product of modular ratio and the stress in concrete at the level of the tendons.

In the case of a post-tensioned beam, the elastic shortening would occur prior to anchorage and thus would not affect the prestress if all the prestress is applied in one operation. Usually, there will be a number of tendons to be stressed one by one, and so any tendon being stressed would cause elastic shortening loss to the tendons stressed and anchored earlier. It is tedious to compute the exact shortening of the fibres at the steel level for each of the tendons. For design purposes, an approximate procedure is adopted. The resultant loss of prestress in tendons tensioned one by one is calculated on the basis of half the product of the modular ratio and the stress in concrete adjacent to the tendons averaged along the length. For example, in the case of a simply supported beam, the mean of the stresses at the end section and at midspan may be multiplied by the modular ratio to arrive at the loss due to elastic shortening.

(b) *Creep of concrete*

The strain due to creep of concrete is defined as the time dependent deformation of concrete due to a sustained stress. The creep deformation is a complex phenomenon depending on many factors, such as: (i) stress level; (ii) the duration of stress; (iii) maturity of concrete; and (iv) environmental conditions. For purposes of design, IRC code specifies the values of creep strain as in Table 5.9 for concrete with ordinary portland cement. The stress for the calculation of the loss due to creep shall be taken as the stress in concrete at the centroid of the prestressing steel.

Table 5.9 Creep Strain in Concrete.

Maturity of concrete at the time of stressing as a percentage of f_{ck}	Creep strain per 10 MPa
40	0.00094
50	0.00083
60	0.00072
70	0.00061
75	0.00056
80	0.00051
90	0.00044
100	0.00040
110	0.00036

(c) *Shrinkage of concrete*

Concrete undergoes gradual diminution in volume during the early age due to drying and chemical reactions in cement. This change is referred as shrinkage. The extent of shrinkage depends on the water-cement ratio, weather conditions and size of the member. The bridge code specifies that the loss of prestress in steel, due to residual shrinkage of concrete, may be estimated from the values of strain due to residual shrinkage given in Table 5.10 for concrete with ordinary portland cement. The loss of prestress in post-tensioned members is smaller than for pretensioned members in view of the longer time interval between casting of concrete and transfer of prestress in the former.

Table 5.10 Shrinkage Strain.

Age of concrete at the time of stressing in days	Strain due to residual shrinkage
3	0.00043
7	0.00035
10	0.00030
14	0.00025
21	0.00020
28	0.00019
90	0.00015

(d) *Relaxation of steel*

When a prestressing tendon is stretched and maintained at constant length, the stress in the tendon decreases with time due to relaxation. The magnitude of relaxation depends on the characteristics of the steel and the level of prestress. The value of relaxation can be determined in a Laboratory by applying an initial stress of 0.7 f_p for 1000 hours under constant length and then adopting value of three times the 1000 hour value as the final relaxation at 0.5 million hours. Here f_p denotes the ultimate strength of the prestressing steel. The above value is the relaxation corresponding to an initial stress level of 0.7 f_p, which is assumed to reduce to zero at a stress level of 0.5 f_p. The intermediate values may be interpolated linearly. When certified values are not available, the Bridge Code specifies that the relaxation losses at 1000 hours may be assumed as 0, 2.5%, 5.0% and 9.0% for initial prestress of 0.5 f_p , 0.6 f_p , 0.7 f_p and 0.8 f_p , respectively, for normal relaxation steel. Corresponding values for low relaxation prestressing steel (LRPS) may be assumed as half the above values. For calculating the temporary stress at transfer, the Code specifies the relations between relaxation losses and time up to 1000 hours.

(e) *Seating of anchorages*

Depending on the system of prestressing, losses in prestress occur due to slip of wires and draw-in of male cones at the anchorages. When using bars and threaded nuts (as in Dywidag system), slipping at anchorages can be totally eliminated. The anchorage take-up is a property of the system and is independent of the length of the tendon. For systems using wires and wedges, the anchorage take-up is about 2 to 3 mm. The loss of prestress due to slip is computed as the product of E_s and the strain obtained as slip divided by the total length of the tendon.

(f) *Frictional losses*

Losses in prestress occur along the span due to (i) curvature effect due to friction between the tendon and its supports; and (ii) wobble effect along the length. The steel stress in prestressing tendon σ_{px} at any distance x from the jacking end can be computed from Equation (5.10).

$$\sigma_{po} = \sigma_{px} \cdot e^{(\mu\theta + kx)} \quad \text{... (5.10)}$$

where σ_{po} = steel stress at the jacking end
σ_{px} = steel stress at a point distant x from the jacking end
e = base of Naperian logarithm
μ = coefficient of friction
θ = angle change in the direction of the tendon in radians
k = wobble coefficient per metre length of steel
x = distance between the points of operation of σ_{po} and σ_{px} in metres.

The values of the coefficients k and μ may be assumed as in Table 5.11 for calculating the friction losses.

Table 5.11 Values of Coefficients k and μ.

Type of High Tensile Steel	Type of Sheath or Duct	Values for Design	
		k	μ
Wire Cable	Bright metal	0.0091	0.25
	Galvanised	0.0046	0.20
	Lead Coated	0.0046	0.18
	Unlined Duct in concrete	0.0046	0.45
Strand (uncoated stress relieved)	Bright metal	0.0046	0.25
	Galvanised	0.0030	0.20
	Lead Coated	0.0030	0.18
	Unlined Duct in concrete	0.0046	0.50
	Corrugated HDPE	0.0020	0.17

5.12.8 COVER AND SPACING OF PRESTRESSING STEEL

The clear cover measured from the outside of sheathing should be at least 75 mm or the diameter of the duct, whichever is greater. When grouping of cables is not involved, the minimum clear distance between individual cables should be maintained as 50 mm or 10 mm in excess of the largest size of aggregate used or the diameter of the duct, whichever is greater. Grouping of cables should be avoided altogether in case of aggressive environment and to the extent possible in moderate exposure conditions. If unavoidable, cables may be grouped vertically up to two cables only, maintaining a minimum spacing of 50 mm between groups. Individual cables or ducts of grouped cables shall be draped or deflected in the end portions of members, ensuring a minimum clear spacing of 50 mm between cables in the end one metre of the members.

5.13 Notation for Detailing Concrete Bridges

At the present time, the detailing practice varies from place to place, and no standard notation is followed in the field. The aim should be to evolve a notation which would convey the ideas of details and dimensions with a minimum of writing on the drawings. A concise notation, suggested by the author[13] and followed in the illustrations included in this text, is described below:

(a) All linear dimensions should be shown in mm. The unit need not be written with every dimension; it will be understood. For example, a slab thickness of 25 cm will be noted as '250'.

(b) No comma is necessary to indicate thousands in number; but a space will be left in its place. Thus, a span of 25 m will be shown on the drawing as '25 000'. A space is optional with a four digit number.

(c) The size of reinforcement will be indicated in terms of its nominal diameter in mm and ϕ. For instance, a round bar of 25 mm diameter will be denoted by 25 ϕ. A square bar of 25 mm size will be shown as '25 □'.

(d) If a group of bars is to be referred to, the number of bars will be shown ahead of the size with a hyphen. Thus, a group of 12 bars of 28 mm diameter in the tensile zone of a T-beam will be designated as '12–28 ϕ'.

(e) The spacing of bars, as in a slab, can be shown by a number next to the bar size. For example, '16 ϕ 150' would indicate slab bars of 16 mm diameter placed with a centre-to-centre spacing of 150 mm.

(f) When it is desired to name the bars, so as to facilitate preparation of bar bending schedule, the designation may be placed ahead of the bar size with a hyphen. For instance, 'A –16 ϕ 240' refers to 'A-bars', which are of 16 mm diameter and are spaced at 240 mm centres.

(g) In case of stirrups, the number of legs may be included ahead of the size of the bar. Thus, a four-legged stirrup of 12 mm diameter will be indicated as '4 –12 ϕ' on the cross section. However, on the longitudinal section of a beam, the stirrup spacing may be included by adding the spacing dimensions such as '4 –12 ϕ 200' indicating a spacing of 200 mm for the above stirrups.

(h) Prestressing cables may be shown by the number of wires in the cable and size of individual wires, e.g., '12–7ϕ' would denote a 12-wire cable of 7 mm diameter wires. If a group of cables is to be referred to, an integer denoting the number of cables may be placed ahead of the cable size enclosed in parenthesis, as '2 (12 –7ϕ)' for two cables.

5.14 Steel Construction

5.14.1 Materials

Steel used in bridge construction is usually structural steel conforming to IS: 2062 (steel for general purposes) or IS: 8500 (structural steel microalloyed) or IS: 11587 (structural weather resistant steel)[14-16]. Steel covered by the above specifications contain: iron, a small percentage of carbon and manganese, and very small quantities of alloying elements to improve specific properties of the final product, such as copper, nickel, chromium, molybdenum and zirconium. The strength of the steel increases with increase in carbon content to some extent, but ductility and weldability decrease.

The important properties of structural steel are the following: (a) Yield stress and Ultimate strength in tension; (b) Ductility; (c) Weldability; and (d) Weather resistance. The yield strength in tension is the most important strength parameter for structural steel. The yield stress may be clearly defined in the case of mild steel with low carbon content or may have to be derived from a stress-strain diagram for high strength steel. Ductility in steel permits a structure to undergo perceptible deflection in case of overload, thus providing advance warning of impending failure. Structural steel normally has a minimum elongation of 17 to 26 %. Weldability is an important property for modern steel bridges. This is ascertained from the 'carbon equivalent' test as defined in Equation (5.4). Corrosion resistance is improved by adding chromium, copper and nickel as alloys. The resulting steel is known as 'weathering steel'.

For purposes of design, steel is assumed to have the following properties:

Modulus of Elasticity, $E_s = 2.11 \times 10^5$ MPa
Shear modulus = 77×10^3 MPa
Poisson's ratio = 0.30
Coefficient of thermal expansion = 11.7×10^{-6}/ °C

Connections in steel members can be by welding, riveting or bolting. The welding electrodes should comply with the requirements of IS: 814 for welding of mild steel and of IS: 1442 for welding of high tensile steel[17–18]. The rivets and rivet bars should conform to IS: 1148 and IS: 1149 respectively, for mild steel and high tensile steel members [19,20]. Mildsteel bolts and nuts should have a minimum tensile strength of 440 MPa and a minimum elongation of 14 per cent on a gauge length of 5.65 $\sqrt{area}$. High tensile steel bolts are to be manufactured from high tensile steel having a minimum tensile strength of 580 MPa.

5.14.2 Permissible Stresses

(a) *Basic permissible stresses*

The basic permissible stresses under dead load, live load with impact and centrifugal forces should be limited to those given in Table 5.12 for structural steel. These values may be exceeded to specified extent for certain combinations of loads as given in Clause 506.2.1

of the Code[2], subject to the limitation that the increased stress should not exceed 0.9 of the yield stress. The permissible stress in axial and flexural compression should take into account the effects of buckling.

Table 5.12 Basic Permissible Stresses in Steel.

Sl No.	Description	Permissible Stress
1.	Axial tension on net area	$0.60\ f_y$
2.	Axial compression on effective section	$0.60\ f_y$
3.	Flexure	
	In plates, flats and tubes	$0.66\ f_y$
	In girders and rolled sections	$0.62\ f_y$
4.	Shear	
	Maximum stress	$0.43\ f_y$
	Average stress	
	For $f_y \le 250$ MPa	$0.38\ f_y$
	For $f_y > 250$ MPa	$0.35\ f_y$
5.	Bending stress on flat surface	$0.80\ f_y$

(b) *Combined stresses*

Members subjected to both axial and bending stresses should be so proportioned that the quantity ($f_a / F_a + f_b / F_b$) does not exceed unity,

where f_a = calculated axial stress
F_a = appropriate allowable axial stress
f_b = calculated bending stress
F_b = appropriate allowable bending stress.

Under combined bending, bearing and shear stresses, the equivalent stress f_e is computed from Equation (5.11).

$$f_e = \sqrt{f_b^2 + f_p^2 + f_b . f_p + 3 f_s^2} \qquad \text{.... (5.11)}$$

where f_b = bending stress
f_p = bearing stress
f_s = shearing stress

The equivalent stress under any combination of loads should not exceed 92 per cent of the yield stress.

5.14.3 General Details

(a) *Effective span:* The effective span for main girders will be distance between centres of bearings, and for cross girders the distance between the centres of main girders.

(b) *Effective depth:* The effective depth of plate or truss girder is taken as the distance between the centres of gravity of the upper and lower flanges or chords.

(c) *Minimum depth*: The minimum depth is not to be less than the following:

For trusses: 0.1 effective span

For R.S. joists and plate girders: 0.04 effective span

For composite steel and concrete bridge:

Overall depth: 0.04 effective span

Steel beam or girder: 0.033 effective span

(d) *Spacing of main girders:* The centre to centre spacing of main girders should not be less than 1/20 of the span.

(e) *Minimum sections:* The minimum thickness of plates used in bridge construction is 8 mm for main members and 6 mm for floor plates and parapets. If one of the sides is not accessible for painting, the minimum thickness should be increased by 2 mm. The minimum size of angle to be used is ISA 7550. End angles connecting cross girders to main girders should be not less than three-quarters of the thickness of web of cross girders.

(f) *Deflections*: For rolled steel beams, plate girders and lattice girders, the total deflection due to dead load, live load and impact should be limited to 1/600 of span and that due to live load and impact should be limited to 1/800 of span. For cantilever arms, the above deflections at the tip should be limited to 1/300 and 1/400 of span, respectively, In the calculations for deflections, gross moment of inertia and gross area will be used.

(g) *Camber:* Beams and plate girder spans up to 35 m need not be cambered. In open web spans, the lengths of members are adjusted such that when the girders are loaded with full dead load and 75 per cent of the live load without impact producing maximum bending moment, they will take up the true geometrical shape assumed in their design.

(h) *Effective length of struts*: For the purpose of calculating the value of L/r ratio, the effective length shall be taken as below:

(i) Effectively held in position and restrained in direction at both ends ...0.7 *L*

(ii) Effectively held in position at both ends and restrained in direction at one end ...0.85 *L*

(iii) Effectively held in position at both ends but not restrained in direction ...1.0 *L*

(iv) Effectively held in position and restrained in direction at one end, and at the other end partially restrained in direction but not held in position ...1.5 *L*

(v) Effectively held in position and restrained in direction at one end, but not held in position or restrained in direction at the other end ... 2.0 *L*

where *L* = the length of strut from centre to centre of intersection with supporting members or the cantilever length in the case of (*v*).

(i) *Effective sectional area*: The effective sectional area of a member or flange in tension is the gross sectional area with deduction for rivet and bolt holes. The area to be deducted is the sum of the sectional areas of the maximum number of holes in any cross-section at right angles to the direction of stress in the member. The diameter of the hole for this calculation is taken as the actual diameter of the hole for shop rivets and turned and fitted bolts and 3 mm over the diameter of the bolt or rivet for countersunk bolts or rivets.

(j) *Plates in compression:* The unsupported width of a plate measured between adjacent lines of rivets, bolts or welds connecting the plate to other parts of the sections shall preferably not exceed 45 *t*, where *t* is the thickness of single plate or the aggregate thickness of two or more plates provided these plates are adequately tacked together. The unsupported projection of any plate measured from its edge to the line of rivets, bolts or weld connecting the plate to the other parts should not exceed 16 *t*.

(k) *Pitch of rivets and bolts*: The minimum distance between centres of rivets or bolts will be 2.5 times the diameter of the rivet or bolt hole. The maximum distance between the centres of any two adjacent rivets or bolts connecting together elements in contact of compression or tension members shall not exceed the smaller of 32 *t* or 300 mm, where *t* is the thickness of the thinner outside element. The maximum pitch in the direction of stress shall not exceed 16 *t* or 200 mm in tension members, and 12 *t* or 200 mm in compression members. The pitch of rivets or bolts in a line adjacent to and parallel to an edge of an outside plate should not exceed 100 mm plus 4 *t* or 200 mm, whichever is the smaller value.

(l) *Edge distance*: The minimum distance from the centre of any rivet or bolt to the edge should be 1.5 times the diameter.

(m) *Curtailment of flange plates*: Each flange plate should be extended beyond its theoretical cut-off point. The extension should be adequate to accommodate rivets, bolts or welds to develop the loads in the plate for the computed moment at the theoretical cut-off point.

(n) *Web thickness*: The minimum thickness of web plate is 10 mm. The ratio of the clear depth between the flange angles to the web thickness should not exceed 200 for mild steeland 180 for high tensile steel for vertically stiffened webs.

(o) *Web stiffeners*: Stiffeners should be provided at the points of support and concentrated loads, on both sides of the web plate. The outstanding legs of each pair of stiffeners should be proportioned such that the bearing stress does not exceed the permissible bearing stress.

5.15 Traffic Aspects of Highway Bridges

(a) *Minimum widths of carriageway*

The minimum clear widths of carriageway to be adopted for various types of traffic are as below:

(i) Vehicular traffic:

Single lane bridge	4.25 m
Two-lane bridge	7.5 m
Multi-lane bridge	7.5 m plus 3.5 m for every additional lane over two lanes.

Three-lane bridges with two-directional traffic are not to be constructed.

(ii) Additional width for cycle traffic:

Without overtaking	2.0 m
With overtaking	3.0 m

As a guide for computing the required roadway, the following capacities may be adopted:

(i) Vehicles: 1000 vehicles per hour per lane width of 3.75 m

(ii) Cycles : 3600 cycles per day for two-lane (2 m)

(b) *Median*

In case of bridges with four or more number of lanes, it would be desirable to have a raised median along the centre line of the bridge. The median serves to physically separate the traffic from opposing directions and improve road safety. For economic reasons, the width of the median may be kept low, but should not be less than 1.2 m.

(c) *Kerbs*

It is desirable to provide a kerb of 600 x 225 mm on either side of the roadway on the bridge. The roadside edge of the kerb will have a slope of 1 in 8 for 200 mm height and a curved edge with a radius of 25 mm at the top.

(d) *Sidewalk (Footpath)*

Since the majority of road users in our country are pedestrians and since pedestrians form the major single category of victims in accidents, the author would advocate the provision of sidewalk on either side of the bridge on all bridges, in the interest of traffic safety. The width may be a minimum of 1.5 m on bridges in rural areas and may be increased suitably in urban areas. The capacity of the sidewalk may be taken as 108 persons per minute. The width should be increased in steps of 0.6 m for every additional capacity of 54 persons per minute.

(e) *Parapets and hand rails*

Parapets are provided for culverts with a minimum height of 450 mm above road surface. Railings should be for a height of 1.1 m less one half of the horizontal width of the top rail. The clear distance from the lower rail to the top of the kerb should not be more than 150 mm and the space between the bottom rail and the top rail should be filled by means of closely spaced horizontal or inclined members.

(f) Bridge width

Bridges are often constructed with a narrower width than the road on either side from consideration of reducing the initial cost. Studies in USA have shown that the accident rate when the bridge width is 0.3 m less than the approach width rises to about eight times that for bridge width of 0.6 m over that of the approaches. Except in the case of very long bridges, it is desirable to provide the full width of roadway over the bridge also.

IRC:5-1998 has stipulated the following requirements for the width between the outermost faces of the bridge: (i) For minor bridges of total length less than 60 m, the full roadway width of the approaches, subject to a minimum of 10 m for hill roads/other district roads and 12 m for other cases; (ii) For two-lane bridges having a total length more than 60 m in non-urban situations, minimum 10.5 m to provide for 7.5 m wide carriageway and 1.5 m wide footpath on either side; (iii) For two-lane bridges in urban situations, the full roadway width of the approaches; and (iv) For multi-lane bridges in urban and non-urban situations and for bridges on expressways, the full roadway width of approaches.

(g) Overpasses and underpasses

In case of a traffic intersection involving two levels, a detailed study should be made to determine which road should be taken over or under the structure. Principal considerations are: economics, a layout to fit the topography, the major traffic movements, the type and character of highways, and aesthetics.

An overpass gives drivers less feeling of restriction and confinement and has less problems of drainage. On the other hand, an underpass may be more advantageous where the major road can be built close to the existing ground. For an underpass clearances as specified in Section 3.4 should be provided.

(h) Sunken and elevated roads

Sunken and elevated roads will be required when it is desired to increase the traffic capacity along existing roads in built up urban areas. Great care will have to be bestowed in their location and planning to avoid destruction of amenity or dimunition of values of historic monuments on the surface.

Sunken roads in tunnels can be planned so as not to interfere with surface roads or developments. But their initial and maintenance costs are high. Elevated roads on viaducts are easier to construct and maintain. Their design is similar to that of any normal bridge.

(i) Pedestrian bridges

Pedestrian bridges over roads offer opportunities for imaginative architectural design. They should permit a vertical clearance of 5 m for the roadway below. The minimum width of a footway should be 2.5 m with a minimum vertical clearance of 2.25 m. Where possible, ramped approaches may be provided in preference to steps. The gradient should preferably be about 1 in 30.

(j) *Sight distance*

Sight distance is measured between points 1.2 m above the carriageway along the centre lines of both the near side and offside lanes of the carriageway. The sight distance should not be less than the stopping distance, which varies with the design speeds, e.g., it is 60, 80, 110 and 150 m for design speeds of 50, 65, 80 and 100 km/hr, respectively. Care should be taken to ensure that the sight distance is not reduced below the stopping distance as mentioned above by shrubs in verges, piers, abutments, etc., especially in the case of underpasses.

(k) *Bridge lighting*

The lighting of bridges should be designed tastefully to provide light to the traffic as well as to enhance the beauty and visibility of the bridge at night. This is particularly important for cable stayed bridges. Sodium vapour lamps are preferable to incandescent bulbs. The light standards should be aligned along the railing and should match with the railing design. Provision of rail level lighting by fluorescent tube lights recessed into the hand rails will not be effective even in the case of urban flyovers. Besides being wasteful in energy and ineffective in lighting the full area of bridge carriageway, these fittings are vulnerable to juvenile vandalism. The provision of high mast area lighting is nowadays adopted for urban bridges and grade separations in the place of fluorescent lamps along the railing. The arrangement of the masts (height and spacing) should be so designed as to ensure a minimum level of illumination of 30 lux.

In the case of cantilever bridges and balanced cantilever bridges, the light posts should not be located at articulations or at the central joints. The author has come across a major prestressed concrete cantilever bridge, where one row of mercury vapour lamps located at every hinge location (midspan) went out of use due to severe vibrations.

5.16 Aesthetics of Bridges

Bridge construction is a fascinating field, which calls for the integrated application of expertise in many areas of Civil Engineering and also an appreciation of the art of aesthetics. As mentioned by D.B. Steinman, “from its foundation rested in bedrock to its towering pylons and vaulting span, a bridge is a thing of wonder and beauty”.

Bridges exert a strong influence on the environment. Once built, a bridge is an aesthetic statement reflecting the quality of life in the community; and it should be the endeavour of the bridge engineer to make it an appealing statement. If we take an aerial view of a city, the first and foremost items meeting the eye are the bridges and the grade separated road intersections (e.g. London, Cologne and Sydney). These bridges become landmarks, if not monuments of contemporary achievement. Hence, we should devote particular attention to aesthetics of bridges besides their functional aspects, so that these structures enhance the environment.

Aesthetic design should precede the structural design of any major bridge. Aesthetics, referred by D. P. Billington as structural art, should be a primary design objective, and emphasis on overall form should have precedence over analysis of the parts. The three leading ideals of structural art are efficiency, economy and elegance[21]. The engineer should consider aesthetics in the determination of the type of bridge, and in the selection of spans, cross section of deck, girder depth, piers and abutments. Aesthetics is not something that is added to a design, but it is the outward manifestation of a sound design. While considerations of function, fitness and truth should predominate, other qualities to be sought are symmetry, harmony, proportion, expressiveness, simplicity, style, feeling, repose, grace and conformity to environment. Aesthetic effects are enhanced by attention to colour, light, texture, shape and proportions. For a skillfully designed bridge, the additional cost required to achieve aesthetic elegance is not likely to exceed 2 to 3 percent of the overall cost. The reader may refer to Reference 22 for a detailed treatment of bridge aesthetics.

Aesthetics is a matter of taste and hence it is difficult to codify the rules for judgment of the visual quality in bridge design. However, a few basic criteria can be stated as guidelines, and these are illustrated with examples.

(a) The bridge structure should blend with the landscape. For example, an open spandrel deck type arch span will suit a gorge with rocky abutments or an area near a waterfall, such as the Rainbow bridge near Niagara Falls.

(b) The systems, lines and edges employed in the bridge structure should be in good order, e.g. concrete arch bridge such as the Gladesville bridge in Sydney. Easily perceivable order in a bridge enhances its aesthetic quality, while complexity in design decreases the visual delight.

(c) Bridge design is a form of structural art. Bridges have the capacity to induce delight in the viewer. The design should highlight the flow of forces and should also conform to the needs of economy. A fine example is Maillart's Salginatobel bridge. A more recent application is Christian Menn's Sunniberg bridge.

(d) Mixing of several systems, such as beams, arches, suspension or frames, or decks of different materials in any one bridge may be avoided. However, this mix may be acceptable in a very long bridge like the 38.6 km long Pontchartrain bridge, New Orleans, USA.

(e) In a long multi-span bridge, there should be order and rhythm, e.g. the Tunkannock Viaduct in Pennsylvania, USA. The viewer follows the repeating arches and feels the flow with the regularity of the heartbeat.

(f) The use of members and components in all possible directions and sizes usually results in an ugly structure. While it is difficult to achieve aesthetic elegance with steel truss bridges, we have examples of good design in the Runcorn Widnes bridge and the Godavari Second bridge.

(g) An elegant bridge should have simplicity of line, expressing emphatically the structural form, e.g. the Fehmarnsund bridge and the Rodenkirchen bridge in Germany, and the Sunshine Skyway bridge in Florida, USA.

(h) Extraneous ornamentations serve to diminish rather than enhance the elegance of the bridge. A classical example is the comparison between the designs put forward by Brunel and Telford in a competition for the Clifton bridge at Bristol in England[23]. Brunel's design with simple lines anchoring the suspension chains directly in rock on each side blends well with the surroundings, while Telford's design with Gothic-style piers clashes disastrously with the landscape.

(i) The proportions of the various components of a bridge should be in harmony with each other. The depth of deck should have a pleasing relation with the span. For a bridge across a river or gorge, it is preferable to have an odd number of spans. In a layout with three spans, the centre span should be longer than the outer span.

(j) The final shape of the structure should highlight the special qualities of the materials of construction. Thus a spandrel filled arch would express gracefully the robust qualities of stone masonry. On the contrary, a prestressed concrete bridge should highlight the high-strength materials by its slender proportions, e.g. the Gambhirkhad bridge in India, and the Fisherstrasse bridge in Dusseldorf, Germany.

(k) A bridge should reflect the contemporary state of technical advancement in materials and construction. Thus a marked difference is readily discernible between well-designed bridges of 1880 and 1998, e.g. the Brooklyn bridge, 1883 and the Akashi Kaikyo bridge, 1998.

(l) For flyovers and elevated highways within urban areas, the supporting piers should be slender, so that the street life below is least disturbed and the view through the bridge is least obstructed. A good example in this connection is the Jan Wellam Platz elevated road in Dusseldorf, Germany. For a multi-span bridge, it is better to have the width of a pier less than one-eighth of the span. When the pier consists of a number of columns, the total width of the pier between the extremes of the columns may be less than one-third of the span to avoid a feeling of clutter of columns.

(m) Landspans may be preferred to high embankments in case of bridges across navigable rivers, e.g. the Zuari bridge.

(n) External renderings should be done with taste. The impression of a beam-and-slab prestressed concrete deck can be enhanced by using a near white fascia beam and a dark grey colour for the webs of the main girders.

(o) The lighting of bridges should be designed tastefully so as to provide adequate light to the traffic, and also to contribute to an aesthetic setting to the environment at night. Mercury vapour or sodium vapour luminairres are preferable to incandescent bulbs.

(p) The sloping ground from the abutment to the edge of the roadway or the stream below the bridge may be paved with concrete paving blocks or rubble stones, so as to secure a neat appearance besides affording slope protection.

(q) A little effort on securing pleasant landscaping around bridges in urban areas will go a long way to make life more pleasant to the community. Lawns, trees and footpaths may be provided around the bridges.

The aesthetic analysis of a bridge should start with the preliminary sketching of the bridge form and should be repeated at the various stages of detailing and proportioning the constituent elements of the bridge. At every stage, the visual quality of the design should be checked from different angles. Designs embodying structural clarity and fitness to the

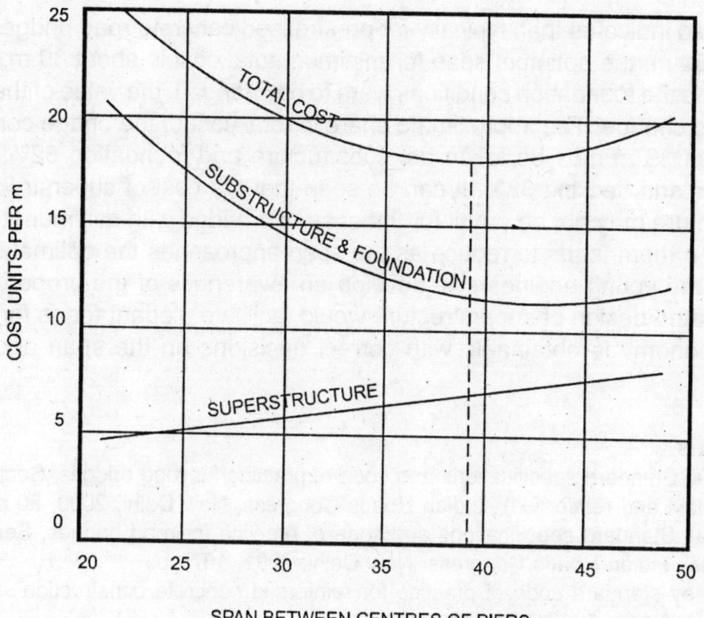

Figure 5.3 Typical Cost Components for Prestressed Concrete Road Bridges.

environment are likely to result in elegant structures. Sketches, perspective view of the structure set in the environment and models are to be used for major constructions. Aesthetics could be learnt only by practice and by a careful study of successful bridges designed by such stalwarts as Eiffel, Freyssinet, Leonhardt, Maillart, Menn, Muller and Nervi. In order to foster an awareness of the importance of aesthetics, bridge engineers should be imparted initial instruction on bridge aesthetics during their study at the University.

5.17 Relative Costs of Bridge Components

The total cost of a bridge is the sum of the costs on foundation, substructure, girders of the superstructure and decking. No general proportions of costs of the various components can be specified, as they depend on the type of structure and the site conditions.

Typical cost components for prestressed concrete road bridges in the span range 20 to 50 m are shown in Fig. 5.3, as adapted from Reference 24. The plots are based on computations of alternative designs with varied spans for a major bridge across Betwa river, adopting a two-girder prestressed concrete-reinforced concrete composite deck of two-lane carriageway, cellular R.C. pier and well foundations.The cost is indicated in relative units, as the absolute values vary also with time. The cost of decking per m is nearly constant in the span range considered. The cost of prestressed concrete girders per m increases with increase in span as can be expected. The cost of superstructure per m, which is the sum of the costs due to prestressed concrete girders, deck slab, kerbs and railings, thus increases with increase in span as shown in the figure. The cost per m due to substructure and foundation decreases rapidly with increase in span up to about 40 m and then increases gradually.

Fig. 5.3 also indicates that, typically for prestressed concrete road bridges in the span range 20 m to 50 m, the optimum span for minimum total cost is about 39 m. If the bridge cross section and the foundation conditions were to be changed, the value of the economical span would also change. The approximate share in total cost of the bridge components for a bridge of span 39 m may be taken as: substructure and foundation 62%, prestressed concrete girders and decking 38%. It can be seen that the cost of superstructure and the cost of substructure may not be equal for the case of a bridge with minimum total cost, but the gap between them tends to reduce as the span approaches the optimal solution. The author would urge young engineers to develop an awareness of the proportionate costs. While sophisticated design of superstructure would facilitate elegant forms for the required span, major economy is obtainable with correct decisions on the span and the type of foundation.

5.18 References

1. 'IRC:21-2000 Standard specifications and code of practice for road bridges, Section III Cement concrete (plain and reinforced)', Indian Roads Congress, New Delhi, 2000, 80 pp.
2. 'IRC:24-2001 Standard specifications and code of practice for road bridges, Section V - Steel road bridges', Indian Roads Congress, New Delhi, 2001, 157 pp.
3. 'Indian railway standard code of practice for reinforced concrete construction', Govt. of India, Ministry of Railways, 1962, 48 pp.
4. 'Indian railway standard code of practice for the design of steel or wrought iron bridges carrying rail, road or pedestrian traffic', Govt. of India, Ministry of Railways, 1962, 87 pp.
5. 'IS:456-1978, Indian standard code of practice for plain and reinforced concrete', Bureau of Indian Standards, New Delhi, 1983, 148 pp.
6. 'IRC:18-2000 Design criteria for prestressed concrete road bridges (Post-tensioned concrete)', Indian Roads Congress, New Delhi, 2000, 61 pp.
7. Reddi, S.A., 'Value engineering for bridge projects', In Rajagopalan, N., 'Advances and innovations in bridge engineering', Allied Publishers Limited, Chennai, 1999, pp. 535-560.
8. 'IRC:44-1976 Tentative guidelines for cement concrete mix design for road pavements', Indian Roads Congress, New Delhi, 1996, 15 pp.
9. 'IS:432-1966, Indian standard specifications for mild steel and medium tensile steel bars and hard drawn steel wire for concrete reinforcement', Bureau of Indian Standards, New Delhi, 1966, Part I, 12 pp.
10. 'IS:1786-1966, Indian standard specification for cold twisted steel bars for concrete reinforcement', Bureau of Indian Standards, New Delhi, 1966, 8 pp.
11. Victor, D.J., 'Effect of bar cutoff on shear strength of reinforced concrete beams with web reinforcement', M.S. Thesis, University of Texas, Austin, Texas, 1965, 107 pp.
12. 'IS:1343-1980, Indian standard code of practice for prestressed concrete', Bureau of Indian Standards, New Delhi, 1988, 62 pp.
13. Victor, D.J., 'Concise notation for detailing reinforced concrete bridges', Transport Communications Quarterly Review, Annual Number, No. 244, Dec. 1970, pp. 109-110.
14. 'IS:2062-1992, Indian standard specification for steel for general structural purposes', Bureau of Indian Standards, New Delhi, 1992.
15. 'IS:8500-1992, Indian standard specification for structural steel microalloyed (Medium and high strength qualities)', Bureau of Indian Standards, New Delhi, 1992.
16. 'IS:11587-1986, Indian standard specification for structural weather resistant steels', Bureau of Indian Standards, New Delhi, 1986.
17. 'IS:814-1991, Indian standard specification for covered electrodes for metal arc welding of carbon and carbon manganese steel', Bureau of Indian Standards, New Delhi, 1991.

18. 'IS:1442-1964, Indian standard specification for covered electrodes for metal arc welding of high tensile structural steel', Bureau of Indian Standards, New Delhi, 1964, 10 pp.
19. 'IS:1148-1982, Indian standard specification for hot rolled rivet bars for structural purposes', Bureau of Indian Standards, New Delhi, 1982.
20. 'IS:1149-1982, Indian standard specification for high tensile rivet bars for structural purposes', Bureau of Indian Standards, New Delhi, 1982.
21. Billington, D.P., and Gottemoeller, F., "Bridge aesthetics – Structural art", <u>In</u> Chen, W.F., and Duan, L., (Eds.), "Bridge Engineering Handbook", CRC Press, Boca Raton, Florida, 2000, pp. 3-1 to 3-19.
22. Leonhardt, F., 'Bridges - aesthetics and design', Deutsche Verlags-Anstalt, Stuttgart, 1982, 308 pp.
23. Beckett, D., 'Great buildings of the world - Bridges', Paul Hamlyn, London, 1969, 191 pp.
24. Ojha, S.K., 'The economics of highway bridge planning and construction', Jl. Indian Roads Congress, New Delhi, Vol. 40-1, Paper No. 326, Oct. 1979, pp. 131-152.

Chapter 6

Culverts

6.1 General

Culvert is a cross drainage work whose length (total length between the inner faces of dirtwalls) is less than 6.0 m. In any highway or railway project, the majority of cross drainage works fall under this category. Hence culverts collectively are important in any project, though the cost of individual structures may be relatively small.

Culverts may be classified according to function as highway or railway culvert. The loadings and structural details of the superstructure would be different for these two classes.

Based on the construction of the structure, they can be of the following types:

(a) Reinforced concrete slab culvert
(b) Pipe culvert
(c) Box culvert
(d) Stone arch culvert
(e) Steel girder culvert for railways.

Though this Chapter deals mainly with single vent culverts, it is possible to adopt the above types of superstructure for multiple vents and thus for bridges with a wider linear waterway between abutments. The first three types are described in this Chapter. The other two types are dealt with under steel and masonry arch bridges.

As stipulated in Clause 112.1 of IRC:5-1998, for culverts and minor bridges of total length less than 60 m, the width between the outermost faces of the bridge should be the full roadway width of the approaches, subject to a minimum of 12 m for roads other than hill roads and other district roads. Though this stipulation is intended to be applicable mainly to National Highways, it may be adopted for State Highways if sufficient funds are available. Kerbs of minimum width 500 to 600 mm and height 300 mm above top of deck slab are desirable. The waterway required for the culvert is designed in such a manner that the estimated discharge can be passed with a low velocity, e.g. 1.5 metre per second, and the HFL is adequately below the bottom of deck.

Raised footpaths need not be provided for culverts and minor bridges having total length of less than 30 m, unless such raised footpaths are existing on the approaches. For bridges with footpaths, footpath load of 5 kN/m^2 is to be considered.

Submersible bridges usually have reinforced concrete slabs for decking and adopt multiple spans up to 8 m each. The deck slab and the substructure are designed on lines similar to those for reinforced concrete slab culverts. Hence the consideration of submersible bridges is included under this Chapter.

6.2 Reinforced Concrete Slab Culvert

6.2.1 General

Reinforced concrete slab culverts are economical for spans up to about 8 m, though the slab bridge type can be used for spans up to about 10 m. The thickness of the slab and hence the dead load are quite considerable as the span increases. However, the construction is relatively simpler due to easier fabrication of formwork and reinforcements and easier placing of concrete. This type of culvert can be used both for highway and railway. Though the discussion in this section refers mainly to highway construction, the concepts can be suitably adapted for a railway culvert.

The components of a culvert with R.C. deck slab are the following:

(a) Deck slab
(b) Abutments, wing walls and approach slabs
(c) Foundations
(d) Kerbs and railings.

6.2.2 Deck Slab

The deck slab should be designed as a one-way slab to carry the dead load and the prescribed live load with impact and still to have stresses within the permissible limits. For a culvert on a State Highway, the width of the bridge may be adopted as 12 m to permit two-lane carriageway. The deck slab should be designed for the worst effect of either one lane of IRC 70R/Class AA tracked vehicle, or one lane of 70R/Class AA wheeled vehicle, or two lanes Class A load trains. Thus, according to the present practice, it is necessary to compute the maximum live load bending moment for three different conditions of loading, and then adopt for design the greatest of the three values.

The author has shown in Reference 1 that, for the computation of live load bending moment, only one loading condition need to be considered, namely Class AA wheeled vehicle for spans up to 4 m and Class AA tracked vehicle for spans exceeding 4 m. If the shear is desired to be computed, Class AA wheeled vehicle is to be considered for spans up to 6 m and tracked vehicle beyond 6 m for single lane bridges. However, for two-lane bridges, the shear due to Class AA wheeled vehicle controls the design for all spans from 1 m to 8 m. The design moment for distributors is taken as 0.3 of the live load moment plus 0.2 of the dead load moment.

The Ministry of Road Transport & Highways, Government of India, referred herein as MORTH, has published a set of plans for 3.0 m to 10.0 m span reinforced concrete deck slabs[2]. The overall width of the bridge is assumed as 12.0 m. Concrete grade M25 (M30 for 10.0 m span) and steel reinforcement to Fe 415 grade of IS:1786 are adopted. Allowing for 'moderate' conditions of exposure, the minimum cover to reinforcement is taken as 50 mm. Approach slabs are provided for 3.5 m length on both approaches. Salient details of the general arrangements are shown in Figs. 6.1 and 6.2 for the case of a slab deck bridge with and without footpath applicable for spans 3.0 m to 10.0 m. The thickness and the principal reinforcements of the deck slab for the above span range are summarized in Table 6.1. Though a culvert strictly refers to bridge of length less than 6.0 m, solid slab decks up to 10.0 m are covered here for convenience. For additional details, the reader may see Reference 2.

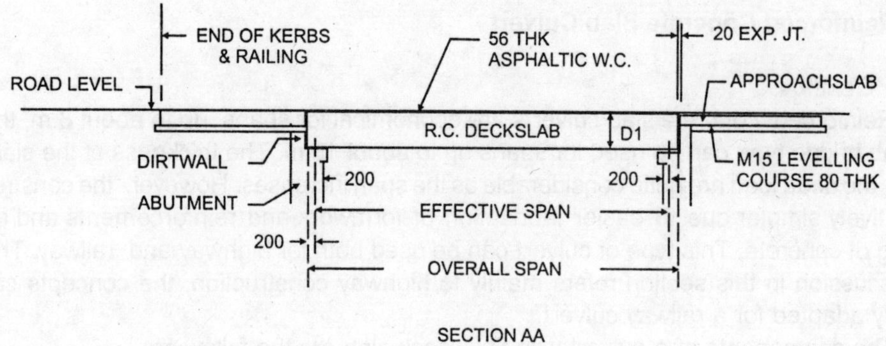

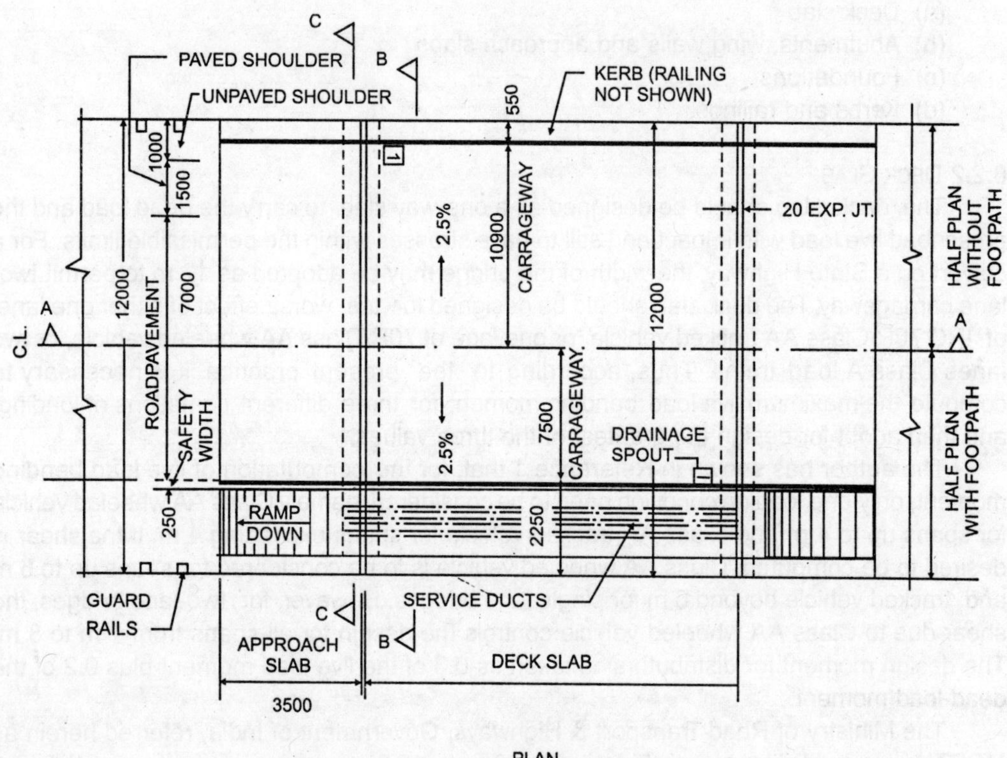

Figure 6.1 General Arrangement for R.C. Slab Culvert.

Wearing course (also referred as wearing coat) over the deck slab for culverts may be of asphaltic concrete with uniform thickness of 56 mm or of cement concrete to a thickness of 75 mm. The details of the wearing course are discussed in Section 14.14. The top of the deck slab is sloped in the transverse direction to provide the camber (2.5%) for storm water drainage. It may be noted that until recently, the top of the deck slab has been kept horizontal and the camber was provided by varying the thickness of the wearing course. The dead load of wearing course may be assumed as 2 kN/m^2.

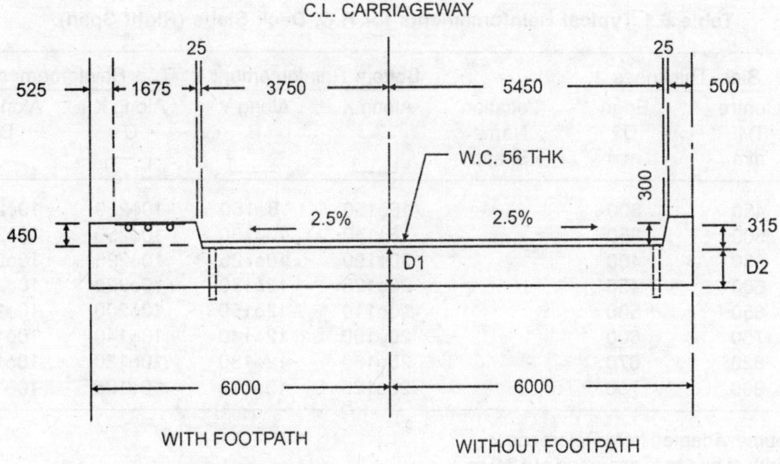

SECTION BB

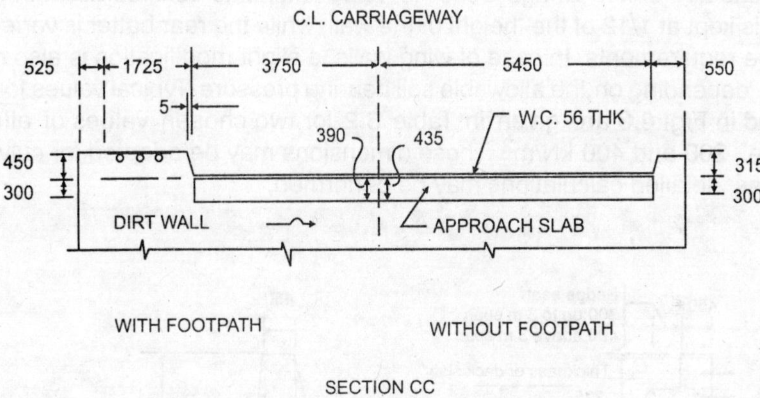

SECTION CC

Figure 6.2 General Arrangement for R.C. Slab Culvert - Details.

6.2.3 Effective Width of Dispersion

The bending moment per unit width of slab caused by a concentrated load can be calculated by estimating the width of slab that may be taken as effective in resisting the bending moment due to the concentrated load. The method of assessment recommended in clause 305.16 of IRC Bridge Code[3] is described in Appendix A.

6.2.4 Abutments and Wing Walls

Typical sections of abutments and wing walls for slab bridges are shown in Fig. 6.3. The sections should be so dimensioned that the walls can sustain the prescribed dead and live loads, besides the earth pressures without overturning or sliding. Further, the pressure

Table 6.1 Typical Reinforcements for R.C. Deck Slabs (Right Span).

Effective Span	Slab Thickness		Location	Bottom Reinforcement		Top Reinforcement		Side
	Centre D1	Edge D2	Name	Along X A	Along Y B	Along X C	Along Y D	Along Y E
m	mm	mm	Shape	⌴	⌴	⊓	∧	—
3.0	450	300		16ϕ150	8ϕ150	10ϕ250	10ϕ250	10ϕ
4.0	500	350		16ϕ140	10ϕ150	10ϕ250	10ϕ250	10ϕ
5.0	550	400		20ϕ150	10ϕ125	10ϕ225	10ϕ225	10ϕ
6.0	600	450		20ϕ125	12ϕ150	10ϕ220	10ϕ220	10ϕ
7.0	650	500		20ϕ110	12ϕ150	10ϕ200	10ϕ200	10ϕ
8.0	750	600		20ϕ100	12ϕ140	10ϕ140	10ϕ140	10ϕ
9.0	820	670		25ϕ130	12ϕ130	10ϕ130	10ϕ130	10ϕ
10.0	900	750		25ϕ125	12ϕ115	10ϕ125	10ϕ125	10ϕ

Note: 1. Source: Adapted from Reference 2
2. Width of bridge is assumed at 12.0 m
3. Direction X is along the roadway and Y is in transverse direction. Rod names as in Fig. 6.9.

on the soil should be within the allowable pressure. Equivalent surcharge for live load as given in clause 217 of IRC Bridge Code[4] is to be taken into consideration. In Fig. 6.3, the front batter is kept at 1/12 of the height of the wall, while the rear batter is varied to suit the soil pressure requirements. In case of wing walls, a slight modification is also made in the front offset, depending on the allowable soil bearing pressure. Typical values for A, B and C as indicated in Fig. 6.3 are given in Table 6.2 for two chosen values of allowable soil pressure, i.e., 200 and 400 kN/m^2. These dimensions may be adopted for culverts and, in special cases, detailed calculations may be performed.

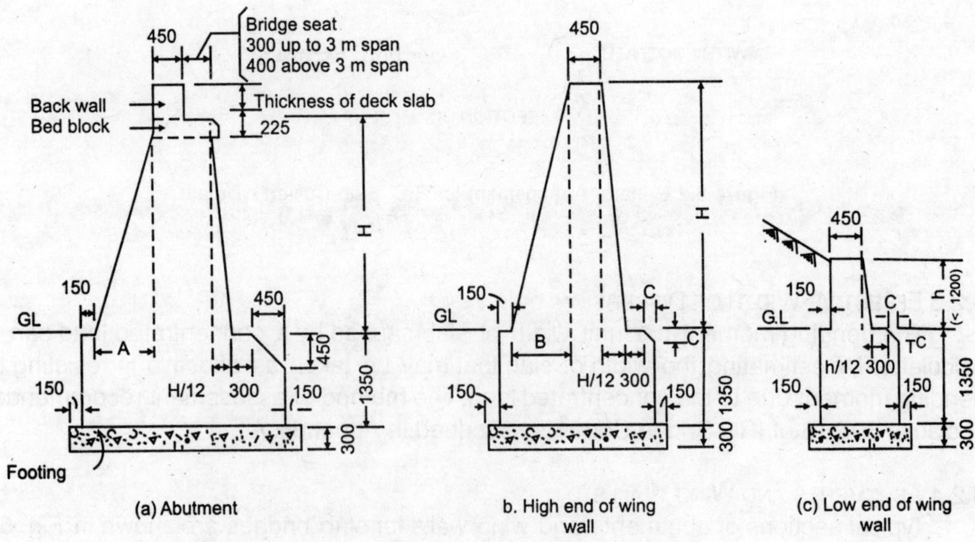

Figure 6.3 Sections of Abutments and Wing Walls for Slab Bridges.

Table 6.2 Dimensions of Abutments and Wing Walls for Slab Bridges. (See Fig. 6.3)

Maximum pressure on soil kN/m^2	Dimension	Value in metres for height H in metres				
		1.5	2.0	2.5	3.0	3.5
200	A	0.75	0.90	1.20	1.65	2.60
	B	0.20	0.30	0.45	0.65	1.06
	C	0.45	0.45	0.45	0.45	0.45
400	A	0.30	0.45	0.55	0.75	0.95
	B	0.20	0.30	0.45	0.65	0.85
	C	0.15	0.15	0.15	0.15	0.15

The foundations should be taken below the scour depth and to a depth where the variation in the moisture content of the subsoil will be absent. The depth of foundation shown should be considered the minimum, unless rock is encountered at a higher level. In the latter case, the masonry should be heeled on to the rock for a depth of about 300 mm, or connected by dowels to the rock if the rock surface is sloping.

The purpose of the footing is to serve as a seal in poor soil or as a bedding course in water bearing strata for the main masonry above. The footing is not taken as effective in designing the abutment section. When the clear span is less than H/6 + 1.5 m, the bottom footings of the two abutments of a single span culvert should be combined.

If the entire abutment is of mass concrete, no special treatment except leveling is required for the bridge seat. If the main abutment is of brick or stone masonry, a bed block of concrete is provided for a thickness of about 225 mm. Wing walls may be either splayed at an angle of 45° or taken perpendicular to the abutment. If splayed, the height of wall will reduce as it extends away from the abutment. At the low end, the height of wall should not be more than 1.2 m. The splayed wing wall is particularly suited for high embankments. If the wall is at right angles to the abutment, it is known as return type wing wall or simply return wall. In the latter case, suitable revetment should be provided to retain the earth of the embankment.

The earth filling behind the abutments and wings should be specially consolidated in order to avoid an excessive depression immediately clear of the bridge deck. A porous backfill of about 500 mm thickness should be provided immediately behind the abutment, with 150 mm diameter weep holes placed at suitable intervals with a gentle slope so as to be 150 mm above normal water level on the vent side, to ensure drainage for the filling behind the abutments.

Hand rails or parapets should be provided over the kerbs on either side of the road over the deck slab. For small culverts, parapet walls of height 750 mm will be adequate. For spans above 4 m, a combination of R.C. posts and rails may be more pleasing.

6.3 Example for R.C. Slab Culvert

6.3.1 Given Data

Culvert to be on State Highway.

Width of bridge : 12.0 m

No footpath provided

Conditions of exposure: 'moderate'

Materials: Concrete grade M25

Steel-Deformed bars to IS : 1786 (grade Fe 415)

Clear span = 5.0 m

Height of vent = 3.0 m

Depth of foundation = 1.35 m

Wearing course: 56 mm thick asphaltic concrete

The design will be made in this section following the current practice for the determination of live load moment.

6.3.2 DESIGN OF DECK SLAB

(i) *Preliminary dimensions:*

Slab thickness (assumed) = 550 mm

Using 20 mm diameter main bars and clear cover of 50 mm,

effective depth = 550 −(50 + 10) = 490 mm

Width of bearing = 400 mm

Effective span is the smaller of (5.0 + 0.4) or (5.0 + 0.49) m; hence adopt 5.4 m.

(ii) *Dead load B.M. per metre width of slab:*

Weight per m^2 of slab = $(0.55 + 0.40)/2 \times 24 = 11.40$ kN

Wearing course per m^2 of slab = 2.00 kN

Total dead load per m^2 = 13.40 kN

$$\text{Dead load B.M.} = \frac{13.4 \times 5.4^2}{8} = 48.8 \text{ kN.m}$$

(iii) *Live load B.M. - Class AA tracked vehicles:*

Impact factor = 23.5%

The tracked vehicle is placed symmetrically on the span (Fig. 6.4).

Effective length of load = 3.6 + 2 (0.550 + 0.056) = 4.812 m

Effective width of slab is computed using Equation (A.1)

$$b_e = k.\, x\left(1 - \frac{x}{L}\right) + b_w$$

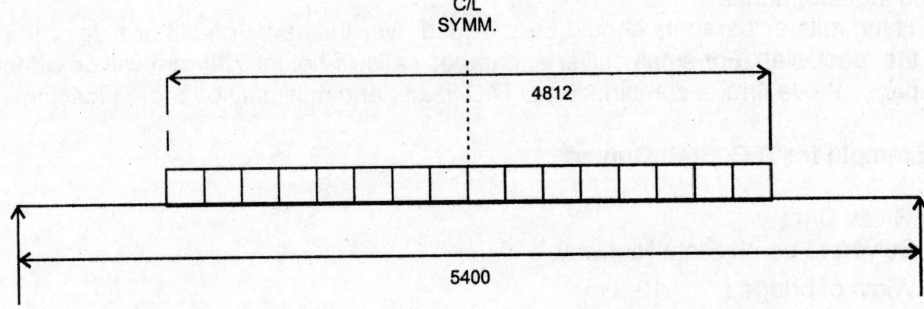

Figure 6.4 Longitudinal Placement of Track for Maximum Bending Moment.

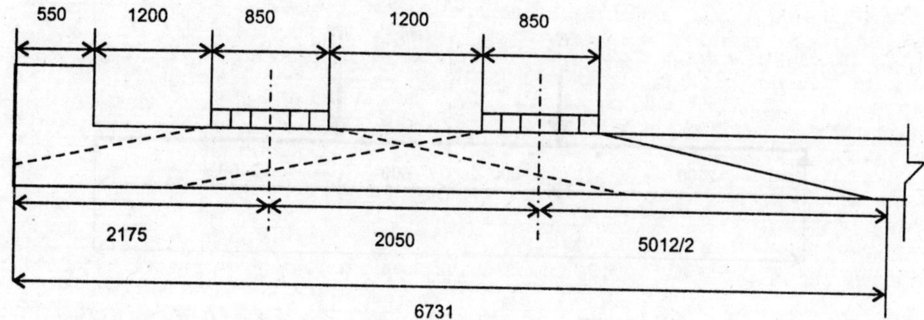

Figure 6.5 Disposition of Tracked Vehicle for Minimum Dispersion Width and Hence Maximum Bending Moment.

Here $x = 2.7$ m; $L = 5.4$ m; $L' = 12.0$ m; $L'/L = 2.22$; From Table A.1, $k = 3.00$

$b_w = 0.85 + 2 \times 0.056 = 0.962$ m

Substituting these values in Equation (A.1), $b_e = 5.012$ m.

The tracked vehicle is placed close to the kerb with the required minimum clearance, as in Fig. 6.5. The spread of load to the left is limited to the actual width available, i.e., to 2.175 m from the centre of the left track instead of half the effective width. There is an overlap of effective widths for the two tracks. The dispersion to the right of the right side track is allowed as computed, i.e., half the effective width beyond the centre of right track. The net effective width of dispersion for the two tracks, after allowing for the overlap, is 6.731 m as shown in Fig. 6.5.

Total load of two tracks including impact = $700 \times 1.235 = 864.5$ kN

$$\text{Average intensity of load} = \frac{864.5}{4.812 \times 6.731} = 26.7 \text{ kN/m}^2$$

Maximum B.M. due to live load (from Fig. 6.2).

$$= \frac{26.7 \times 4.812}{2} \times 2.7 - \frac{26.7 \times 4.812^2}{8} = 96.2 \text{ kN.m}$$

(iv) *Live load B.M. - Class AA wheeled vehicle*

Impact factor = 25%

The loads are to be arranged in such a manner as to produce maximum bending moment in the slab. First, it is necessary to determine the width of dispersion parallel to span. If the dispersion areas overlap, then the loads are to be placed symmetrically with reference to the centre of span. Width of dispersion parallel to span is computed as below:

$$t_p = t_e + 2\,(t_w + t_s)$$

where t_p = width of dispersion parallel to span

t_e = width of tyre contact area parallel to span

t_w = thickness of wearing course

t_s = effective depth of slab

Here t_e = 150 mm, t_s = 490 mm, t_w = 56 mm

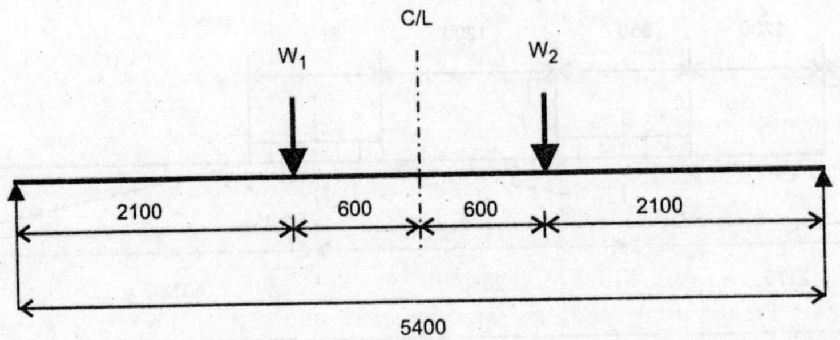

Figure 6.6 Placement of Loads of Class AA Wheeled Vehicle for Maximum Bending Moment.

Substituting, $t_p = 1242 \text{ mm} > 1200 \text{ mm}$

Arrange the wheel loads symmetrically with respect to the centre of span as shown in Fig. 6.6.

Length of load along the span = 1.2 + 1.242 = 2.442 m

Effective width of slab perpendicular to span for load W_1 or W_2 is computed using Equation (A.1),

$$b_e = 3.0 \times 2.1\left(1 - \frac{2.1}{5.4}\right) + (0.3 + 2 \times 0.056) = 4.262 \text{ m}$$

The wheeled vehicle is placed close to the kerb with the minimum required clearance as shown in Fig. 6.7. The net effective width for all the four wheels of axle shown as W_1 is 6.231 m. Intensity of loading per square metre (q) due to W_1 (or W_2) including impact is computed as below :

$$q = \frac{400 \times 1.25}{6.231 \times 2.442} = 32.9 \text{ kN/m}^2$$

Maximum bending moment at centre of span

$$= 32.9 \times \frac{2.442}{2}\left(2.7 - \frac{2.442}{4}\right)$$

$$= 83.9 \text{ kN.m}$$

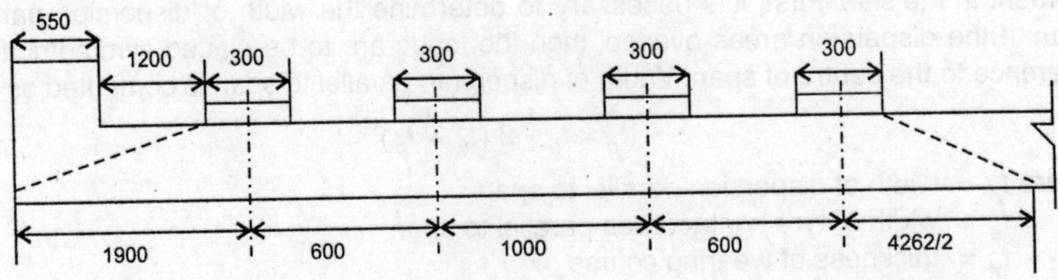

Figure 6.7 Location of Class AA Wheeled Vehicle for Minimum Dispersion Width.

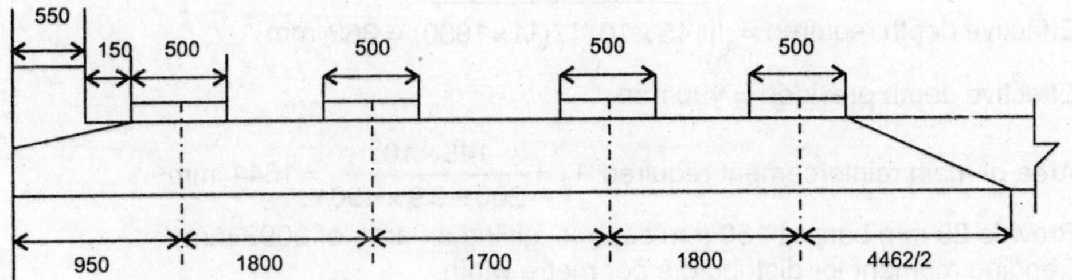

Figure 6.8 Disposition of Two Lanes of Class A Wheel Loads for Minimum Dispersion Width.

(v) *Live load B.M. — Class A two-lane loading*

The two axle loads of 114 kN each are to be placed in the span so as to produce maximum bending moment.

Width of dispersion parallel to span = 250 + 2 (490 + 56) = 1342 mm

This is greater than the spacing of 1200 mm between the 114 kN axle loads. The dispersion areas overlap. Hence the axles are to be placed symmetrically with respect to centre line of span to produce maximum bending moment.

$$\text{Impact factor} = \frac{4.5}{6+L} = 0.395$$

The wheels are arranged across the width of the deck as shown in Fig. 6.8.

Using Equation (A.1), effective width of dispersion perpendicular to span

$$= 3.0 \times 2.1\left(1 - \frac{2.1}{5.4}\right) + (0.5 + 2 \times 0.056) = 4.462 \text{ m}$$

Allowing for overlaps, the effective width of dispersion is computed as 8.481 m.

Intensity of loading including impact per metre width

$$= \frac{2 \times 114 \times 1.395}{8.481} = 37.5 \text{ kN/m}$$

$$\text{Maximum B.M. at midspan} = 37.5\left(2.7 - \frac{2.542}{4}\right) = 77.4 \text{ kN.m}$$

(vi) *Design B.M.*

B.M. produced by Class AA tracked vehicle is the greatest of the three conditions. Hence this is adopted for design as the live load bending moment.

Design max. B.M = 96.2 + 48.8 = 145.0 kN.m

(vii) *Structural design of deck slab*

Use M25 concrete and Fe 415 deformed bars.

Permissible compressive stress in concrete = 8.3 MPa

Permissible tensile stress in steel = 200 MPa

Coefficient of resistance = 1.10

Lever arm factor = 0.90

Effective depth required = $\sqrt{(145\times10^6)/(1.1\times1000)}$ = 363 mm

Effective depth provided = 490 mm

Area of main reinforcement required $A_s = \dfrac{145\times10^6}{200\times0.9\times490} = 1644\ \text{mm}^2$

Provide 20 mm bars at 150 mm centres, giving an area of 2093 mm^2

Bending moment for distributors per metre width

$$= 0.3\ M_L + 0.2\ M_D$$

where M_L = live load bending moment including impact per metre width

M_D = dead load being moment per metre width

Bending moment for distributors = $0.3\times93.3 + 0.2\times48.8$ = 37.8 kN.m

Using 12 mm bars, depth to centre of bars = 490 – (10 + 6) = 474 mm

$$\text{Area of distributors required} = \frac{37.8\times10^6}{200\times0.9\times474} = 443\,\text{mm}^2$$

Provide 10 mm dia. bars at 150 mm centres. The reinforcement details are shown in Fig. 6.9.

(viii) *Check for shear*

Usually shear is not critical in one way slabs. However, this may be checked.

Maximum shear force occurs at the support when the live load is nearer to a support. For R.C. slabs, the critical section for shear may be taken at a distance equal to effective depth from the support. Here the design shear is conservatively taken as that at the support.

$$\text{Dead load shear} = 13.4\times\frac{5.4}{2} = 36.2\ \text{kN}$$

To compute the live load shear due to Class AA tracked vehicle, the track is placed in such a way that the longitudinally dispersed zone will just be at the support. For this condition, using the effective width method, maximum shear including effect of impact = 71.5 kN.

By similar procedure, the maximum shear due to Class AA wheel loading including effects of impact = 79.1 kN.

The shear due to two lanes of Class A loading can also be computed. It will be found that it is smaller than the above two values.

Shear from Class AA wheeled loading governs the design.

Total design shear = 36.2 + 79.1 = 115.3 kN

By Equation (5.5)

$$\text{Shear stress}\ \tau = \frac{115.3\times1000}{1000\times490} = 0.26\ \text{MPa}$$

Reinforcement is 20ϕ150

$$A_s = 314\times1000/150 = 2093\ \text{mm}^2$$

$\rho = (100\times2093)/(1000\times490) = 0.427$

Form Table 12 B of IRC : 21-2000,

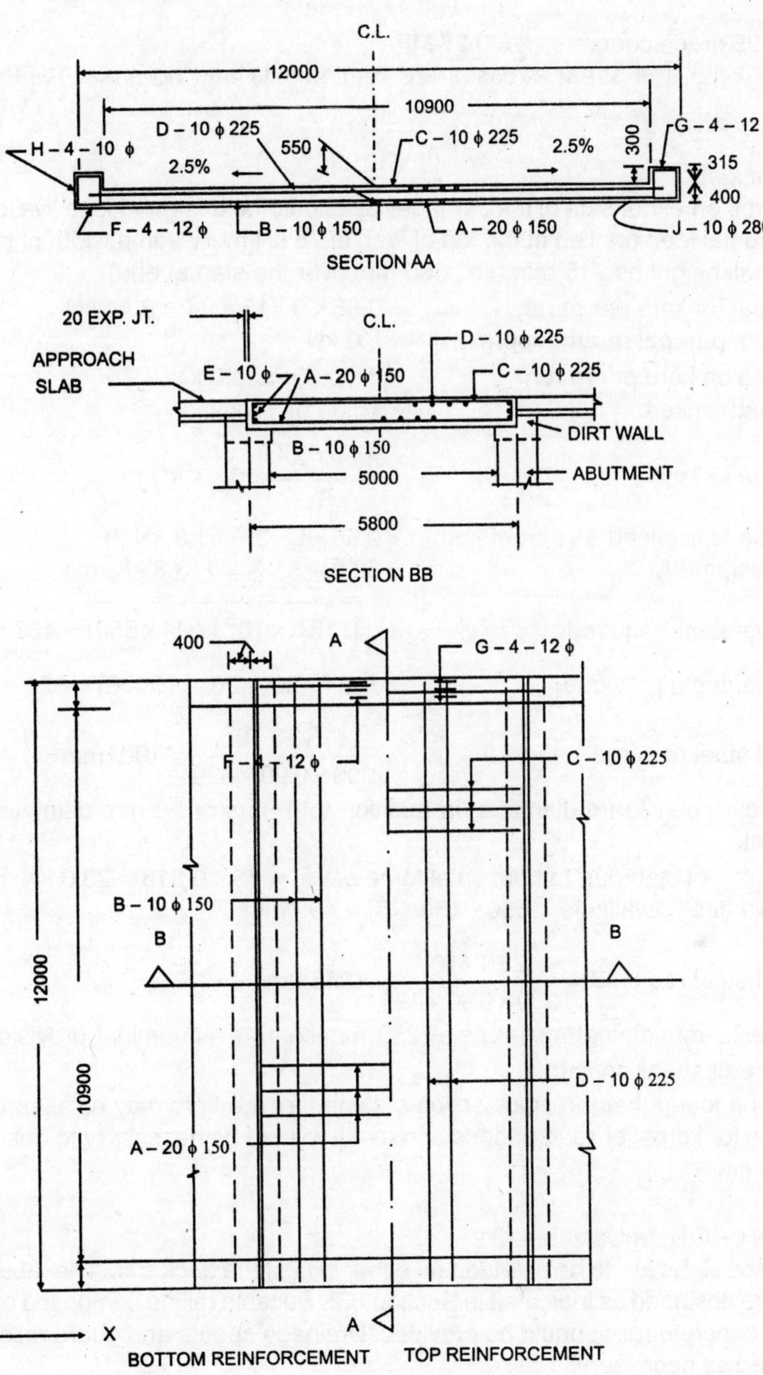

Figure 6.9 Reinforcement Details for Deck Slab.

For M25 grade concrete, $\tau_c = 0.28$ MPa

Since $\tau < \tau_c$, the shear stresses are within limits and no shear reinforcement is necessary.

(ix) *Design of kerb*

The kerbs on either side of the carriageway should be designed for a live load of 4 kN/m^2 and horizontal load applied at the top of kerb of 75 kN/m. Assume width of kerb as 550 mm and overall height as 715 mm (i.e., 300 mm over the slab at end).

Dead load of kerb per m run $= 0.55 \times 0.715 \times 24 = 9.44$ kN

Weight of parapet per m (approx.) $= 5.0$ kN

Live load on kerb per m run $= 4 \times 0.55 = 2.20$ kN

Total load on kerb $= 16.64$ or 17 kN

B.M. due to kerb $= \dfrac{17 \times 5.4^2}{8} = 62.0$ kN.m

B.M. due to live load as part of kerb $= 0.55 \times 93.3 = 51.3$ kN.m

Total design B.M. $= 62.0 + 51.3 = 113.3$ kN.m

Effective depth required $= \sqrt{(113.3 \times 10^6)/(1.1 \times 550)} = 433$ mm

Effective depth provided $= 715 - (50 + 20 + 10 + 6) = 629$ mm

Area of steel required $= \dfrac{113.3 \times 10^6}{200 \times 0.90 \times 629} = 1001\,\text{mm}^2$

Use 4 bars of 12 mm diameter, in addition to 4 bars of 20 mm diameter from slab reinforcement.

B.M. at top of slab due to horizontal force per m $= 75 \times 0.315 = 23.6$ kN.m

Effective depth available $= 550 - (50 + 5) = 495$ mm

Area of steel required $= \dfrac{23.6 \times 10^6}{200 \times 0.90 \times 495} = 265\,\text{mm}^2$

Provide 10 mm diameter stirrups at 280 mm centres as nominal provision. This will also take care of shear on kerb.

Note: It will be found that the final section obtained for the kerb may be assumed as type design for kerbs for road bridges and that it will not be necessary to design the kerb every time.

6.3.3 Design of Other Components

Approach slabs are to be provided on either side of the deck slab. The abutments and wing walls are designed as indicated in Section 6.2. Suitable railing composed of R.C. posts and precast concrete rails should be provided. Drainage spouts and guard rails/stones are to be provided as necessary.

6.4 Author's Charts for Design of Deck Slab of Slab Bridges

The author has developed design charts to enable rapid design of deck slab of highway

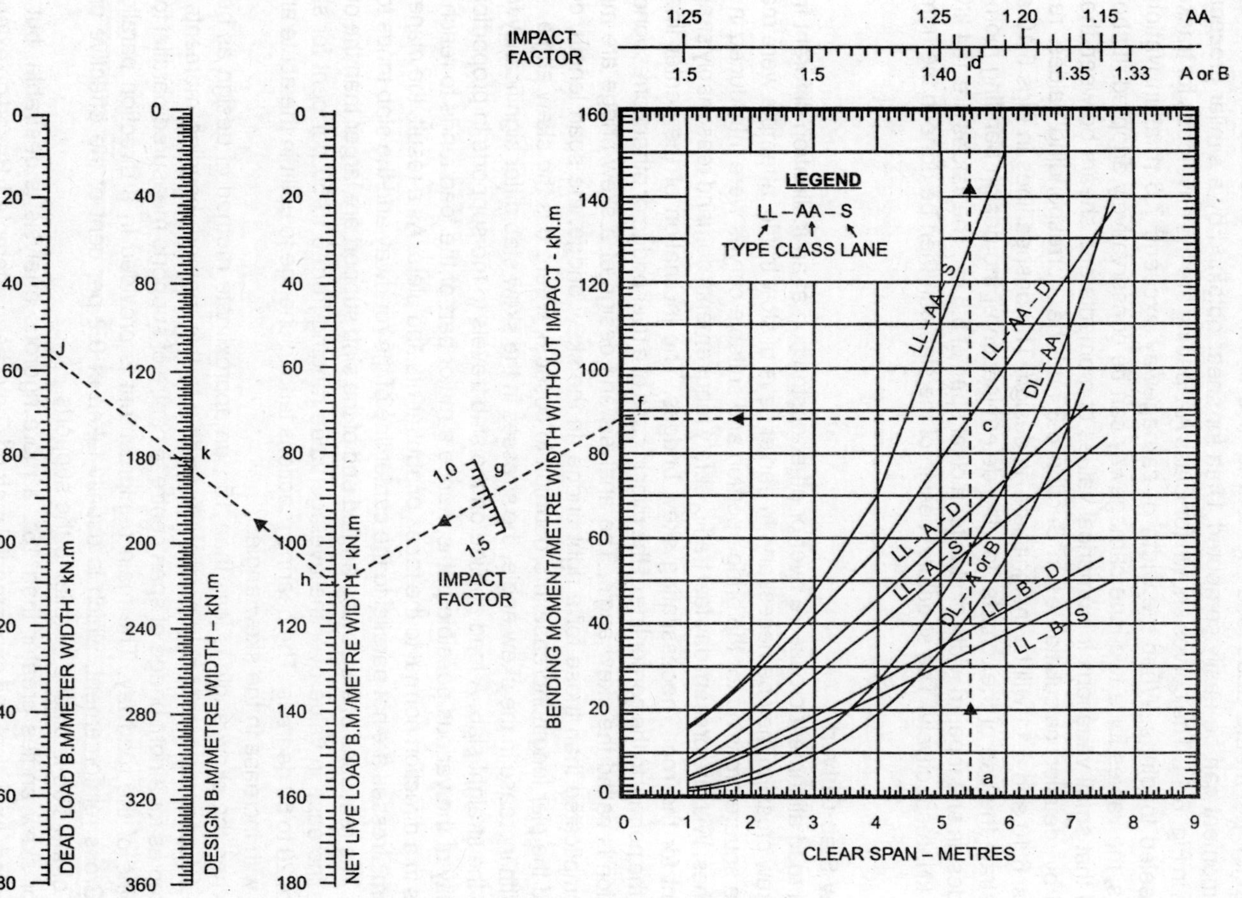

Figure 6.10 Chart for Design Bending Moment Per Metre Width of Deck Slab.

slab bridges[1]. Figure 6.10 gives curves for dead load moment and live load moment without impact per m width of slab. The curves for live load moment are based on the effective width of dispersion method discussed in Section 6.2. The horizontal scale at top gives the impact factor. Estimated dead load moment is also shown. With these data, the design bending moment can be easily computed. Data for shear obtained by a similar procedure are given in Fig. 6.11. The curves shown for Class AA loading are the envelopes for tracked and wheeled loadings. When the width of carriageway exceeds 7.5 m as in National Highways, the values for a two-lane carriageway can be conservatively adopted. It should be noted that small variations in numerical values of moments and shears between those obtained by detailed calculation as in Section 6.3 and the corresponding values read from Figs. 6.10 and 6.11 will be insignificant in design. The dashed lines in Figs 6.10 and 6.11 illustrate the use of the charts for the determination of the design bending moment and the design shear per m width of slab for a clear span of 5.5 m . The values are 154 kN.m and 132 kN, respectively. The structural design of the slab can then be done in the normal manner.

6.5 Skew Slab Culvert

If a road alignment crosses a river or other obstruction at an inclination different from 90°, a skew crossing may be necessary. In earlier days of slow traffic, attempts were made to have a square crossing for the bridge portion and suitable curves were introduced in the approaches. With the present day fast traffic, safety requirements demand reasonably straight alignment for the road, necessitating skew bridges. The inclination of the center line of traffic to the normal to the center line of the river in case of a river bridge or other corresponding obstruction is called the skew angle. The analysis and design of a skew bridge are much more complicated than those for a right bridge. In the skew bridge, the span length, deck area and the pier length increase in proportion to cosec θ, where θ is the skew angle.

With increase in the skew angle, the stresses in the skew slab differ significantly from those in the straight slab. A load applied on the slab travels to the supports in proportion to the rigidity of the various possible paths; hence a major part of the load tends to reach the supports in a direction normal to the faces of abutments and piers. As a result, the planes of maximum stresses are not parallel to the center line of the roadway and the slab tends to be warped. The reactions at the obtuse angled end of the slab support are larger than the other end, the increase in value over the average value ranging from 0 to 50 per cent for skew angles of 20 to 50 degrees. The bearing reactions tend to change to uplift in the acute angle corners with increase in the skew angle.

When the skew angle is less than 15°, an approximate method of design as below may be adopted as reasonably correct. According to this method, bending moments are calculated as for a right bridge of span centre to centre of supports measured parallel to the center line of the roadway. The main reinforcement is provided in a direction parallel to traffic. Cross reinforcement, which is usually taken at 0.2 per cent of the effective cross section of the slab is placed parallel to the supports.

For skew angles greater than 15°, a more rigorous analysis is desirable, but it is complicated. Analytical and experimental methods have been attempted. Based on extensive tests on skew slab models made of gypsum plaster, Rüsch and Hergenroder[5] have presented influence surfaces for bending moments and torsional moments at critical points of skew slabs under a concentrated load placed anywhere on the slabs for various span/width ratios

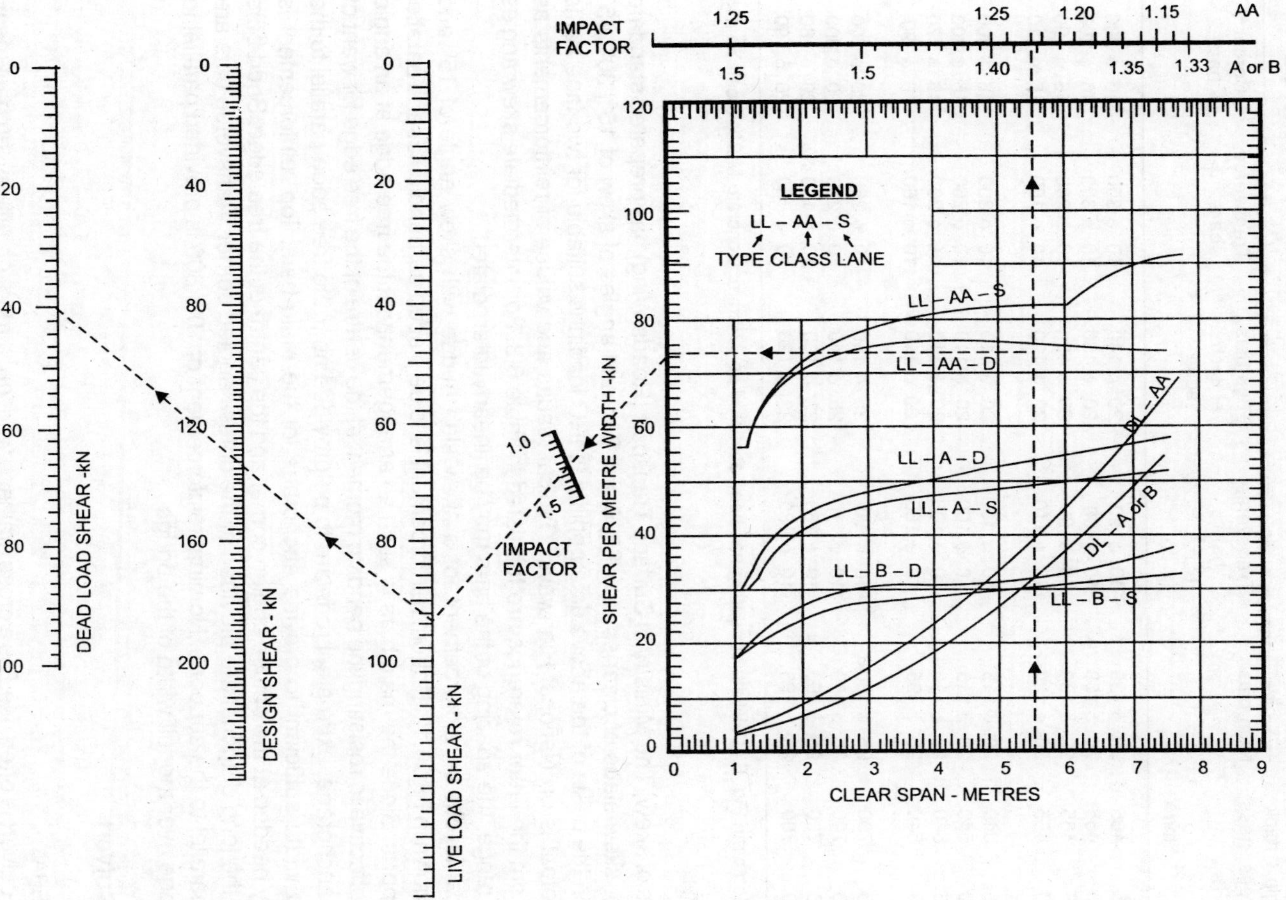

Figure 6.11 Chart for Design Shear Per Metre Width of Deck Slab.

Table 6.3 Typical Reinforcements for Skew Slab Culverts.

Clear Span m	Skew angle degrees	Slab thickness mm	Bottom reinforcement			Top reinforcement	
			Main bars	Transverse bars	Edge bars at each edge	Transverse bars	Corner bars
	15	465	20 ϕ 30	10 ϕ 90	20 ϕ 140	10 ϕ 300	16 ϕ 300
	30	465	20 ϕ 130	10 ϕ 90	20 ϕ 130	10 ϕ 300	16 ϕ 300
5	45	465	20. ϕ 130	10 ϕ 90	20 ϕ 120	10 ϕ 200	16 ϕ 190
	60	465	20 ϕ 120	10 ϕ 110	20 ϕ 120	10 ϕ 180	16 ϕ 110
	15	540	20 ϕ 115	10 ϕ 100	20 ϕ 110	10 ϕ 300	16 ϕ 300
	30	540	20 ϕ 115	10 ϕ 100	20 ϕ 110	10 ϕ 260	16 ϕ 300
6	45	540	20 ϕ 115	10 ϕ 100	20 ϕ 110	10 ϕ 200	16 ϕ 170
	60	540	20 ϕ 95	10 ϕ 160	20 ϕ 110	10 ϕ 160	16 ϕ 90
	15	700	25 ϕ 150	10 ϕ 90	25 ϕ 150	10 ϕ 260	20 ϕ 300
	30	700	25 ϕ 150	10 ϕ 110	25 ϕ 140	10 ϕ 230	20 ϕ 300
8	45	700	25 ϕ 150	10 ϕ 150	25 ϕ 130	10 ϕ 160	20 ϕ 180
	60	700	25 ϕ 95	10 ϕ 180	25 ϕ 120	10 ϕ 90	20 ϕ 90

Note: Notation as in Fig. 6.12. Width of bearing has been assumed as 370 mm. Concrete M 20. Steel Fe 415 to IS: 1786.

and angles of skew. The Ministry of Surface Transport (Roads Wing) has prepared standard designs for skew slabs of clear spans of 5, 6 and 8 m and angles of skew of 15°, 30°, 45° and 60°, on the basis of the above data applied to IRC loadings suitable for two-lane traffic without footpaths on National Highways[6]. Typical results and values of reinforcements as adapted from the latter reference are tabulated in Table 6.3. For intermediate skew angles within the range, the spacing of the bars may be linearly interpolated.

Typical layout of reinforcement for a skew slab bridge with skew angle of 15° and above is shown in Fig. 6.12. Bottom reinforcement is to be placed perpendicular and parallel to the supports. Since the main bars in such an arrangement cut the free edge at an angle and are ineffective in resisting the bending moment at the centre of the free edge for want of adequate anchorage, extra steel rods are to be provided near the free edge parallel to the edge for a width sufficient to provide anchorage for the main bars. Top reinforcement is particularly needed at the obtuse angled corner and the centre of the free edge. Short bars parallel to the long diagonal are provided at the obtuse angled corner. Transverse bars are provided parallel to the supports. Nominal reinforcement of 10 ϕ 300 is provided parallel to the free edge over the full width of the bridge.

6.6 Pipe Culvert

6.6.1 General

Pipe culvert is often used as a cross drainage work on a road or railway embankment, when the discharge through the culvert is small, and when there is no defined channel as in flat country. According to IS:458[7], concrete pipes are classified as non-pressure (NP) and pressure pipes (PP). There are four types of non-pressure pipes (NP1 to NP4), and

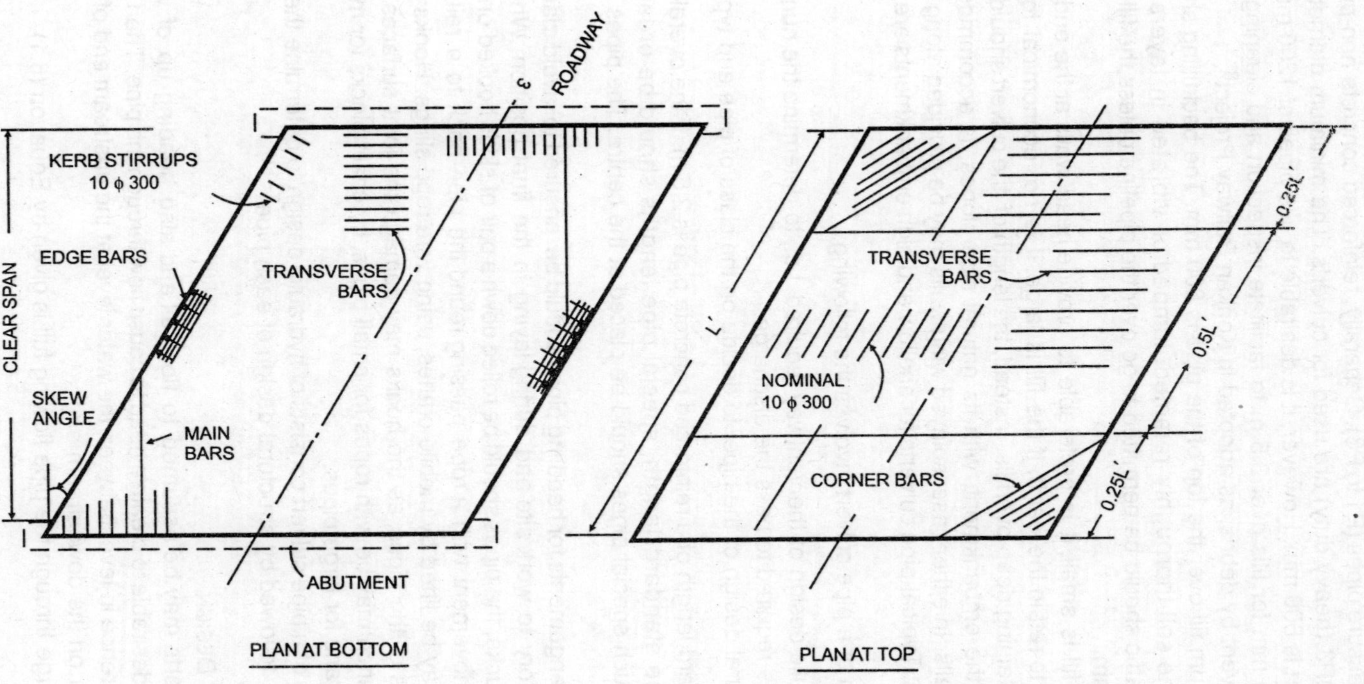

Figure 6.12 Layout of Reinforcement for Skew Slab Bridges.

three types of pressure pipes (P1 to P3). Generally, reinforced concrete non-pressure pipes designated as NP3 (heavy duty) are used for culverts. The minimum diameter of the pipe used for a culvert is 600 mm. However, it is desirable to use pipes of 1200 mm for fills upto 3.5 m and 1800 mm for fills over 3.5 m to facilitate inspection and cleaning and to avoid blocking of the vent by debris, as adopted in Konkan Railway Project[8].

The minimum fill over the top of the pipe is 600 mm. The backfilling should be done carefully and the soil thoroughly rammed, tamped or vibrated in layers of 150 mm thickness. No traffic should be permitted to go over the pipeline unless the fill over the pipe is at least 600 mm.

When the fill is small, it is preferable to provide head walls at the ends of the road formation width to retain the earth. If the fill is large, it is not economical to provide high head walls for retaining deep overfills. Instead, the length of the culvert should be increased suitably so that the embankment, with its natural side slopes, is accommodated without high retaining walls. In either case, splayed wing walls may be provided along with the head walls at the ends. Typical pipe culverts for shallow and high embankments are shown in Fig. 6.13.

The design of a pipe culvert involves the following:

(a) Hydraulic design of the ventway required, i.e., to determine the number and size of pipes required to pass the discharge.
(b) Structural design of the pipe, to decide on the class of pipe and type of bedding.

The standard length of a reinforced concrete pipe is 2.5 m. If the overall length is not a multiple of the standard length, special pipe lengths should be ordered from the manufacturer. Such special pipes should be placed at the centre. The pipes are joined by collar joints.

Proper arrangements for handling pipes should be ensured to avoid damages during transit from factory to work site and during laying in the final position. While unloading manually from trucks, the pipes should be rolled down a pair of skids hooked on to the trucks, controlling the movement with a rope passing round the pipe back to a rail on the truck. Heavier pipes may be lifted by mobile cranes, using wire rope slings. Hooks are not to be used in the ends of the pipes, as the hooks may damage the joint surfaces. Lowering of pipes into the trench may be with ropes for small pipes, by chain blocks for medium pipes, and by mobile crane for large pipes.

The design of a pipe culvert consists of hydraulic design to determine the size and the number of pipes, followed by structural design of each pipe[7–11].

6.6.2 Hydraulic Design

Pipe culverts may be assumed to flow full and also heading up of water on the upstream side is desirable to create a head to cause flow through the pipe. Then the operating head is the difference in level between the water level at the upstream end of the pipe and the level of water on the downstream end.

The discharge through the pipe flowing full is given by Equation (6.1).

$$Q = AK\sqrt{2gH} \qquad \dots (6.1)$$

where Q = discharge in m^3 per second

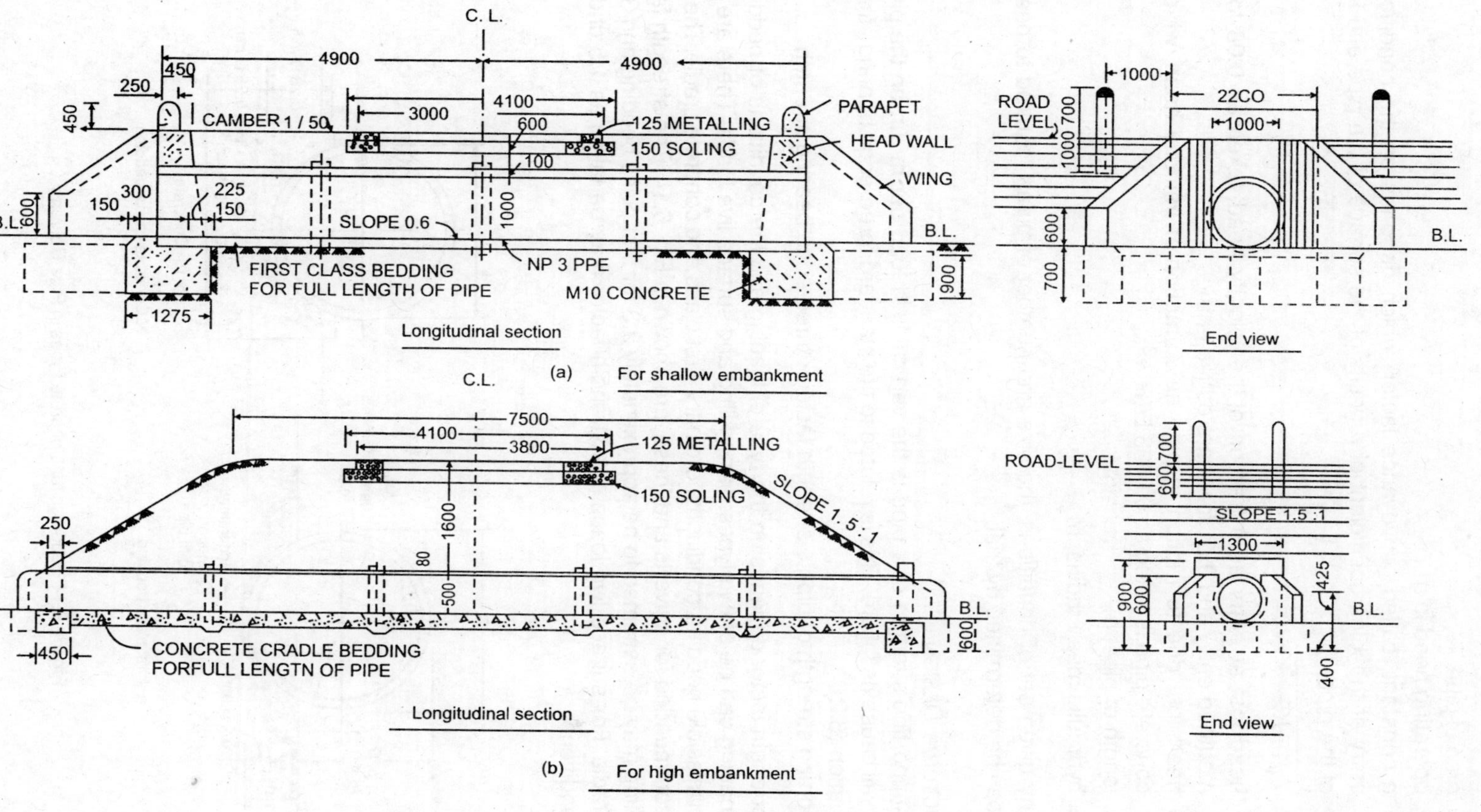

Figure 6.13 Typical Details of Pipe Culverts.

A = area of pipe
H = operating head in m
K = a constant, called conveyance factor, which depends on the conditions of entry at the inlet, coefficient of rugosity of the surface of the pipe and length of the pipe

$$= 1/\sqrt{1 + K_e + K_f}$$

K_e = head lost at entry as a fraction of the velocity head, taken as 0.08 for bell mouthed entry and 0.51 for sharp edged entry
K_f = head lost due to friction in pipe as a fraction of velocity head, which for concrete pipe is taken as $0.0033\ L/(R)^{1.3}$
L = length of pipe, m
R = hydraulic mean radius, m

For any given set of conditions, the size and number of pipes required to pass the discharge can be determined by trial.

6.6.3 Structural Design

The supporting strength of a pipe is the vertical field load per m run on the pipe as laid, which will cause its failure. This is related to the standard three-edge bearing strength test by Equation (6.2).

Supporting strength of pipe = strength factor × three-edge bearing strength ... (6.2)

The strength factor depends on the type of loading and the conditions of bedding of the pipe. Loading can be of two types: earth filling and surface live loads. There are three types of permissible bedding conditions: ordinary, first class and concrete cradle. The latter two are recommended for culverts and these are shown in Fig. 6.14. The strength factors for earth filling may be assumed to be approximately 2.3 for first class bedding and 3.7 for concrete cradle. For surface live loads, the strength factor may be taken as 1.5 under all bedding conditions.

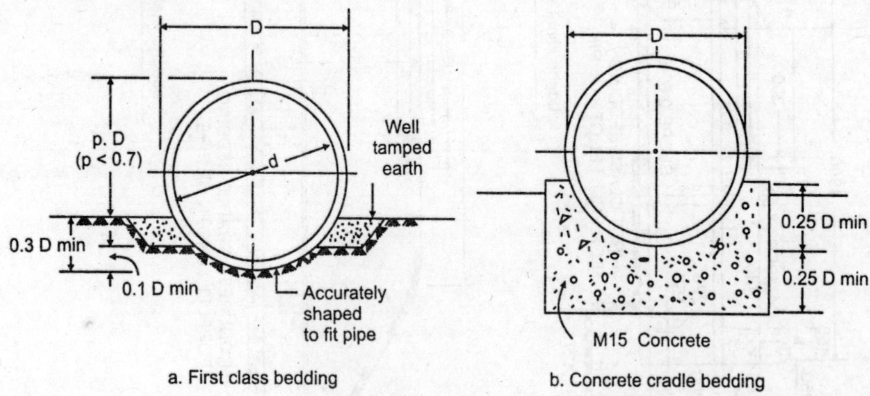

Figure 6.14 Beddings for Concrete Pipes in Pipe Culverts.

The type of non-pressure pipe and bedding should be so chosen that under the worst combination of field loading, a factor of safety of 1.5 is available, as in Equation (6.3).

$$\frac{\text{Three - edge bearing strength in kN/m}}{\text{Factor of safety}} = \frac{\text{W due to filing material in kN/m}}{\text{Corresponding strength factor}} + \frac{\text{W due to surface load in kN/m}}{\text{strength factor for surface load}} \quad ...(6.3)$$

The load on a pipe due to embankment material can be calculated from Equation (6.4).

$$W = C_c . w . D^2 \quad ... (6.4)$$

where W = vertical external load in kN/m of pipe due to embankment material

C_c = coefficient depending on the ratio of height of embankment H to external diameter of pipe, and condition of laying (See IS: 783-1959)

w = density of embankment material in kN/m^3

D = external diameter of pipe in m.

For average conditions, the loads due to earthfill may be obtained from Table 6.4, which is based on a value of 18 kN/m^3 for w.

Table 6.4 Load on Pipe due to Earthfill.

Pipe size d	D	Embankment loading on pipe in kN/m for various depths H in m									
mm	mm	1	2	3	4	5	6	7	8	9	10
NP3-											
500	650	16.8	34.9	54.0	65.3	99.0	118.5	150.0	171.0	191.0	212.0
600	760	18.7	42.6	62.4	75.5	103.0	133.0	160.5	186.0	197.0	230.0
700	860	20.0	48.0	68.0	86.5	118.5	154.5	174.0	209.0	228.0	246.0
800	980	24.2	55.3	79.5	107.0	136.5	163.0	202.0	230.0	260.0	288.0
900	1100	28.3	58.7	91.5	122.0	150.0	185.0	209.0	257.0	297.0	338.0
1000	1200	28.6	59.5	101.0	132.0	163.0	202.0	228.0	259.0	324.0	357.0
1200	1430	33.4	74.0	115.0	159.0	200.0	226.0	274.0	315.0	352.0	427.5

The load on a pipe due to highway wheel load P can be computed from Equation (6.5).

$$W = 4\, C_s\, .\, I.P. \quad .... (6.5)$$

where W = vertical external load in kN/m due to concentrated surface load

C_s = influence coefficient depending on D and H, as given in Table 6.5

H = vertical depth of top of pipe below the surface in m

D = external diameter of pipe in m

P = concentrated (wheel) load in kN

I = impact factor (recommended value 1.5 for highways).

Table 6.5 Influence Coefficient C_s for Concentrated Surface Load for Highways.

Pipe size d	D	Embankment loading on pipe in kN/m for various depths H in m									
mm	mm	0.1	0.2	0.3	0.4	0.6	0.8	1	2	3	4
NP3-											
500	650	0.246	0.228	0.198	0.169	0.117	0.083	0.060	0.017	0.008	0.005
600	760	0.247	0.234	0.210	0.182	0.131	0.094	0.068	0.022	0.010	0.006
700	860	0.247	0.236	0.215	0.186	0.140	0.102	0.075	0.024	0.010	0.006
800	980	0.249	0.240	0.220	0.196	0.149	0.110	0.083	0.027	0.013	0.007
900	1100	0.249	0.241	0.225	0.202	0.156	0.117	0.089	0.029	0.014	0.008
1000	1200	0.249	0.242	0.228	0.205	0.162	0.123	0.095	0.032	0.015	0.010
1200	1430	0.249	0.242	0.230	0.209	0.171	0.131	0.104	0.036	0.020	0.011

The railway load occurs as a uniformly distributed load because of the sleepers and ballast. The load on a buried pipe in a railway embankment is given by Equation (6.6).

$$W = 4\ C_s\ U.D. \qquad \text{... (6.6)}$$

where W = load on pipe in kN/m

C_s = influence coefficient depending on the length of the sleeper, distance between two axles and depth of the top of pipe below the surface

U = uniformly distributed load in kN/m^2 on the surface directly above the pipe

$$= \frac{Pl}{4AB} + 2W_t.\ B$$

P = axle load in kN (229 kN for broad gauge)

A = half the length of the sleeper in m (1.35 m for broad guage)

B = half the distance between the two driving axles in m (0.92 m for broad gauge)

W_t = weight of track structure in kN/m (generally 3 kN/m)

D = outside diameter of the pipe in m.

For broad guage loading, Equation (6.6) will reduce to Equation (6.6a)

$$W = 339\ C_s.D \qquad \text{... (6.6a)}$$

The values of the coefficient C_s for broad gauge are given in Table 6.6.

Table 6.6 Influence Coefficient C_s for Broad Gauge Railway Loading.

H m	C_s	H m	C_s
0.1	0.250	1.0	0.183
0.2	0.249	2.0	0.094
0.3	0.245	3.0	0.052
0.4	0.240	4.0	0.032
0.5	0.233	5.0	0.021
0.6	0.224	6.0	0.015
0.7	0.218	7.0	0.011
0.8	0.205	8.0	0.009
0.9	0.193	9.0	0.007
		10.0	0.005

6.6.4 Examples

(a) *Example 1*

Check the adequacy of NP3 class for pipe with first class bedding for a culvert pipe of 600 mm diameter under 1 m depth of cover across a road.

Load due to earthfill, from Table 6.4 = 18.7 kN/m

C_s for 1 m cover, from Table 6.5 = 0.068

Assuming IRC Class AA wheel load of 62.5 kN to be directly above the pipe,

Loading on pipe $= 4\ C_s$ P.I.

$= 4 \times 0.068 \times 1.5 \times 62.5 = 25$ kN/m

Three-edge bearing strength of NP3 class pipe is 70.8 kN/m (vide IS: 458-1971). The strength factor required for the bedding is obtained by using Equation (6.3).

$$\frac{70.8}{1.5} = \frac{18.7}{\text{S.F.}} + \frac{25}{1.5}$$

$$\text{S.F.} = 0.83$$

The strength factor for first class bedding is 2.3 and hence the provision is adequate.

(b) *Example 2*

Determine the class of pipe and type of bedding required for a concrete pipe of 1200 mm diameter to be laid with its invert level at 7 m under a broad gauge railway embankment.

Assuming the external diameter of the pipe to be 1430 mm, the height of embankment over the pipe will be 7.0 – (1.43 – 0.11) = 5.68 m.

Load on the pipe due to earthfill (by interpolation from Table 6.4) = 215.7 kN/m.

Using Equation (6.6a) and the value for C_s from Table 6.6, load on the pipe due to railway track = $339 \times 0.017 \times 1.43 = 8.2$ kN/m

Choosing NP3 class pipe, three-edge bearing strength is 135 kN/m (vide IS: 458-1971).

Using Equation (6.3)

$$\frac{135}{1.5} = \frac{215.7}{\text{S.F.}} + \frac{8.2}{1.5}$$

$$\text{S.F. required} = 2.55$$

Plain concrete cradle bedding which gives a strength factor of 3.7 is suitable.

6.7 Reinforced Concrete Box Culverts

6.7.1 Components

Reinforced concrete rigid frame box culverts are used for square or rectangular openings with spans up to about 4 m. The height of vent rarely exceeds 3 m. The top of the box section can be at the road level or can be at a depth below road level with a fill depending on the site conditions.

The box culvert consists of the following components:

(i) Barrel of box section of sufficient length to accommodate the carriageway and the kerbs.

(ii) Wing walls splayed at 45° to retain the embankments and also to guide the flow of water into and out of the barrel.

6.7.2 Loading Cases

The loading conditions to be considered in the design of the barrel (per unit length of the barrel) may be classified into the following six categories:

(1) *Concentrated vertical load due to wheel loads*

This is computed from Equation (6.7)

$$W = \frac{P.I}{e} \quad \text{... (6.7)}$$

where P = wheel load

I = impact factor

e = effective width of dispersion

The reaction at the foundation is assumed to be uniform.

(2) *Uniform vertical load*

The track load and the weight of wearing coat and deck slab occur as uniform load. The foundation reaction is uniform.

(3) *Weight of walls (Also for the case of uplift)*

The weights of the two side walls are assumed to cause uniform reaction at foundation. This loading case may also be used for consideration of uplift on the bottom slab.

(4) *Pressure from contained water*

The barrel is assumed to be full with water at the top of the opening. A triangular distribution of pressure is assumed.

(5) *Triangular lateral load*

The earth pressure computed according to Coloumb's theory is applied to both sides. The earth pressure is applied alone when the live load surcharge is neglected, or in combination with case 6, when considering live load surcharge also.

(6) *Uniform lateral load*

The effect of live load surcharge when acting alone will be a uniform lateral load. This loading is considered uniform on both sides. When combined with case 5, the effect of trapezoidal loading will be obtained.

The loading cases are shown schematically in Fig. 6.15. The notation used in subsequent discussion is also shown in this figure. The fixing moments for the six loading conditions are given in Table 6.7.

6.7.3 Hydraulic Design

The design of ventway would depend on the discharge to be catered for. Except in the case of buried barrel, the maximum flood level will be below the bottom of top slab allowing

Table 6.7 Moments in Box Culverts*

Loading case	Fixing moments	
	$M_A = M_{A'}$	$M_D = M_{D'}$
1	$-\frac{WL}{12}\left[\frac{2K+4.5}{(K+3)(K+1)}\right]$	$-\frac{WL}{24}\left[\frac{K+6}{(K+3)(K+1)}\right]$
2	$-\frac{wL^2}{12(K+1)}$	$-\frac{wL^2}{12(K+1)}$
3	$+\frac{WL}{6}\left[\frac{K}{(K+3)(K+1)}\right]$	$-\frac{WL}{6}\left[\frac{3+2K}{(K+3)(K+1)}\right]$
4	$+\frac{pH^2}{60}\left[\frac{K(2K+7)}{(K+3)(K+1)}\right]$	$+\frac{pH^2}{60}\left[\frac{K(2K+8)}{(K+3)(K+1)}\right]$
5	$-\frac{pH^2}{60}\left[\frac{K(2K+7)}{(K+3)(K+1)}\right]$	$-\frac{pH^2}{60}\left[\frac{K(3K+8)}{(K+3)(K+1)}\right]$
6	$-\frac{p.KH^2}{12(K+1)}$	$-\frac{p.KH^2}{12(K+1)}$

*See Fig. 6.15 for notation and details of loading cases.
Positive moment indicates tension on inside face.

a vertical clearance. In this case, the design of ventway will be similar to that for a culvert with R.C. slab deck. The design of ventway for a buried barrel will be similar to that for a pipe culvert. Usually, the ratio of span to height of vent lies between 1 : 1 and 1.5 : 1. The top of bottom slab will be at bed level.

6.7.4 Structural Design

The governing moments, thrusts and shears at the critical section of a box culvert are tabulated in Table 6.8 for the span to height ratios of 1 : 1 and 1.5 : 1. In this table, the walls and slabs are assumed to have the same thickness. Moments, thrusts and shears are computed, preferably using a tabular form, for the six cases and are algebraically added to get the net effect. Reinforcements are provided and detailed to provide adequate resistance to the effects of the applied forces.

Details of a typical reinforced concrete box culvert of span 3.6 m and height of vent of 2.4 m are given in Fig. 6.16.

6.8 Submersible Bridges

6.8.1 Definition

A submersible bridge is a bridge designed to be overtopped during heavy floods, but having its road formation level fixed in such a manner that the duration and frequency of interruption caused to traffic in a year during floods will not exceed specified amounts.

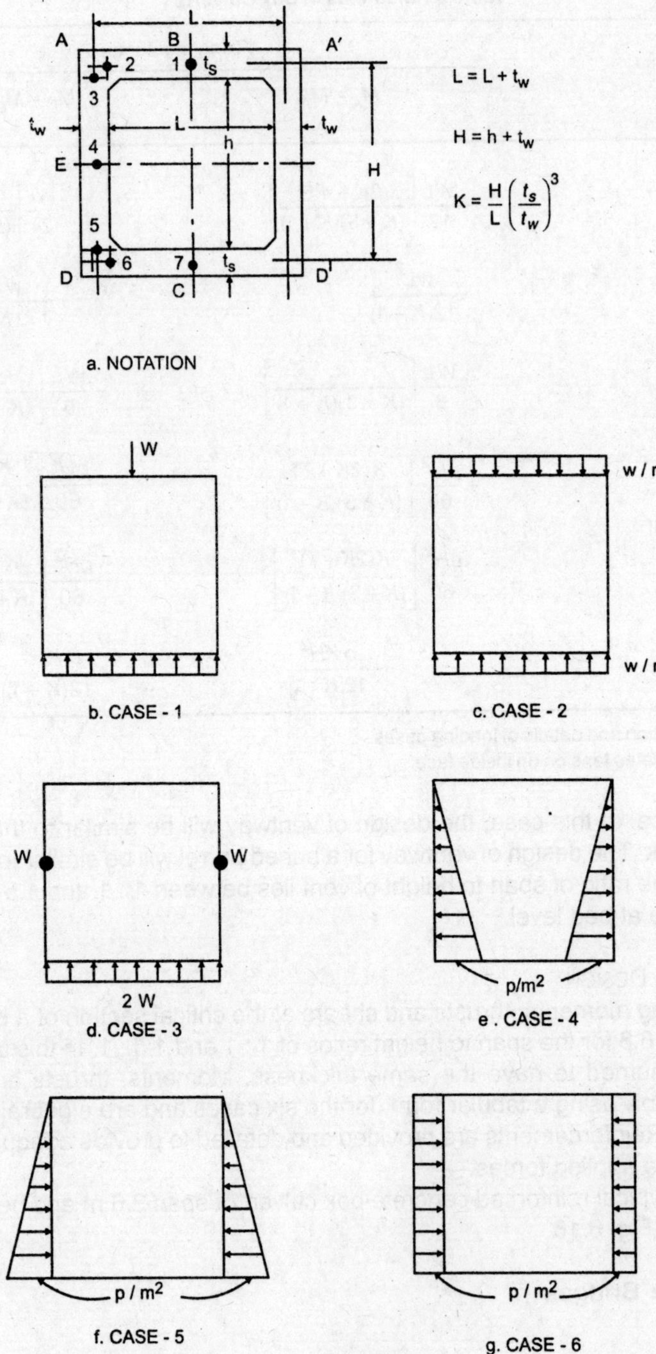

Figure 6.15 Loading Cases for Box Culverts.

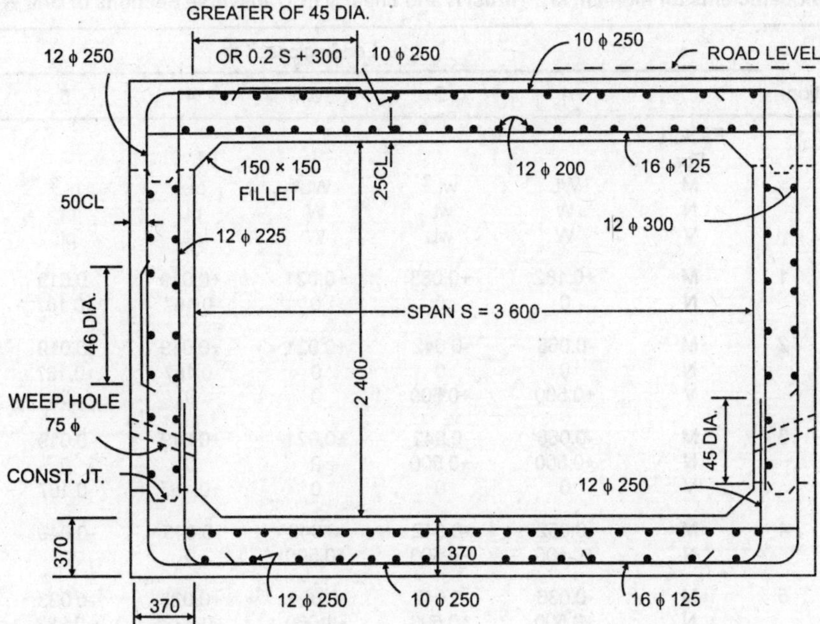

Figure 6.16 Typical Cross-Section of Box Culvert.

6.8.2 Applicability

The submersible bridge is essentially a low-cost bridge and is a compromise between a high-level bridge and a bed-level causeway or paved dip. It is best suited when the difference between the ordinary flood level (OFL) and the highest flood level (HFL) is considerable and the site is subject to wide fluctuations of floods of short duration. The ordinary flood is defined as the maximum flood which is expected normally to occur every year, whereas the highest flood is the highest flood ever recorded or calculated as possible. The submersible bridge can be adopted only on roads of minor importance with low traffic, where interruption to traffic for short durations of up to three days at a time and eighteen days in a year will not be of much consequence. Thus we find that submersible bridges are most suitable for roads in hilly terrain where the streams are near their origin, for minor district roads and village roads and for roads sponsored by community projects where low cost is of great importance.

When the cost of a high level bridge at a particular site on a major district road of low traffic intensity is such as to delay the provision of a crossing indefinitely, and where a submersible bridge at a lower cost could serve the traffic requirements well enough, then, as a special case, adoption of a submersible bridge on a major district road may be recommended. Though less frequent, such cases can still be met with in many States. Submersible bridges are not to be adopted on State Highways and National Highways.

6.8.3 Selection of Site

The considerations for site selection for high level bridges generally apply to submersible bridges. Due to the possibility for outflanking, a straight reach of river and adequate distance from confluence of large tributaries are important. Cutting rather than embankment is to be

Table 6.8 Coefficients for Moment M_1, Thrust N and Shear V in Transverse Sections of Unit Width*.

L : H	Section			Loading case 1	2	3	4	5	6
			Factors For						
			M	WL	wL^2	WL	pL^2	pL^2	pL^2
			N	W	wL	W	pL	pL	pL
			V	W	wL	W	pL	pL	pL
1:1	B	1	M	+0.182	+0.083	+0.021	+0.019	-0.019	-0.042
			N	0	0	0	-0.167	+0.167	+0.500
	A	2	M	-0.068	-0.042	+0.021	+0.019	-0.019	-0.042
			N	0	0	0	-0.167	+0.167	-0.500
			V	+0.500	+0.500	0	0	0	0
	A	3	M	-0.068	-0.042	+0.021	+0.019	-0.019	-0.042
			N	+0.500	+0.500	0	0	0	0
			V	0	0	0	+0.167	-0.167	-0.500
	E	4	M	-0.052	-0.042	-0.042	-0.043	-0.043	+0.083
			N	+0.500	+0.500	+0.500	0	0	0
	D	5	M	-0.036	-0.042	-0.004	+0.023	-0.023	-0.042
			N	+0.500	+0.500	+1.000	-0.333	+0.333	0
			V	0	0	0	0	0	+0.500
	D	6	M	-0.036	-0.042	-0.104	+0.023	-0.023	-0.042
			N	0	0	0	0	0	+0.500
			V	-0.500	-0.500	-1.000	-0.333	+0.333	0
	C	7	M	+0.088	+0.083	+0.146	+0.023	-0.023	-0.042
			N	0	0	0	-0.333	+0.333	+0.500
1.5 : 1	B	1	M	+0.170	+0.075	.0.018	+0.015	-0.015	-0.033
			N	0	0	0	-0.167	+0.167	+0.500
	A	2	M	-0.079	-0.050	+0.018	+0.015	-0.015	-0.333
			N	0	0	0	-0.167	+0.167	+0.500
			V	+.500	+0.500	0	0	0	0
	A	3	M	-0.079	-0.050	+0.018	+0.015	-0.015	-0.033
			N	+0.500	+0.500	0	0	0	0
			V	0	0	0	+0.167	-0.167	-0.500
	E	4	M	-0.062	-0.050	-0.050	-0.047	+0.047	+0.092
			N	+0.500	+0.500	+0.500	0	0	0
	D	5	M	-0.045	-0.050	-0.118	+0.018	-0.018	-0.033
			N	+0.500	+0.500	+1.000	0	0	0
			V	0	0	0	-0.333	+0.333	+0.500
	D	6	M	-0.045	-0.050	-0.118	+0.018	-0.018	-0.033
			N	0	0	0	-0.333	+0.333	+0.500
			V	-0.500	-0.500	-1.000	0	0	0
	C	7	M	+0.079	+0.075	+0.132	+0.018	-0.018	-0.033
			N	0	0	0	-0.333	+0.333	+0.500

*See Fig. 6.15 for notation and details of loading cases.

Note : Positive moment indicates tension on inside face.
Positive thrust indicates compression on the section.
Positive shear indicates that the summation for forces at the left of the section acts outward when viewed from within.

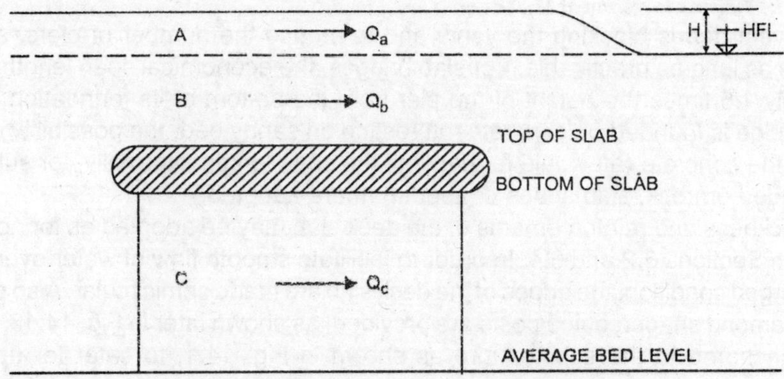

Figure 6.17 Discharge through a Submersible Bridge.

desired for approaches. Hence well defined high banks are desirable. As scour of bed is particularly more severe in the case of submersible bridges, a site having good foundation facilities is to be chosen.

6.8.4 Hydraulic Considerations

Reinforced concrete slab spans are most suitable for submersible bridges, as these offer the least obstruction to water flow in submerged condition. Pipes are suitable only for small causeways where the ordinary flood is very small. Arch spans are not suitable.

The design discharge at the bridge site is to be determined as explained in Chapter 2. The total discharge (Q) passed through a submersible bridge can be visualised as the sum of three parts as in Fig. 6.17 and in Equation (6.8)

$$Q = Q_a + Q_b + Q_c \qquad \ldots (6.8)$$

where $Q_a = A_a \cdot \frac{2}{3} C \cdot \sqrt{2g} \cdot \frac{(H+h_a)^{1.5} - h_a^{1.5}}{H}$

$Q_b = A_b \cdot C \cdot \sqrt{2g(H+h_a)}$

$Q_c = A_c \cdot C \cdot \sqrt{2g\,(H+h_a)}$

A_a, A_b, A_c = area of waterway in parts A, B, and C, respectively

H = difference in water level between upstream and downstream sides of the bridge

h_a = head due to velocity of approach = $\frac{V^2}{2g}$

g = acceleration due to gravity

C = constant equal to 0.625 for flow through parts A and B and 0.9 for flow through part C.

To prevent debris blocking the vents and to reduce the number of piers, spans are chosen to be as long as practicable. For slab bridges, the economical span length would be approximately 1.5 times the height of the pier from the bottom of its foundation to its top. When the bridge is founded on concrete raft resting on sandy bed, the possibility of tension occurring in the concrete raft would further limit the span length. Normally, for submersible bridges founded on rock, slab spans of about 8 m are adopted.

The thickness and reinforcements of the deck slab may be adopted as for normal slab bridges, as in Sections 6.2 and 6.4. In order to facilitate smooth flow of water over the deck in the submerged condition, the edges of the deck slab are made semicircular. Also perforated kerbs and diamond shaped guide posts are provided, as shown later in Fig. 14.13. The deck should be anchored to the substructure as shown in Fig. 14.1, to cater for uplift under submersion.

The linear waterway is so designed that the OFL discharge could be passed through the vents. It is advisable to allow the full bank to bank waterway without restricting it, so that the afflux is small and the velocity in the vents is kept within safe limits. Restriction of waterway by guide banks which is common in the case of high level bridges is not suitable for submersible bridges. The sides of the stream may be provided with stone pitching for a short distance on the upstream and downstream sides up to 300 mm above HFL where convenient or at least up to 300 mm above OFL.

6.8.5 Determination of Road Level At Bridge

Proper determination of the road level is very important in the case of submersible bridges and is the prime criterion for the successful functioning of the bridge so far as road usage is concerned. The decision on the extent of permissible interruption to traffic is critical in fixing the road level and hence would affect the cost. It has been normal to permit interruption to traffic for a maximum period of three days at a time for upto six times in a year. The author had recommended in 1959 that the road level should be so fixed that the bottom of the deck slab is at OFL[12]. The thickness of the slab would take care of the afflux and the water level upstream of the bridge will be close to but less than the road level for ordinary flood conditions. This is particularly suitable when there is a substantial difference between OFL and HFL. Needless to say, the above recommendation would be found successful only if the OFL had been determined correctly.

However, if the peak of the ordinary flood occurs for a very short time and if that peak is much higher than the succeeding next higher levels, it may then be desirable to fix up the road level lower than the OFL. In hilly areas where floating trees may be expected under HFL conditions, the road level should be sufficiently low to allow the floating trees over the deck slab without causing damage to the deck slab.

6.8.6 Foundation

When hard rock is available at shallow depths above the calculated scour level, the structure should be secured to the rock by means of adequate dowel bars. If hard soil other than rock, which is not erodible at the velocity obtaining at the site, is met with, the foundation can be rested in such hard soil with a grip length of about 1 m.

Sandy strata is generally not a favourable condition for submersible bridges. If the sand strata extends for considerable depth, resort is usually made to a monolithic base construction or a cement concrete raft which runs continuously over the entire length. Sloped aprons and cut off walls on both upstream and downstream sides are provided to guard against scour and undermining. Pitching with large diameter boulders is provided in the bed on either side beyond the cut off walls.

6.9 References

1. Victor, D.J., and Chettiar, C.G., 'Design charts for highway bridge slabs', Transport-Communications Quarterly Review, Indian Roads Congress, September 1969, pp. 105-109. (Also reprinted in M.P. Subordinate Engineers' Magazine, Feb. 1970, pp. 7-14).
2. 'Standard plans for 3.0 m to 10.0 m reinforced cement concrete solid slab superstructure with and without footpaths for highways', Ministry of Surface Transport (Roads Wing), Government of India, 1991.
3. 'IRC: 21-2000 Standard specifications and code of practice for road bridges: Section III- Cement concrete (plain and reinforced)', Indian Roads Congress, 2000, 80 pp.
4. 'IRC: 6-2000 Standard specifications and code of practice for road bridges: Section II- Loads and Stresses', Indian Roads Congress, 2000, 55 pp.
5. Rüsch, H., and Hergenroder, A., 'Influence surfaces for moments in skew slabs', Cement and Concrete Association, London, 1964.
6. Drg. No. BD/12-74 and BD/13-74, Ministry of Shipping and Transport (Roads Wing), Govt. of India, July 1974.
7. 'IS: 458-1971 Indian standard specifications for concrete pipes (with and without reinforcement)', Bureau of Indian Standards, New Delhi 1971, 23 pp.
8. Konkan Railway Corporation, 'A treatise on Konkan Railway', Konkan Railway Corporation, 1999, 490 pp.
9. IRC:SP:13-2004, 'Guidelines for the design of small bridges and culverts', Special Publication 13, Indian Roads Congress, New Delhi.
10. 'IS: 783-1959, Indian Standard Code of practice for laying of concrete pipes', Bureau of Indian Standards, New Delhi, 1959, 34 pp.
11. Rau, V.V.S., 'Structural design of concrete pipes in accordance with IS: 783-1959', Journal of Institution of Engineers (India), Vol. 50, No.5, Part CI3, Jan. 1970, pp. 99-107.
12. Victor, D.J., 'The investigation, design and construction of submersible bridges', Journal of Indian Roads Congress, Vol. XXIV-Part 1, Oct. 1959, pp. 181-213.

Chapter 7

Reinforced Concrete Bridges

7.1 General

Reinforced concrete is well suited for the construction of highway bridges in the small and medium span range. The usual types of reinforced concrete bridges are:

Slab bridges;
Girder and slab (T-beam) bridges;
Hollow girder bridges;
Balanced cantilever bridges;
Rigid frame bridges;
Arch bridges; and
Bow string girder bridges.

Design of solid slab bridges has been described in detail in Chapter VI. For slab bridges of spans longer than 10 m, the dead load can be reduced by adopting voided slab design using circular polystyrene void formers. In this case, it is important to ensure that the void formers and the reinforcement are held firmly in the formwork during construction. The void diameter is usually less than 0.6 of the slab thickness. In order to cater to shear stresses, the voids are stopped some distance away from the supports to leave a solid section at the supports.

The other types are explained in this Chapter. A complete design is worked out for a typical T-beam bridge of 14.5 m effective span, as this is the most frequently used type. The Ministry of Road Transport & Highways, Government of India (also referred as MORTH) has specified that for bridges on National Highways with total length less than 30 m the overall width between the outermost faces of the railing kerb be adopted the same as the roadway width of the adjoining road, i.e. at 12.0 m for two-lane carriageway plus shoulders[1]. Since the purpose of the illustrative example is to acquaint the beginner with the principles and techniques of design, the worked example uses a carriageway of 7.5 m with kerbs of 600 mm on either side, giving a total width of the bridge of 8.7 m. Such a width will be applicable for State Highways and also for NH in rural sections for a T-beam bridge of multiple spans resulting in a total length in excess of 30 m. The procedure used therein could be adapted for other types with suitable modifications. Only the design of superstructure for one span is considered herein.

The total cost is usually the governing factor in the selection of the proper type of concrete bridge in any particular case. However, the problem is sometimes complicated by special requirements, such as aesthetics, navigational or traffic clearance below the bridge, limited time for construction, and restrictions on provision of formwork.

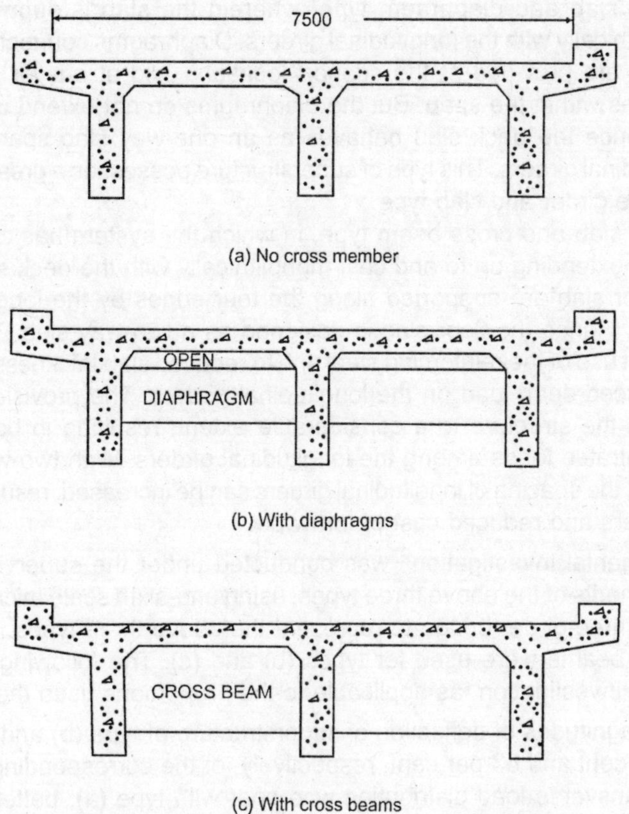

(a) No cross member

(b) With diaphragms

(c) With cross beams

Figure 7.1 Typical Cross Sections of T-beam Bridges.

7.2 T-beam Bridges

7.2.1 General

The T-beam bridge is by far the most commonly adopted type in the span range of 10 to 25 m. The structure is so named because the main longitudinal girders are designed as T-beams integral with part of the deck slab, which is cast monolithically with the girders. T-beam bridges are usually cast-in-situ on fasework resting on ground. The formwork required may be complex to accommodate the bulb for the tensile zone in the T-rib, provision for cross girders and the various chamfers. Simply supported T-beam spans of over 25 m are rare as the dead load then becomes too heavy. However, the author has noted a single span of 35 m for Advai bridge in Goa, which is probably the longest span simply supported reinforced concrete T- beam bridge in India.

The superstructure may be arranged to conform to one of the following three types, as also shown in Fig. 7.1.

(a) Girder and slab type, in which the deck slab is supported on and cast monolithically with the longitudinal girders. No cross beams are provided. In this case, the deck slab is designed as a one-way slab spanning between the longitudinal girders. The system does not possess much torsional rigidity and the longitudinal girders can spread laterally at the bottom level. This type is not adopted in recent designs.

(b) Girder, slab and diaphragm type, wherein the slab is supported on and cast monolithically with the longitudinal girders. Diaphragms connecting the longitudinal girders are provided at the support locations and at one or more intermediate locations within the span. But the diaphragms do not extend up to the deck slab and hence the deck slab behaves as an one-way slab spanning between the longitudinal girders. This type of superstructure possesses a greater torsional rigidity than the girder and slab type.

(c) Girder, slab and cross beam type, in which the system has at least three cross beams extending up to and cast monolithically with the deck slab. The panels of the floor slab are supported along the four edges by the longitudinal and cross beams. Hence the floor slab is designed as a two-way slab. This leads to more efficient use of the reinforcing steel and to reduced slab thickness and consequently to reduced dead load on the longitudinal girders. The provision of cross beams stiffens the structure to a considerable extent, resulting in better distribution of concentrated loads among the longitudinal girders. With two-way slab and cross beams, the spacing of longitudinal girders can be increased, resulting in less number of girders and reduced cost of formwork.

An experimental investigation[2] was conducted under the supervision of the author into the relative merits of the above three types, using one-sixth scale micro concrete models of a 20 m span three-girder bridge to represent the above three types. Three diaphragms and three cross beams were used for types (b) and (c). The following conclusions were derived from the investigation, as applicable to the proportions used therein:

(i) The magnitudes of deflection of superstructure of type (b) and type (c) were only 74 per cent and 63 per cent, respectively, of the corresponding value of type (a).

(ii) The transverse load distribution was poor with type (a), better with type (b) and best with type (c)

(iii) The magnitudes of ultimate load capacity for the superstructure of type (b) and type (c) were 132 per cent and 162 per cent, respectively, of the corresponding value for type (a).

It is evident from the above, that the arrangement of type (c) is to be recommended for adoption, wherever possible.

The standard drawings of MORTH[1] for bridges on National Highways provide the cross camber of the roadway at the top portion of the deck slab. In such a design, the top of the slab provides for a slope (– 2.5%) from the centre line to the sides, and the wearing course is of uniform thickness. In the earlier designs, the cross camber was adjusted in the wearing course and the top of the deck slab was horizontal. For the purpose of learning the principles of design of the deck, the earlier procedure is simpler and adequate. The reader is advised to peruse the above reference and to be aware of the modifications.

7.2.2 Components of a T-Beam Bridge

The T-beam superstructure consists of the following components as also indicated in Fig. 7.2.

(i) Deck slab

(ii) Cantilever portion

(iii) Footpaths, if provided, kerbs and handrails

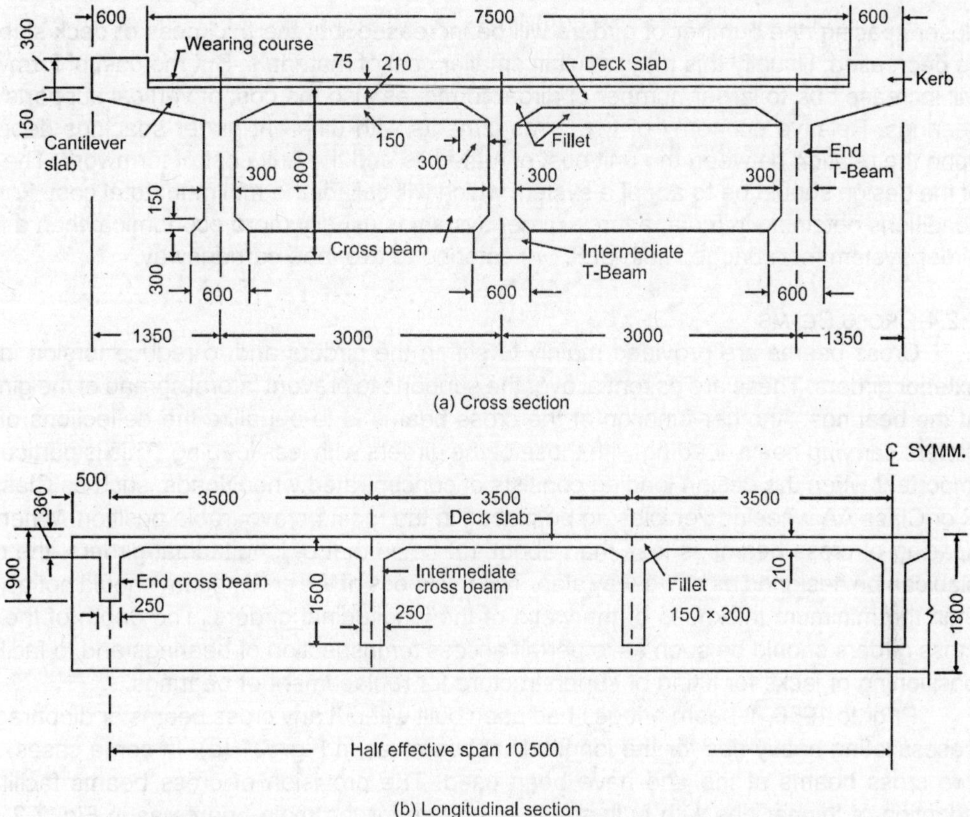

Figure 7.2 Components of T-beam Bridge with Typical Dimensions for 20 m Clear Span.

(iv) Longitudinal girders, considered in design to be of T-section
(v) Cross beams or diaphragms
(vi) Wearing course.

Standard details are used for kerbs and hand rails. The width of the kerb may vary from 475 mm to 600 mm. Wearing course can be of asphaltic concrete of average thickness 56 mm or of cement concrete of M30 grade for an average thickness of 75 mm. Footpaths of about 1.5 m width are to be provided on either side for bridges located in municipal areas; these may be omitted for bridges on rural stretches of roads. It is, however, desirable to provide footpaths even for a bridge on a rural section, if the overall length of the bridge is large. Typical details of footpath, kerb, handrails and wearing course are discussed in chapter 14.

7.2.3 Number And Spacing of Main Girders

The illustration shown in Fig. 7.2 features three main girders, applicable for a two-lane carriageway of 7.5 m width. If the width of the bridge is adopted as 12.0 m, at least four main girders will be necessary. The lateral spacing of the longitudinal girders will affect the cost of the bridge. Hence in any particular design, the comparative estimates of several alternative arrangements of girders should be studied before adopting the final design. With closer spacing, the number of girders will be increased, but the thickness of deck slab will be decreased. Usually this may result in smaller cost of materials. But the cost of formwork will increase due to larger number of girder forms, as also the cost of vertical supports and bearings. Relative economy of two arrangements with different girder spacings depends upon the relation between the unit cost of materials and the unit cost of formwork. The aim of the design should be to adopt a system which will call for the minimum total cost. For the conditions obtaining in India, a three-girder system is usually more economical than a four-girder system for a bridge width of 8.7 m catering to two-lane carriageway.

7.2.4 Cross Beams

Cross beams are provided mainly to stiffen the girders and to reduce torsion in the exterior girders. These are essential over the supports to prevent lateral spread of the girders at the bearings. Another function of the cross beams is to equalize the deflections of the girders carrying heavy loading with those of the girders with less loading. This is particularly important when the design loading consists of concentrated wheel loads, such as Class 70 R or Class AA wheeled vehicles, to be placed in the most unfavourable position. When the spacing of cross beams is less than about 1.8 times that of longitudinal girders, the deck slab can be designed as a two-way slab. The thickness of the cross beam should not be less than the minimum thickness of the webs of the longitudinal girders. The depth of the end cross girders should be such as to permit access for inspection of bearings and to facilitate positioning of jacks for lifting of superstructure for replacement of bearings.

Prior to 1956, T-beam bridges had been built without any cross beams or diaphragms, necessitating heavy ribs for the longitudinal beams as in Fig. 7.1 (a). In some cases, only two cross beams at the end have been used. The provision of cross beams facilitates adoption of thinner ribs with bulb shape at bottom for the main beams as in Fig. 7.2. The current Indian practice is to use one cross beam at each support and to provide one to three intermediate cross beams. Diaphragms have been used instead of cross beams in some cases in the past. The author's tests referred to in Section 7.2.1 indicate that provision of one cross beam at each end and one at the centre is definitely advantageous in reducing deflection and increasing ultimate load capacity, though the additional benefit in providing more than three cross beams is not significant. Similar inferences have also been reported by Kostem[3] based on extensive analytical tests.

7.2.5 Design of Deck Slab

If the deck slab is spanning in one direction as in Fig. 7.1 (a) and (b), the bending moments for dead load may be computed as in a continuous slab, continuous over the longitudinal girders. For concentrated loads, the bending moment per unit width of slab may be computed using the effective width formula given in clause 305.16.2 of I.R.C. Bridge Code for each concentrated load. The formula is given as Equation (A.1) in Appendix A.

When the slab is supported on four sides, as for the case shown in Fig. 7.1 (c), the deck slab may be designed as a two-way slab using Pigeaud's theory as explained in Appendix A. Curves useful for design by this method are available in many text books[4] and are also included in Appendix A. The curves are intended for slabs simply supported at the four sides. In order to allow for continuity, the values of maximum positive moments are multiplied by a factor of 0.8. In design computations, the effective span is taken as the clear span.

7.2.6 Cantilever Portion

The cantilever portion usually carries the kerb, handrails, footpath if provided and a part of the carriageway. The critical section for bending moment is the vertical section at the junction of the cantilever portion and the end longitudinal girder. For the computation of bending moment due to live load, the effective width for cantilever is assessed from the formula given in clause 305.16.2 of the Bridge Code as also included as Equation (A.2) in Appendix A.

The reinforcement should be so detailed that the cranked bars from the deck slab could be used as half of the main reinforcement for the cantilever. The top bars of deck slab may be extended to the cantilever to provide the other half. This step in design would facilitate easier placing of reinforcement. The distributors for the cantilever portion are computed as corresponding to a moment of 0.2 times dead load moment plus 0.3 times the live load moment.

7.2.7 Design of Longitudinal Girders

In Fig. 7.1, the longitudinal girders are shown with straight T-ribs. Such ribs are necessary when cross beams are not used. When multiple cross beams are used, the rib is made thinner and the bottom of T-rib is widened to an extent sufficient to accommodate the tensile reinforcing bars, as shown in Fig. 7.2. However, straight ribs are convenient for cranking of main bars and would facilitate easier formwork. Hence straight ribs are preferred for spans less than 18 m.

For the computation of the bending moment due to live load, the distribution of the live loads between longitudinals has to be determined. When there are only two longitudinal girders, the reactions on the longitudinals can be found by assuming the supports of the deck slab as unyielding. With three or more longitudinal girders, the load distribution is estimated using any one of the rational methods. Three of these are as below:

(a) Courbon's method
(b) Hendry-Jaegar method[5]
(c) Morice and Little version of Guyon and Massonnet method [4,6].

The salient features of these methods are described in Appendix B. By using any one of the above methods, the maximum reaction factors for intermediate and end longitudinal girders are obtained. The bending moments and shears are then computed for these critical values of reaction factors. The procedure will be clear from the illustrative example given in Section 7.3.

7.2.8 Design of Cross Beams

Since the purpose of the cross beams is to stiffen the longitudinal beams and to provide a stiff superstructure, an approximate design is usually adequate. However, if a more rigorous design is required, the following procedure may be adopted. Dead load B.M. is computed considering a trapezoidal distribution of the weight of deck slab and wearing course, besides including the self weight. The cross beam is considered continuous over two spans. Suitable weighted moment factors may be computed, considering different dispositions of the standard loading over the cross beams. Similarly, shears may also be computed. Reinforcements may then be provided to suit the values of moment and shear. Additional cranked bars (usually two bars of 20 ϕ or 22 ϕ) may be provided to cater to diagonal tension.

Using the approximate method, the depth of the intermediate cross beam may be arranged such that the bottom of the cross beam is at the top of the bottom flange of the longitudinal beam when a bulb is provided or to a depth of at least 0.75 of overall depth when straight T-ribs are adopted. Width of cross beam may be adopted nominally as 250 mm. The reinforcements may be provided at 0.5% of gross area at bottom and 0.25% of gross area at top. The same reinforcements may also be used for the end cross beam. Nominal shear reinforcement consisting of 12 ϕ two-legged stirrups or 10 ϕ four-legged stirrups at 150 mm centres will usually be adequate.

The design of the end cross beam may be performed on the same lines as for the intermediate cross beam. In earlier days, the depth of the end cross beam had been reduced to about 0.6 of that for intermediate beam. With the use of elastomeric bearings, provision has to be made for the possibility of lifting the deck to replace the bearings. Hence the present practice is to keep the depth of the end cross beam the same as for the intermediate cross beam. The bottom reinforcement may be taken as half the bottom reinforcement for the corresponding intermediate cross beam. The top reinforcement is kept the same. In addition, two bars of 20 ϕ or 22 ϕ are provided at top as cranked bars to cater to diagonal tension occurring during the lifting operation. The locations of jacks for lifting have to be indicated, and additional mesh reinforcement should be provided at these locations. The details are indicated in para 7.3.7.

7.3 Illustrative Example of T-beam Bridge

Problem: To design the superstructure for one span for a T-beam bridge to be built on a rural section of a State Highway. The bridge consists of five spans of 14.5 m. Assume 'moderate' exposure, and cement concrete wearing course.

Solution:

7.3.1 Preliminary Design

Clear roadway = 7.5 m
Assume three T-beams spaced at 2.5 m intervals
Effective span of T-beam = 14.5 m
Assume five cross beams at 3.625 m intervals.

The preliminary dimensions may be assumed (based on experience) as shown in Fig. 7.3. These may be checked later and modified, if necessary. M25 grade concrete and high yield deformed bars of Fe 415 grade to IS: 1786 will be used. Clear cover to reinforcement is taken as 40 mm.

7.3.2 Deck Slab

The slab is supported on four sides by beams.
Thickness of slab H = 215 mm
Thickness of wearing course D = 75 mm
Span in the transverse direction = 2.5 m
Effective span in transverse direction = 2.5 - 0.3 = 2.2 m
Span in the longitudinal direction = 3.625 m
Effective span in the longitudinal direction = 3.625 - 0.25 = 3.375 m

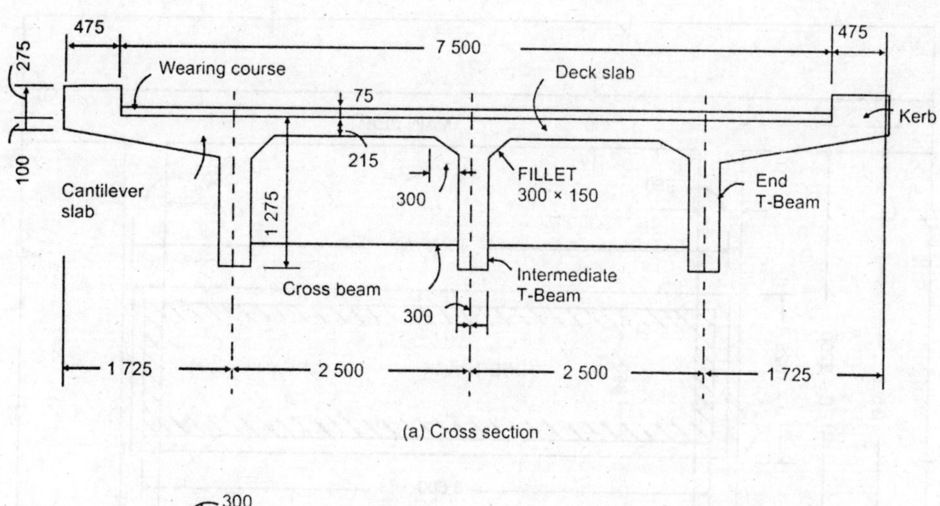

(a) Cross section

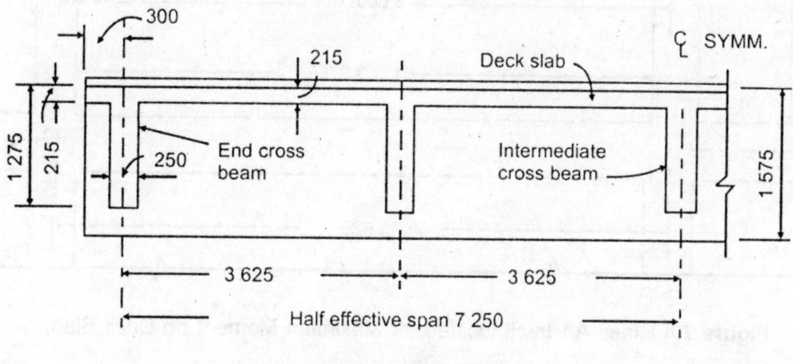

(b) Longitudinal section

Figure 7.3 Preliminary Dimensions for T-beam Deck.

(i) Maximum B.M. due to dead load

Weight of deck slab = 0.215 x 24 = 5.16 kN/m²
Weight of wearing course = 0.075 x 22 = 1.65 kN/m²
Total weight = 6.81 kN/m²

Since the slab is supported on all four sides and is continuous, Pigeaud's curves will be used to get influence coefficients to compute moments.

Ratio K = Short span / Long span = 2.2 / 3.375 = 0.652

1 / K = 1.53

From Pigeaud's curves, m_1 = 0.047 and m_2 = 0.0175
Total dead weight = 6.81 x 2.2 x 3.375 = 50.56 kN
Moment along short span = (0.047 + 0.15 x 0.0175) 50.56 = 2.51 kN.m
Moment along long span = (0.0175 + 0.15 x 0.047) 50.56 = 1.24 kN.m

(ii) *Live load B.M. due to IRC Class AA tracked vehicle*

Size of one panel of deck slab = 2.5 m x 3.625 m

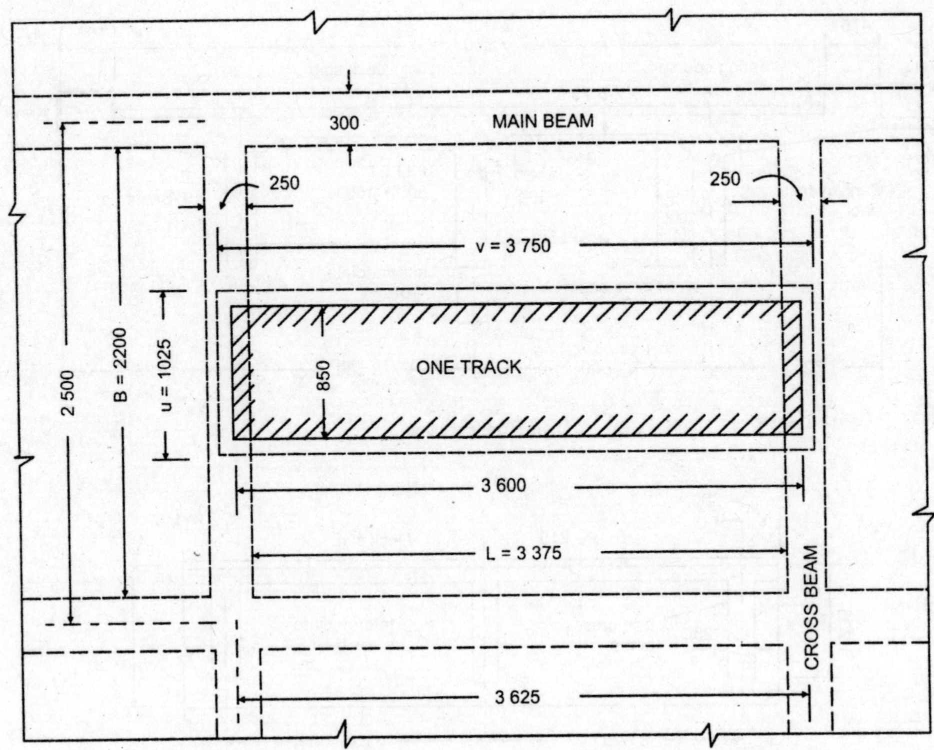

Figure 7.4 Class AA track located for Maximum Moment on Deck Slab.

One track of the tracked vehicle is placed symmetrically on the panel as shown in Fig. 7.4. Track contact dimensions are taken from Fig. 3.3.

Impact factor fraction = 25%

Width of load spread along short span

$$u = \sqrt{(0.85 + 2 \times 0.075)^2 + 0.215^2} = 1.025 \text{ m}$$

Similarly width of load spread along longitudinal direction

$$v = \sqrt{(3.6 + 2 \times 0.075)^2 + 0.215^2} = 3.756 \text{ m limited to } 3.375 \text{ m}$$

$$K = 0.652\ ;\ \frac{u}{B} = \frac{1.025}{2.2} = 0.466\ ;\ \frac{v}{L} = \frac{3.375}{3.375} = 1.0$$

Using Pigeaud's curves

$$m_1 = 7.75 \times 10^{-2}$$

$$m_2 = 2.675 \times 10^{-2}$$

Total load per track including impact = 1.25 x 350 = 437.5 kN

$$\text{Effective load on the span} \quad = 437.5 \times \frac{3.375}{3.756} = 393.1 \text{ kN}$$

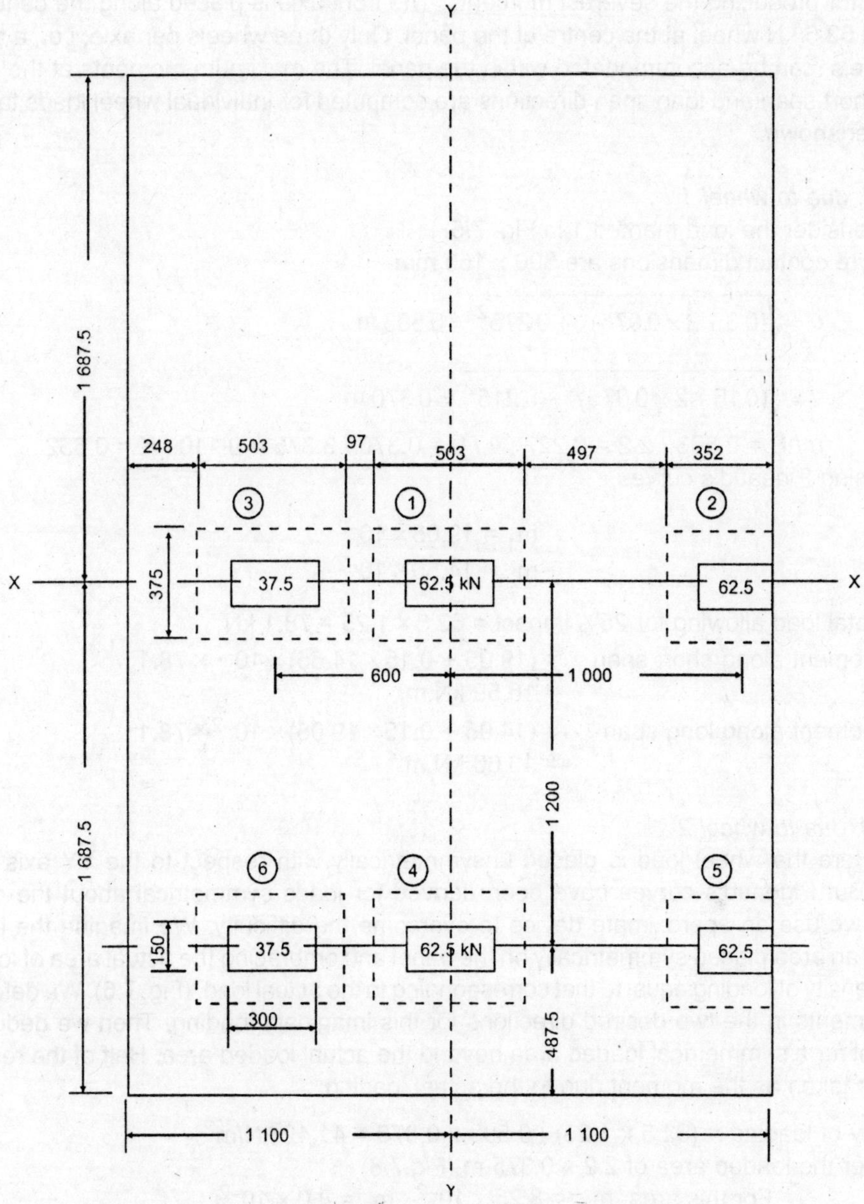

Figure 7.5 Disposition of Class AA Wheeled Vehicle for Maximum Moment.

Moment along shorter span = $(7.75 + 0.15 \times 2.675) \times 10^{-2} \times 393.1 = 32.04$ kN.m

Moment along longer span = $(2.675 + 0.15 \times 7.75) \times 10^{-2} \times 393.1 = 15.09$ kN.m

(iii) *Live load B.M. due to IRC Class AA wheeled vehicle*

The Class AA wheeled vehicle should be placed on the deck slab panel as shown in

Fig. 7.5 for producing the severest moments. The front axle is placed along the centre line with the 62.5 kN wheel at the centre of the panel. Only three wheels per axle, i.e., a total of six wheels, can be accommodated within the panel. The maximum moments at the centre in the short span and long span directions are computed for individual wheel loads taken in the order shown.

(a) *B.M. due to wheel 1*

Consider the load marked 1 in Fig. 7.5.

Tyre contact dimensions are 300 x 150 mm

$$u = \sqrt{(0.3 + 2 \times 0.075)^2 + 0.215^2} = 0.503 \text{ m}$$

$$v = \sqrt{(0.15 + 2 \times 0.075)^2 + 0.215^2} = 0.370 \text{ m}$$

u / B = 0.503 / 2.2 = 0.227 ; v / L = 0.370 / 3.375 = 0.110 ; K = 0.652

Using Pigeaud's curves,

$$m_1 = 19.05 \times 10^{-2}$$
$$m_2 = 14.55 \times 10^{-2}$$

Total load allowing for 25% impact = 62.5 × 1.25 = 78.1 kN

Moment along short span = $(19.05 + 0.15 \times 14.55) \times 10^{-2} \times 78.1$
= 16.59 kN.m

Moment along long span = $(14.05 + 0.15 \times 19.05) \times 10^{-2} \times 78.1$
= 13.60 kN.m

(b) *B.M. due to wheel 2*

Here the wheel load is placed unsymmetrically with respect to the YY axis of the panel. But Pigeaud's curves have been derived for loads symmetrical about the centre. Hence we use an approximate device to overcome the difficulty. We imagine the load to occupy an area placed symmetrically on the panel and embracing the actual area of loading, with intensity of loading equal to that corresponding to the actual load. (Fig. 7.6). We determine the moments in the two desired directions for this imaginary loading. Then we deduct the moment for a symmetrical loaded area beyond the actual loaded area. Half of the resulting value is taken as the moment due to the actual loading.

Intensity of loading = (62.5 x 1.25) / 0.503 x 0.375 = 41.42 kN/m^2

Consider the loaded area of 2.2 x 0.375 m (Fig.7.6)

For this area, $m_1 = 8.25 \times 10^{-2}$; $m_2 = 8.0 \times 10^{-2}$

Moment along short span = $(8.25 + 0.15 \times 8.0)\ 10^{-2} \times 2.2 \times 0.375 \times 41.42$
= 32.29 kN.m

Moment along long span = 31.56 kN.m

Next, consider the area between the real and the 'dummy' load, i.e., 1.496 m x 0.375 m.

For this area, $m_1 = 11.3 \times 10^{-2}$; $m_2 = 10.35 \times 10^{-2}$

Moment along short span = $(11.3 + 0.15 \times 10.35)\ 10^{-2} \times 1.496 \times 0.375 \times 41.42$
= 27.89 kN.m

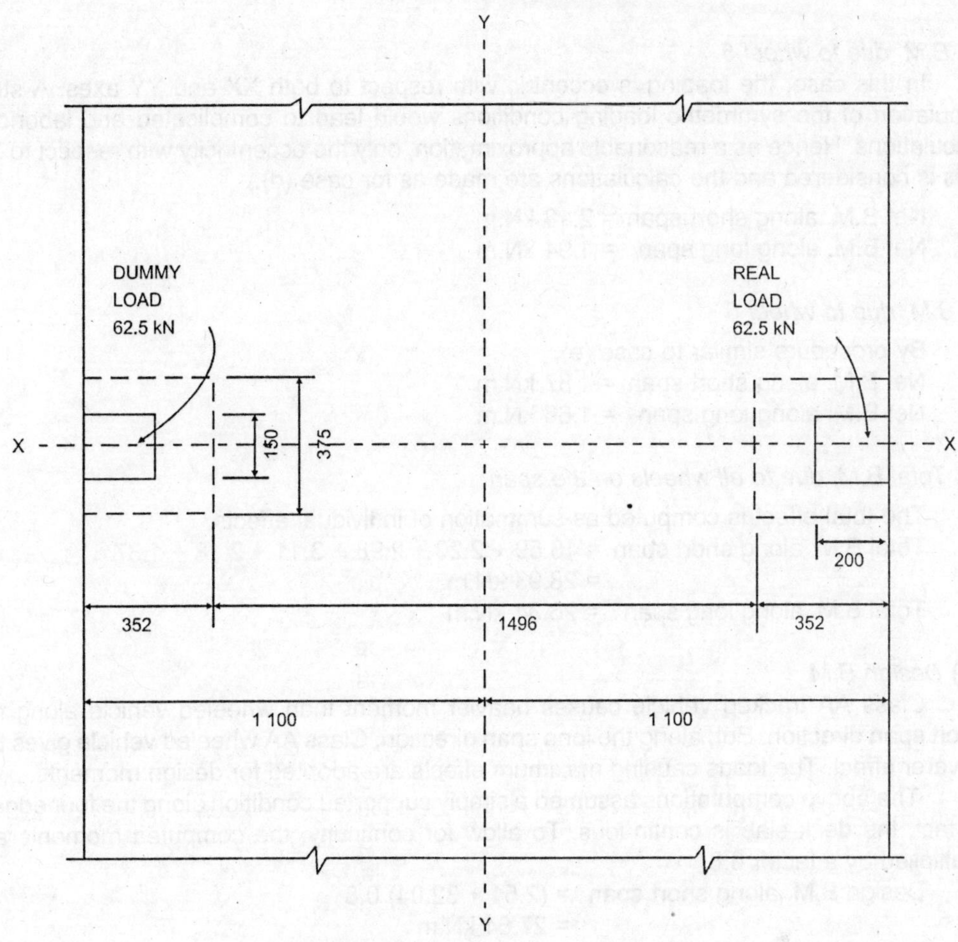

Figure 7.6 Details of Disposition of Wheel 2.

Moment along long span = 26.14 kN.m
Net B.M. along short span = 1/2 (32.29 – 27.89) = 2.20 kN.m
Net B.M. along long span = 1/2 (31.56 – 26.14) = 2.71 kN.m

(c) *B.M. due to wheel 3*

By similar procedure as for case (b), we get
B.M. along short span = 2.98 kN.m
B.M. along long span = 3.66 kN.m

(d) *B.M. due to wheel 4*

Here the load is eccentric with respect to XX axis. By similar procedure as for case (b) but with the load area extended with respect to XX axis, we get

B.M. along short span = 3.11 kN.m
B.M. along long span = 2.77 kN.m

(e) *B.M. due to wheel 5*

In this case, the loading is eccentric with respect to both XX and YY axes. A strict simulation of the symmetric loading conditions would lead to complicated and laborious calculations. Hence as a reasonable approximation, only the eccentricity with respect to XX axis is considered and the calculations are made as for case (d).

Net B.M. along short span = 2.18 kN.m
Net B.M. along long span = 1.94 kN.m

(f) *B.M. due to wheel 6*

By procedure similar to case (e),
Net B.M. along short span = 1.87 kN.m
Net B.M. along long span = 1.66 kN.m

(g) *Total B.M. due to all wheels on the span*

The total effect is computed as summation of individual effects.

Total B.M. along short span = 16.59 + 2.20 + 2.98 + 3.11 + 2.18 + 1.87
= 28.93 kN.m

Total B.M. along long span = 26.34 kN.m

(iv) *Design B.M.*

Class AA tracked vehicle causes heavier moment than wheeled vehicle along the short span direction. But, along the long span direction, Class AA wheeled vehicle gives the severer effect. The loads causing maximum effects are adopted for design moments.

The above computations assumed a simply supported condition along the four edges. In fact, the deck slab is continuous. To allow for continuity, the computed moments are multiplied by a factor 0.8.

Design B.M. along short span = (2.51 + 32.04) 0.8
= 27.64 kN.m

Design B.M. along long span = (1.24 + 26.34) 0.8
= 22.07 kN.m

(v) *Reinforcement*

σ_{cb} = 8.3 MPa ; σ_{st} = 200 MPa ; j = 0.90 ; and R = 1.10

$$\text{Effective depth required} = \sqrt{\frac{27.64 \times 1000 \times 1000}{1.10 \times 1000}} = 159 \text{ mm}$$

Effective depth provided assuming 12 mm diameter main bars
= 215 - 40 - 6 = 169 mm

Hence the provision of total depth of 215 mm is safe.

$$\text{Area of main reinforcement} = \frac{27.64 \times 1000 \times 1000}{200 \times 0.90 \times 169} = 909 \text{ mm}^2$$

Adopt 12 mm dia. bars at 110 mm centres, giving an area of 1028 mm^2

$$\text{Area of longitudinal reinforcement} = \frac{22.07 \times 1000 \times 1000}{200 \times 0.90 \times 157} = 781 \text{ mm}^2$$

Adopt 12 mm dia. bars at 140 mm centres, giving an area of 808 mm^2

7.3.3 CANTILEVER SLAB

(i) *Moment due to dead load*

The total maximum moment due to dead load per metre width of cantilever slab is computed as in the following table, using details from Fig. 7.7.

S. No.	Description	Load kN	Lever arm	Moment kN.m
1.	Handrails (approx.)	1.74	1.40	2.44
2.	Kerb 0.475 × 0.275 × 24	3.13	1.34	4.19
3.	Wearing course			
	1.1 × 0.075 × 22	1.81	0.55	1.00
4.	Slab 1.575 × 0.1 × 24	3.78	0.79	2.99
	0.5 × 0.25 × 1.575 × 24	4.72	0.52	2.45
	Total			13.07

(ii) *Moment Due to live load*

Due to the specified minimum clearance, Class AA loading will not operate on the cantilever slab. Class A loading is to be considered and the load will be as shown in Fig. 7.7.

Effective width of dispersion b_e is computed by Equation (A.2).

$$b_e = 1.2\ x + b_w$$

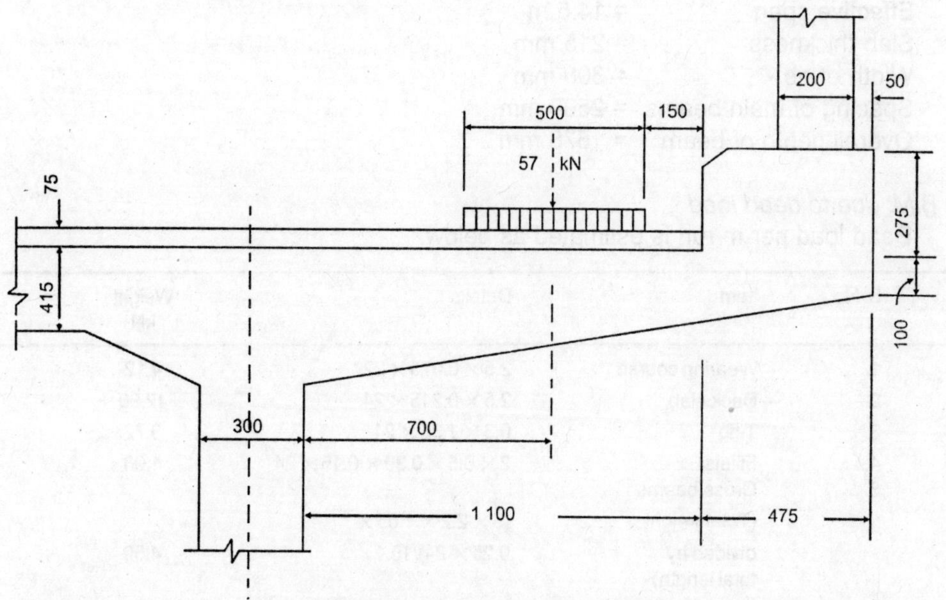

Figure 7.7 Cantilever Slab with Class A Wheel.

Here x = 0.70 m

b_w = 0.25 + 2 × 0.075 = 0.40 m

Hence b_e = 1.24 m

Live load per m width including impact = 57 × 1.5/1.24 = 68.95 kN

Maximum moment due to live load = 68.95 × 0.7 = 48.26 kN.m

(iii) *Reinforcement*

Total moment due to dead load and live load = 13.07 + 48.26 = 61.33 kN.m

Effective depth required $= \sqrt{\dfrac{61.33 \times 1000 \times 1000}{1.10 \times 1000}} = 236 \text{ mm}$

Effective depth provided = 350 – 40 – 8 = 302 mm

Area of main reinforcement required $= \dfrac{61.33 \times 1000 \times 1000}{200 \times 0.90 \times 302} = 1128 \text{ mm}^2$

Adopt 16 mm dia. bars at 220 mm centres plus 12 mm dia. bars at 220 mm centres giving a total area of 1428 mm^2

B.M. for distributors = 0.2 × 13.07 + 0.3 × 48.26 = 17.09 kN.m

Area of distributors $= \dfrac{17.09 \times 1000 \times 1000}{200 \times 0.90 \times 289} = 329 \text{ mm}^2$

Provide 10 mm dia. bars at 220 mm centres giving an area of 357 mm^2

7.3.4 Intermediate Longitudinal Girder

(i) *Data*

Effective span	= 14.5 m
Slab thickness	= 215 mm
Width of rib	= 300 mm
Spacing of main beams	= 2500 mm
Overall depth of beam	= 1575 mm

(ii) *B.M. due to dead load*

Dead load per m run is estimated as below:

S. No.	Item	Details	Weight kN
1.	Wearing course	2.5 × 0.075 × 22	4.12
2.	Deck slab	2.5 × 0.215 × 24	12.90
3.	T-rib	0.3 × 1.35 × 24	9.72
4.	Fillets	2 × 0.5 × 0.30 × 0.15 × 24	1.08
5.	Cross beams (Total weight divided by total length)	(5 × 2.2 × 1.05 × 0.25 × 24)/15.1	4.59
		Total	32.41

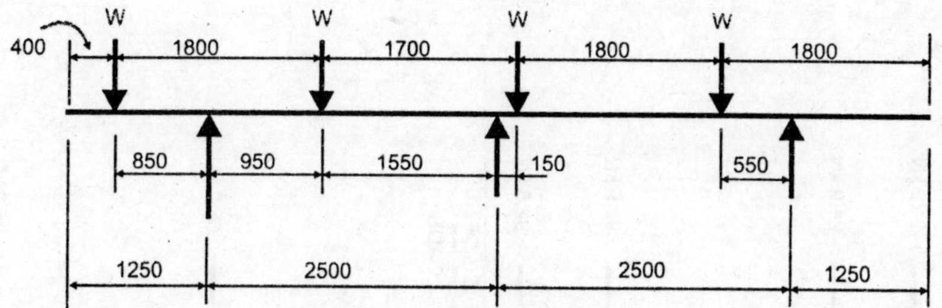

Figure 7.8 Transverse Disposition of Two Trains of Class A Loading for Determination of Reactions on Longitudinal Beams.

$$\text{Maximum B.M.} = \frac{32.41 \times 14.5 \times 14.5}{8} = 852 \text{ kN.m}$$

(iii) *B.M. due to live load*

Maximum live load B.M. would occur under Class A two lane loading.

$$\text{Impact factor fraction} = \frac{4.5}{6 + 14.5} = 0.22$$

The loading is arranged in the transverse direction as shown in Fig. 7.8, allowing the minimum clearance near the left kerb. All the four wheel loads are of equal magnitude.

Live load B.M. can be determined by using any one of the following methods:

(a) Courbon's method
(b) Hendry-Jaegar method
(c) Morice-Little method

The use of the first two methods is detailed here. The third method is outlined in Appendix B. In any practical design, it is adequate to use only one of the above three methods.

(iv) *Live load B.M. - Courbon's method*

The conditions for the applicability of Courbon's method are satisfied. Equation (B.1) is used to determine the reaction factors.

Here, P = 4 W; n = 3; e = 0.7 m

It is assumed that the values of I for all the three girders are equal.

Reaction factor for girder A

$$R_a = \frac{4W}{3}\left[1 + \frac{3I}{2\left(I \times 2.5^2\right)} \times 2.5 \times 0.7\right] = 1.89W$$

Similarly

$$R_b = \frac{4W}{3}(1 + 0) = 1.33W$$

$$R_c = 4W - (1.89 + 1.33)\,W = 0.78\,W$$

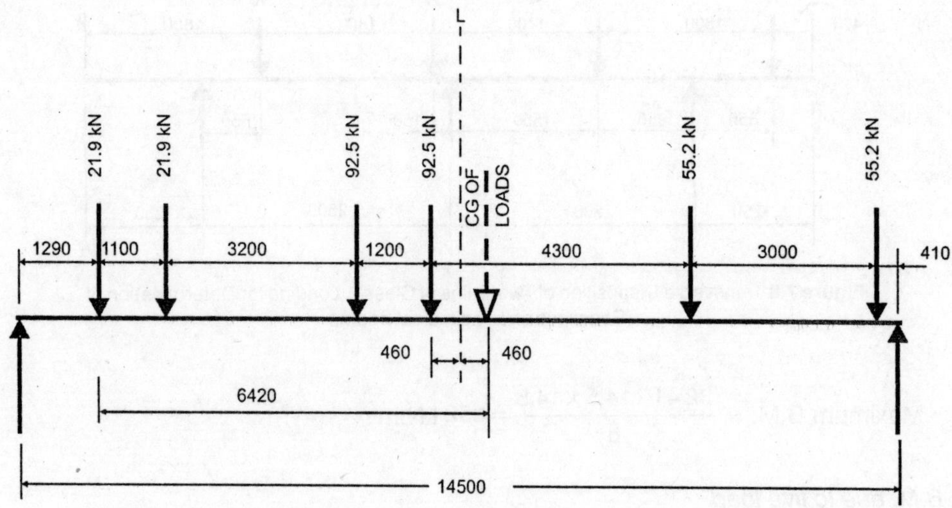

Figure 7.9 Arrangement of Wheel loads in the Longitudinal Direction for Maximum Moment.

In the longitudinal direction, the first six loads of Class A train can be accommodated on the span. The centre of gravity of this load system will be found to be located at a distance of 6.42 m from the first wheel. The loads are arranged on the span as shown in Fig. 7.9, such that the maximum moment will occur under the fourth load from left. The loads shown in the figure are corresponding Class A train loads multiplied by 1.33, the reaction factor at the intermediate beam obtained as above, and further multiplied by the impact factor of 1.22. For example, the first load of 21.9 kN is the product of the first train load of 13.5 kN and the factors 1.33 and 1.22.

For the conditions shown in Fig. 7.9, the maximum bending moment occurring under the fourth load from left is found to be 750.4 kN.m.

(v) *Live load B.M. - Hendry-Jaegar method*

To evaluate the parameter A by Equation (B.2), the moment of inertia of transversals I_t and longitudinal girders I should be computed. The sections of intermediate longitudinal girder and intermediate cross beam can be taken from Figs. 7.2 and 7.3 assuming the top flange width to be the spacing of the respective beam.

Values of moments of inertia about the neutral axis (taken as axis through centre of gravity of gross section) will be obtained as below:

M.I. of longitudinal girder $I = 0.4347 \text{ m}^4$

M.I. of cross beams $I_t = 0.1229 \text{ m}^4$

From Equation (B.2),

$$\text{Parameter A} = \frac{12}{\pi^4}\left(\frac{14.5}{2.5}\right)^3 \frac{5 \times E \times 0.1229}{E \times 0.4347} = 34$$

For the purpose of design, coefficients for F = ∞ will be adopted. Using Figs. B.3 and B.4, the distribution coefficients for A = 34 will be obtained as in Table 7.1.

Table 7.1. Distribution Coefficients.

Unit load on girder	Distribution Coefficients for Girder A	Girder B	Girder C
A	0.40	0.32	0.28
B	0.32	0.36	0.32
C	0.28	0.32	0.40

Treating the deck slab as continuous in the transverse direction, the support moments at the locations of the three longitudinal girders due to the loading shown in Fig. 7.8 are computed using the method of moment distribution. The moments at A, B and C will be found to be – 0.85 W, – 0.19 W and 0, respectively. Reactions R_A, R_B and R_C are determined from the support moments.

$$R_A = 1.884\ W$$
$$R_B = 1.352\ W$$
$$R_C = 0.764\ W$$

These reactions are treated as loads on the interconnected girder system and multiplying these by the respective distribution coefficients and adding the results under each girder, the final reaction at each girder is obtained as shown in Table 7.2.

Maximum bending moment on the intermediate beam is obtained by proportion from the value computed by Courbon's method.

$$\text{Maximum B.M.} = \frac{750.4}{1.33} \times 1.334 = 753 \text{ kN.m}$$

Table 7.2 Determination of Reaction Factors.

	Load	Reaction Girder *A*	Girder *B*	Girder *C*
1.	1.884 W on girder A	0.40 × 1.884 *W* = 0.754 *W*	0.32 × 1.884 *W* = 0.603 *W*	0.28 × 1.884 *W* = 0.528 *W*
2.	1.352 W on girder B	0.32 × 1.352 *W* = 0.433 *W*	0.36 × 1.352 *W* = 0.487 *W*	0.32 × 1.352 *W* = 0.433 *W*
3.	0.764 W on girder C	0.28 × 0.764 *W* = 0.213 *W*	0.32 × 0.764 *W* = 0.244 *W*	0.40 × 0.764 *W* = 0.305 *W*
	Net reaction	1.400 *W*	1.334 *W*	1.266 W

(vi) *Design maximum B.M.*

Live load B.M. obtained from Hendry-Jaegar method will be adopted.

Design B.M. = Moment due to D.L. + Moment due to L.L.

= 852 + 753 = 1605 kN.m

(vii) *Design of section*

Effective flange width for the T-beam section will be determined as per Clause 305.15.2 of IRC Bridge Code.

Effective flange width = thickness of web + 0.2 x 0.7 x effective span

= 0.3 + 0.14 x 14.5 = 2.33 m

Allowing a distance of 120 mm from the bottom of T-beam to the centre of gravity of rods, and assuming the centre of compression to be 120 mm below the top, and allowing a stress of 180 MPa as permissible stress at the centre of gravity of the steel area, area of steel required

$$A_s = \frac{1605 \times 10^6}{180 \times (1575 - 120 - 120)} = 6679 \text{ mm}^2$$

Provide 12 bars of 28 mm dia. in three rows of four bars each.

A_s provided = 7389 mm

For the arrangement of bars adopted, the actual stresses in the extreme steel fibre and the extreme concrete fibre should be computed and checked to be within permissible values.

In this case, the neutral axis is located at 122 mm below the top. The c.g. of reinforcement is at 120 mm above the bottom of rib.

Effective depth = 1455 mm

The actual stresses will be found to be within limits.

Assuming the stress at bottom row of steel to be 200 MPa, stresses at other rows are found by proportion of the distance from neutral axis.

Stress at middle row = 191 MPa

Stress at top row = 182 MPa

Resisting moment of each 28 mm diameter bar at bottom row

$$= 615 \times 200 \left(1455 - \frac{122}{3}\right) = 173.6 \text{ kN.m}$$

R.M. of each 28 mm bar at middle row = 165.8 kN.m

R.M. of each 28 mm bar top row = 158.0 kN.m

R.M. at centre of span = 4 (173.6 + 165.8 + 158.0) = 1990 kN.m

The beam is divided into 6 equal parts and the values of B.M. and shear are calculated at each section using influence lines. The values are tabulated in Table 7.3.

Table 7.3 Moments and Shears at Different Sections.

Section No.	Distance from support m	Maximum moment, kN.m			Max. shear, kN			Required spacing of 4-10 ϕ stirrups mm
		D.L.	L.L.	Total	D.L.	L.L.	Total	
1.	0	0	0	0	223	235	458	120
2.	2.4	446	341	787	149	180	329	120
3.	4.8	815	392	1207	75	128	203	135
4.	7.25	987	439	1426	0	77	77	200

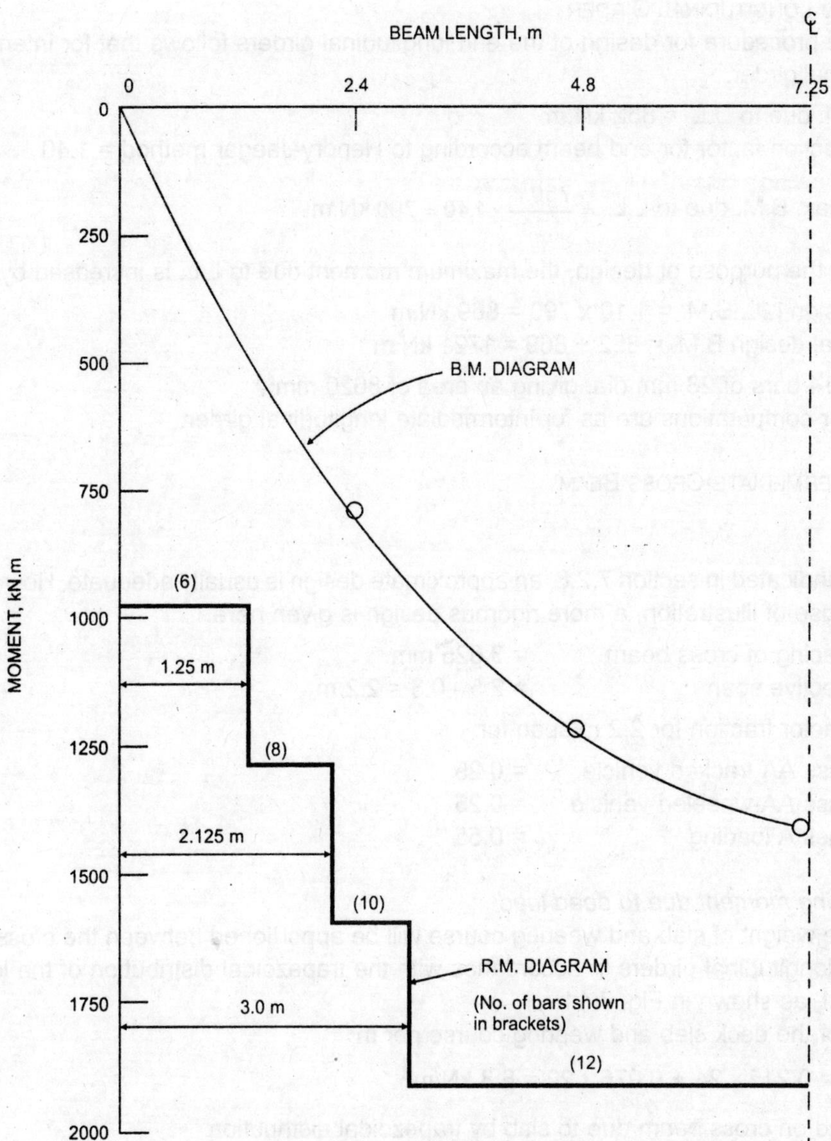

Figure 7.10 Bar Curtailment Diagram for Intermediate T-beam.

Tensile reinforcement bars may be curtailed as the moment reduces. An extension of at least 12 diameters or 300 mm whichever is greater should be provided beyond the points of theoretical cut off. The curtailed bars should be bent up and anchored in the compression zone. The bar curtailment diagram is shown in Fig. 7.10.

Web reinforcement should be provided to cater for shear, in accordance with the rules for shear reinforcement. In this case, 10 mm dia. 4-legged stirrups are provided and the required spacings are shown in Table 7.3.

7.3.5 End Longitudinal Girder

The procedure for design of the end longitudinal girders follows that for intermediate longitudinal girder.

B.M. due to D.L. = 852 kN.m

Reaction factor for end beam according to Hendry-Jaegar method = 1.40

Hence max. B.M. due to L.L. $= \frac{750.4}{1.33} \times 1.40 = 790$ kN.m

For the purpose of design, the maximum moment due to L.L. is increased by 10%.

Design L.L. B.M. = 1.10 x 790 = 869 kN.m

Total design B.M. = 852 + 869 = 1721 kN.m

Provide 14 bars of 28 mm dia. giving an area of 8620 mm².

The other computations are as for intermediate longitudinal girder.

7.3.6 Intermediate Cross Beam

(i) *Data*

As indicated in section 7.2.8, an approximate design is usually adequate. However, for the purpose of illustration, a more rigorous design is given here.

Spacing of cross beam = 3.625 mm

Effective span = 2.5 - 0.3 = 2.2 m

Impact factor fraction for 2.2 m span for

Class AA-tracked vehicle = 0.25

Class AA-wheeled vehicle = 0.25

Class A loading = 0.55

(ii) *Bending moment due to dead load*

The weight of slab and wearing course will be apportioned between the cross beams and the longitudinal girders in accordance with the trapezoidal distribution of the loads on the panel, as shown in Fig. 7.11.

Weight of the deck slab and wearing course per m^2

$$= 0.215 \times 24 + 0.075 \times 22 = 6.8 \text{ kN/m}^2$$

Total load on cross beam due to slab by trapezoidal distribution

$$= 2 \times 0.5 \times 2.2 \times (2.2/2) \times 6.8 = 16.5 \text{ kN}$$

Self weight of cross beam and weight of wearing course over the cross beam

$$= 2.2 \times 0.25 \times 1.25 \times 24 + 2.2 \times 0.25 \times 0.075 \times 22 = 17.4 \text{ kN}$$

Total dead load on cross beam in one span = 33.9 kN

The cross beam is continuous over two spans. The exterior girders restrain the cross beam at the ends, and at the middle girder, the beam approaches a fixed condition. The exact degree of restraint at the girder locations is difficult to determine, and is somewhere

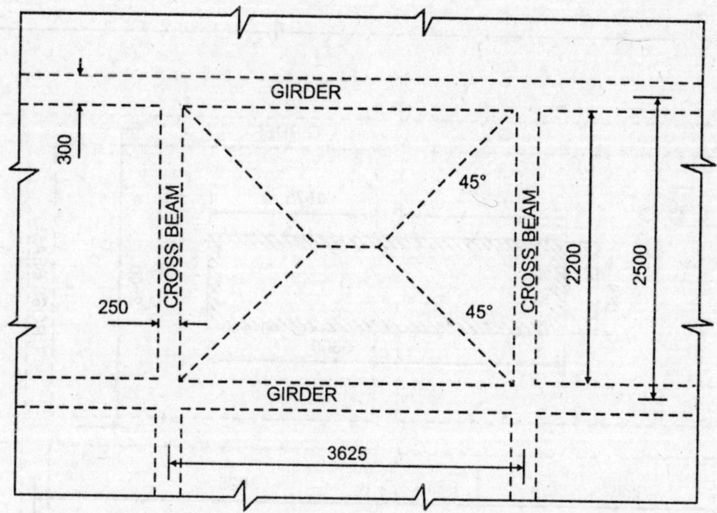

Figure 7.11 Deck Panel Showing Trapezoidal Distribution of Dead Load.

intermediate between the free and fixed condition. Moment coefficients are listed in standard texts[6] for free and fixed ends and for uniform and concentrated loads, for multiple spans. Weighted coefficients are chosen as the sum of one-third of the value corresponding to the case of two-span continuous beam with free ends and two-thirds of the value corresponding to a single fixed ended span.

Coefficient for maximum positive bending moment

$$= (1/3) \times 0.07 + (2/3) \times 0.042 = 0.051$$

Coefficient for maximum negative bending moment

$$= (1/3) \times 0.125 + (2/3) \times 0.083 = 0.097$$

Positive B.M. $= 0.051 \times 33.9 \times 2.2 \times 1 = 3.8$ kN.m

Negative B.M. $= 0.097 \times 33.9 \times 2.2 \times 1 = 7.2$ kN.m

(iii) *Bending moment due to live load*

Class AA tracked vehicle produces severer effect than the other loadings. Fig. 7.12 shows the disposition of one track on a cross beam.

$$\text{Load on cross beam} = 2\left[350 \times \frac{1.675}{3.6} \times \frac{2.662}{3.625}\right] + 350 \times \frac{0.25}{3.6} = 263 \text{ kN}$$

Coefficient of maximum positive B.M. due to concentrated load

$$= (1/3) \times 0.203 + (2/3) \times 0.125 = 0.151$$

Coefficient of maximum negative B.M. due to concentrated load

$$= (1/3) \times 0.188 \times (2/3) \times 0.125 = 0.146$$

Positive B.M. including impact $= 0.151 \times 263 \times 2.2 \times 1.25 = 109$ kN.m

Negative B.M. including impact $= 0.146 \times 263 \times 2.2 \times 1.25 = 106$ kN.m

(iv) *Design of section*

Design positive B.M. $= 3.8 + 109 = 112.8$ kN.m

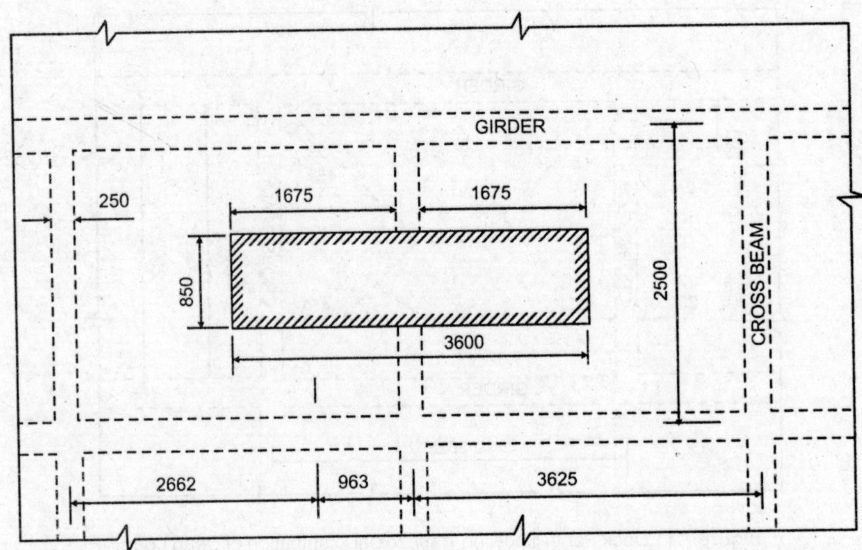

Figure 7.12 Disposition of Class AA Track for Maximum Bending Moment on Cross Beam.

Effective depth $= 1275 - 73 = 1202$ mm

Area of steel required $= \dfrac{112.8 \times 10^6}{200 \times 0.90 \times 1202} = 521\ \text{mm}^2$

Add 0.3% of area of the beam to give additional stiffness to the beam.

Additional area of steel $= \dfrac{0.3}{100} \times 250 \times 1202 = 901\ \text{mm}^2$

Total area of the steel required $= 521 + 901 = 1422\ \text{mm}^2$

Provide 6 bars of 22 ϕ giving an area of 2281 mm²

Design negative B.M. = 7.2 + 106 = 113.2 kN.m

Area of steel required $= \dfrac{113.2 \times 10^6}{200 \times 0.90 \times 1202} = 523\ \text{mm}^2$

Provide 3 bars of 22 ϕ giving an area of 1140 mm²

(v) *Provision for shear*

Arrange Class AA tracked vehicle as shown in Fig. 7.13.

Total load on cross beam due to track as found in (iii) = 263 kN

Shear near girder *C* $= 263 \times \dfrac{2.0}{2.2} + \dfrac{263 \times 0.375}{0.85} \times \dfrac{0.375}{2 \times 2.2} = 249$ kN

Shear including impact of 25% = 311 kN

Shear due to dead load = 17 kN

(computation not shown here)

Total shear = 328 kN

Shear stress $= \dfrac{328 \times 1000}{250 \times 0.90 \times 1202} = 1.2$ MPa

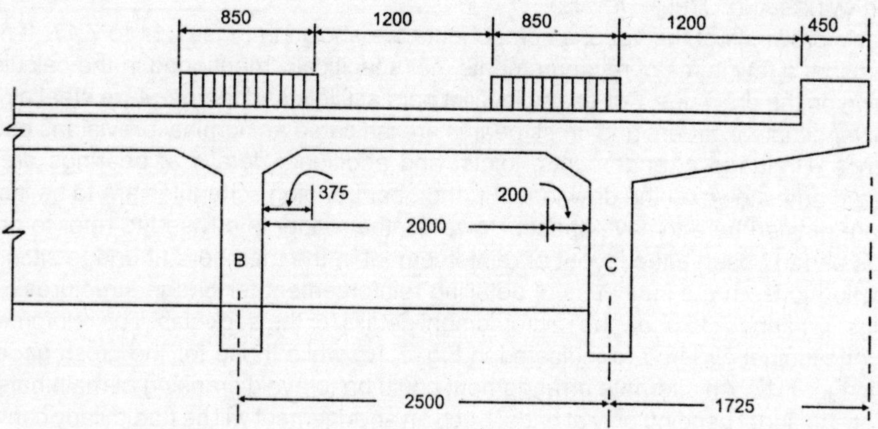

Figure 7.13 Disposition of Class AA tracked vehicle for maximum shear on cross beam.

4-legged stirrups of 10 mm diameter rods at spacing of 150 mm will be adequate.
Two bars of 22ϕ are provided as cranked bars as additional provision to cater to diagonal tension.

(vi) *By nominal provisions*

For comparison, the nominal provisions indicated in section 7.2.8 may be tried.

$$\text{Area of steel required at bottom} = \frac{0.5}{100} \times 250 \times 1275 = 1594 \text{ mm}^2$$

$$\text{Area of steel required at top} = \frac{0.25}{100} \times 250 \times 1275 = 797 \text{ mm}^2$$

Use 6-22ϕ at bottom and 3-22ϕ at top.
Nominal shear reinforcement: 4-legged stirrups 10 ϕ 150
These provisions are same as those derived in (iv) and (v) above.

7.3.7 End Cross Beam

The design of the end cross beam can be done on the same lines as for the intermediate cross beam. The depth of the end cross beam is adopted the same as for the longitudinal girder for 1000 mm at the supports and as for the intermediate cross beam for the remaining portion.

The bottom reinforcement at midspan of the cross beam between two longitudinal girders may be taken as half of that for the intermediate cross beam, i.e. three bars of 22 mm diameter, in addition to the two cranked bars of 22 mm diameter. The top reinforcement may be taken as three bars of 22 mm diameter, plus two cranked bars of 22 mm diameter.

The locations of four jacks for lifting provision are indicated on the end cross beams. At each of these locations, two layers of 6 mm mesh reinforcement of suitable size are provided, one layer at 20 mm and the other at 100 mm from the concrete surfaces. Similar mesh reinforcement of suitable size in two layers are provided at the supports above the bearings.

7.3.8 Drawings and Other Details

Basic drawings of the various components are shown in Figs. 7.14 to 7.17. It may be observed that a few items of reinforcements, not specifically mentioned in the calculations, are shown on the drawings. For example, fillet bars at fillets and temperature steel along the depth of longitudinal girders and cross beams are indicated as nominal provisions based on experience. Drainage arrangements, joints, and additional details at bearings are to be considered and shown on the drawings. Further, bar bending schedules are to be included. These are omitted here to save space. However, the reader is advised to refer to practical drawings and to observe the layout of reinforcement in the members at bridge sites during construction to learn the subtleties of detailing reinforcement for bridge structures.

Figs. 7.14 and 7.15 show the reinforcement details for the deck slab. The reinforcements for the longitudinal girders are indicated in Fig. 7.16, while those for the cross girders are shown in Fig. 7.17. An alternate arrangement could be to avoid cranking of main bars in the girders and to adopt bending only at ends. Such an arrangement will be particularly convenient for longer spans using 32 ϕ bars. The drawings shown use 28ϕ and 22ϕ bars for which cranking is manageable.

7.4 Hollow Girder Bridges

Reinforced concrete hollow girder bridges are economical in the span range of 25 to 30 m. The closed box shape provides torsional rigidity, and the depth can be varied conveniently along the length as in continuous deck or in balanced cantilever layout. The cross-section can consist of a single cell or can be multi-cellular. The extra torsional stiffness of the section makes this form particularly suitable for grade separations, where the alignment is normally curved in plan. The cells can be rectangular or trapezoidal, the latter being used increasingly in prestressed concrete elevated highway structures. Reinforced concrete hollow girder bridges are currently not favoured.

A typical cross section of a reinforced concrete hollow girder superstructure suitable for two-lane traffic on a National Highway for a simply supported clear span of 30 m is shown in Fig. 7.18. The components of the girder are: (i) the cantilever portion including the kerb; (ii) the top slab carrying the roadway; (iii) the webs, in this case two exterior webs and one central web; and (iv) diaphragms, typically two end diaphragms and three intermediate diaphragms.

The design of the simply supported hollow girder can be performed on similar lines as for a T-beam superstructure with a few modifications. The tensile bars are mainly spread over a larger area in the soffit slab. A few of the rods can be accommodated in the webs as shown in Fig. 7.18. If curtailment of bars is desired, the curtailed portion may be provided with nominal (smaller diameter) bars. In situations of higher labour costs and relatively lower material costs as in developed countries, it may even be desirable to extend the bars straight avoiding addition of different sized bars for part length. The webs are designed to carry the shear. The main bars in higher rows provided in the webs may be cranked as per bar curtailment from bending moment consideration and anchored at the top. Vertical stirrups are provided to cater to the requirements of shear. Usually two-legged stirrups of 12 mm or 16 mm diameter are adopted with variable spacing.

Rectangular openings are provided in the diaphragms to enable removal of formwork from inside the cells after casting. Detailing of reinforcement should ensure that the edges are duly strengthened. It is desirable to provide one access opening of 750 mm diameter in

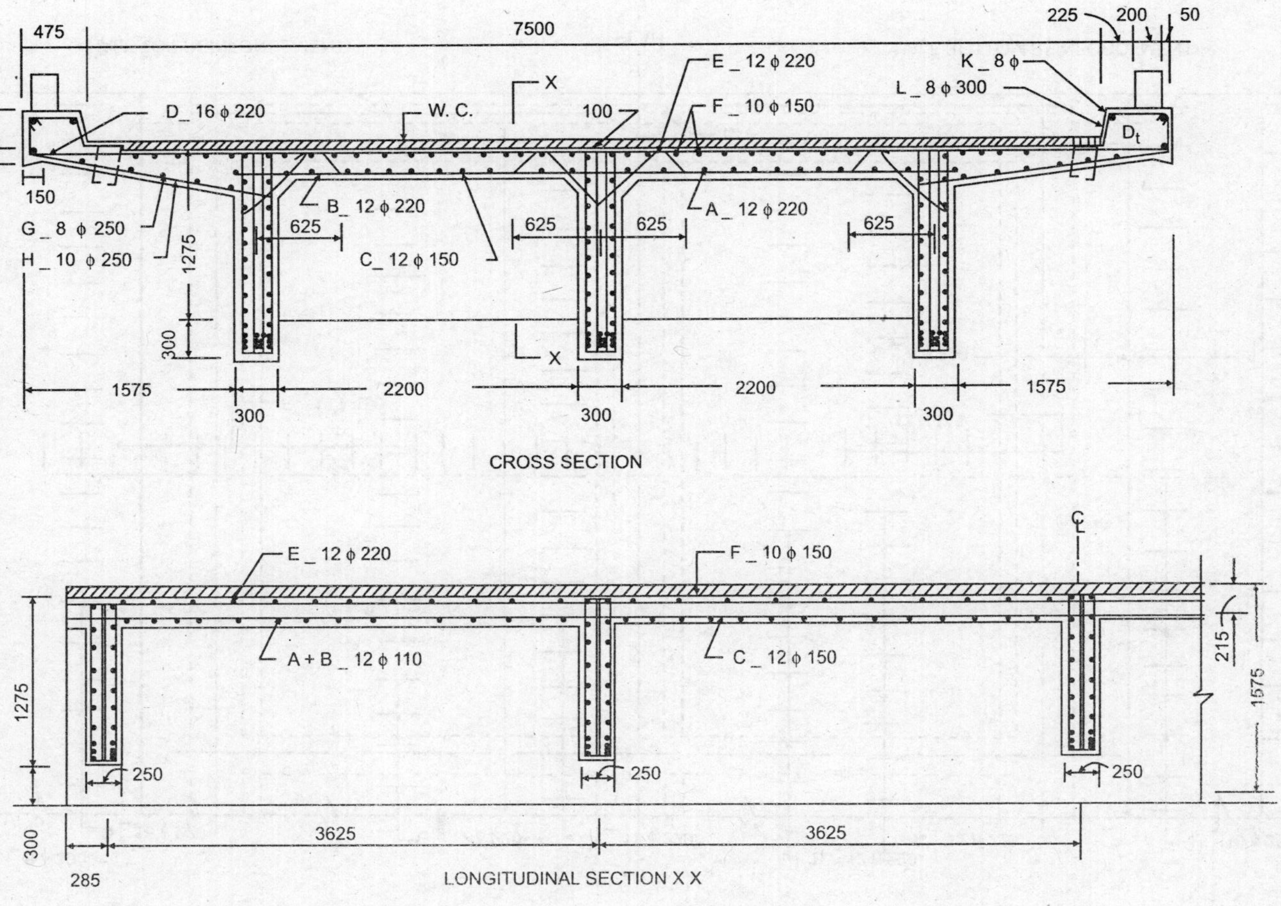

Figure 7.14 T-beam Bridge - Reinforcement Details for Deck Slab.

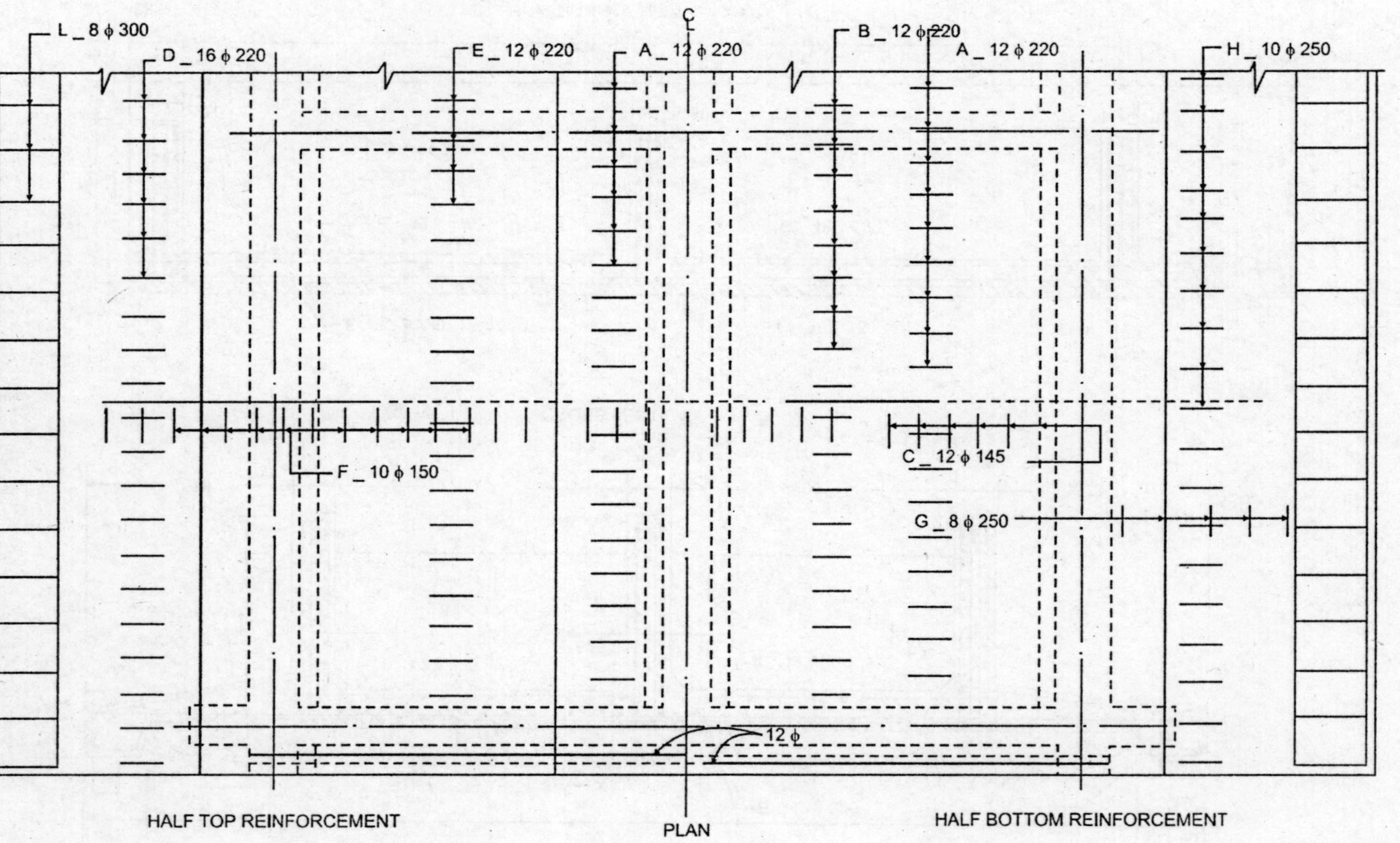

Figure 7.15 T-beam Bridge - Plan of Deck Slab Reinforcement.

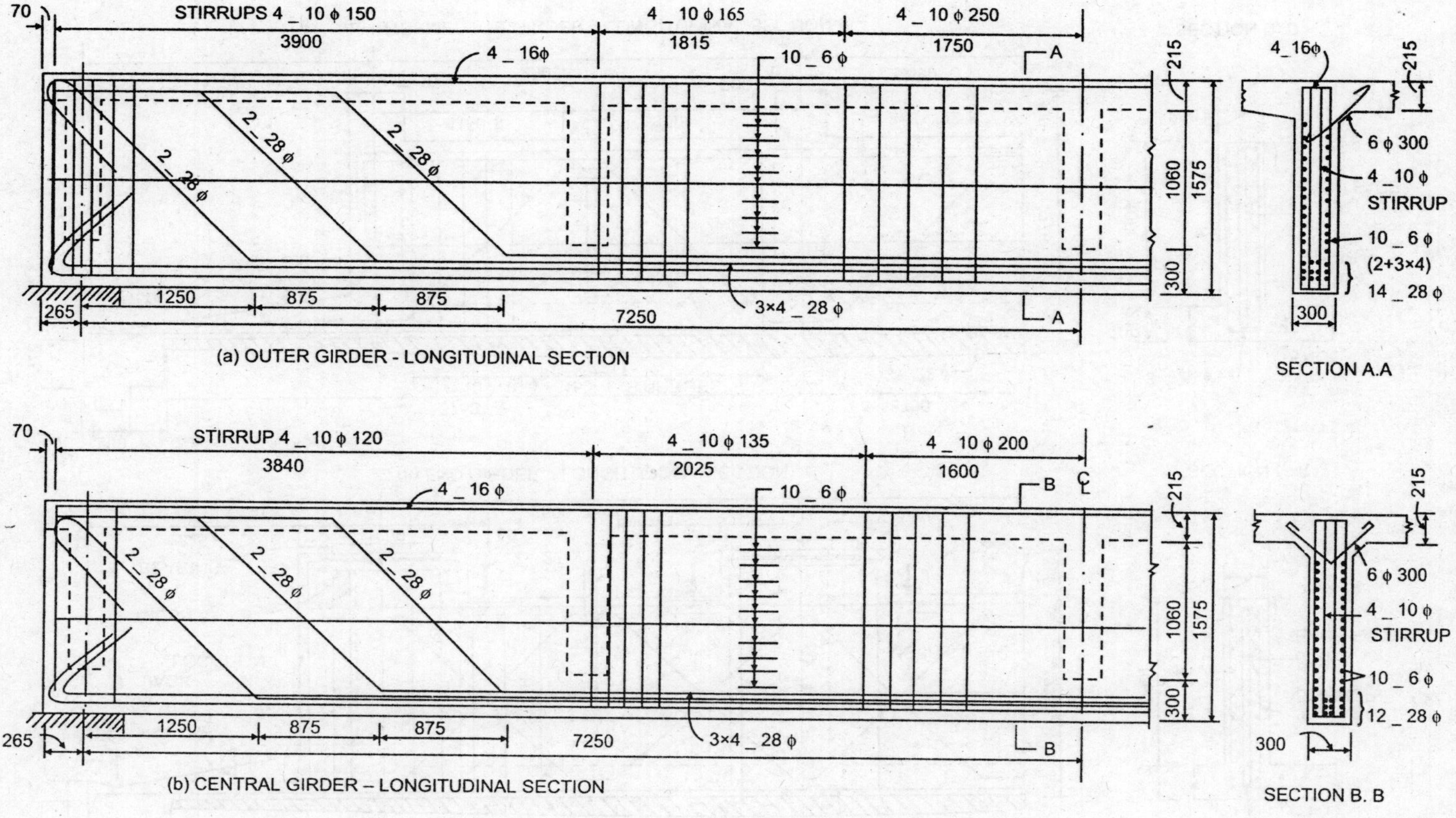

Figure 7.16 T-beam Bridge - Longitudinal Girders.

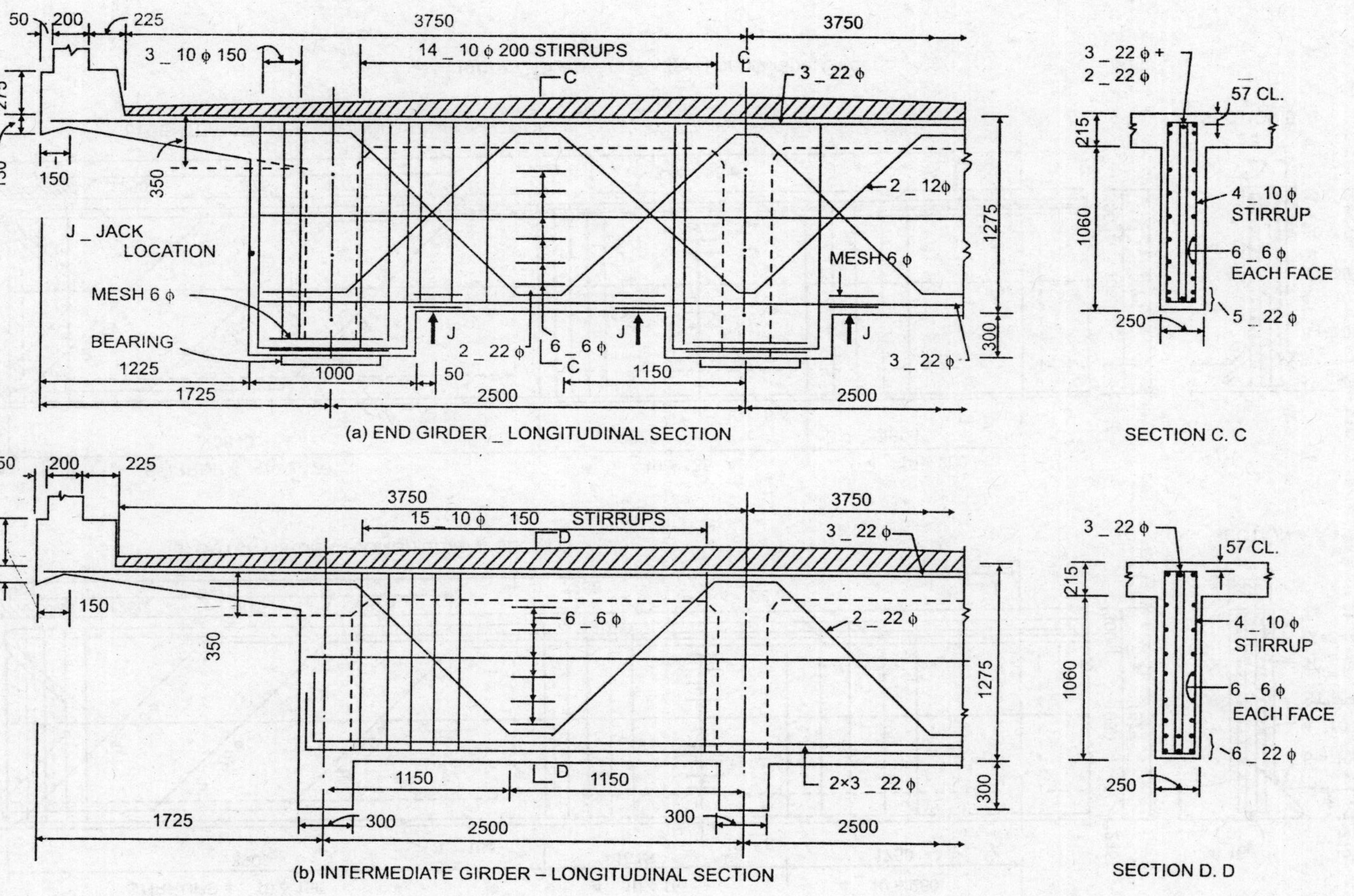

Figure 7.17 T-beam Bridge - Cross Girders.

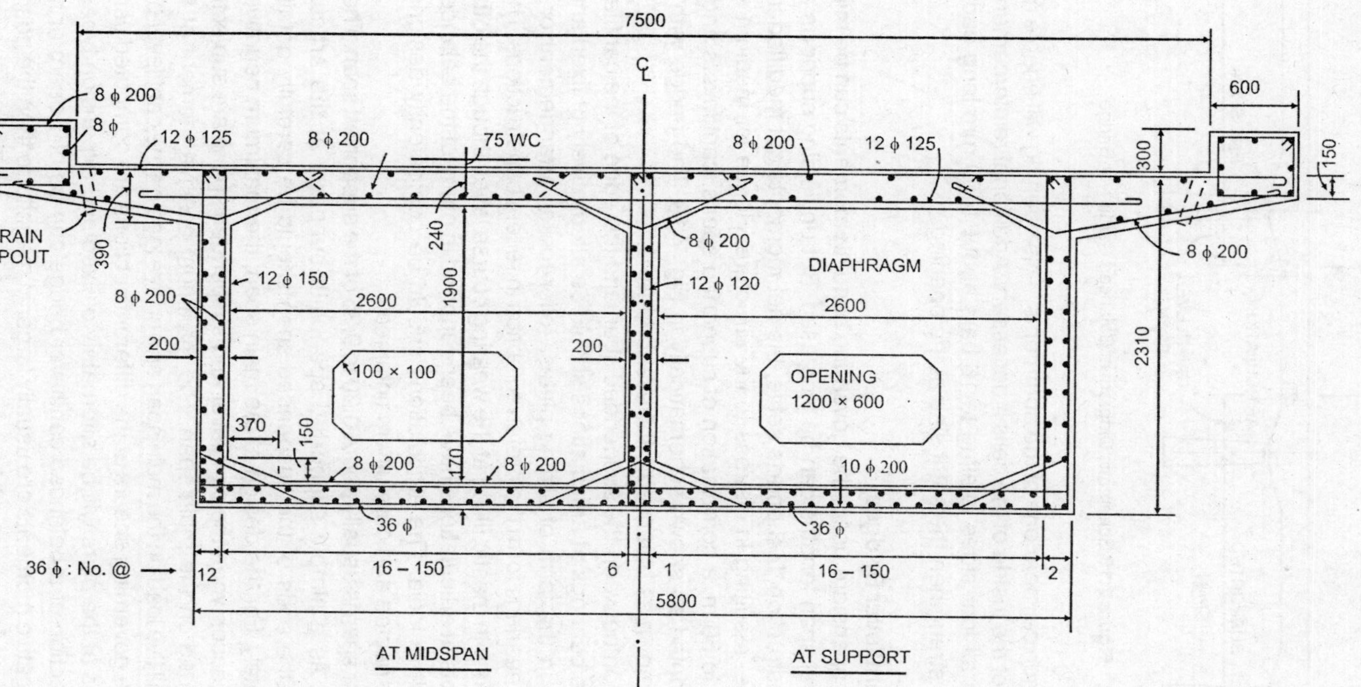

Figure 7.18 Typical Cross Section of a Hollow Girder Deck of Clear Span of 30 m.

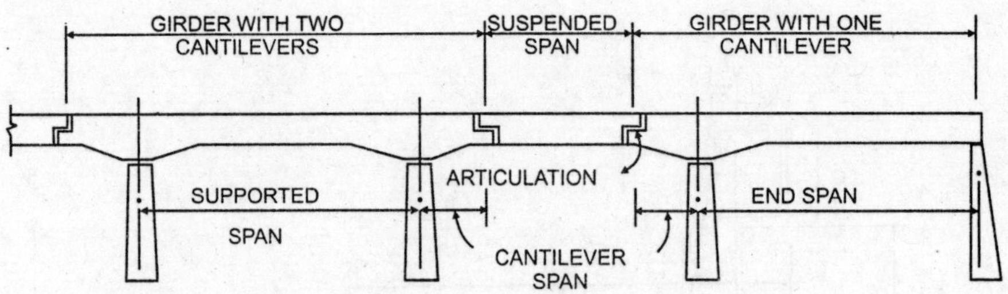

Figure 7.19 Schematic Diagram of Balanced Cantilever Bridge.

the soffit slab for each cell near one of the abutments. This opening will enable maintenance personnel to inspect the inside of the cells if necessary. Additional reinforcement of 2-12ϕ at top and bottom on all four sides totaling to 16 bars about 1400 mm long each should be provided to locally strengthen the soffit slab at the opening.

7.5 Balanced Cantilever Bridges

If continuous spans are used, the governing bending moments can be minimised and hence the individual span lengths can be increased. But unyielding supports are required for continuous construction. If supports settle, the net moments get modified in magnitude as well as in sense, resulting in distress to the structure. Hence for medium spans in the range of about 35 to 60 m, a combination of supported spans, cantilevers and suspended spans may be adopted as shown schematically in Fig. 7.19. The bridge with this type of superstructure is known as balanced cantilever bridge.

The connection between the suspended span and the edge of the cantilever is called an articulation. The bearings at articulations should be alternatively of fixed and expansion types and can be in the form of sliding plates, roller-rocker arrangement or elastomeric pads. Elastomeric bearings form the preferred option in recent constructions. The expansion joint should be filled with mastic filling at the wearing course level, though the other parts can be left open. Defects at articulations have been reported in many of the balanced cantilever bridges so far built in India. The articulations should be competently designed, properly built, frequently inspected and carefully maintained.

The cantilever span is usually about 0.20 to 0.25 of the supported span. The suspended span is designed as a simply supported span with supports at the articulations. The reinforcements at the ends of the suspended span should be carefully detailed so as to carry the shear safely. For the design of the main span, the maximum negative moment at the support would occur when the cantilever and suspended spans are subjected to full live load with no live load on the main span. The maximum positive moment at the midspan would occur with full live load on the main span and no live load on the cantilever or suspended span. Similarly, the governing shears at the different sections are computed using influence lines. The bearings at the piers will be alternately of fixed and expansion types.

The cross section of a balanced cantilever bridge can be of T-beam or hollow girder type. Since the negative moments are usually larger in magnitude than the positive moment at midspan, the depth at support will be greater than at midspan. The soffit can be arranged to be on a parabolic profile or as two inclined lines with a central horizontal line.

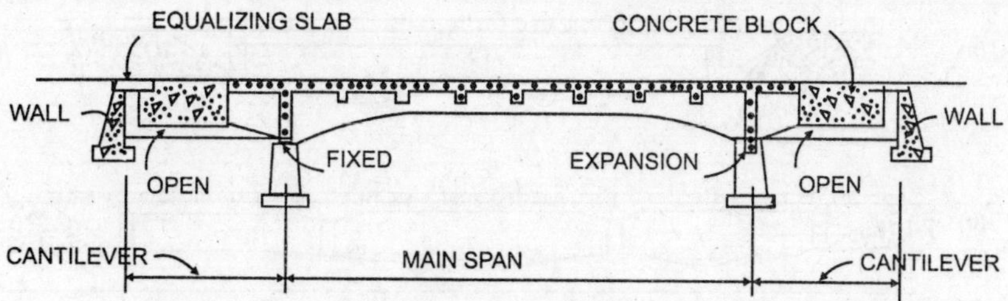

Figure 7.20 Bridge with Counter-weighted Cantilevers - General Arrangement.

A variation of the above scheme, though less commonly adopted, is the case of the bridge with counter-weighted (sometimes concealed) cantilevers as shown in Fig. 7.20. The arrangement is suitable for bridges with one opening where head-room is restricted and where a rigid frame bridge is not convenient. Such situations arise in the case of highway overbridge across railway tracks in plain areas. Applications in Germany have resulted in central depth to main span ratios as low as 1 in 35 with spans up to 45 m. Since this structure is statically determinate with vertical reactions at the supports, relatively light piers can be used. Normally, the depth of girder at the support will be about 2 to 4 times that at the midspan and the ratio of main span to the depth at midspan will be between 15 and 35.

The size and shape of the counter-weighted cantilever are determined after trying several combinations. The usual length of the cantilever varies from one-fifth to one-third of the main span. The cantilever can be in the form of a short and heavy block or a long and relatively lighter block. It is advisable to provide an equalizing slab at either end as shown in Fig. 7.20, so that any vertical movement of the end of the cantilever is smoothened out for traffic at the road level.

The advantages of balanced cantilever designs over simply supported girder designs are as below:

(a) Less concrete, steel and formwork are required for cantilever designs.
(b) The reactions at the piers are vertical and central permitting slender piers.
(c) The cantilever design requires only one bearing at every pier while the simply supported design needs two bearings. Hence the width of the pier can be smaller.
(d) Fewer expansion bearings are needed for the full structure, resulting in lower first cost and maintenance.

A disadvantage of this type of structure is that it requires a little more skill on the part of the designer and a more elaborate detailing of the reinforcements. The half-joint arrangement for cantilever and drop-in span decks present a difficult situation for access for inspection. Rundown and ponding of stormwater (salt-laden in situations using de-icing salts) tend to hasten corrosion in the members. Hence current design practice discourages the use of balanced cantilever spans from the point of view of durability.

7.6 Continuous Girder Bridges

Continuous girder bridges, not connected monolithically to supports, are suitable when unyielding supports are available. Typical shapes are shown in Fig. 7.21. The spans can be equal, but usually the end spans are made about 16 to 20% smaller than the intermediate spans. The decking can be in the form of slab, T-beam or box section. The bending moments

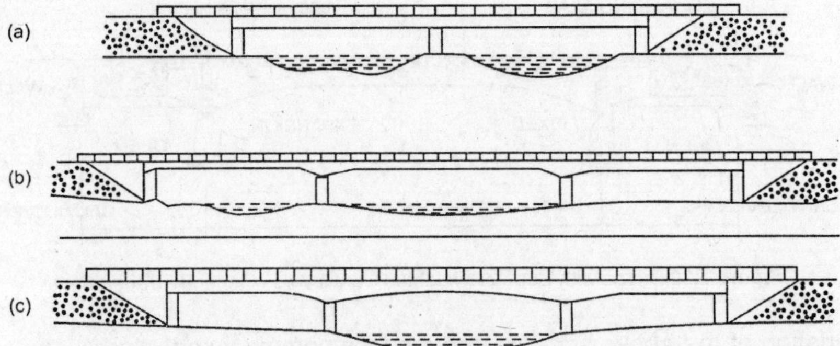

Figure 7.21 Typical Continuous Reinforced Concrete Bridges.

and shears at the various sections of the bridge are evaluated by using influence lines. The negative bending moment at the support induces tension at top and hence renders the top slab of a T-beam bridge ineffective as the flange for a T-beam. Further, the bending moment will in general be larger at the support than at midspan. In view of the above, a haunched profile or a curved soffit is normally used, and the section at the support is strengthened with compression reinforcement besides provision of thickened webs and a cross beam. In the case of continuous slab bridges, the thickness of slab at support will be approximately 1.3 to 1.8 times the minimum thickness at midspan and the length of haunched portion will be about 0.20 to 0.25 of the span. All but one of the bearings will be of the expansion type.

Continuous girder bridges have the following advantages over simply supported girder bridges:

(a) The depth of decking at midspan will be much smaller. This is particularly important in the case of overbridges where the headroom available is generally restricted.
(b) As a corollary to the above, the quantities of steel and concrete will be less, resulting in reduced cost. Also reduced depth of deck leads to decrease in cost of approach ramps and earthwork.
(c) Fewer bearings are required. At each pier, only one bearing is needed, as against two bearings required for simply supported designs. Hence the piers can be narrower. Although the cost of individual bearings will be higher, the total cost on bearings will be lower.
(d) Fewer expansion joints will be required. For a continuous girder design, only two joints are needed at the ends, while the simply supported girder design will require one joint on each abutment and pier. Elimination of joints enhances the riding quality over the bridge.
(e) Since the bearings are placed on the centre lines of the piers, the reactions of the continuous girder are transmitted centrally to the piers.
(f) The continuous girder bridge suffers less vibration and deflection.

The disadvantages of continuous girder designs over the simply supported girder designs may be listed as below:

(a) Uneven settlement of foundations may lead to disaster. Hence this type of structures should not be used in situations where unyielding foundations cannot be obtained at a reasonable cost.

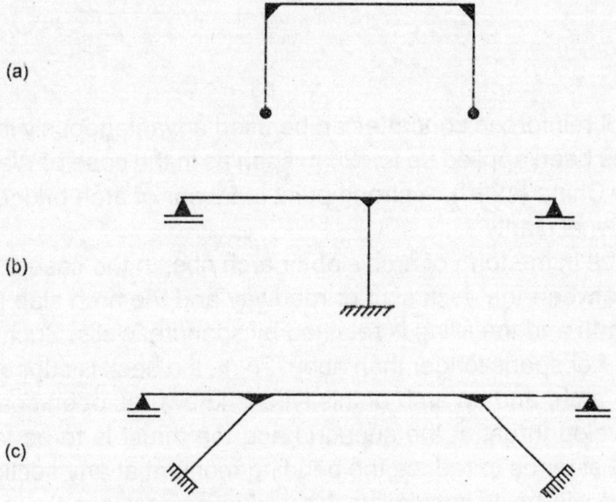

Figure 7.22 Types of Rigid Frame Bridges.

(b) The detailing and placing of reinforcements need extra care.
(c) The sequence of placing concrete and the sequence of removing formwork have to be carefully planned.
(d) Being statically indeterminate, the design is more complicated than simple beams.

7.7 Rigid Frame Bridges

Rigid frame bridges are structures consisting of a number of parallel girders (or slab instead of girders) which are rigidly connected to the supporting columns or piers. Usually the decking and substructure are cast monolithically.

The types of rigid frames normally used in bridge construction are shown in Fig. 7.22. The arrangement of Type (a) is suitable for single span openings as in the case of bridges over railway tracks. Type (b) shows a two-span bridge with the base of the column fixed. If the base is hinged, which is a more common condition, the column is tapered downward. This type can also be used for bridges with greater number of spans by adding the required number of intermediate columns. Type (c) offers an aesthetically pleasing structure over restricted access highways and has been used extensively over expressways in USA and Germany.

Rigid frames possess the advantages listed for continuous bridges and have the following additional advantages:

(a) No bearings are needed at the supports.
(b) The rigid connections result in more stable supports, than possible with independent piers of comparable dimensions.
(c) In view of the slender dimensions, the supporting piers or columns cause the least obstruction to view for the traffic below the bridge.

7.8 Arch Bridges

7.8.1 General

Arch bridges of reinforced concrete can be used advantageously in the span range of 35 to 200 m, and has been applied up to 420 m span as in the case of Wanxiang bridge over the Yangtze river in China (1996). A strong point in favour of arch bridges is their aesthetic elegance and functional clarity.

The arch can be in the form of arch slab or arch ribs. In the case of slab type for short spans, the space between the deck slab or roadway and the arch slab (called spandrel) is usually filled with earth and the filling is retained by spandrel walls. Such an arch is called a spandrel filled arch. For spans longer than about 25 m, the deck is supported on columns or walls resting on the arch, and an arch of this type is known as open spandrel arch.

All arches develop thrust at the supports and the thrust is to be taken by unyielding abutments. This thrust tends to reduce the bending moment at any section of the arch. The aim of the designer will be to maximize this reduction, so that the arch will have only compression in the section. While it is possible to nearly eliminate bending moments due to dead load by choosing the arch axis to coincide with the thrust line for bending moment, live load will invariably cause net bending moments.

7.8.2 Design Considerations

The arch axis is generally governed by three considerations: (a) span and rise from the road gradient and navigation or traffic clearances below, (b) the economical shape from point of view of saving of materials, and (c) the beauty of the intrados. The most important parameter is the rise-span ratio, the economical value of which lies between 0.30 and 0.20. A large rise reduces the thrust and leads to thinner arch section. Flatter arches lead to more aesthetic structures. The usual profiles adopted in practice are parabolic, segmental and elliptical. Parabolic arches will be admirably suited in rugged country with exposed solid rock for abutments. In plains, and particularly for a spandrel filled arch, a segmental profile may be more satisfactory. Elliptical shape is not much favoured, except in cases where clearance requirements need an almost vertical surface of the soffit near the springing. A parabolic profile is first assumed and the thrust lines are drawn for the dead load and for dead load plus live load. The final profile is adjusted to result in minimum flexural stresses in the arch section.

Arches are designed by trial and error. First, the preliminary dimensions are assumed. For solid ribbed arches, the span-depth ratio is generally in the range of 70 to 80. Influence lines for horizontal thrust and bending moments are constructed using first principles. The resulting stresses are then checked against allowable stresses, and the sections are redesigned, if necessary. The main characteristics of the three types of arches are discussed below. For a detailed discussion of the analysis of arches, the reader may refer to standard treatises in structural analysis.

For arches of large spans, the arch cross section is typically of box type with two or three cells. The arch width and depth are chosen from stability considerations. For ease in construction, the outer dimensions of the arch cross section are kept constant. The diaphragms under the spandrel columns are kept vertical.

7.8.3 Types of Arches

Three types of arches are used: (a) the fixed arch, (b) the two-hinged arch, and (c) the three-hinged arch. A variation of the two-hinged arch bridge is the bowstring girder bridge, in which a tie member connecting the two ends of the arch carries the horizontal thrust from the arch and permits the reactions to be vertical as in the case of girder bridges. The bowstring girder bridge is discussed in section 7.9.

A special type of concrete arch bridges has been developed by Robert Maillart, as exemplified in the Salginatobel bridge (1930), where the integrated structural action of thin arch slabs with monolithically cast stiffening beams results in light arches with light superstructure This model was later followed by Christian Menn in the design of several elegant arch bridges including the Rhine bridge at Reichenau (1964).

A fixed arch is statically indeterminate and its analysis is based on the use of the following assumptions:

(a) The span length remains unaltered.
(b) The abutments remain fixed in position, and there is no vertical displacement or rotation of arch axis at the abutments.

It is therefore evident that settlement, spread or rotation of abutments, as well as temperature change, shrinkage and rib shortening will introduce additional stresses. For a detailed discussion on the analysis of reinforced concrete fixed arches, the reader may refer to Reference 8.

In a two-hinged arch, the thrust line must pass through the two hinges, placed at the springings. This arch is indeterminate to the first degree. Its analysis is based on the following assumptions:

(a) The span length remains unchanged.
(b) The abutments remain in position, but the arch axis may rotate, as the hinge is assumed to be incapable of offering resistance to bending.

Thus rotation and moderate vertical settlement will not affect the stresses. But spread of abutment and temperature change, shrinkage and rib shortening will induce additional stresses.

The horizontal thrust in a two-hinged arch including the effect of rib shortening is given by Equation (7.6).

$$H = \frac{\int \frac{1}{EI} My\,ds}{\int \frac{1}{EI} y^2\,ds + \frac{L}{EA_c}} \qquad \text{... (7.6)}$$

where M = bending moment at a section considering the arch as a simply supported beam
y = rise of the arch at the section
I = moment of inertia of the section
E = modulus of elasticity
L = span of the arch
A_c = area of the arch at the section

The effect of a rise of temperature T on the horizontal thrust can be computed by substituting the numerator of Equation (7.6) by + E α TL, where α is the coefficient of linear expansion. The effect of shrinkage is generally taken as equivalent to that of a fall of temperature of about 8 degrees C, resulting in a negative thrust at the abutments.

The three-hinged arch is statically determinate and its analysis is quite simple. This type is generally chosen in situations where the abutments are not suitable for a fixed arch but where an arch profile would fit the surroundings. Temperature, shrinkage and rib shortening are not considered in design. Under any system of loading, the thrust line passes through the three hinges. For dead load, the parabolic shape may be advantageous as the thrust line will very nearly coincide with the arch axis. However, live load will induce moments. When a bridge spans a gorge with exposed rock faces, the parabolic shape may be more appropriate.

7.8.4 Construction Methods

Concrete arches of moderate spans have been constructed on staging. For spans of 200 m and above, the current trend is to use the free cantilever method. After constructing the springing in situ, spandrel columns/piers are constructed at the springings. These piers are extended by auxiliary steel pylons for facilitating successive cantilevering with stay cables in fan form. The pylons are provided with anchor stays at the land side. The stay cables support the arch segments, which are connected by prestressed steel bars inside the hollow arch box.

The Wanxiang bridge over the Yangtze river, which holds the record as the longest concrete arch span, is actually a steel-composite truss encased in concrete.

7.9 Bow String Girder Bridge

A bow string girder bridge consists of two arches which are tied horizontally at the springings by tie members, so that the reactions at the supports will be only vertical. Thus the structure combines the advantages of two-hinged arches, at the same time eliminating the main difficulty of arch supports, i.e., the horizontal thrust at each support. This type of bridge is normally used in the span range of 30 to 35 m and is applicable in situations where unyielding abutments required for arches are not available and good headroom is required under the bridge adjacent to the abutments. Usually, the girder is supported on pin bearings at one end to permit rotations and roller bearings at the other end to provide for expansion.

The decking is usually between and monolithic with the tie members. Footpaths, if provided, are cantilevered outside the ties. The deck slabs are cast monolithic with cross girders, spanning from tie to tie. Each cross girder would be carried directly by a pair of hangers (also known as suspenders). Kerbs of at least 150 mm height and 300 mm width should be provided to prevent damage to hangers from vehicles. Bracings between the two arches are to be provided, where possible, to cater to the wind load. The bow string girder bridge uses less quantities of materials than an alternate design with a reinforced concrete girder bridge, but it would require more costly formwork. Hence the overall economy is not quite apparent. However, the enhanced aesthetic appearance of a bow string girder bridge merits attention.

For the design of a bowstring girder, preliminary dimensions are first assumed for all the members. These are checked and adjusted after analysis. The analysis follows that for a two-hinged arch. The horizontal thrust is given by Equation (7.7).

$$H = \frac{\int \frac{1}{E_c I} . My\, ds}{\int \frac{1}{E_c I} . y^2 ds + \frac{L}{E_c A_c} + \frac{L}{E_s A_s}} \qquad \text{... (7.7)}$$

where A_s = area of steel reinforcement in the section (both in arch rib and tie)
A_c = area of concrete in the section

Eliminating E_c and writing in summation form, we get Equation (7.8).

$$H = \frac{\Sigma_0{}^L \frac{1}{I} Myds}{\Sigma_0{}^L \frac{1}{I} y^2 ds + \frac{L}{A_c} + \frac{L}{mA_s}} \quad \text{... (7.8)}$$

where m = modular ratio.

The form of Equation (7.8) is convenient for performing computations by treating the arch rib as composed of a number of segments of equal width. Here A_c is substituted by A_{av}, i.e., average area of concrete section.

The bending moment M_x at any section is obtained from Equation (7.9).

$$M_x = M - H.y \quad \text{...(7.9)}$$

This bending moment M_x is assumed to be shared by the arch rib and tie in proportion to their moments of inertia, since the arch and tie are connected together by hangers.

The bending moment taken by the arch section is given by Equation (7.10).

$$M_R = \left\{ \frac{M_x}{I_R + I_T} \right\} . I_R \quad \text{... (7.10)}$$

where M_R = moment taken by arch rib
I_R = moment of inertia of the arch rib
I_T = moment of inertia of the tie.

The tie is designed for a bending moment M_T given by Equation (7.11), the tension due to vertical load, and the tension due to bending moment caused by wind load.

$$M_T = M_x - M_R \quad \text{... (7.11)}$$

The design procedure for a bowstring girder bridge is illustrated with an example giving a brief outline of the design.

7.10 Example of Bow String Girder Bridge

Problem

Design a bow string girder superstructure for a clear span of 30 m for two-lane traffic on a rural section of a National Highway.

Solution

(a) *Data*

Assuming width of bearing at each support to be 1.2 m, the effective span centre to centre of bearings will be 31.2 m. Width of roadway is taken as 7.5 m. Select eleven bays with cross beams and hangers at 2.85 m centres.

For the purpose of this example, M20 grade concrete is selected. Mild steel reinforcement to IS: 432 will be used, with the following permissible tensile stresses: 95

MPa for bracings as stress reversals are possible, 115 MPa for hangers and 125 MPa for other members.

The components to be designed are: (i) deck slab, (ii) cross beams, (iii) suspenders, (iv) braces, (v) arch rib, and (vi) tie beams.

(b) *Deck slab*

Due to monolithic construction, the effective span for design of deck slab in the longitudinal and transverse directions may be assumed as the respective clear spans. The design would follow similar procedure as done under Section 7.3.2. It would be seen that an overall thickness of 220 mm would be adequate. The reinforcements required would be:

For positive moment	: 16ϕ 120
For negative moment at end cross beam	: 16ϕ 240 + 12ϕ 240
At intermediate cross beam	: 16ϕ 240
Distributors	: 12ϕ 200
Nominal temperature steel	: 12ϕ 300 both ways.

(c) *Cross beams*

The cross beams are designed as T-beams supported at the centre lines of suspenders. The procedure to be followed in design is similar to that for T-beam superstructure as in Section 7.3. Typical details of an intermediate cross beam are included in Fig. 7.23, which gives the details of reinforcements for the superstructure.

(d) *Length of segments and heights of hangers*

The rise of the arch is usually between 0.15 to 0.3 of span for an economical design. Here we can adopt a rise/span ratio of 0.2.

Rise = 0.2 x 31.2 = 6.24 m.

Assuming a parabolic shape, the equation of the arch with origin at left springing can be expressed by

$$y = kx\,(L - x)$$

Substituting $y = 0.2\,L$ at $x = 0.5\,L$, and $L = 31.2\ m$, we get

$$y = \frac{x(31.2 - x)}{39}$$

The ordinate at any point along the arch axis can be computed from the above equation. The span is divided into 44 equal parts. The width of each part = 31.2 / 44 = 0.709 m.

A total of 10 hangers are provided symmetrically; thus they occur at the 4th, 8th, 12th....40th stations. The length of each segment along the arch axis is computed from

$$ds = \sqrt{dx^2 + dy^2}$$

$dx = 0.709$ m

dy = difference in ordinates at the two ends of the segment.

The summation of the lengths of 22 segments for one half of the arch axis gives the half length of the arch as 17.14 m. In a complete design, the ordinates and lengths of segments at the various stations will have to be tabulated for use in later computations.

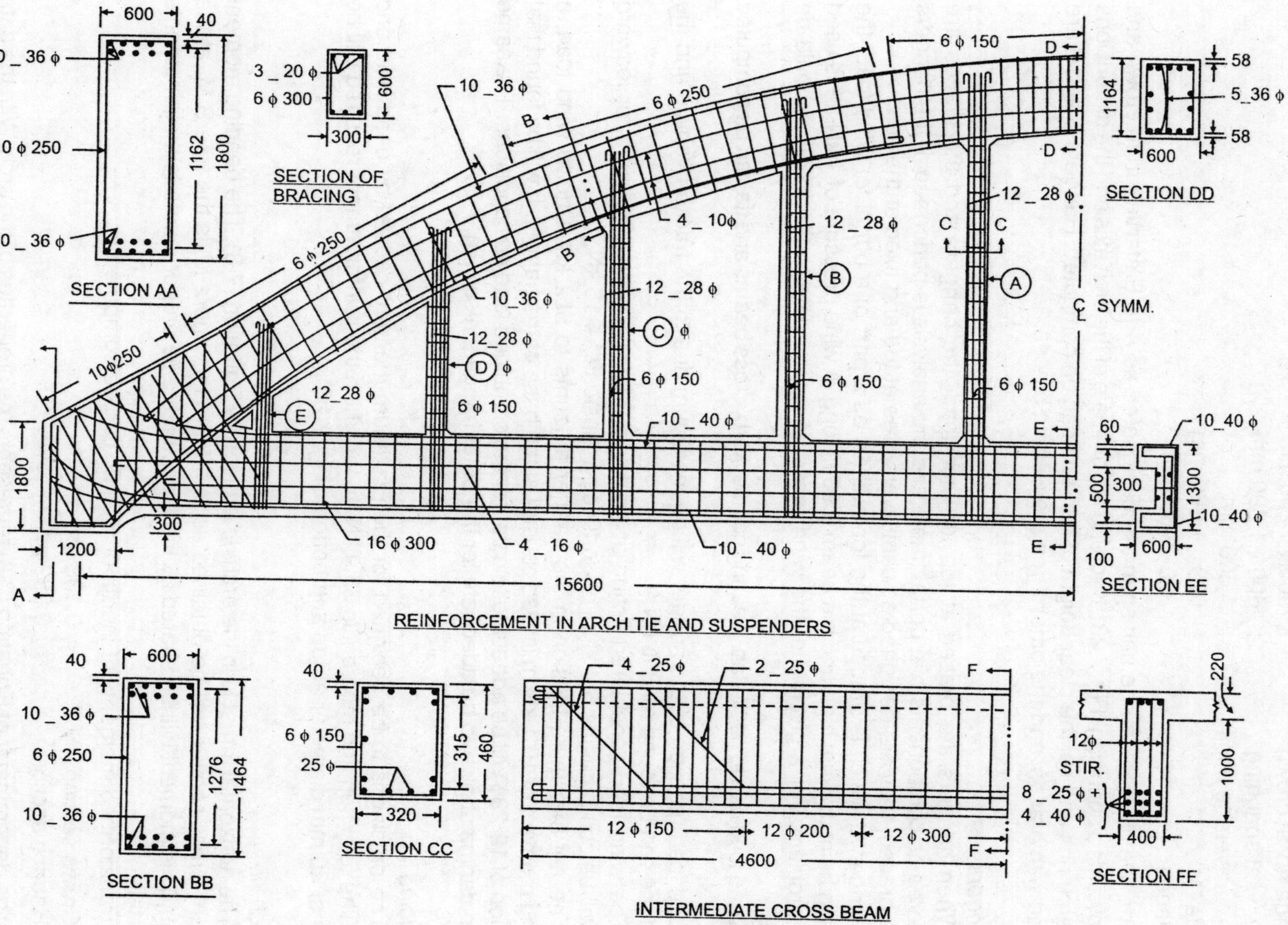

Figure 7.23 Bow String Girder Bridge 30 m Clear Span - Reinforcement Details.

The following sections are assumed for the different members:

Arch rib, at crown	:	600 × 1 200 mm
at springing	:	600 × 1 800 mm
average depth	:	1 500 mm
Tie	:	600 × 1 300 mm
Hanger	:	320 × 460 mm

The five hangers in one half span are designated as A to E starting from crown and going towards a springing (Fig. 7.23). Knowing the ordinate of the arch axis and the dimensions of the arch ribs and tie, the clear height of hangers can be computed. Hangers A and B are provided with braces, and the other three are without braces.

(e) *Hangers*

When hangers are braced at top for wind load, the hangers and the braces are considered as components of a portal frame and the analysis is performed. If the hangers are not braced, they are designed as cantilevers, free at top and fixed at the tie.

Hanger A, which is braced at the top, will act as a part of a portal for wind load, the portal leg being 5.54 m high and the beam 8.56 m long. A wind pressure of 1 kN/m² is used in this problem. For a given site, the wind pressure appropriate to the locality should be used.

Dead load due to deck slab, wearing course, kerb, cross beams and tie can be computed as 192 kN.

Live load reaction of the hanger will be maximum due to the tracked vehicle and the magnitude can be found as 356 kN.

The wind load considered acting at the knee (after allowing 50% extra for leeward side) will be computed as 6.4 kN. This induces a tension of 4.15 kN.

The net maximum tension in the hanger amounts to 552 kN. Due to wind load, a bending moment of 10.1 kN.m is also caused at the base of the hanger. It will be found that provision of 12-28ϕ will be adequate to cater for the combined loading conditions. The same section can be shown to be adequate for the other hangers also (Fig. 7.23).

(f) *Bracing*

The bracing is to be designed for the maximum moment caused by dead load and wind. The total bending moment is 35.9 kN.m, for which the reinforcement required will be 3 bars of 20 mm diameter at top and bottom (Fig. 7.23).

(g) *Arch rib*

The horizontal thrust H in the arch rib is given by Equation (7.8). The bending moment M at any cross section will be obtained as $M = M_f - H.y$, where M_f is the free B.M. at the section. This moment will be shared by the arch and the tie proportionate to their moments of inertia.

Transformed moments of inertia of arch rib can be computed as:

Crown section	:	0.138 m^4
Haunch section	:	0.299 m^4

The denominator of Equation (7.8) represents the properties of the arch. With the data available from complete calculations of the previous steps, it would be possible to obtain the value of the denominator as 34.47 m units. Assuming loads at different sections, conveniently at the locations of hangers, the values for the numerator of Equation (7.8) for the corresponding locations of unit load can be obtained. With these values, the ordinates for influence lines for horizontal thrust can be determined. The final values are shown in Table 7.4.

The influence line ordinates for bending moment at crown of the arch rib are computed by considering a unit load applied at the hanger points. Thus, if the unit load is applied at a distance 'a' from the crown, the ordinate at x of the influence line for moment at the crown will be R(L/2) - 1.a - H.y, where R is the reaction of the unit load on the side of the load, and H is the horizontal thrust, whose value can be obtained from part (a) of Table 7.4. By a similar procedure, the values of ordinates of the bending moment at quarter point (hanger C) are also obtained. These ordinates are given in Table 7.4.

The critical sections and effects to be considered are: section at the crown for horizontal thrust and moment; section at the quarter point for horizontal thrust, shear and moment; and the section at the springing for shear and thrust. Thrust at any point on the arch can be computed as H. cos α - V. sin α, where α is the inclination of the arch with the horizontal, and V is the shear at the section considered.

For the computation of dead load effects, the dead load can be considered to be acting at the locations of the suspenders. Dead load is due to the weight of deck, hanger, part of arch and brace, if any. The effects will be found as below:

B.M. at crown = 362.4 kN.m
B.M. at quarter point = 320.0 kN.m
Horizontal thrust = 1799 kN

For computing the bending moment at crown section, the Class AA tracked vehicle is arranged as in Fig. 7.24.

Induced load at centre of arch axis, allowing for an impact of 14.5%

$$= 700 \times 1.145 \times \frac{5.38}{8.56} = 511 \text{ kN}$$

$$\text{L.L. per m length of girder} = \frac{511}{3.6} = 142 \text{ kN}$$

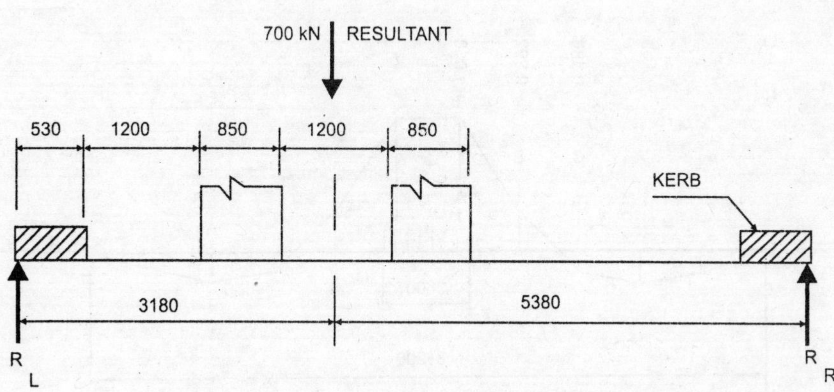

Figure 7.24 Disposition of Class AA Tracked Vehicle for Maximum Moment at Crown Section.

Table 7.4 Influence Line Ordinates for Horizontal Thrust and Moments at Crown and Quarter Point.

	Locations of unit load at hangers									
	E	D	C	B	A	A'	B'	C'	D'	E'
(a) *Horizontal thrust*										
$\Sigma_0^L \frac{1}{I} My\ ds$	886	1714	2452	2973	3346	3346	2973	2252	1714	886
I.L. Ordinate	0.257	0.497	0.712	0.862	0.971	0.971	0.862	0.712	0.497	0.257
(b) *Moment at crown*										
I.L. Ordinate	–0.187	–0.264	–0.186	0.292	1.030	1.030	0.292	–0.186	–0.264	–0.187
(c) *Moment at quarter point*										
I.L. Ordinate	0.784	1.659	2.660	1.130	–0.170	–0.942	–1.182	–1.222	–0.917	–0.469

Applying this load to the influence line diagram as shown in Fig. 7.25, L.L. bending moment = 521.1 kN.m.

Total B.M. at crown = 362.4 + 521.1 = 885.5 kN.m. Similarly, the total thrust at crown can be found as 2204 kN.

The total B.M. as computed above is shared by the arch and tie in proportion to their moments of inertia, since the arch and tie are connected together by hangers.

M.I. of arch rib, I_R = 0.138 m^4
M.I. of tie, I_T = 0.137 m^4

$$\text{B.M. taken by arch} = \frac{M.I_R}{I_{R+}\,I_T}$$

$$= \frac{883.5 \times 0.138}{0.138 + 0.137} = 443\ \text{kN.m}$$

B.M. taken by tie = 440.5 kN.m

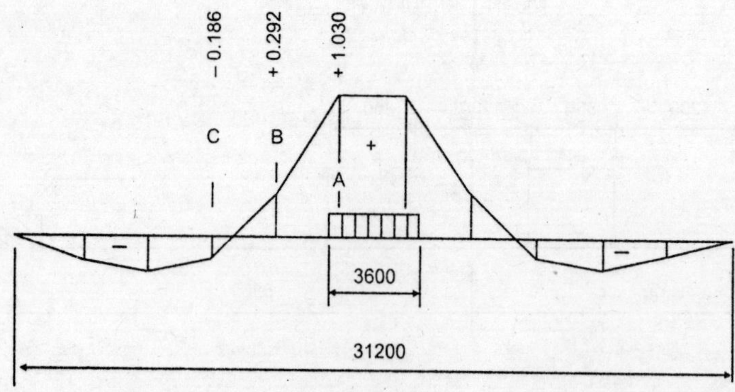

Figure 7.25 Influence Line for Moment at Crown Section with Class AA Tracked Vehicle.

The arch rib is designed for a moment 443 kN.m and a thrust of 2204 kN. The section DD shown in Fig. 7.23 will be adequate.

By a similar procedure, the section at quarter point is to be designed for a B.M. of 955 kN.m and a thrust of 2287 kN and vertical shear of 1037 kN. It can be checked that the section BB as shown in Fig. 7.23 is adequate.

(h) *Tie member*

The tie member is subjected to direct tension due to the horizontal thrust in the arch, bending moment shared with the arch rib as computed above and also stresses due to wind load caused by the ties bending in a horizontal plane.

Horizontal thrust = 2204 kN

B.M. shared with arch rib = 440.5 kN.m

Area exposed to wind, including 50% allowance for the leeward side = 156 m^2 .

Assuming a wind pressure of 1 kN/m^2 over this area, equivalent wind load per m of span = 156 x 1 / 31.2 = 5 kN/m

Length of moving load for Class AA tracked vehicle = 7.25 m. The vehicle is considered to be located with its centre at midspan. Wind load on moving vehicle is 3 kN/m.

Hence the maximum B.M. at centre due to wind load

$$= \frac{5 \times 31.2^2}{8} + \frac{7.25 \times 3}{2}\left(\frac{31.2}{2} - \frac{7.25}{4}\right) = 760.3 \text{ kN.m}$$

Tension in leeward tie beam due to this wind moment = 760.3 / 8.56 = 89 kN

Total tension in tie beam = 2204 + 89 = 2293 kN.

The tie member will then be designed for a bending moment of 440.5 kN.m and a direct tension of 2293 kN. The section CC as shown in Fig. 7.23 will be adequate.

The additional stresses caused in the tie by temperature and shrinkage will be usually very small with properly functioning bearings. They can be normally neglected in the design of the tie member of a bow string girder bridge.

(i) *Detailing*

The principal details have been worked out as above and in a complete design all the requirements of reinforcements would be worked in detail and checked at a number of locations along the various members. In a monolithic structure of this kind, special care is necessary to detail the connections of various members to facilitate proper placement of the reinforcement and the correct sequence of the casting operations. The reinforcement at the junction of the tie and arch rib should be carefully detailed to ensure adequate anchorage for tie reinforcement and for arch rib steel, besides providing sufficient stirrups to prevent shearing off of the arch rib from tie at this junction. Some details of the reinforcement for the bow string girder can be seen from Fig. 7.23.

7.11 References

1. "Standard plans for highway bridges: R.C.C. T-beam and slab superstructure', Ministry of Surface Transport (Roads Wing), New Delhi, 1993, 159 pp.
2. Victor, D.J., and Lakshmanan, N., 'Reactions in three-girder bridge decks', Journal of Indian Roads Congress. Vol. 36-1, Oct. 1975, Paper No. 299, pp. 41- 68.
3. Kostem, C.N., 'Lateral live load distribution in prestressed concrete highway bridges', In"

Yilmaz, C., and Wasti, S.T. (Ed.), 'Analysis and design of bridges', Martinus Nijhoff Publishers, The Hague, 1984, pp. 213-223.

4. Rowe, R.E., 'Concrete bridge design', C.R. Books Ltd., London, 1962, First Edition, 336 pp.
5. Hendry, A.W. and Jaegar, L.G., 'The analysis of grid frameworks and related structures', Chatto and Windus, London, 1958, 308 pp.
6. Morice, P.B., and Little, G., 'Analysis of right bridge decks subjected to abnormal loading', Cement and Concrete Association, London, July 1956, 43 pp.
7. Taylor, F.W., Thompson, S.E., and Smulski, E., 'Reinforced concrete bridges', John Wiley and Sons, New York, 1955, 456 pp.
8. Chettoe, C.S., and Adams, H.C., 'Reinforced concrete bridge design', Chapman and Hall Ltd., London, 1952, 416 pp.

Chapter 8

Prestressed Concrete Bridges

8.1 Types of Prestressed Concrete Bridges

Based on the method of construction, prestressed concrete bridges may be classified into four categories: (a) cast-in-situ bridges; (b) bridges with precast girders; (c) bridges with segmental cantilever construction; and (d) incrementally launched bridges. The method of construction determines the design and computations for the superstructure.

The cast-in-situ method of construction is applicable for short spans below 50 m in locations where centering can be erected without difficulty. Bridges using precast girders placed in position by cranes are suitable for spans in the range of 30 to 40 m, particularly bridges in urban areas requiring least obstruction to traffic below. Segmental cantilever construction enables the elimination of centering and falsework, and is therefore eminently suited for construction across wide and deep valleys involving very tall piers, and for structures requiring uninterrupted passage of traffic or navigation during construction. Incremental launching technique is used for bridges with a series of short spans in the range of 35 to 45 m, and this technique is discussed in Section 15.7.2.

8.2 Types of Prestressing

Basically there are two types of prestressing:

(a) Pretensioning
(b) Post-tensioning.

The application of these types to bridge construction is explained in this Chapter. The selection of a particular type of prestressing method for any given bridge will depend on many factors, including, the proximity and availability of pretensioning plant and equipment, availability with the contractor of the post-tensioning know-how and equipment, size of members, and the number of units of the same type. Pretensioning with straight tendons can usually be economical, while post-tensioning with draped tendons can be more advantageous structurally. The above two advantages can be combined in certain circumstances by using a combination of pretensioned and post-tensioned tendons. The concrete can be either fully prestressed or partially prestressed. In the former case, the longitudinal stresses are always under compression throughout the section. In case of partial prestressing, tensile stresses may be allowed to occur to a limited extent in part of the section under certain loading conditions.

There are several different systems seeking to impart the prestressing force. The systems which are popular in this country include: Freyssinet system, Magnel-Blaton system, Gifford-Udall system and BBRV system. The hardware and exact details pertaining to these systems can be had from the original patent holders or from standard texts on prestressed concrete [1-6]. In this text, it is assumed that the reader is familiar with these systems as well as the basic theory of prestressed concrete. A brief description of prestressed concrete as applicable to bridge design has been included in Section 5.12 of this book.

8.3 Pretensioning

Pretensioning is a method of prestressing in which the steel tendons are tensioned before the concrete has been placed in the moulds. In this technique, the tendons (wires or strands) are tensioned by hydraulic jacks bearing against strong abutments between which the moulds are placed. After the setting and hardening of the concrete, the tendons are released from the tensioning device and the forces in the tendons are transferred to concrete by bond. Mould for pretensioned work should be sufficiently strong and rigid to withstand, without distortion, the effects of placing and compacting concrete, as well as those of prestressing.

Special attention is needed from the point of view of production technology. Steam curing is often used to accelerate the hardening process so that the release of wires may be advanced and the formwork reused. The steam curing cycle has to be evolved at the particular site. A 12-hour cycle has been used at Pamban bridge for accelerated curing of precast girders[7] (post-tensioned in this case). The cycle consisted of 1.5 hours presteaming period, raising of temperatures from atmospheric to 70° C in 2 hours, keeping temperature constant at 70° C for about 6 hours and cooling from 70°C to atmospheric temperature in 2 hours. It was possible to attain a concrete strength of about 35 MPa within 12 hours of concreting with high strength ordinary portland cement. The application of this technique to bridge construction will be economical only if a large number of identical girders are to be cast.

The tendons used in pretensioned girders must be of small diameter, since the transfer of stress from the tendon to the concrete is by bond. For a given cross sectional area, which determines the force possible in a tendon, the bond area available per unit length increases with decrease in diameter. It is preferable to use seven-wire strands (10 to 13 mm nominal diameter) instead of wires as tendons for bridge girders. Where wires are used, they should invariably be of the indented type. The increase in bond resistance for indented wire of 7 mm diameter over a smooth wire of the same diameter is, however, not very significant. Crimping of the wires at the ends (approximately 2 mm crimp with a pitch of 40 mm) will improve the bond characteristics very significantly over the smooth wire. The author has found crimping to be very effective in imparting pre-tension to short beam specimens used for torsion research in the laboratory; there was no detrimental effect on ultimate strength.

It is usual to attempt placing of the prestressing wires or strands spaced closely together in regular grid pattern on a cross section and in a straight profile. Such arrangement would lead to correct positioning at midspan, but unfavourable placement at the supports causing tensile stresses at top. This situation can be got over by providing a few prestressing tendons at the top of the beam.

Deflected strands are employed by manufacturers of precast girders in USA. The use of deflected tendons leads to reduced concrete sections and hence reduced dead load. However, additional investment on the plant is necessary to provide for hold-downs and special equipment for raising the tendons. It is possible to avoid tensile stresses at the top of supports by preventing bond for some of the tendons for a computed length near the ends by covering the tendons with plastic tube or by greasing. But this latter procedure is not generally favoured.

End blocks can be omitted for pretensioned girders, if straight tendons are used. When deflected tendons are used, care should be taken to avoid distinct concentration of the top cables at a distance away from the bottom cables. The occurrence of two large concentrations of prestress at the ends have been known to induce hairline cracks in the end block even with additional mild steel reinforcement[1]. Nominal transverse reinforcement should be provided for a length of 0.4 of depth of girder, the minimum being 0.5 per cent of the cross section of the web.

8.4 Post-tensioning

A member is referred as post-tensioned member, if the tendons are stressed and anchored at each end of the member after the concrete has been cast and has attained sufficient strength to withstand the prestressing force. The post-tensioning method basically requires the following steps: (i) the prestressing tendon is assembled in a flexible metal sheath and anchor fittings are attached to its ends; (ii) the tendon assembly is placed in the form and tied in place, along with other untensioned and auxiliary reinforcement; (iii) concrete is placed in the form and allowed to cure to the specified strength; (iv) tendons are stressed to computed extent and anchored; (v) the space around the tendon within the sheath is grouted under pressure with cement grout; and (vi) anchor fittings are covered with a protective coating. The tendon provides a precompression force to reduce cracking under service load and also serves as tension reinforcement under the ultimate load condition. The integrity of the grout duct and the surrounding concrete governs the corrosion protection of the high-strength, low ductility steel tendon. Grouting also helps to avoid fatigue failure in the steel at the anchorages. A significant part of the prestressing force can be imparted using external tendons, with the duct grouted with grease or petroleum wax to give á soft, flexible filler.

From the point of view of bridge construction, the basic differences between pretensioning and post-tensioning can be listed as below:

(a) Post-tensioning is well suited for prestressing at a construction site without the need for costly factory-type installations.
(b) Cast-in-place structures can be conveniently stressed by post-tensioning, which would not be possible with pretensioning.
(c) With post-tensioning, tendons can have curved trajectories, which lead to structural advantages, particularly for shear resistance.
(d) The need for individual tensioning, special anchorages, sheath and grouting results in a higher unit cost (cost per kN of effective prestressing force) for post-tensioning than for pretensioning.
(e) Many of the post-tensioning devices are covered by patents, restricting the user to purchase materials and equipment from the patent holders. This difficulty is not present in pretensioning.
(f) It is possible to fabricate a beam with a number of precast elements, which are post-tensioned together to form one structural unit.

The prestressing tendon may be of the following types:

(a) Cable with a fixed number of (parallel) wires, e.g., 12-5ϕ or 12-7ϕ in Freyssinet system, 32-7ϕ in Magnel-Blaton system. (Numbers and sizes can be varied within certain ranges).
(b) Seven-wire strands, e.g., Gifford-Udall strand system (13 mm nominal diameter as currently available), Freyssinet system with 13 mm and 15 mm diameter strands.
(c) High tensile bars, e.g., Lee-McCall bars 28ϕ , Dywidag bars 26ϕ.

In order to allow for the necessary movement of the tendons inside the concrete section during tensioning, the tendons are installed in special ducts (also called sheathing) that are made of galvanized steel or polyethylene. The sheathing can be of a flexible galvanized metal hose of interlocking construction having sheet of thickness ranging from 0.20 mm to 0.30 mm and of size from 25 mm to 70 mm. This can be made at site using a light special machine or can also be purchased in coils. Alternatively, the ducts can be of HDPE. The duct must be strong enough to retain its shape during handling. The ducts are tied to untensioned reinforcement to prevent misalignment during placing and vibrating the surrounding concrete. The duct should have a low coefficient of friction with steel tendons and should be impervious to cement paste. The duct area is usually 2 to 2.5 times the prestressing steel area.

Steel forms are invariably used for casting post-tensioned bridge decks. In view of the slender dimensions of the girders and particularly the webs, it is often necessary to use form vibrators in addition to needle vibrators to ensure adequate compaction.

8.5 Pretensioned Prestressed Concrete Bridges

8.5.1 Application

Typical cross sections of bridges with precast pretensioned members are shown in Fig. 8.1. The construction may follow one of the following methods:

(a) Members which are cast in the final shape and just assembled together at site. The solid and hollow precast slab segments shown in Fig. 8.1 (a) and (b) and the beam members shown in Fig. 8.1 (c) relate to this method. The openings shown in Fig. 8.1 (b) may also be rectangular. This method is suitable for short spans.

(b) Beams of inverted T-shape placed side by side and the gaps filled with in-situ concrete to form an integral slab, as in Fig. 8.1 (d). Transverse bond rods are provided just above the bottom flange passing through preformed holes in the precast beams. This method can be used up to a span of about 20 m. In view of the good transverse distribution of load possible in this type of construction, transverse prestressing can be omitted. Other advantages include the elimination of propping and soffit shuttering, use of small cranes or hoisting devices for placing the beams in position, and the economy achieved by the use of relatively small prestressed members in conjunction with larger quantities of concrete cast in place. The sections of the beams have been standardised in UK and USA facilitating factory production.

(c) Composite deck consisting of precast girders placed at definite spacing and connected with cast-in-place diaphragms and deck slab, as in Fig. 8.1 (e). In USA, these bridge girder sections are standardised and are known as AASHTO-PCI standard (Types I to VI) bridge sections. This procedure constitutes one of the most economical types of prestressed concrete bridges. Special attention should be given to the provision of adequate shear connectors.

The first application of pretensioning in highway bridge construction in India was in the reconstruction of Wallajah Bridge and in the widening of Wellington Bridge in Chennai by Tamil Nadu Highways Department. The section adopted is similar to the one shown in Fig. 8.1 (d). In this construction, the long-line method was followed for casting the beams, using 7 mm diameter high-tensile indented wires. Steam curing was utilised to facilitate early

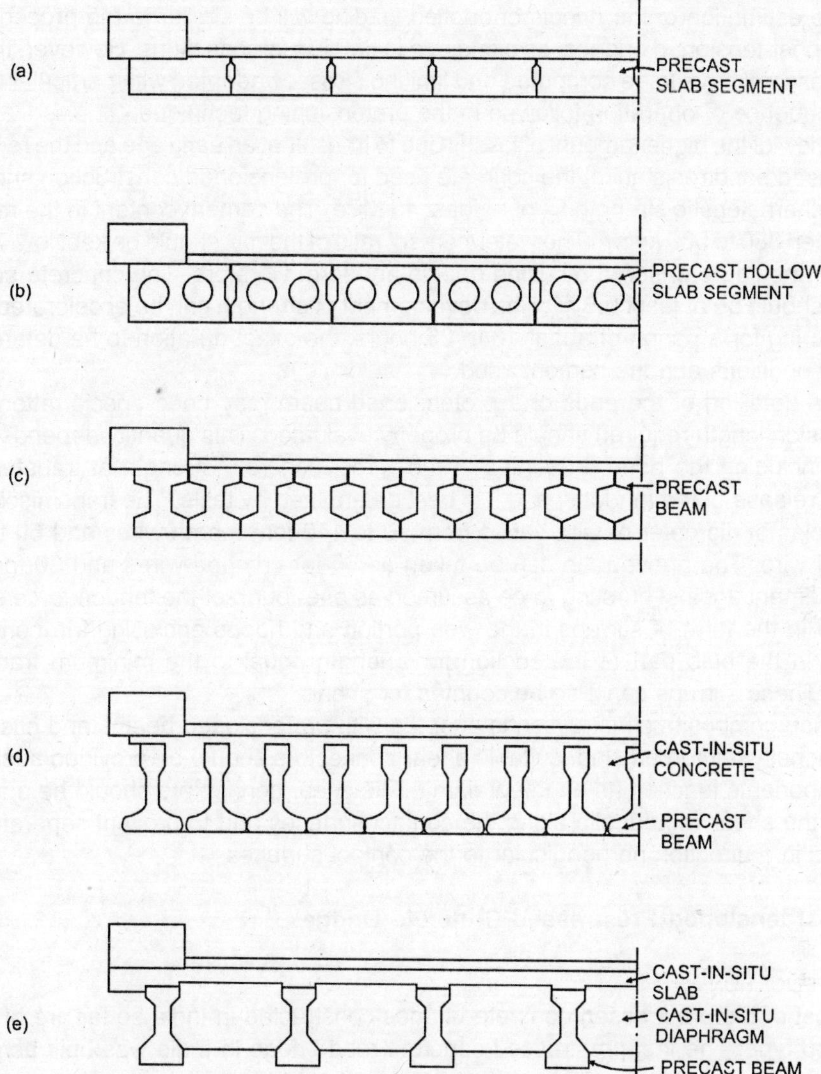

Figure 8.1 Cross Sections of Bridges with Precast Pretensioned Members.

release of tendons and quick turnover for steel moulds. The Delhi Noida Toll Bridge Project has evolved an innovative design of precast pretensioned beams for spans of 12, 15 and 24 m using a common geometrical cross section and varying number of prestressing tendons to suit the span.

8.5.2 Principles of Design

The Indian Roads Congress has not yet formulated a code for the design of pretensioned concrete bridges. The provisions of the Indian Standard IS:13438 can be used to the extent applicable.

The estimation of the effects of applied loading will be similar to the procedure to be used for post-tensioned bridges, as explained in detail in later sections. However, the effects of prestress will have to be computed and limiting stresses adopted with particular reference to the sequence of operation followed in the pretensioning technique.

In view of the higher amount of losses due to transfer at an early age and the requirement of increased bond resistance, the concrete used for pretensioned construction should have 28- day characteristic strength f_{ck} of at least 45 MPa. The cement content in the mix should be between 380 to 530 kg/m³. The water/cement ratio of the mix should be kept low. Adequate vibration should be ensured by using needle and form vibrators. The concrete strength at transfer should be at least 0.5 f_{ck}. The development of strength can be accelerated by using steam curing for a period of about 18 to 25 hours, the exact duration to be determined for the site conditions and the cement used.

The detailing of the ends of a pretensioned beam may need special attention. The transmission length required should be properly evaluated. This quantity depends on many factors, including the bond properties of the wires, concrete strength at transfer and the mode of release of the tendons; and it is best determined by tests. The transmission length as multiples of diameter of wire varies from 70 to 140 for smooth wires and 50 to 100 for indented wire. The above ratio can be taken as 65 for crimped wires and 30 for strands. Reinforcement against bursting force assumed as one-fourth of the tendon force should be provided in the form of stirrups in the web portion and hoops enclosing the concentrated tendons in the bulb part of the section for a length equal to the minimum transmission length[9]. These stirrups can also be counted for shear.

When composite construction is adopted with pretensioned beams and cast-in-place conventionally reinforced slab, suitable shear connectors should be provided such that the two components function as an integral unit. The shear connectors should be adequate to transfer the shear without slip along the contact surfaces and to prevent separation of the elements in a direction perpendicular to the contact surfaces.

8.6 Post-tensioned Prestressed Concrete Bridges

8.6.1 Application

Most of the prestressed concrete bridges constructed in India so far are of the post-tensioned type. The first prestressed concrete road bridge in India was built across Palar river near Chingleput in Tamil Nadu in 1954 using the Magnel-Blaton system, though three railway bridges of smaller span had been built in Assam in 1948. The Palar bridge consisted of 23 spans of 27 m each. Since then, a large number of bridges have been built using different systems of prestressing.

Typical arrangements of bridge decks with post-tensioned girders suitable for simply supported constructions are shown in Fig. 8.2. Basically the arrangements may be one of the following types:

(i) Fully cast-in-situ construction, as in Fig. 8.2 (a), e.g., Ranganadi bridge near Lakshimpur town in Assam consisting of seven spans of 30 m each[10]. This type is suitable when site conditions permit putting up the centering for the deck from the river bed and the river is dry for a major part of the year. It cannot be used if the river is perennial or if there is a considerable depth of water.

(ii) Bridge deck with precast prestressed girders, assembled together and transversely prestressed as in Fig. 8.2(b), e.g., the first Koratalayar bridge at km 26/4 of NH-5 near Chennai each span being 18 m. This arrangement is convenient for spans in the range of 15 to 20 m. The crossbeams are also precast with the main girders. The gap between two precast girders is about 25 mm. It is packed with 1:1.5: 3 mix fine concrete before applying transverse prestress. After the transverse prestressing, the wearing coat is laid. Special care should be taken during casting of the precast girders to ensure proper matching of the ducts for transverse prestress so that the slab alignment is truly set as intended in the drawing from design.

(iii) Deck consisting of precast girders similar to Fig. 8.2 (b), but with wider spacing of girders and a cast-in-situ 'gap slab' and cross beam portion between precast girders as in Fig 8.2 (c), e.g. Kovalam bridge near Chennai. Transverse prestressing is essential to ensure stability of the cast-in-place deck slab and functioning of the full deck as one unit. This arrangement can be adopted for spans up to about 30 m.

(iv) Deck with composite construction, consisting of precast prestressed concrete girders and cast-in-place deck slab. This type is particularly suited for spans in the range of 30 to 50 m. The advantage here is that the weight of components for launching will be reduced and transverse prestressing can be avoided leading to savings in cost. The precast girder can be of T or U shape as in Figs. 8.2 (d) or (e).The type shown in Fig. 8.2 (d) will permit the use of bonded tendons and the dimensions indicated correspond to a span of 30 m. Torsional rigidity is secured by cast-in-situ cross girders connecting the short sloping precast diaphragms. Particular care is required to avoid lateral buckling of the precast girder during launching for erection. Fig. 8.2 (e) represents the scheme adopted for Palar Bridge for a span of 27 m, using U-shaped precast girders with unbonded tendons (though covered with cement mortar for resistance against corrosion). Special attention should be given to detailing the shear connection between the precast and cast-in-situ components.

The trend in recent years is more towards the use of pre-cast girders than towards cast-in-situ construction. The arrangements shown in Fig. 8.2 (b) and (c) require transverse prestressing in the deck slab and in the diaphragms. Besides causing difficulties regarding alignment of transverse cable ducts, transverse prestressing involves additional costs on anchorages and tendons. The arrangement shown in Fig. 8.2 (b) results in heavier loads for erecting the precast girders than for the arrangement in Fig. 8.2 (c). Some engineers would like to avoid doubts about the structural integrity of the 'gap slab' in Fig. 8.2 (c). Hence recent constructions favour the composite construction shown in Fig. 8.2 (d). With the use of high tensile steel strands for prestressing tendons, the cables are reduced in number and are handled more conveniently.

8.6.2 PRINCIPLES OF DESIGN

(a) *Design criteria*

The criteria to be followed for the design of post-tensioned prestressed concrete road bridges are specified by the Indian Roads Congress in IRC:18 [11]. Certain additional aspects of design are discussed briefly in the following subsections.

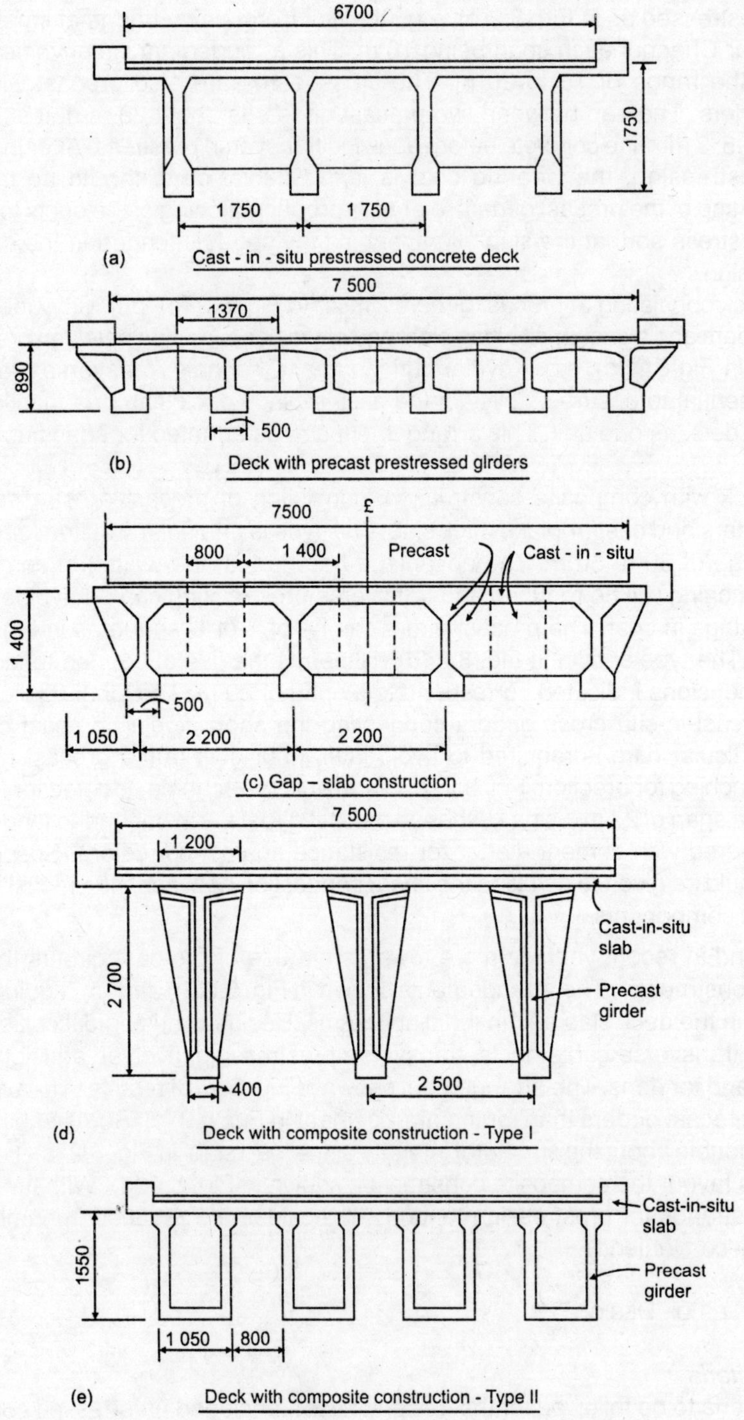

Figure 8.2 Typical Bridge Decks with Post-tensioned Girders.

The detailed calculations of the various effects would depend on many factors, including, the type of cross section, the system and sequence of prestressing, and the method of construction. The application of the design criteria and some aspects of the design would be evident from the following description, the illustrative example of gap-slab type given in Section 8.7 and from the drawings relating to the composite type bridge discussed in Section 8.8.

(b) *Outline of design of simply supported decks*

The steps involved in the design generally include the following:

(i) List the properties of the materials used such as grade of concrete, high tensile steel and untensioned steel. Usually, concrete of grade M40 is used for post-tensioned girders.

(ii) Assume preliminary dimensions, based on experience. The overall depth is usually about 75 to 85 mm for every metre of span. The thickness of deck slab is about 150 to 200 mm with transverse prestressing and about 200 to 250 mm in composite construction. The minimum thickness of web of precast girder is 150 mm plus the diameter of cable duct. The bottom width of the precast beam may vary from 500 to 800 mm.

(iii) Compute section properties. It is permissible to compute these based on the full section without deducting for the cable ducts.

(iv) Compute dead load moments and stresses for girders.

(v) Calculate live load moments and stresses for girders for the severest applicable condition of loading. Any rational method may be used for load distribution among the girders.

(vi) Determine the magnitude and location of the prestressing force at the point of maximum moment. The prestressing force must meet two conditions :

1. It must provide sufficient compressive stress to offset the tensile stresses which will be caused by the bending moments.
2. It must not induce either tensile or compresive stresses which are in excess of those permitted by the specifications.

(vii) Select the prestressing tendons to be used and work out the details of their locations in the member. Avoid grouping of tendons, as grouting of some after first stage prestressing may lead to leakage of grout into the remaining as yet ungrouted ducts if the sheaths are not leak proof. If unavoidable, use group of two cables one vertically above the other. Allow a minimum clear cover of 50 mm to the cables. Ensure that a normal needle vibrator can reach almost to the bottom row of cables while placing the concrete for the girder.

(ix) Determine the profile of the tendons, and check the stresses at critical points along the member under initial and final conditions. The following two combinations of conditions should be considered:

1. Initial prestress plus dead load only.
2. Final prestress plus full design load.

The profile of the tendons may be parabolic for the full length from one anchorage to the other. Alternatively, a tendon may have a central straight portion with parabolic profile at either end. The draping of the tendons helps to bring the stresses in concrete at every section within the permissible range. This also serves to reduce the net shear at any section.

While it is possible to anchor some of the draped tendons on the deck surface, recent practice is to avoid such deck-anchored tendons. If necessary, the sizes of the tendons and the webs are suitably increased and the tendons are anchored at the ends of the girder in the end blocks. The moment due to applied load is zero at the end of the girder. The anchorages should be spread evenly over as large an area as possible. At the ends of the girders, the c.g. of steel should ideally coincide with the c.g. of concrete. At every other section, the c.g. of steel must lie in such a position that the internal moments of resistance balance the applied moments, so that the concrete stresses lie within the permissible range.

(x) Check the ultimate strength to ensure that the requirements of IRC:18 are met. The ultimate load under moderate exposure condition is to be taken as 1.25 G + 2 SG + 2.5 Q where G, SG and Q denote permanent load, superimposed load and live load including impact, respectively. For severe exposure condition, the ultimate load is to be taken as 1.5 G + 2 SG + 2.5 Q.

(xi) Work out the shear stresses at different sections and design the shear reinforcement.

(xii) Check the stresses at the jacking end due to cable forces and these should be within permissible limits.

(xiii) Design the end block. The end block should be rectangular in section of width equal to the width of the bottom flange of the girder.

(xiv) Ensure that adequate untensioned reinforcement is provided as specified in Section15 of IRC:18. The bars of such reinforcement should have a diameter not less than 8 mm and the spacing not more than 200 mm. The reinforcement in the vertical direction should be not less than 0.18 per cent of the plan area of web/bulb of the girder. The longitudinal reinforcement should be not less than 0.15 per cent of the cross sectional area. The spacing should be uniform to the extent possible.

(xv) Design the deck slab. Any rational method may be used. If the deck slab is transversely prestressed as in Fig. 8.2 (c), the method as indicated in Section 8.7 (r) may be acceptable. If the deck is of composite type as in Fig. 8.2 (d) with reinforced concrete slab, the design may follow the procedure as for deck slab for R.C. T-beam bridge.

(xvi) Design the cross beams for the transverse bending moment. Any rational method may be used. In case of end cross beam, provision should be made for possible jacking for replacement of bearings.

(c) *Design of end block*

The portion of a prestressed concrete girder surrounding the anchorages of the tendons at an end of the girder is called an end block. It is usually made rectangular in section with width equal to the width of the bottom flange of the girder. The purpose of the end block is to distribute the concentrated prestressing forces at the anchorages and to facilitate gradual transmission of the forces to the basic cross section. The length of the end block should be about one-half the depth of the girder, but not less than 600 mm or its width. The thickening of the web of the girder towards the end block should be achieved gradually with a splay in plan of not more than 1 in 4. The portion housing the anchorages should as far as possible be precast.

The concentrated force at an anchorage causes bursting tensile force and spalling tensile force in the end block. The bursting tensile force is estimated from tabular values given in IRC:18 as a function of the force in the tendon and the ratio of the loaded width to the total width at the anchorage without overlap. The stress is maximum at a distance of about half the width of the end block behind the anchorage. In rectangular end blocks, the bursting tensile forces should be assessed in two principal directions. The reinforcement for the bursting tensile force should be provided in the region 0.1 b_2 to b_2 behind the anchorage where b_2 is the side of effective anchor area.

Spalling tensile forces occur immediately behind the anchorage and at the end face of the anchorage zone. The reinforcement is provided by two layers of meshes of small diameter. Also the end face of the anchorage zone is continually reinforced to prevent edge spalling, the reinforcement being placed as close to the end as possible. Additionally, spiral reinforcement is provided for embedded anchors.

Typical details of reinforcements provided for the end block may be seen from Fig. 8.15 and Fig. 8.22.

(d) *Deck slab*

Experiments on prestressed concrete slabs have shown that they assure a large factor of safety against both cracking and failure. It may therefore be permissible to disregard the condition of ensuring permanent compression which is often considered as a base for designing members of prestressed concrete structures.

The security of the slab lies in the formation of an arch in its interior, 'an arch' bearing on lateral beams that are fastened together by prestressing cables serving as ties. It is admitted that the tension in the concrete may attain as limit the level of prestressing cables in section at support as well as at midspan. The centre of thrust is then situated at the third points of the reduced sections (from the most compressed fibres). Hence the rise of the line of thrust. On the side beams the centre of thrust is assumed to coincide with the point of passage of the cables. The thrust Q is taken equal to the permanent prestressing force.

In practice, these assumptions lead to the following calculating procedure:

(i) Find out the bending moment *M* at midspan for a unit strip of slab assuming it to be simply supported along its edges.
(ii) Determine the allowable rise *f* of the assumed arch.
(iii) Find thrust Q of arch as *M*/*f*.
(iv) Determine the prestressing force *F* per unit length of slab. The application of this method can be seen from the example in Section 8.7.

8.6.3 Construction Sequence

An acceptable sequence of construction activities for post-tensioned bridge deck is as below[12]:

Day after casting main girders	Activity
14	Stressing of first stage cables
21	Casting of deck slab and cross girders
49	Casting of superimposed dead loads other than wearing course
56	Stressing of second stage cables
After 56	Installation of expansion joint, wearing course.

Stressing of first stage cables can be done earlier on achieving concrete strength of 35 MPa. Subsequent activities can also be suitably advanced keeping the same time intervals.

8.7 Illustrative Example of Design

The example briefly indicates the steps required in the design of a prestressed concrete deck of gap-slab construction. In order to conserve space, many details of routine calculations are omitted and final values are given to enable an understanding of the principles behind the various steps. For a practical design, a more complete working of all the details and preparation of many additional drawings will be required.

(a) DATA

(i) *Characteristics of the bridge*

Clear span	= 18.0 m
Effective span	= 18.8 m
Total length of girder	= 19.6 m
Clear roadway	= 7.5 m

Design is required for one span.

(ii) *Basis of calculations*

(a) Live load: IRC Class AA or Class A two-lane traffic.
(b) Regulations: IRC:5; IRC:6; IRC:18; IRC:21.

(iii) *Materials*

(a) Tensioned steel:
High tensile steel, 5 mm and 7 mm dia. wires, conforming to clause 4.3.2 of IS:1343-1960. Ultimate tensile strength is 1600 MPa for 5 mm wires and 1500 MPa for 7 mm wires.
Modulus of Elasticity = 2.1×10^5 MPa.

(b) Untensioned steel:
HYSD bars Grade Fe 415 conforming to IS:1786.

(c) Concrete:
Controlled concrete : M40
Permissible stresses as per Section 7 of IRC: 18
Maximum permanent compressive stress = 13.2 MPa
Temporary compressive stress at transfer = 0.50 f_{cj}
f_{cj} being the concrete strength at transfer
Permissible permanent tensile stress = 0
Permissible temporary flexural tensile stress = 0.05 f_{cj}
Modulus of Elasticity $E_c = 5000\sqrt{40} = 31620$ MPa

(b) PRELIMINARY DIMENSIONS

Bridge deck consisting of precast prestressed girders of T-section with cast-in-place gap slabs will be used.

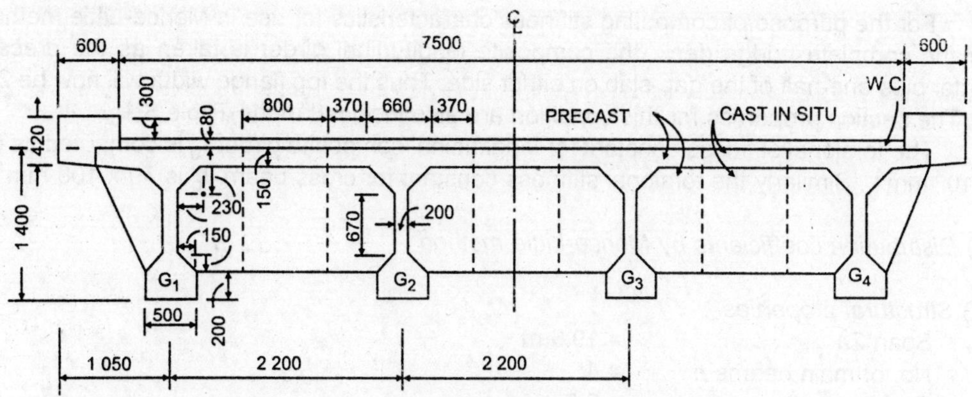

Figure 8.3 Preliminary Dimensions for the Example in Section 8.7.

Overall depth required	$= 75 \times 18.8 = 1410$ or 1400 mm
Thickness of slab	$= 150$ mm
Thickness of web	$= 150 + 50 = 200$ mm

Preliminary dimensions are assumed as shown in Fig. 8.3.

Spacing of precast girders	$= 2.2$ m
Spacing of cross beams	$= 4.7$ m

(c) Design of Precast Girders

(i) *Section properties*

No deductions for holes are made in calculating the properties of girder and cross beam. This is permissible, since no allowance for the transformed area of the prestressing tendons is included in the calculation.

The cross-sectional dimensions of a precast longitudinal girder can be obtained from Fig. 8.3. The section properties are then computed and tabulated as in Table 8.1. Here y_t and y_b denote the distance to extreme fibre of concrete at top and bottom, respectively, from the neutral axis. I denotes the second moment of area. The cross section for a cross beam is as in Fig. 8.18.

Table 8.1 Section Properties.

	Section	Area $\times 10^3$ mm²	y_t mm	y_b mm	I $\times 10^8$ mm⁴	I/y_t $\times 10^8$ mm³	I/y_b $\times 10^8$ mm³
Precast Girder	Midspan	596	546	854	1407	2.57	1.65
	Support	829	602	798	1575	2.62	1.97
Cross beam	Midspan	905	202	948	695	3.44	0.73
Precast Girder with cast-in-place slab G_2, G_3	Midspan	715	471	929	1632	3.46	1.76
–do– G_1, G_4	Midspan	656	461	939	1619	3.51	1.72

For the purpose of computing stiffness characteristics for use in Morice-Little method for the complete bridge deck, the composite longitudinal girder is taken as the precast girder plus one-half of the gap-slab on either side. Thus the top flange width will now be 2.2 m. The section properties for this condition are also included in the Table 8.1.

The torsional stiffness constant for longitudinal composite girder I_o is computed as 67 x 10^8 mm^4 . Similarly the torsional stiffnéss constant for cross beam J_o is 79 x 108 mm^4 .

(ii) *Distribution coefficients by Morice-Little method*

(1) *Structural properties*

Span 2*a*	= 19.6 m
No. of main beams *n*	= 4
Girder spacing *p*	= 2.2 m
Effective width 2*b* = np	= 8.8 m
No. of cross beams *m*	= 5
Cross beam spacing *q*	= 4.7 m

Second moment of area of composite girder $I = 1632 \times 10^8$ mm^4

Distributed longitudinal stiffness $i = I/p$

$$i = \frac{1632 \times 10^8}{2200} = 0.742 \times 10^8 \text{ mm}^4/\text{mm}$$

Second moment of area of cross beams $J = 695 \times 10^8$ mm^4

Distributed transverse stiffness $j = J/q$

$$j = \frac{695 \times 10^8}{4700} = 0.148 \times 10^8 \text{ mm}^4/\text{mm}$$

Bending stiffness parameter $\theta = \left(\frac{b}{2a}\right)\left(\frac{i}{j}\right)^{0.25}$

$$\theta = \frac{4.4}{19.6} \cdot \left[\frac{742}{148}\right]^{0.25} = 0.34$$

Distributed torsional stiffness $i_o = \frac{I_0}{p}$

$$i_o = \frac{\left(67 \times 10^8\right)}{2200} = 3.05 \times 10^6 \text{ mm}^4/\text{mm}$$

Distributed transverse torsional stiffness $j_0 = \frac{J_0}{q}$

$$j_0 = \frac{\left(79 \times 10^8\right)}{4700} = 168 \times 10^6 \text{ mm}^4/\text{mm}$$

Torsional stiffness parameter $\alpha = \dfrac{G(i_o + j_0)}{2E\sqrt{ij}}$

$$\frac{G}{2E} = \frac{1}{4(1+\mu)} = \frac{1}{4.6}$$

$$\alpha = \frac{(3.05 + 168)10^6}{4.6\sqrt{0.742 \times 10^8 \times 0.148 \times 10^8}} = 0.03$$

(2) *Unit load distribution coefficients*

The unit load distribution coefficients for the bridge deck for θ of 0.34 and α equal to 0.03 are computed using Morice-Little curves and the procedure described in Appendix B. The final distribution coefficients are shown in Table 8.2.

Table 8.2 Unit Load Distribution Coefficients K_α for $\theta = 0.34$ and $\alpha = 0.03$.

Loading position	Reference station									Row integral
	–b	*–3b/4*	*–b/2*	*–b/4*	*0*	*+b/4*	*+b/2*	*+3b/4*	*+b*	
–b	3.79	3.00	2.24	1.54	0.82	0.27	–0.33	–0.83	–1.33	7.95
–3b/4	3.00	2.53	1.95	1.43	0.92	0.47	–0.02	–0.39	–0.83	8.01
–b/2	2.24	1.95	1.64	1.31	0.99	0.69	0.33	–0.02	–0.33	7.85
–b/4	1.54	1.43	1.31	1.22	1.08	0.90	0.69	0.47	0.27	8.02
0	0.82	0.92	0.99	1.08	1.14	1.08	0.99	0.92	0.82	7.96
+b/4	0.27	0.47	0.69	0.90	1.08	1.22	1.31	1.43	1.54	8.02
+b/2	-0.33	-0.02	0.33	0.69	0.99	1.31	1.64	1.95	2.24	7.85
+3b/4	-0.83	-0.39	-0.02	0.47	0.92	1.43	1.95	2.53	3.00	8.01
+b	-1.33	-0.83	-0.33	0.27	0.82	1.54	2.24	3.00	3.79	7.95

(3) *Equivalent load λP at the nine standard positions*

Five loadings are considered as shown in Fig. 8.4 - (i) Loads from kerbs, (ii) Class AA tracked vehicle placed symmetrically (Position 1), (iii) Class AA tracked vehicle placed at

Table 8.3 Equivalent Loads λP at the Standard Positions.

Loading condition	Reference station								
	–*b*	–3*b*/4	–*b*/2	–*b*/4	0	+*b*/4	+*b*/2	+3*b*/4	+*b*
Kerb Position	0.742	0.258	...	...	...	...	...	0.258	0.742
Class AA Position 1	...	...	...	0.932	0.136	0.932	...	...	...
Class AA Position 2	...	...	0.932	0.136	0.932	...	...	...	...
Class A Position1	...	0.410	0.590	0.773	0.454	0.773	0.590	0.410	...
Class A Position 2	0.045	0.955	0.410	0.590	0.864	0.909	0.227	...	...

extreme side (Position 2), (iv) Class A two-lane placed symmetrically, and (v) Class A two-lane placed to one side. The equivalent load coefficients λ for the standard positions corresponding to these loads are listed in Table 8.3.

(4) *Actual distribution coefficients K'*

The equivalent load multipliers λ as obtained from Table 8.3 are applied to the corresponding coefficients of Table 8.2 to obtain the coefficients K'. The detailed working for Class AA - Position 2 only is shown in Table 8.4. The actual distribution coefficients K' for the five cases are tabulated in Table 8.5.

Table 8.4 Distribution Coefficient *K'* for Class *AA*-Position 2.

Loading position	E.L.M. λ	Reference station								
		$-b$	$-3b/4$	$-b/2$	$-b/4$	0	$+b/4$	$+b/2$	$+3b/4$	$+b$
$-b$	0									
$-3b/4$	0									
$-b/2$	0.932	2.09	1.82	1.53	1.22	0.92	0.64	0.31	–0.02	–0.31
$-b/4$	0.136	0.21	0.19	0.18	0.17	0.15	0.12	0.09	0.06	0.04
0	0.932	0.77	0.86	0.92	1.01	1.06	1.01	0.92	0.86	0.07
$+b/4$	0									
$+b/2$	0									
$+3b/4$	0									
$+b$	0									
$\sum \lambda K_\alpha$		3.07	2.87	2.63	2.40	2.13	1.77	1.32	0.90	0.50
$K' = \dfrac{\sum \lambda K_\alpha}{2}$		1.54	1.44	1.32	1.20	1.07	0.89	0.66	0.45	0.25

Table 8.5 Distribution Coefficients *K'* for Various Loadings.

Loading condition	Reference station								
	$-b$	$-3b/4$	$-b/2$	$-b/4$	0	$+b/4$	$+b/2$	$+3b/4$	$+b$
Kerb position	1.19	1.08	0.96	0.92	0.85	0.92	0.96	1.08	1.19
Class AA Position1	0.90	0.95	1.00	1.07	1.09	1.07	1.00	0.95	0.90
Class AA Position 2	1.54	1.44	1.32	1.20	1.07	0.89	0.66	0.45	0.25
Class A Position 1	0.95	1.00	0.99	1.02	1.03	1.02	0.99	1.00	0.95
Class A Position 2	1.46	1.36	1.24	1.15	1.04	0.90	0.73	0.60	0.45

(5) *Distribution coefficients for actual beam positions*

In this problem, the actual beam positions coincide with the reference points, i.e., -3b/4, - b/4, + b/4, and 3b/4, respectively. The values of the coefficients for the girder positions for the five loadings under consideration are listed in Table 8.6.

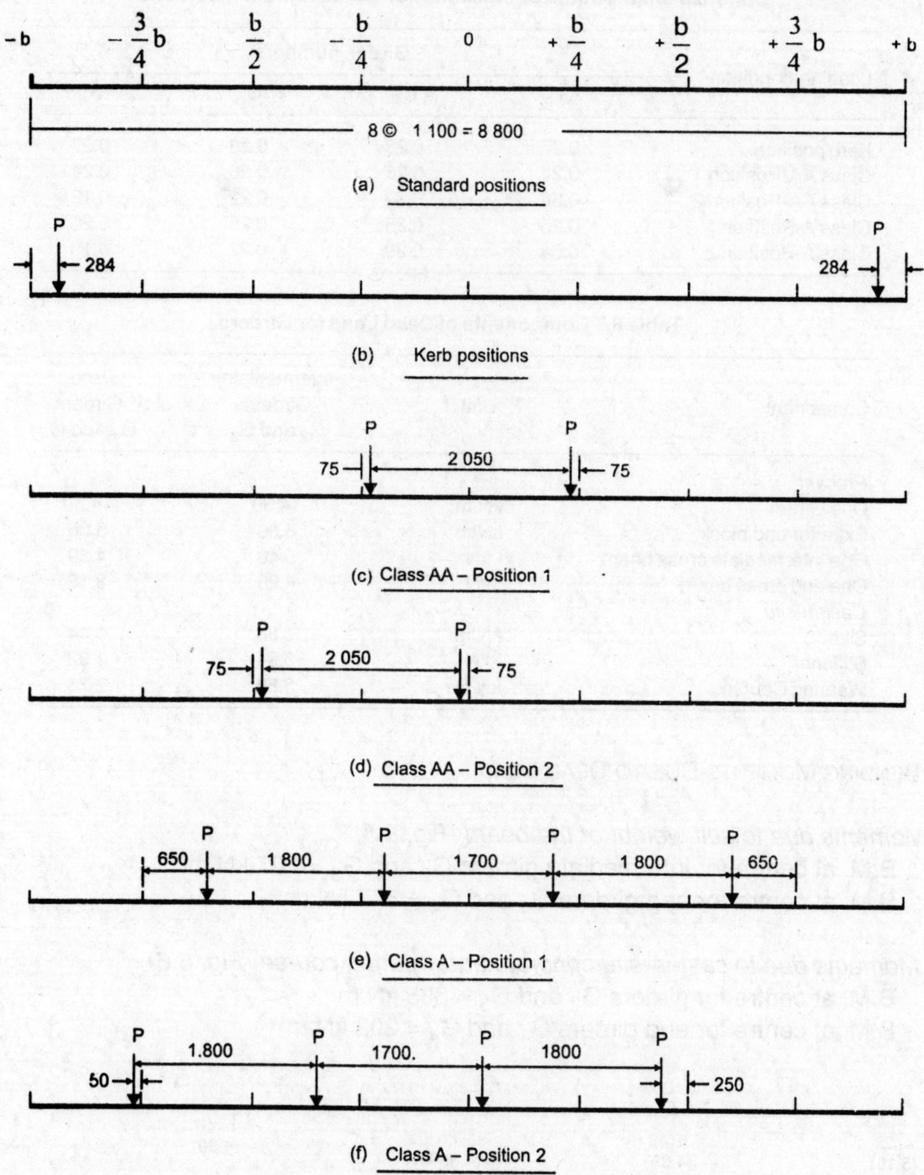

Figure 8.4 Various Load Positions for Calculating λ P.

(d) Dead Load

The components of dead load for girders are computed and tabulated as in Table 8.7. The dead load components for the girders are shown in Fig. 8.5 for precast portion and in Fig. 8.6 for cast-in-situ portion, respectively. In these figures, the values for end girders are shown in parantheses. In addition, the dead load due to kerbs and handrails in this case will be 7.32 kN/m acting at 284 mm from edge of kerb.

Table 8.6 Distribution Coefficients for Actual Girder Positions.

Loading condition	Girder numbers			
	G_1	G_2	G_3	G_4
Kerb position	0.27	0.23	0.23	0.27
Class AA-Position 1	0.24	0.26	0.26	0.24
Class AA-Position 2	0.36	0.30	0.22	0.12
Class A-Position 1	0.25	0.25	0.25	0.25
Class A-Position 2	0.34	0.29	0.22	0.15

Table 8.7 Components of Dead Load for Girders.

Component	Unit	Intermediate Girders G_2 and G_3	End Girders G_1 and G_4
Precast			
One Girder	kN/m	14.30	14.30
Extra for end block	kN/m	6.06	6.06
One Intermediate cross beam	kN	5.46	4.39
One end cross beam	kN	4.26	3.18
Cast-in-situ			
Slab	kN/m	2.88	1.44
Stiffener	kN	3.84	1.92
Wearing Course	kN/m	3.88	2.73

(e) Bending Moments Due to Dead Load

(i) *Moments due to self weight of the beam (Fig. 8.5)*

B.M. at centre for intermediate girders G_2 and G_3 = 687 kN.m

B.M. at centre for end girders G_1 and G_4 = 677 kN.m

(ii) *Moments due to cast-in-situ concrete and wearing course (Fig. 8.6)*

B.M. at centre for girders G_2 and G_3 = 335 kN.m

B.M. at centre for end girders G_1 and G_4 = 202 kN.m

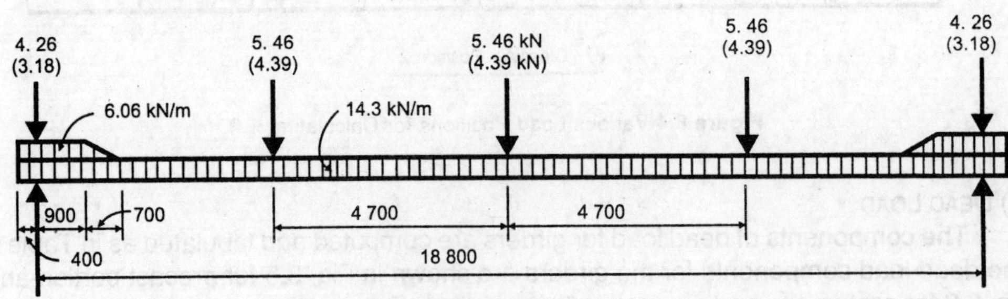

Figure 8.5 Dead Load due to Precast Girders.

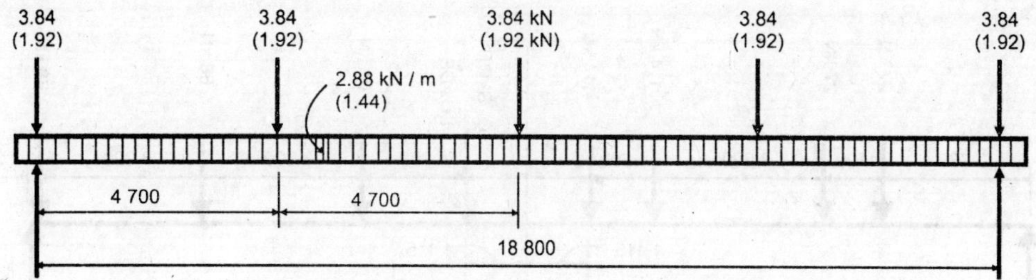

Figure 8.6 Dead Load due to Cast-in-situ Portions.

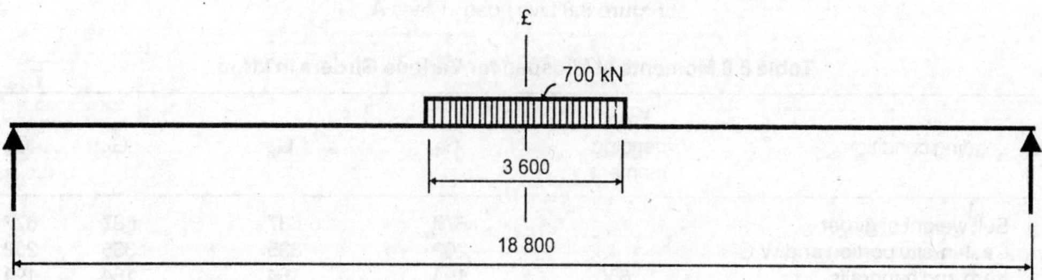

Figure 8.7 Live Load - Class AA.

(iii) *Moments due to kerb and handrails*

Total moment due to the kerbs and handrails = 650 kN.m

(f) MOMENT DUE TO LIVE LOAD

(i) *Live load Class AA*

Impact factor = 1.10

The track is placed at the centre for maximum B.M. (Fig. 8.7).

$$\text{B.M. at midspan including impact} = 1.10 \times 350\left(\frac{18.8}{2} - \frac{3.6}{4}\right) = 3270 \text{ kN.m}$$

(ii) *Live load Class A*

$$\text{Impact factor fraction} = \frac{4.5}{6+L} = \frac{4.5}{6+18.8} = 0.18$$

The condition for maximum bending moment is as shown in Fig. 8.8. The maximum B.M. occurs under the 114 kN load nearest to midspan.

Max. B.M. due to one lane of Class A = 1376 kN.m

Max. B.M. for two lanes with impact = 2 x 1.18 x 1376 = 3260 kN.m

Moments at midspan for various girders are now listed in Table 8.8. Items 3 to 7 include moment correction factor of 1.10.

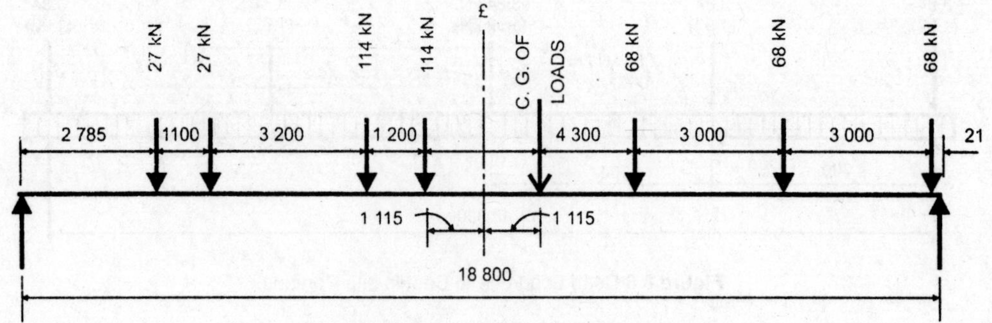

Figure 8.8 Live Load - Class A.

Table 8.8 Moments at Midspan for Various Girders in kN.m.

No.	Loading condition	Free bending moment	G_1	G_2	G_3	G_4
1.	Self weight of girder	...	677	687	687	677
2.	Cast-in-situ portion and W.C.	...	202	335	335	202
3.	Kerb and handrails	650	194	164	164	194
4.	Class AA position 1	3270	864	935	935	865
5.	Class AA position 2	3270	1295	1079	791	433
6.	Class A position 1	3260	896	897	897	896
7.	Class A position 2	3260	1219	1040	789	538
8.	Total live load-highest vaiue of 4 to 7	...	1295	1079	935	896

Table 8.9 Bending Stresses at Midspan (+ Indicates Compression).

Loading condition	Girder G_1 Bending moment kN.m	Girder G_1 Stress, MPa top	Girder G_1 Stress, MPa bottom	Girder G_2 Bending moment kN.m	Girder G_2 Stress, MPa top	Girder G_2 Stress, MPa bottom
Self weight of girder	677	+2.63	−4.10	687	+2.67	−4.16
Cast-in-situ portion	202	+1.22	−0.80	335	+1.30	−2.03
Kerb and handrails	194	+0.55	−1.12	164	+0.47	−0.93
Live load	1295	+3.69	−7.50	1079	+3.11	−6.26
Total		+8.09	−13.60		7.55	−13.38

(g) Stresses Due to Bending Moment

Due to symmetry, it is sufficient to design and check only the girders G_1 and G_2 . The bending stresses due to the different loading conditions are tabulated in Table 8.9. The stresses are obtained by dividing the moments by the appropriate values of section modulus.

(h) Prestress For Intermediate Girders G_2 and G_3

(i) *Cables*

Cables of 12-8φ high tensile wires are used.

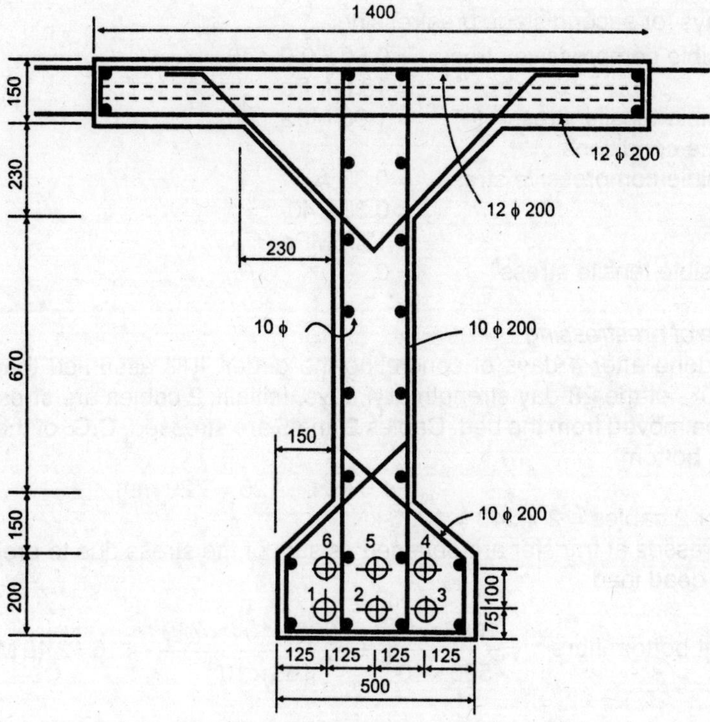

Figure 8.9 Cross Section at Midspan for Intermediate Girder.

Ultimate stress of 8 mm wires = 1500 MPa

Allowable stress in steel before allowing for creep and shrinkage

$$= 1500 \times 0.6 = 900 \text{ MPa} = 0.9 \text{ kN/mm}^2$$

Allowable force at transfer for one 12-8ϕ cable

$$= 12 \times \frac{\pi}{4} (8)^2 \times 0.9 = 543 \text{ kN}$$

Allowable force at jacking end for one cable

$$= \frac{0.7 \times 1.5}{0.9} \times 543 = 633 \text{ kN}$$

6 cables are provided, as shown in Fig. 8.9.

(ii) *Permissible stresses in concrete*

At 4 days for first stage prestressing

Permissible compressive stress	$= 0.50 f_{cj}$
	$= 0.50 \times 0.6 \times 40$
	$= 12.0$ MPa
Permissible tensile stress	$= 1.20$ MPa

At 21 days for second stage prestressing

Permissible compressive stress $= 0.50 \times 0.9 \times 40$

$= 18.0$ MPa

Permissible tensile stress $= 1.80$ MPa

At service conditions

Permissible compressive stress $= 0.33\ f_{ck}$

$= 0.33 \times 40$

$= 13.2$ MPa

Permissible tensile stress $= 0$

(iii) *First stage of prestressing*

This is done after 4 days of concreting the girder. It is assumed that the concrete would gain 60% of the 28-day strength at 4 days. Initially 2 cables are stressed to enable the girder to be moved from the bed. Cables 2 and 5 are stressed. C.G. of the cables lies at 125 mm from bottom.

$$e = 854 - 125 = 729 \text{ mm}$$

Force for 2 cables = 2 x 543 = 1086 kN

Initial stresses at transfer are obtained as sum of the stress due to prestress and the stress due to dead load.

$$\text{Stress at bottom fibre} = \frac{1086 \times 10^3}{596 \times 10^3} + \frac{1086 \times 10^3 \times 729}{1.65 \times 10^8} - 4.16 = 2.46 \text{ MPa}$$

$$\text{Stress at top fibre} = \frac{1086 \times 10^3}{596 \times 10^3} + \frac{1086 \times 10^3 \times 729}{2.57 \times 10^8} + 2.67 = 1.41 \text{ MPa}$$

Stress at c.g. of the two cables = 1.82 + 4.8 x 729/854

= 5.92 MPa

Hence average stress at centroid of cables

= 0.5(1.82 + 5.92) = 3.87 MPa

Creep strain (assuming strength of concrete at 60% of f_{ck}) during the time of first stage prestressing = 7.2×10^{-4} mm/mm per 10 MPa

Percentage of loss due to creep = $(7.2 \times 10^{-4}/10) \times 3.87 \times (2.1 \times 10^5/900) \times 100 = 6.50\%$

Loss due to relaxation of steel corresponding to average initial stress of 0.6 of ultimate strength is assumed as per IRC: 18.

Percentage loss due to relaxation = 0.55 x 2.5 = 1.38%

Loss due to shrinkage in concrete = $(4.1 \times 10^{-4})\ (2.1 \times 10^5/900) \times 100 = 9.57\%$

Total losses = 17.45% or 18%

Final force after losses due to creep, shrinkage and relaxation of steel

$= 543 \times 0.82 = 445$ kN

Final Stresses:

Stress at bottom fibre = 0.82 (1.82 + 4.80) = 5.43 MPa

Stress at top fibre = 0.82 (1.82-3.08) = −1.03 MPa

(iv) *Second stage of prestressing*

This is done at 21 days. It is assumed that concrete would gain 90% of the 28-day strength at 21 days. Cables 1, 3, 4, and 6 are stressed.

Initial force per cable = 543 kN

c.g. of the 4 cables will be at 125 mm above bottom.

Eccentricity = 854 – 125 = 729 mm

Initial stresses before losses due to second stage cables are obtained by proportion (2 × 729/854 = 1.71)

Stress at bottom fibre $= 1.71 \times 6.62 = 11.32$ MPa – I

Stress at top fibre $= -1.71 \times 1.26 = -2.15$ MPa – II

Stress at c.g. of cables $= 1.71 \times 5.92 = 10.11$ MPa – III

Direct stress at support $= 1.71 \times 1.82 = 3.11$ MPa – IV

(v) *Stress due to first stage cables*

The losses that the first stage cables must have incurred during the interval between first and second stages are calculated and the exact forces of these cables at the time of second stage of prestressing are determined.

Loss due to relaxation in steel = 1.38%

Loss due to shrinkage of concrete that had occurred between the two stages

$$= 9.57 - \frac{2\times10^{-4}.(2.1\times10^{5})}{900}\times 100 = 4.90\%$$

Loss due to creep in concrete calculated as the difference between the values for 4 and 21 days for an average stress of 3.87 MPa

$$= 6.50 - \frac{4.4\times10^{-4}}{10}\times 3.87\times\frac{2.1\times10^{5}}{900}\times 100 = 2.53\%$$

Elastic shortening of first stage cables due to second stage prestressing is computed by approximate method as in Section 11.1 of IRC:18.

$$\text{Modular ratio} = \frac{E_s}{E_c} = \frac{2.1\times10^{5}}{31620} = 6.64$$

Average stress due to second stage prestress = 0.5 (10.11 + 3.11)
= 6.61 MPa

Loss of prestress due to elastic shortening = 0.5 × 6.64 × 6.61
= 21.95 MPa

Percentage loss due to elastic shortening $= \left[\frac{21.95}{900}\right]\times 100$
= 2.44%

Total losses for first stage cables till second stage of prestressing = 13.76% or 13.8%

Actual force in first stage cables at the time of second stage of prestressing = 543 × 0.862
= 468 kN

Stresses due to first stage cables are: (e = 729 mm)

$$\text{Stress at bottom fibre} = \frac{2\times468\times10^{3}}{596\times10^{3}} + \frac{2\times468\times10^{3}\times729}{1.65\times10^{8}}$$

= 1.57 + 4.14 = 5.71 MPa –V

Stress at top fibre = –1.10 MPa –VI

Stress at c.g. of the second stage cables = 1.57 + (4.14 × 729/854)
= 5.10 MPa –VII
Direct stress at support = 1.57 MPa –VIII

(vi) *Net Stresses due to first stage and second stage cables*
Total stress at c.g. of the second stage cables due to first and second stage prestressing
= III + VII = 10.11 + 5.10 = 15.21 MPa
Release in compression at the same level due to self weight of girder

$$= -4.16 \times 729/854 = -3.55 \text{ MPa}$$

Net stress at c.g. at midspan = 15.21 – 3.55 = 11.66 MPa
Total direct stress at support = IV + VIII = 3.11 + 1.57
= 4.68 MPa
Average stress at centroid of cables = 0.5 (4.68 + 11.66) = 8.17 MPa
Creep loss for second stage cables

$$= \frac{4.4\times10^{-4}}{10}\times8.17\times\frac{2.1\times10^{5}}{900}\times100 = 8.39\%$$

$$\text{Loss due to shrinkage} = \frac{2\times10^{-4}\times\left(2.1\times10^{5}\right)}{900}\times100 = 4.67\%$$

Loss due to relaxation in steel = 1.38%
Total losses = 8.39 + 4.67 + 1.38 = 14.44% or 14.5%
Final force in second stage cables after all losses = 543 × 0.855 = 464 kN
Final stresses due to second stage prestressing after all losses
Stress at bottom fibre = 11.32 × 0.855 = 9.68 MPa
Stress at top fibre = – 2.15 × 0.855 = –1.84 MPa

Total initial stresses at transfer (before losses) due to first and second stage cables at the time of second stage prestressing:

Stress at bottom fibre = I + V = 11.32 + 5.71 = 17.03 MPa
Stress at top fibre = II + VI = –2.15 – 1.10 = –3.25 MPa

(vii) *Summary of stresses*

The summary of stresses at midspan for intermediate girders G_2 and G_3 at the various sequences of operation from initial prestress to service condition for the intermediate girders G_2 and G_3 is given in a tabular form in Table 8.10.

The stresses are to be checked particularly at the following stages:

Stage 1 : First stage prestressing and self weight. Steps 4 and 5.
Stage 2 : Second stage prestressing and self weight of girder. Steps 8 and 9.
Stage 3 : Second stage prestressing and weight of cast-in-situ portions, kerbs and hand rails. Steps 14 and 15.
Stage 4 : Second stage prestressing and service loads including live load. Steps 17 and 18.

The stresses at every stage are within permissible limits. Hence the section assumed as in Fig. 8.9 is adequate.

In this example, the effects of two cables of first stage prestressing are taken together.

Table 8.10 Summary of Stresses at Midspan for Intermediate Girders G_2 and G_3 (+ Compression).

Sl. No.	Loading case	Girders G_2 and G_3 Bottom MPa	Top MPa
1.	Self weight	−4.16	2.67
2.	I Prestress initial	6.62	−1.26
3.	I Prestress final	5.43	−1.03
4.	1 + 2	2.46	1.41
5.	1 + 3	1.27	1.64
6.	Total initial stress due to I and II stage cables	17.03	−3.25
7.	Second stage final	9.68	−1.84
8.	6 + 1	12.87	−0.58
9.	7 + 5	10.95	−0.20
10.	Cast-in-situ	−2.03	1.30
11.	10 + 8	10.84	0.72
12.	10 + 9	8.92	1.10
13.	Kerbs and handrails	−0.93	0.47
14.	13 + 11	10.01	1.19
15.	13 + 12	7.99	1.57
16.	Live loads	−6.26	3.11
17.	16 + 14	3.75	4.30
18.	16 + 15	1.73	4.68

instead of strictly examining the stresses at the end of stressing of every cable. Similarly, the effects of four cables of second stage prestressing are taken together. This approximation is permissible for design purposes. Alternatively, the effects may be computed exactly based on the sequence of stressing the cables.

(j) SECTION AT MIDSPAN FOR END GIRDERS G_1 AND G_4

The section at midspan for the end girders G_1 and G_4 may be assumed nearly the same as for the intermediate girders G_2 and G_3. Following a procedure similar to that for intermediate girders, it will be found that the stresses are within limits at every stage of operation.

(k) CABLE PROFILE

The six cables can be conveniently anchored at the end, with a spacing of 200 mm, as shown in Fig. 8.10.

In the present example, the profile of each cable is adopted as parabolic without a central straight portion. The ordinates of the cable profile may be obtained from the equation $y = a + b x^2$, where y is the ordinate above the bottom of the girder at a distance x from the midspan, and a and b are constants to be determined for each cable. The inclination θ of the cable at the anchor is given by

$$\tan \theta = \frac{8r}{L}$$

where r is the rise of the parabola and L is the span.

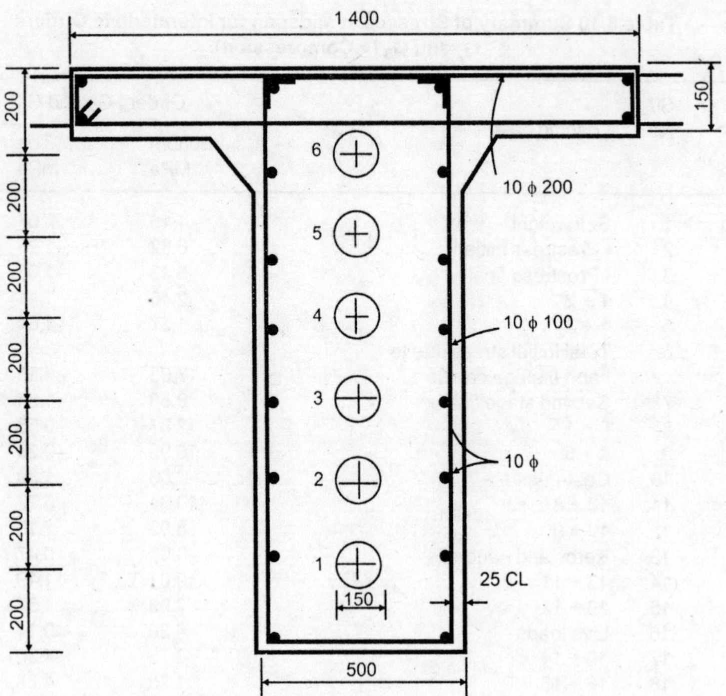

Figure 8.10 Section at Support for Intermediate Girder Showing Cable Anchorage Locations.

The cable profile, the ordinates at different sections and the inclination at the anchorage for each cable for the longitudinal girder are shown in Fig. 8.11.

(l) CHECK FOR BENDING STRESSES AT OTHER SECTIONS

It is necessary to check for stresses at a number of other sections, besides the midspan section. Normally stresses are checked at least at the one-sixth and one-third points along the span. The detailed computations are not furnished here to save space.

(m) CHECK FOR ULTIMATE STRENGTH

The ultimate moment at midspan is checked.

Ultimate moment $M_u = 1.25\ G + 2.0\ SG + 2.5\ Q$

where G, SG and Q denote the effects due to permanent load, superimposed dead load, and live load including impact, respectively.

$$M_u = 1.25\ (687 + 335) + 2\ (164) + 2.5\ (1079)$$
$$= 4303 \text{ kN.m}$$

Ultimate moment of resistance, M_{ult}

(i) *Failure by yield of steel*

$$M_{ult} = 0.9\ d_b .\ A_s .\ f_p$$

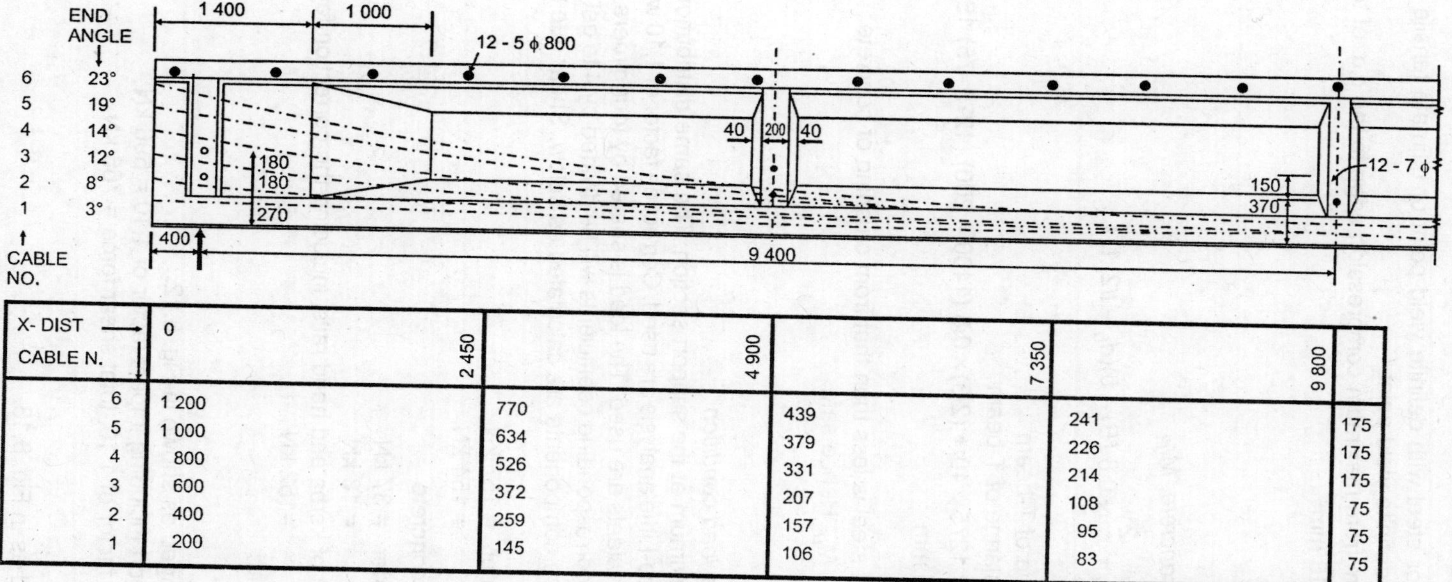

X- DIST → / CABLE N. ↓	0	2 450	4 900	7 350	9 800
6	1 200	770	439	241	175
5	1 000	634	379	226	175
4	800	526	331	214	175
3	600	372	207	108	75
2	400	259	157	95	75
1	200	145	106	83	75

Figure 8.11 Elevation of Precast Girder and Ordinates of Cable Profile.

where f_p = yield stress for steel with definite yield point or ultimate tensile strength for steel without definite yield point

d_b = depth of beam from maximum compression edge to the c.g. of tendons

Here $A_s = 6 \times 603 = 3618 \text{ mm}^2$

$f_p = 1500$ MPa

$d_b = 1250$ mm

Hence $M_{ult} = 6105$ kN.m

(ii) *Failure by crushing of concrete, M_{ult}*

$$M_{ult} = 0.176\, b.\, d_b^2 f_{ck} + \frac{2}{3} \times 0.8\, (B_f - b)(d_b - t/2)\, t.\, f_{ck}$$

where b = width of web

B_f = width of flange of T-beam

t = thickness of flange of T-beam

$M_{ult} = 0.176 \times 200 \times 1275 \times 40 + (2/3) \times 0.8\ (1400 - 200)\ (1275–75)\ 150 \times 40$

$= 6897 \times 10^6$ N.mm

$= 6897$ kN.m

M_{ult} by yielding of steel is less than that from crushing of concrete.

$M_{ult} = 6105$ kN.m $> M_u$. Hence safe.

(n) DESIGN FOR SHEAR

(i) *Shear forces at working load condition*

Shear forces are maximum at the support section. The same distribution coefficients calculated as per the Morice-Little analysis are used. Correction factor of 1.10 will be applied, wherever distribution coefficients are used. The load is shared by four girders. Total values of shear multiplied by the corresponding coefficients will be divided by 4 to get the shear for one girder. The shear force components are calculated as below. Shear due to self weight of girder

For intermediate girder = 157 kN
For end girder = 154 kN

Shear due to cast-in-situ concrete

For intermediate girder = 37 kN
For end girder = 18 kN
Shear due to weight of kerbs and hand rails, including the correction factor of 1.10 =152 kN

Shear force due to live load

(1) *Class AA loading*

The loading is arranged as shown in Fig. 8.12.
Max. Shear at support including impact factor of 1.10 = 696 kN
Applying correction factor of 1.10, total shear force = 766 kN

(2) *Class A loading*

Loading is arranged as in Fig. 8.13.
Shear at support = 348 kN
Shear for two lanes= 696 kN

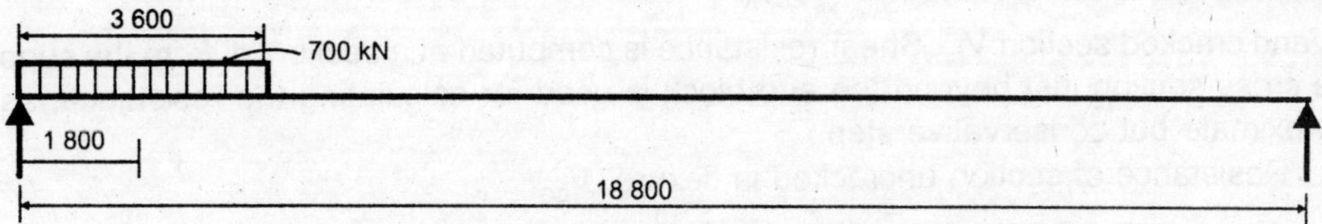

Figure 8.12 Disposition of Class AA Loading for Maximum Shear.

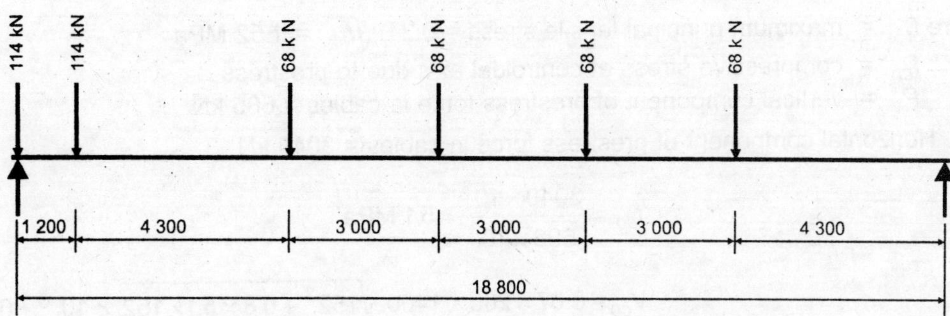

Figure 8.13 Disposition of Class A Loading for Maximum Shear.

Using correction factor of 1.10 and impact factor of 1.181, total shear = 904 kN
The shear forces at the working load condition are listed in Table 8.11.

Table 8.11 Shear Force for Longitudinal Girders in kN.

No.	Loading condition	Total shear force kN	Shear force for G_1 kN	Shear force for G_2 kN
1.	Self weight of girder	...	154	157
2.	Cast-in-situ portion	...	18	37
3.	Kerbs	152	41	35
4.	Total dead load 1+2+3	...	213	229
5.*	Live load Class AA	766	276	230
6.*	Live load Class A	904	309	261
7.	Live load - Max. of 5 and 6	...	309	261
8.	Total shear 4 + 7	...	522	490

*Position 2 which is critical is used for the distribution coefficients.

(ii) *Shear force at ultimate load condition*

Maximum shear force at support at ultimate load condition is obtained as below:

Ultimate shear $V_u = 1.25\,G + 2.0\,SG + 2.5\,Q$

From Table 8.11, $G = 157 + 37 = 194$ kN

$SG = 35$ kN

$Q = 261$ kN

Substituting the above values, $V_u = 965$ kN.

(iii) *Shear resistance*

Shear resistance V_u is taken as the smaller value of resistance of uncracked section

V_{uo} and cracked section V_{cr}. Shear resistance is computed at a section 'd' from the support. The cross section just beyond the end block is used for calculating the resistance, as an approximate but conservative step.

Resistance of section uncracked in flexure, V_{co}

$$V_{co} = 0.67\, bd \sqrt{f_t^2 + 0.8\, f_{cp}.f_t} + P_v$$

where f_t = maximum principal tensile stress = $0.24\sqrt{f_{ck}}$ = 1.52 MPa

f_{cp} = compressive stress at centroidal axis due to prestress

P_v = vertical component of prestress force in cables = 685 kN

Horizontal component of prestress force in cables = 3040 kN

$$f_{cp} = \frac{3040 \times 10^3}{596 \times 10^3} = 5.1 \text{ MPa}$$

$$V_{co} = 0.67 \times 200 \times 1400 \sqrt{1.52^2 + 0.8 \times 5.1 \times 1.52} \times 10^{-3} + 685$$

$$= 1232 \text{ kN}$$

Resistance of section cracked in flexure, V_{cr}

$$V_{cr} = 0.037\, bd_b \sqrt{f_{ck}} + (M_t / M)V$$

where d_b = distance from extreme compression fibre to the centroid of tendons in the section considered

M_t = cracking moment at the section

$= (0.37\sqrt{f_{ck}} + 0.8\, f_{pt})\, I/y$

f_{pt} = stress due to prestress at the tensile fibre distance y from centroid of concrete section whose moment of inertia is I

M = moment due to ultimate load

$$f_{pt} = \frac{3040 \times 10^3}{596 \times 10^3} - \frac{3040 \times 10^3 \times 551}{3.46 \times 10^8}$$

$$= 0.26 \text{ MPa}$$

$$M_t = (0.37\sqrt{40} + 0.8 \times 0.26)\, 3.46 \times 10^8$$

$$= 8.816 \times 10^8 \text{ N.mm} = 882 \text{ kN.m}$$

$$M = 1.25(210 + 10) + 2 \times 185 + 2.5 \times 273$$

$$= 1430 \text{ kN.m}$$

Ultimate shear at distance 'd' from support V = 730 kN

$$V_{cr} = \frac{0.037 \times 200 \times 1152\sqrt{40}}{1000} + \frac{882 \times 730}{1430} = 504 \text{ kN}$$

Available shear resistance $V_c = V_{cr}$ = 504 kN

(iv) *Shear reinforcement*

V is greater than V_c

Hence shear reinforcement is necessary. Shear reinforcement should be obtained from

$$\frac{A_{sv}}{S_v} = \frac{V - V_c}{0.87 f_y . d_t}$$

where A_{sv} = area of two legs of a stirrup

S_v = spacing of stirrup

d_t = depth from extreme compression fibre to the longitudinal bars or to the centroid of tendons

Using 10 mm bars, A_{sv} = 157.2 mm^2

d_t = 1400 – 511 = 889 mm

Substituting values in the above equation, S_v = 223 mm

Provide 10ϕ– 2 legged stirrups at 200 mm centres.

(o) UNTENSIONED REINFORCEMENT

(i) *Vertical direction*

Minimum at midspan = 0.18% of plan area

$$= \frac{0.18}{100} \times 200 \times 1000 = 360 \text{ mm}^2$$

Provide two-legged 10 ϕ stirrups at 200 mm centres

$$\text{Minimum near support} = \frac{0.18}{100} \times 500 \times 1000 = 900 \text{ mm}^2$$

Provide two-legged 10ϕ stirrups at 150 mm centres

(ii) *Longitudinal steel in girder*

$$\text{Minimum area} = \frac{0.15}{100} \times 596 \times 10^3 = 894 \text{ mm}^2$$

Provide 24 bars of 10ϕ as shown in Fig. 8.9.

For end block

$$\text{Minimum area} = \frac{0.15}{100} \times 829 \times 10^3 = 1244 \text{ mm}^2$$

Provide 20 bars of 10ϕ as shown in Fig. 8.10.

(p) STRESSES AT JACKING END

The stresses at the jacking end should be so adjusted that, after allowing for losses due to friction, slip and successive shortening, the net prestress will be as already computed. Slip during seating of anchorages for Freyssinet system is assumed as 2.5 mm. Only a brief indication of the computations involved is included here. For a comprehensive treatment of these losses, the reader is referred to Reference 4.

(i) *Friction losses*

Steel stress σ_{px} in the tendon at a distance x from the jacking end is given by Equation (5.10).

$$\sigma_{po} = \sigma_{px} . e^{(\mu\theta + kx)}$$

where σ_{po} = steel stress at jacking end
μ = coefficient of friction
θ = cumulative angle change in radians
k = wobble coefficient per m length of steel

Here σ_{px} = 900 MPa at x = 9.4 m
μ = 0.25 from Table 7 of IRC: 18
k = 0.0091

For cable 2, θ = 0.131 and σ_{po} = 1017 MPa.

In order to facilitate reasonably uniform distribution of prestress force along the span, it would be desirable to pull half the cables from one end and the remainder from the other end. Thus, cables 2, 1 and 3 will be pulled from one end and cables 5,4 and 6 will be pulled from the other end.

(ii) *Loss due to slip at anchorages*

The loss due to slip at anchorages is computed following a method given in Reference 4.

The tensioning stress diagram for Cable 2 is given in Fig. 8.14. The desired stress is 900 MPa at midspan. After friction losses, the stress at the far end (locked end) will be 783 MPa. The stress required at the jacking end will be 1017 MPa. The variation of stress between the two ends of the cable prior to anchorage slip is assumed to be linear.

During the occurrence of anchorage slip, the friction effects will act in the opposite direction to that during jacking. The friction coefficient μ' for slip is usually taken as 1.5μ. The loss of prestress due to anchorage slip will be maximum at the jacking end and it will not have any effect beyond a distance L_s from the anchorage. The value of L_s is found by trial and error method.

Assume L_s = 5.8 m, Slip s = 2.5 mm, and μ' = 1.5 μ

$$\text{Loss of stress at distance } L_s = (1017 - 900)\left(\frac{5.8}{9.4}\right) = 72 \text{ MPa}$$

$$\text{Slip loss} = \left(1 + \frac{\mu'}{\mu}\right) 72 = 180 \text{ MPa}$$

Area of detensioning diagram due to slip (see Fig. 8.14)

$$= 0.5 \times 180 \times 5.8 = 522 \text{ MPa.m}$$

$$\text{Slip} = \frac{area}{E_s} = \frac{522 \times 10^3}{2.1 \times 10^5} = 2.49 \text{ mm} \approx 2.5 \text{ mm}$$

Hence the length of tendon affected by slip L_s = 5.8 m. Slip loss is taken as 180 MPa as above.

The loss due to slip for every cable may be calculated as above.

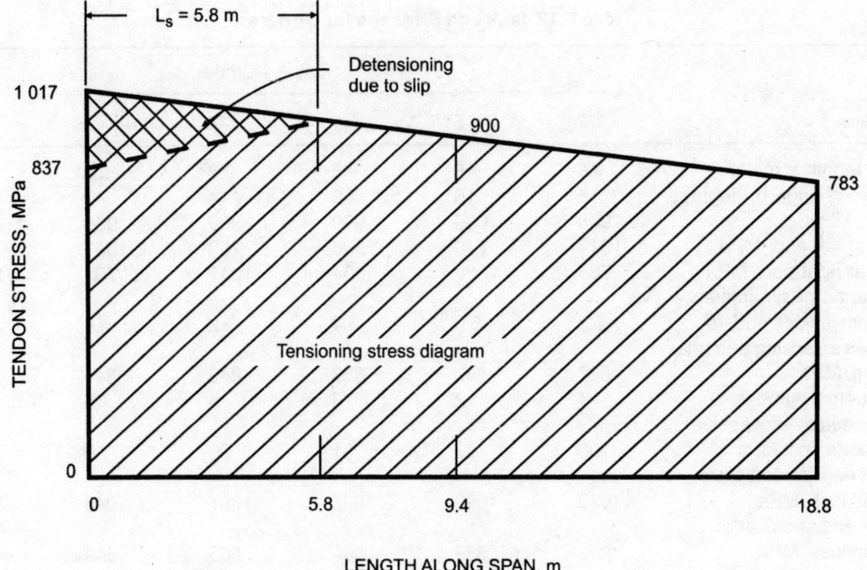

Figure 8.14 Tensioning Stress Diagram for Cable 2.

(iii) *Loss due to successive shortening*

The loss of prestress in any particular cable due to successive shortening depends on the sequence of stressing. The sequence of stressing is as below : Cable numbered 2,5,1,4,3 and 6. The loss of prestress in any cable due to successive stressing can be taken as the summation of the product of modular ratio and the average compressive stress at the concrete adjacent to the tendon due to tensioning of subsequent cables.

(v) *Summary of jack end stresses*

The computations of jack end stresses taking into account the above losses are summarised in Table 8.12.

(q) End Blocks

End blocks are provided at the ends of the girder for a length of 1.4 m. The end block is rectangular with a width of 500 mm. The design is performed as per sections 7.3 and 17 of IRC:18.

Maximum bearing force behind anchor = $603 \times 892 \times 10^{-3}$ = 538 kN

For embedded Freyssinet anchorage, one-third of the anchor force can be assumed to be transmitted by friction.

Force transmitted by bearing = 0.667×538 = 359 kN

Area A_1 for anchorage = 17671 mm^2

Bearing pressure $= \dfrac{359 \times 10^3}{17671} = 20.3$ MPa

Allowable bearing pressure f_b is given by

$f_b = 0.48\, f_{cj} \sqrt{A_2 / A_1}$ or $0.8\, f_{cj}$ whichever is less

Table 8.12 Jack end Stresses for Girders.

Sl. No	Details	Cable Number 2	5	1	4	3	6
1.	Half length x, m	9.8	9.8	9.8	9.8	9.8	9.8
2.	Angle change θ, degrees	8	16	3	14	12	23
3.	σ_{px}, MPa	900	900	900	900	900	900
4.	σ_{po} at left end, MPa	1017	740	996	753	1037	714
5.	σ_{po} at right end, MPa	783	1060	804	1047	763	1086
6.	Distance of slip influence from jacking end, m	5.8	5.0	6.4	5.2	5.4	4.6
7.	Stress at jacking end after slip, MPa	837	847	833	843	840	859
8.	Sequence number	1	2	3	4	5	6
9.	Loss due to successive shortening, MPa	55	39	31	20	9	0
10.	Jack end stress before slip loss, MPa	1072	1099	1027	1067	1046	1086
11.	Jack end stress after slip loss, MPa	892	886	864	863	849	859

A_2 = 40000 mm^2

f_{cj} = 24 MPa for first stage prestressing and 40 MPa for second stage prestressing

Hence f_b = 17.3 MPa for first stage and 28.9 MPa for second stage.

Considering that adequate spiral reinforcement and mesh reinforcement are provided at the anchorage, an increase in bearing pressure of about 30% can be admitted. For first stage prestressing, the allowable bearing pressure can now be increased to 22.5 MPa. Hence provision may be considered safe.

Diameter of anchorage for 12-8ϕ cable = 150 mm

Side of loaded area $2Y_{po}$ = 150 mm

Effective side of end block $2Y_o$ = 200 mm

Y_{po}/Y_o = 0.75

Load in tendon $P_k = 603 \times 0.87 \times 1.5$ = 787 kN

From Table 8 of IRC: 18.

Bursting tensile force F_{bst} = $0.11\ P_k$ = 86.6 kN

$$\text{Reinforcement for bursting tension} = \frac{86.6 \times 10^3}{0.87 \times 415} = 240 \text{ mm}^2$$

This reinforcement should be provided in the region between 0.2 Y_o and 2 Y_o, i.e. between 20 mm and 200 mm from the anchorage.

Provide additional 8ϕ mesh as shown in Fig. 8.15.

Spalling force is taken as 0.04 of the tendon force.

$$\text{Spalling reinforcement} = \frac{0.04 \times 787 \times 10^3}{0.87 \times 415} = 87 \text{ mm}^2$$

Provide 4–10 ϕ bars.

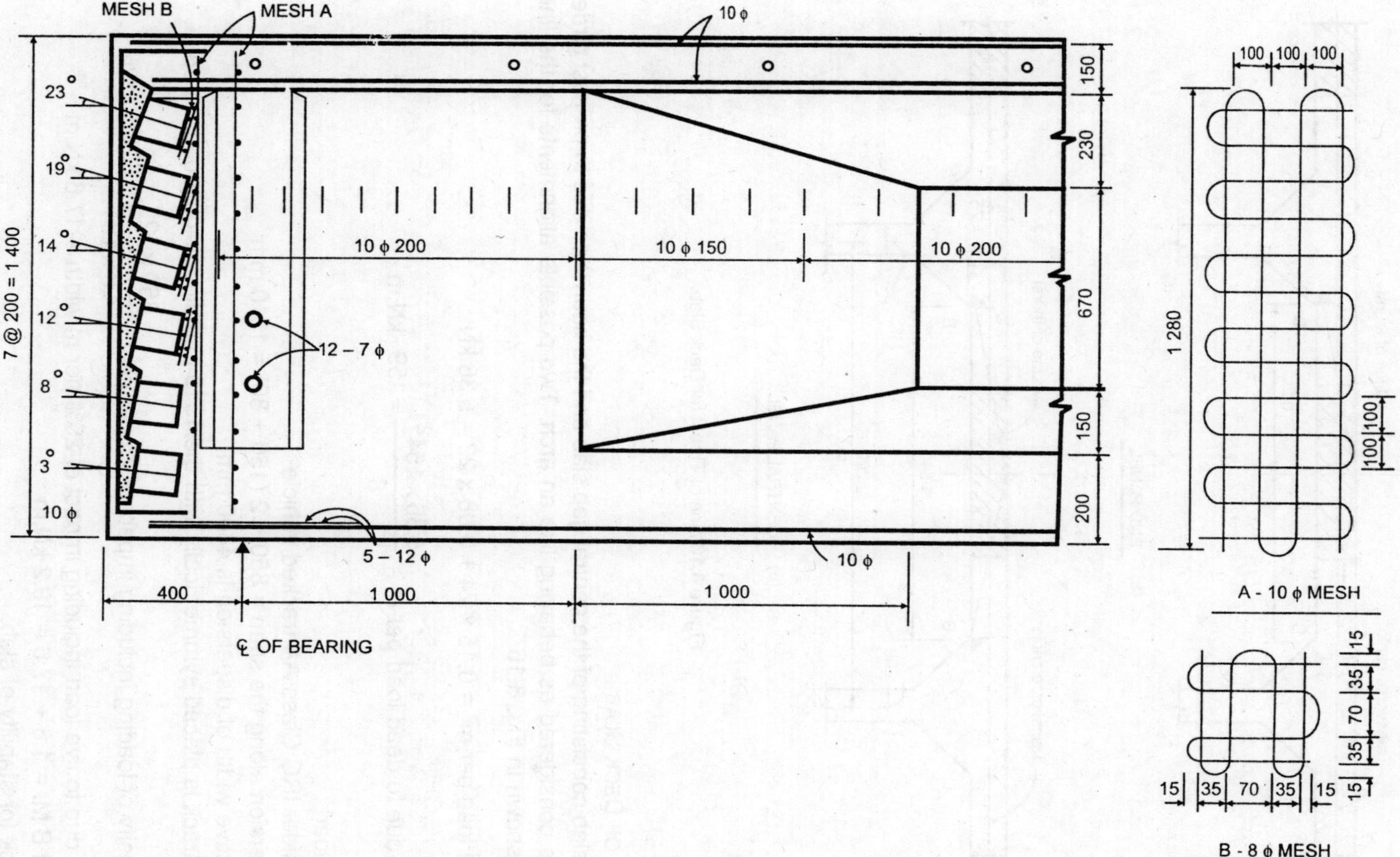

Figure 8.15 End Block Details.

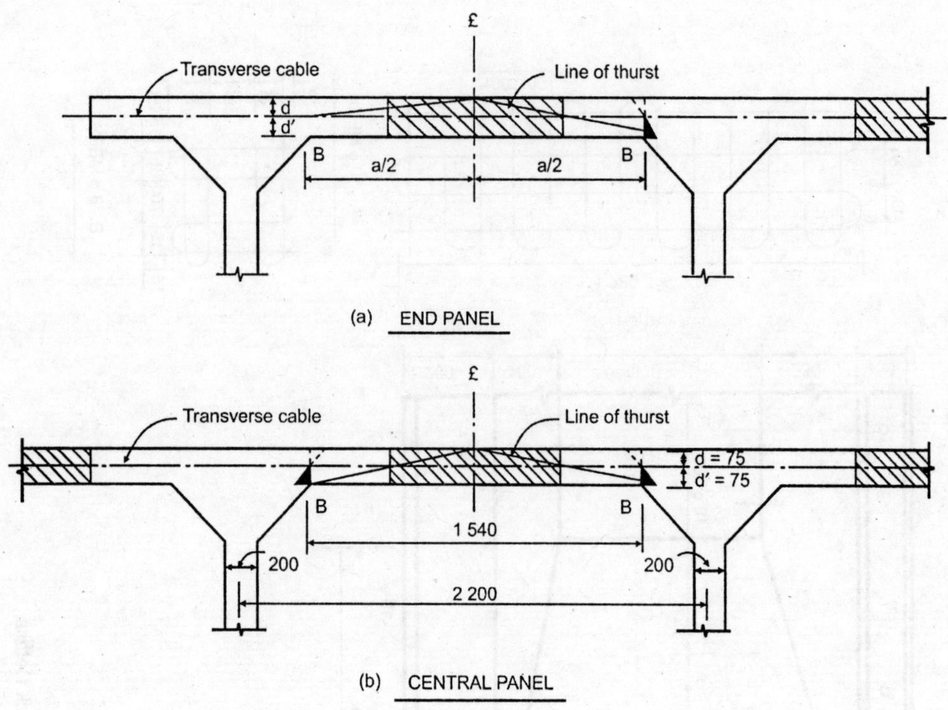

Figure 8.16 Line of Thrust for Deck Slab.

(r) DESIGN OF DECK SLAB

The slab consisting of the central gap slab and the haunches due to precast girder at the ends is considered as behaving like an arch. Two possible alignments for the line of thrust are shown in Fig. 8.16.

Dead load per m² = 0.15 x 24 + 0.08 x 22 = 5.36 kN

B.M. due to dead load per m = $\dfrac{5.36 \times 1.54^2}{8} = 1.59$ kN.m

Live load :
Consider IRC Class AA tracked vehicle
Dispersion along the span = 850 + 2 (150 + 80) = 1310 mm
Effective width of dispersion is 4800 mm
The track is placed symmetrically with the 1310 mm dimension along the span.

Intensity of loading including impact of 25 per cent $= \dfrac{350 \times 1.25}{4.8 \times 1.31} = 69.58$ kN/m²

B.M. due to live load including impact of 25% per m width = 17.6 kN.m
Total B.M. = 1.6 + 17.6 = 19.2 kN.m
Check for stability of slab:

Maximum rise possible for end panel $c = \frac{2}{3}d + \frac{1}{3}d' = \frac{2}{3} \times 75 + \frac{1}{3} \times 75 = 75$ mm

Maximum rise possible for central panel $= \frac{2}{3}\ d + \frac{2}{3}\ d' = 100$ mm

$$\text{Maximum thrust needed per m} = \frac{M}{c} = \frac{19.2 \times 1000}{75} = 256 \text{ kN}$$

Assume use of 12-5ϕ cables at 800 mm centres
Allowable initial prestress for 5ϕ wire = $0.6 \times 1600 = 960$ MPa
Force required per cable $= 0.8 \times 256 = 205$ kN

Final prestress required in cables $= \dfrac{205 \times 1000}{12 \times 19.6} = 872$ MPa

Assume jack stress at $0.7 \times 1600 = 1120$ MPa

Loss due to slip of 2.5 mm $= \dfrac{2.5}{8000} \times 2.1 \times 10^5 = 66$ MPa

Loss due to friction $= 0.0091 \times 8 \times 1100 = 80$ MPa
Final stress in steel at transfer $= 1120 - (66 + 80) = 974$ MPa
Loss due to relaxation (assumed) $= 35$ MPa
Loss due to shrinkage at 28 days $= 1.9 \times 10^{-4} \times 2.1 \times 10^5 = 40$ MPa

Average stress in concrete $= \dfrac{12 \times 19.6 \times 974}{650 \times 150} = 2.3$ MPa

Loss due to creep $= 4.0 \times 10^{-4} \times \dfrac{2.3}{10} \times 2.1 \times 10^5 = 21$ MPa

Final prestress in steel $= 974 - (35 + 40 + 21) = 878$ MPa
This value is less than 960 MPa allowable and is greater than 872 MPa required. Hence safe.

(s) INTERMEDIATE CROSS BEAMS

(i) *Live Load B.M.*

Using Morice-Little method, the transverse bending moment M_y is given by Equation (B.10).

$$M_y = b\,.\,(\mu_\theta \cdot r_1 - \mu_{3\theta} \cdot r_3 + \mu_{5\theta} \cdot r_5)$$

Assuming the track load to be symmetrically placed along the span length 2a, and assuming the length of track as 2c, the r_n terms are given by

$$r_n = \frac{4P_o}{n\pi} \sin \frac{n\pi}{2}.\ \sin \frac{n\pi c}{2a},\quad n = 1, 3, 5, \ldots$$

Here $\theta = 0.34$; $a = 9.8$ m; $c = 1.8$ m; $b = 4.4$ m.
The r_n terms can be evaluated and are determined as

$$r_1 = 0.362\ P_o;\ r_3 = -0.323\ P_o;\ \text{and}\ r_5 = 0.252\ P_o$$

P_o = intensity of loading with impact due to one track

$$= \frac{350 \times 1.25}{3.6} = 121.5 \text{ kN/m}$$

The $\mu\theta$ coefficients relating to the standard load positions for θ, 3θ and 5θ are tabulated as below:

Load position	θ $\times 10^{-4}$	3θ $\times 10^{-4}$	5θ $\times 10^{-4}$
0	2150	934	569
$b/4$	1034	183	13
$b/2$	208	−115	−21
$3b/4$	−536	−134	−15
b	−1230	−144	11

Using these values, the influence lines for transverse moment coefficients are plotted as in Fig. 8.17.

Placing the load as shown in Fig. 8.17, we could get the influence values for maximum positive moments.

$$\mu_\theta = (2150 + 300)\ 10^{-4} = 2450 \times 10^{-4}$$
$$\mu_{3\theta} = (934 - 90)\ 10^{-4} = 844 \times 10^{-4}$$
$$\mu_{5\theta} = (569 - 15)\ 10^{-4} = 554 \times 10^{-4}$$

Substituting

$$M_y = 4.4 \times 10^{-4}\ [2450 \times 0.362 - 844 \times (-0.323) + 554 \times 0.252]\ 121.5$$
$$= 69.5 \text{ kN.m/m}$$

Spacing of cross beams = 4.7 m

B.M. to be resisted by each cross beam = 4.7 × 69.5 = 327 kN.m

Section characteristics of cross beam are available in Table 8.1.

$$\text{Stress at bottom fibre} = -\frac{327 \times 10^6}{0.73 \times 10^8} = -4.48 \text{ MPa}$$

$$\text{Stress at top fibre} = \frac{327 \times 10^6}{3.44 \times 10^8} = 0.95 \text{ MPa}$$

(ii) *Dead load B.M.*

Assuming a triangular area at 45° inclination on either side of the cross beam to contribute to dead load on the cross beam, weight of wearing course and deck slab can be computed as 60.5 kN.

Following a procedure similar to that for live load, maximum transverse bending moment due to cross girders can be evaluated as 6.5 kN.m.

Stresses due to dead load B.M. are:

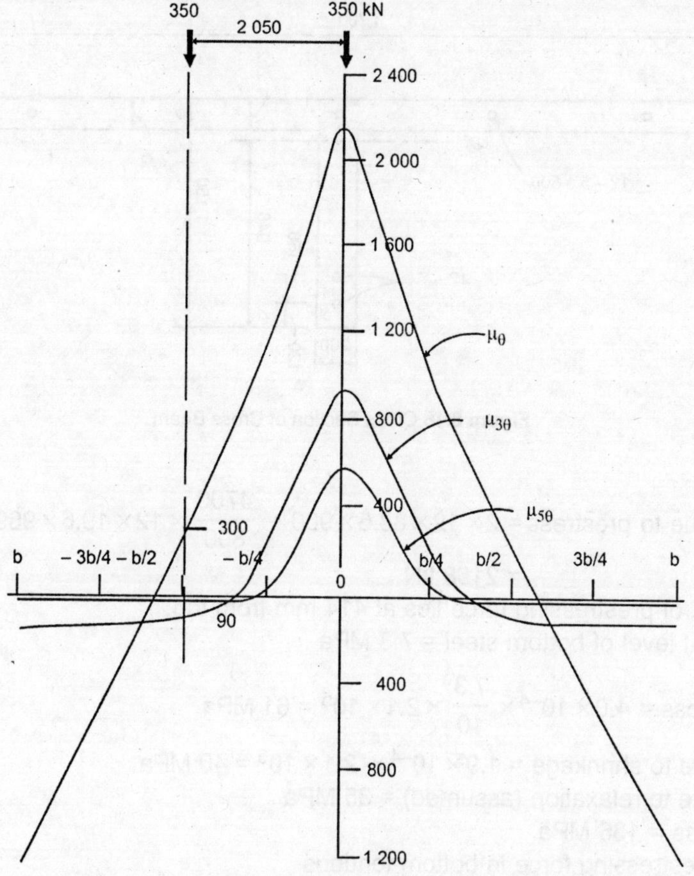

Figure 8.17 Influence Lines for Transverse Moment Coefficients.

$$\text{Stress at bottom fibre} = -\frac{6.5 \times 10^6}{0.73 \times 10^8} = -0.09 \text{ MPa}$$

$$\text{Stress at top fibre} = \frac{6.5 \times 10^6}{3.44 \times 10^8} = 0.02 \text{ MPa}$$

(iii) *Net stresses due to D.L. and L.L.*

Stress at bottom = –4.48 - 0.09 = –4.57 MPa

Stress at top = 0.95 + 0.02 = 0.97 MPa

(iv) *Prestress*

The cross section of a cross beam is shown in Fig. 8.18.

Two cables of 12-7ϕ are provided in the web and the flange has cables of 12-5ϕ at 800 mm centres.

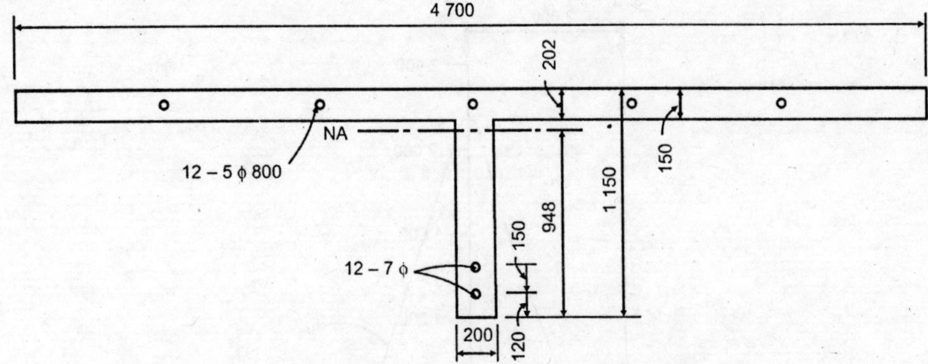

Figure 8.18 Cross Section of Cross Beam.

$$\text{Force due to prestress} = 2 \times 12 \times 38.5 \times 900 + \frac{4700}{800} \times 12 \times 19.6 \times 960$$

$$= 2158 \text{ kN}$$

The c.g. of prestressing force lies at 414 mm from top.
Stress at level of bottom steel = 7.3 MPa

$$\text{Creep loss} = 4.0 \times 10^{-4} \times \frac{7.3}{10} \times 2.1 \times 10^{5} = 61 \text{ MPa}$$

Loss due to shrinkage = $1.9 \times 10^{-4} \times 2.1 \times 10^{5}$ = 40 MPa
Loss due to relaxation (assumed) = 35 MPa
Total loss = 136 MPa
Final prestressing force in bottom tendons

$$= 2 \times 12 \times 38.5\ (900 - 136) \times 10^{-3} = 705 \text{ kN}$$

Final prestressing force in top tendons

$$= \frac{4700}{800} \times 12 \times 19.6\ (960 - 136) \times 10^{-3} = 1139 \text{ kN}$$

Total final prestressing force = 1844 kN
Distance of c.g. of prestressing force from top = 411 mm
The final stress due to prestress can be determined as
Stress at bottom = 7.31 MPa
Stress at top = 0.88 MPa

(v) *Net stresses*

Net stresses due to bending moment and prestress are:
At bottom = – 4.57 + 7.31 = 2.74 MPa
At top = 0.97 + 0.88 = 1.85 MPa

(vi) *Shear*

For maximum shear on cross beam, Class AA track is placed as shown in Fig. 8.19.
Live load shear including impact = 154 kN

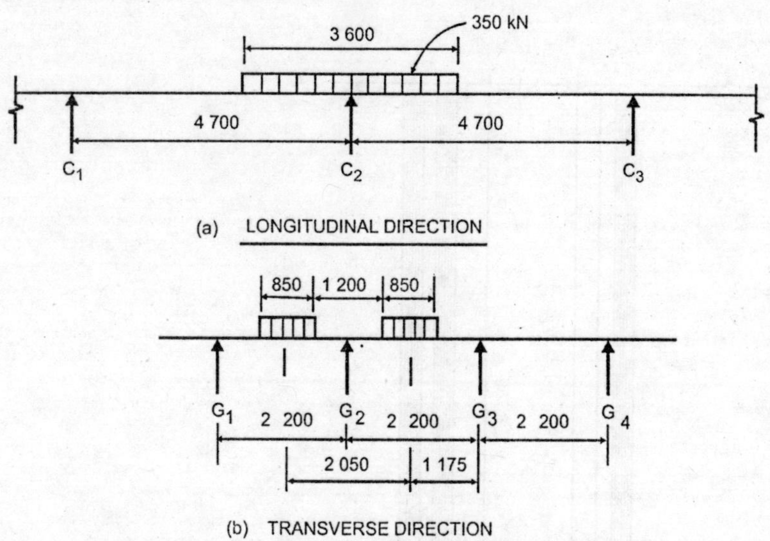

Figure 8.19 Loading for Maximum Shear on Cross Beam.

Dead load shear = 10 kN

Total shear force = 164 kN

Shear stress at level of NA = 1.1 MPa

Compression due to prestress = 2.0 MPa

Provide nominal stirrups (minimum 0.18% of plan area) with 2-legged 10 ϕ stirrups at 300 mm centres.

Provide nominal longitudinal mild steel bars (minimum 0.15% of section) with 6 - 10ϕ bars spaced two at top two at bottom and two at mid depth.

(t) DRAWINGS

Only a few of the sketches required to follow the design calculations are provided here. For a practical design, many other detailed drawings giving complete details, bar bending schedules, etc. will be required.

8.8 Example of Composite Prestressed Concrete Superstructure

Typical details of the composite prestressed superstructure for a span 35 m are shown in Fig. 8.20 to 8.23 as adapted from standard drawings in Reference 12. The deck consists of three precast prestressed girders of T-section with a cast-in-situ reinforced concrete deck slab and cast-in-situ cross girders, with dimensions as shown in Fig. 8.20. The concrete grade adopted is M40. The prestressing cable consists of 12 nos. of 12.7 mm diameter 7-ply Class 2 strand as per IS: 6006.

Each precast girder has eight cables with cable profile and midspan cross section as indicated in Fig. 8.21. The cables are all anchored at the ends of the girder. Cables 1 to 7 each have a straight portion in the middle and curved portions on either side. Cables 1 to 4 and 6 and 7 have curved profile in plan and in elevation. The curved parts are necessary to locate the anchorages at the ends of the girder.

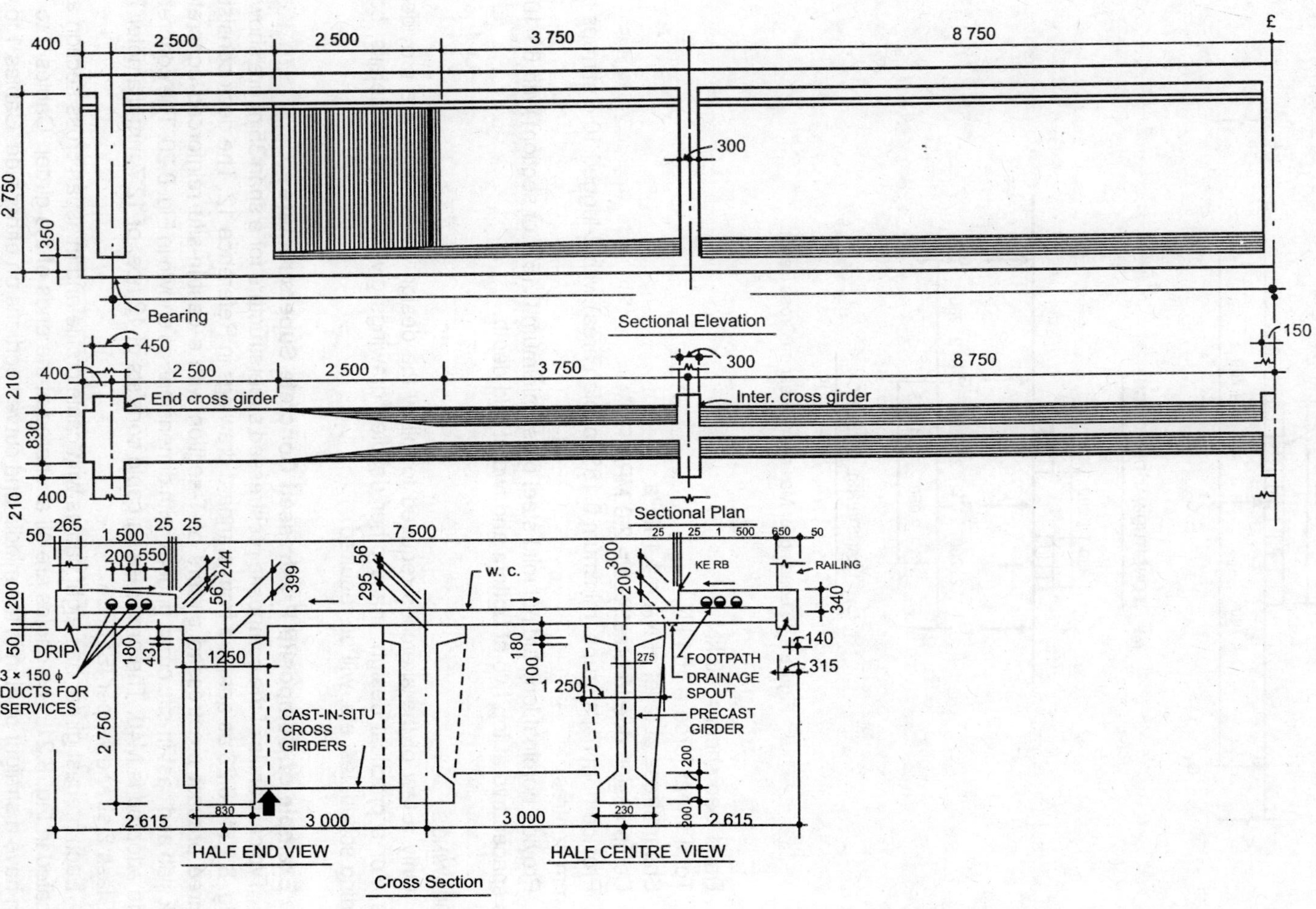

Figure 8.20 Composite P.S.C Superstructure - Dimenstions

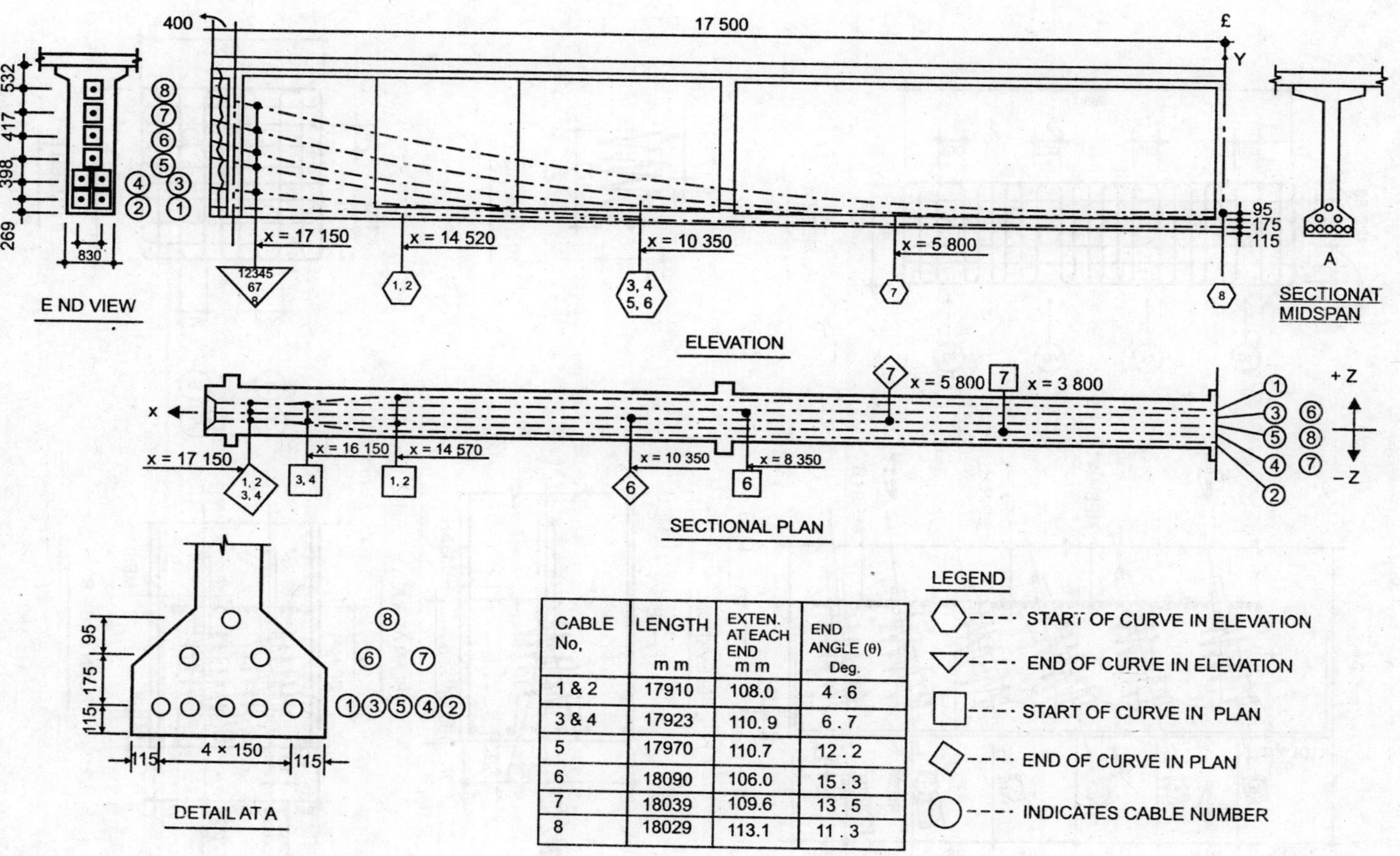

CABLE No,	LENGTH mm	EXTEN. AT EACH END mm	END ANGLE (θ) Deg.
1 & 2	17910	108.0	4 . 6
3 & 4	17923	110. 9	6 . 7
5	17970	110.7	12 . 2
6	18090	106.0	15 . 3
7	18039	109.6	13 . 5
8	18029	113.1	11 . 3

Figure 8.21 Composite P.S.C. Bridge - Cable Profile.

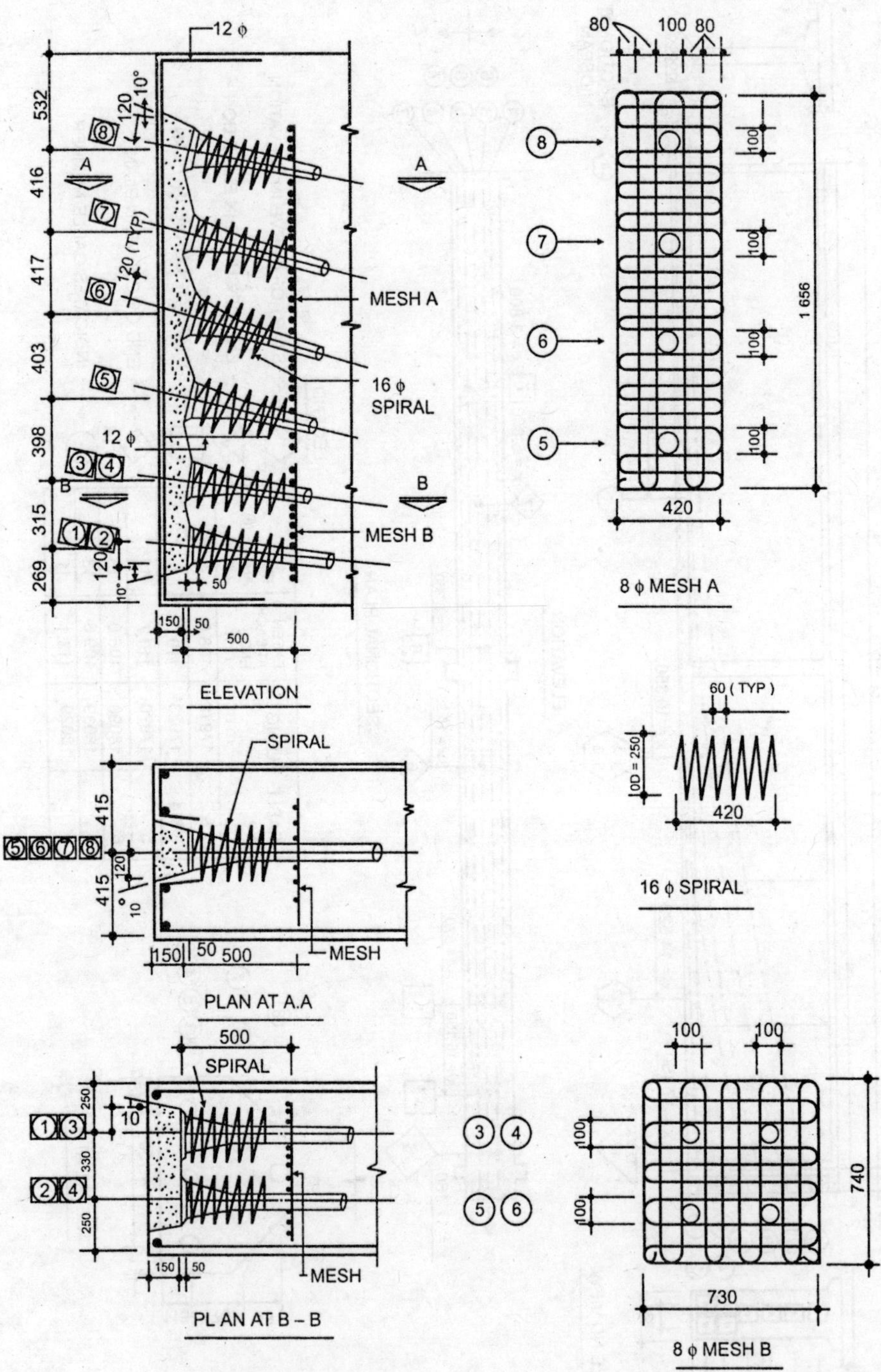

Figure 8.22 Composite P.S.C. Bridge - End Block Details.

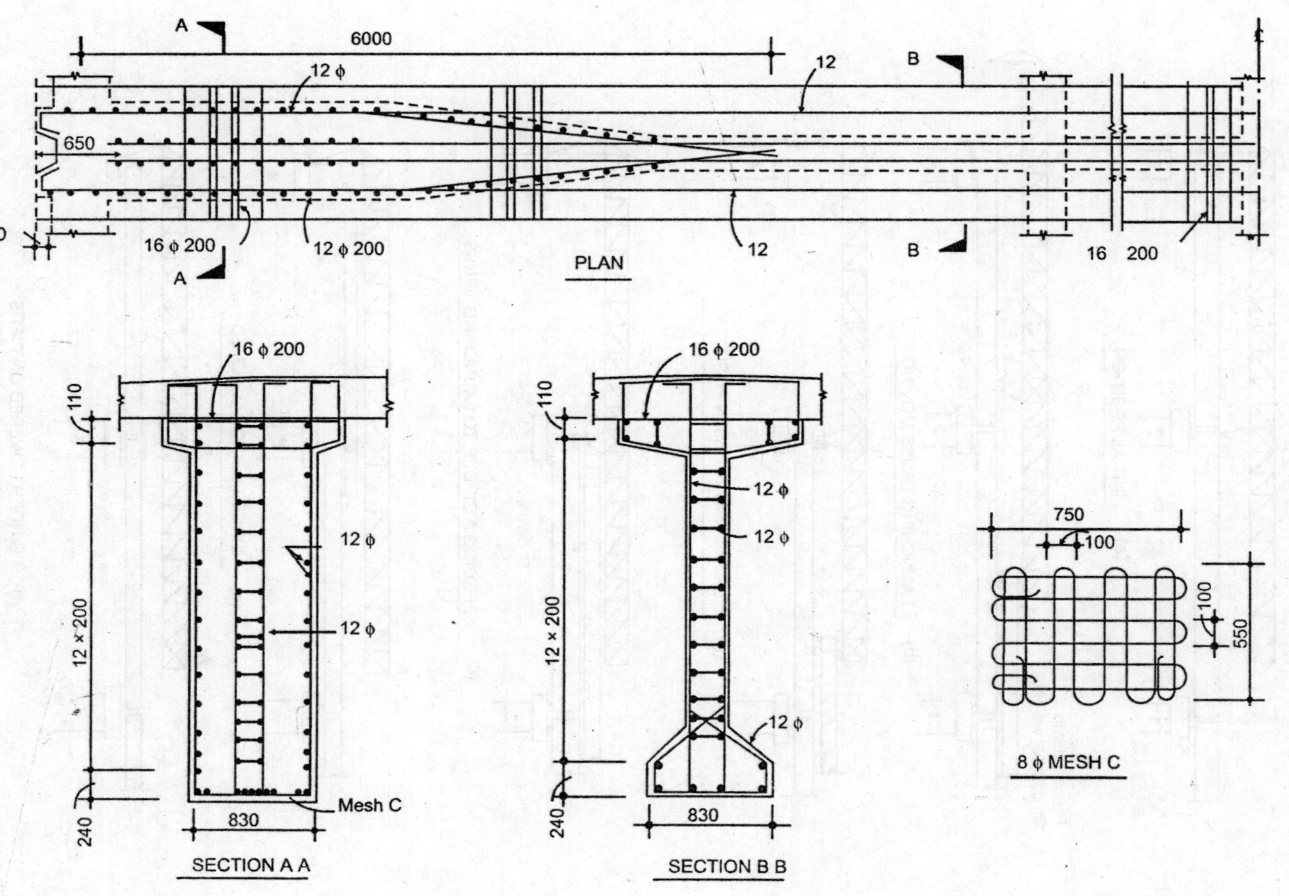

Figure 8.23 Composite P.S.C. Bridge - Reinforcement in Central Girder.

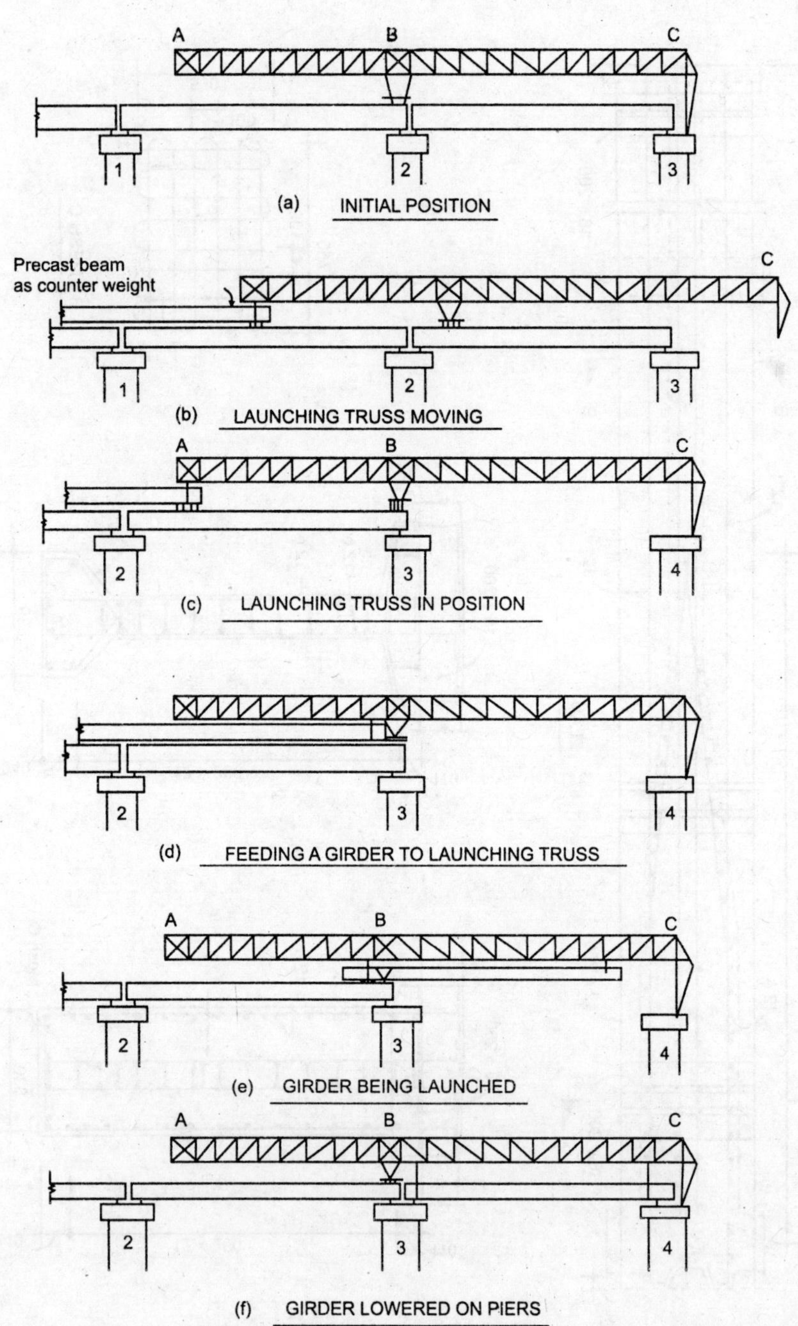

Figure 8.24 Launching of Precast Girders.

The details of the end block are shown in Fig. 8.22. The anchor plates are recessed by 150 mm from the edge of the beam which itself is at 400 mm from the centre line of the bearing. The meshes A and B are provided to cater to bursting tensile force while the spiral reinforcement and 12 mm rods at the edges are intended as protection against spalling tensile forces.

Typical untensioned reinforcements for the central longitudinal girder are shown in Fig. 8.23. For a detailed consideration of composite prestressed bridge superstructure and the additional requirements for actual construction, the reader may see Reference 12.

8.9 Erection of Precast Girders

An essential requirement for the use of precast members in bridge construction is the economic availability of erection equipment. Depending on site conditions, precast bridge members may be erected using truck cranes, crawler cranes, floating cranes or grider launchers. Crane erection is a popular method adopted for short-span or simple-span bridges. In case of river bridges, cranes mounted on floating barges may be used. When the range of tides in a tidal river is considerable, precast griders may be floated on barges during high tide and allowed to rest on the pier supports during low tide.

Typical details of the sequence of operations for erecting a precast prestressed girder using a launching truss are shown in Fig. 8.24. The launching truss is of steel or aluminium alloy, and is approximately 1.75 times the length of the girder to be launched. The truss has a triangular profile and is provided with a central and a front trestle. It moves over rails laid along the centre lines of the webs of beams already launched into position in the previous span. When the launcher is to be moved, a precast girder is attached to the rear end to serve as a counter weight. The launching truss and the girder are moved such that the front trestle rests on the forward pier. The launcher is now in position and ready to erect the girder. The front end of the girder to be launched is then pulled forward with the front end suspended from the truss and the rear end still supported on rail-mounted bogies. As the front end of the girder advances sufficiently, the rear end is also hooked up to the underside of the truss. The girder is then pulled forward further till it is just above the bearings in the span. It is then lowered on to the bearings on the pier.

Precast girders may also be erected using falsework in case of bridges with low heights above dry ground. The girders may be precast in segments, assembled, stressed and grouted on falsework. They are then slid transversely into place.

8.10 Continuous Bridges

8.10.1 Typical Forms of Superstructure

A bridge is called a continuous bridge when the superstructure is longitudinally continuous over intermediate supports on bearings. Continuity in superstructure can be achieved by several methods, including: (a) segmental cantilever construction using cast-in-place or precast segments; (b) span-by-span method; (c) incremental launching method; or (d) precast girders made continuous by cast-in-place slab and diaphragm. The above methods are discussed in Section 8.11 and Section 15.7.

Typical forms of superstructure adopted in continuous spans for bridges and elevated expressways are shown in Fig. 8.25. The applicable ranges of span and depth of deck

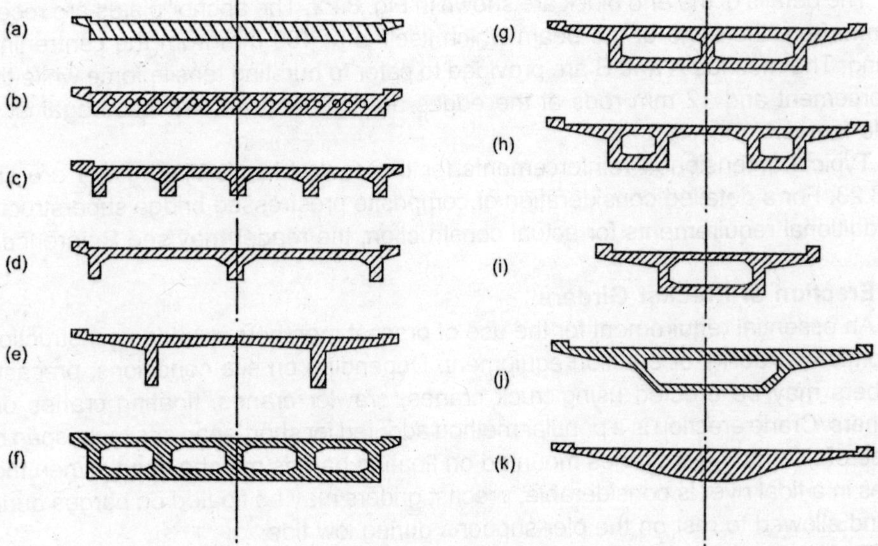

Figure 8.25 Typical Forms of Post-tensioned Prestressed Concrete Bridge Decks.

required for these types are listed in Table 8.13. The numerical values given in the Table are to be taken only as indicative of the order of magnitudes applicable.

Continuity can be used effectively in long span bridges. Due to the increased rigidity resulting from continuous construction, shallower members can be used on long spans without incurring excessive deflections. The vibrations occurring due to traffic loads will also be less.

Table 8.13 Typical forms of Superstructure.

Type Designation in Fig. 8.25	Type	Span Range m	Span / Depth
a	Solid slab	20 to 25	10 to 30
b	Hollow slab	25 to 35	20 to 30
c	T-beam, multi-girder	30 to 45	15 to 25
d	T-beam, three-girder	30 to 45	15 to 25
e	T-beam, two-girder	30 to 40	12 to 15
f	Box, multi-cell	28 to 40	20 to 25
g	Box, two-cell	30 to 40	20 to 25
h	Box, two separated cells	30 to 70	20 to 25
i	Box, single rectangular cell	30 to 50	20 to 50
j	Box, single trapezoidal cell	30 to 50	20 to 30
k	Fish-back pattern	25	25

8.10.2 Continuous Slab Bridges

Prestressed concrete continuous slab type is particularly attractive for elevated expressways passing through urban areas in view of the shallow depth. Slab bridges are easy to construct and they lead to lower construction costs.

The criteria for elastic design are to limit the principal fibre stresses at the top and the bottom surfaces of the slab to prescribed limits. Besides prestressing cables, additional untensioned reinforcement should be provided in the top and bottom surfaces to cater to residual tensile stresses.

8.10.3 Box Girder Arrangements

Box girder construction, of single or multiple cells, has become very popular in recent years. Being a closed section, a box girder has relatively large torsional rigidity compared with other types of beams. The torsional resistance of a box girder can be modified by varying the span/depth ratio and the width/span ratio, as the resistance is proportional to the area enclosed by the median line of the box. Haunches are provided at the junctions of the webs with the top and bottom slabs to facilitate smooth flow of shear stress from one element to another.

For bridge decks with box girder construction, a fairly substantial cantilever flange beyond the outer web of the box is permissible and economical. The width of the webs are to be adequate to cater to flexural and torsional shear, besides accommodating conveniently the prestressing cables and anchorages. Hence the webs are usually thicker than the flanges.

Concrete box sections may not normally require diaphragms between pier locations for preventing box distortion, except in case of box girders having thin concrete webs. The distortional stresses are a minimum when the box is square, but it is inconvenient to adopt square shaped boxes. From consideration of formwork and ease of construction, flanges are horizontal and webs are often vertical. Since a circular tube is the most efficient torsion resisting section, the nearer the shape of the box approaches a circular shape, the more efficient the shape will be with regard to distortional stresses. For practical cases, adjustments are made in the relative thickness of flanges and webs, besides providing haunches at their junctions. Warping in the longitudinal direction due to torsion is usually not a severe problem with practical box girders.

Webs of concrete box girders may be made vertical for convenience in construction. However, when a wide deck is supported on a single cell box girder, it is more convenient to have the outer webs inclined such that the tops of the webs can be widely spaced under the top flange and the span of the bottom flange is reduced. Such an arrangement has many advantages, including: reduced size of substructure, better streamlining for wind forces and, in case of urban elevated highway, better lighting at the road level[13].

8.11 Segmental Cantilever Construction

The economy of continuous construction is enhanced by the adoption of the segmental cantilever construction method pioneered by M/s. Dyckerhof & Widman in Germany [14,15] This method is particularly economical where construction of superstructure using staging from the river bed is not feasible. The segmental cantilever construction may adopt cast-in-place segments or precast segments, and is considered suitable for the span range of 50 to 200 m, though a few bridges have been built beyond 200 m span.

Comprehensive calculations are to be performed for every construction stage to account for time-dependent material behaviour of the segments under successive load stages. The deflection at every stage should be calculated and the needed camber should be determined and applied. Depending on the specific segment configuration and the erection sequence adopted, the superstructure may need to be balanced to ensure stability, by providing temporary supports near the piers to withstand overturning moments from unbalanced loads from the cantilever arms. The piers are to be designed to withstand bending moments due to unfavourable combination of loads on the cantilever arms during construction.

The box section is the best suited cross section for cantilever construction, in view of the torsional rigidity and the availability of the top and bottom slab portions for accommodating the tendons. The single cell box is used in many bridges in India for deck width less than 13 m. The two webs may be vertical or sloping. Box sections with two or three cells are used for wider decks.

8.11.1 Cast-in-Place Segments

The method of cantilever construction using the cast-in-place segments is shown in Fig. 8.26 (a), as used in the Rhine bridge at Worms in Germany built in 1952 (spans 101-114-104 m). The structure is built from a pier or support across the open span without temporary supports in sections of about 3.5 m length. The construction should proceed on both sides of the pier such that at any stage the unbalanced moment is kept minimum. Each section is cast in the cantilever form traveler (CFT) which is attached to the end of the constructed portion. When the concrete attains the specified strength, the segment is pressed against the previous one by means of prestressing tendons ending at the front of the section. Other prestressing tendons are carried through unstressed at a length corresponding to the static requirements of the construction. At each section, the tendons ending there may be bonded by grouting the duct, so that at every stage the construction is safe against rupture. This process is continued till midspan.

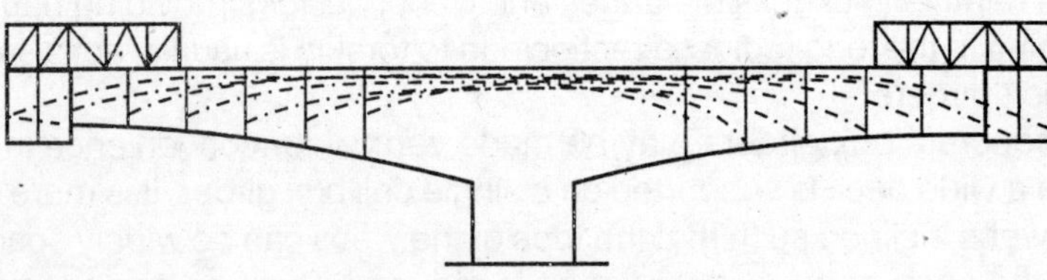

(a) Using cast – in – place segments

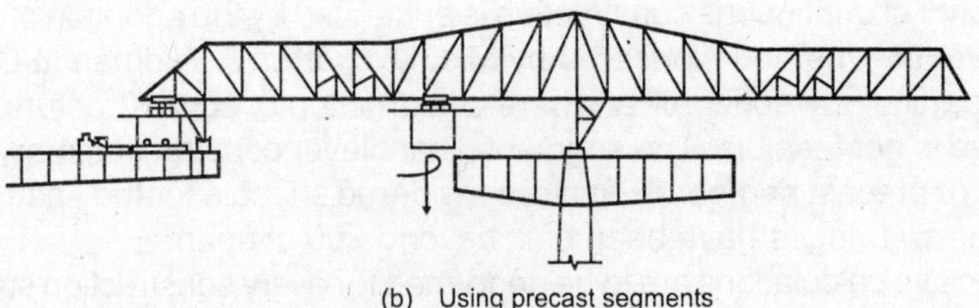

(b) Using precast segments

Figure 8.26 Construction by Segmental Cantilever Method for Long-span Prestressed Concrete Bridges.

The cost of the cantilever form traveler is relatively small. By repeated use of the CFT, the construction of large bridges is facilitated. The avoidance of scaffold from below, the speed of work and the saving in labour cost result in economical construction. The free-cantilever system is ideally suited for haunched girders with a large depth above the pier cantilevering towards the middle of the span.

An improved prestressing technique using short length diagonal prestressing tendons in the webs of the segments was applied in Bendorf bridge in Germany, built in 1964 with three spans of 71-208-71 m adopting variable depth box sections. A record bridge with cast-in-place segments is the Hamana bridge completed in 1976 in Japan with a main span of 240 m.

Many outstanding segmental cantilever bridges have been constructed in India with a single cell box section[16]. The Barak bridge at Silchar in Assam (1961) with a clear span of 122 m was the first prestressed concrete bridge constructed by the segmental cantilever method in India. Other notable bridges include Lubha bridge (1961) with a central span of 130 m, the Bassein Creek bridge (1967) with spans of 114.6 m, the Ganga bridge at Patna (1982) with spans of 121 m, and Jadukata bridge in Meghalaya (1998) with central span of 140 m.

8.11.2 Precast Segments

Long span prestressed concrete bridges can also be built using precast segments. The segments, precast in a nearby casting yard, are about 3.5 m long covering the full width of the bridge. Each segment is transported by a trolley system from the plant to the front of the launching truss, from where it is picked up by a traveling frame hoist suspended from the bottom chord of the truss. The traveling frame has a spreader beam which can be raised, lowered or rotated so that wide sections can pass between the supporting legs of the launching truss and can be lowered into the proper position. The section is then fastened in place with epoxy and prestressing tendons.

The first major bridge using precast segments in cantilever construction was the Choisy-le-Roi bridge over the Seine in France, built in 1962 with 37.5-55-37.5 m spans by M/s. Campenon Bernard[15]. A notable example of this type of construction, shown schematically in Fig. 8.26 (b), is Oleron Viaduct (1966) in France[17]. Other examples include the Osterschelde bridge in the Netherlands, and the Mandovi (first) bridge at Panjim, the Narmada bridge at Zadeshwar, the Ganga bridge at Buxar, and the Pamban bridge at Rameswaram[18,19].

One problem with the use of precast segments is in joining segments prior to stressing. In order to ensure precise matching of the mating faces, each segment is cast directly against the face of the previous segment. This process is known as match-casting. Still the faces may not match precisely at the time of joining, due to differences in temperature and humidity during curing. The joint faces are cleaned by sand blasting and a thin layer of epoxy is applied to the mating faces prior to prestressing. The epoxy serves to lubricate the faces of the joints and in the hardened condition seals the joints against moisture thus protecting the tendons. It may be noted that the epoxy gains a strength higher than the concrete in the segments.

It is not advisable to use a narrow layer of concrete for joining, as this would result in two construction joints with the possibility to permit leak of water from the deck. The use of mortar joints would entail delay due to curing and hardening of the joints. The poor performance of a buttering mortar used between precast segments is cited as a contributory cause for the corrosion of tendons leading to the collapse of the Yns-y-Gwas bridge in UK in 1985[20].

To facilitate shear transfer, it would be desirable to give the face of the joint a saw-toothed profile distributed over the whole depth of the web. Since the precast segments would have matured to almost the full design strength prior to erection, the effects due to shrinkage and creep of concrete would be considerably reduced. Special care is necessary to control the geometry of the segments during fabrication and erection. The design of the completed cantilevers should be checked to carry safely the loads from the launching truss during handling of segments and during the launching of the truss itself from one span to another. Since untensionezd reinforcement is lacking across the joint between any two segments, net tensile stresses cannot be allowed.

8.11.3 Advantages

The segmental cantilever method of construction of prestressed concrete bridges has the following advantages: (a) Centering and falsework are avoided, enabling construction of structures with tall piers and over deep valleys; (b) the speed of construction is enhanced, typically at 1 m per day per CFT for cast-in-place construction and possibly 3 m per day with the use of prefabricated segments; (c) enhanced levels of quality and workmanship are facilitated due to mechanization of repetitive tasks; and (d) the cost of construction permits competition with alternative design of a steel superstructure of long span. Precasting of segments has additional advantages: (i) Shrinkage effects are avoided due to age of concrete at the time of erection; (ii) Creep in concrete is less due to age at time of initial loading; (iii) Saving in construction time as segments can be precast during construction of substructure; and (iv) Protection from weather during concreting as precasting is performed in a factory.

However, the design computations are more complex and voluminous, as a large number of sections have to be checked for safety and stability during the various stages of construction. Further the effects of creep of concrete and relaxation of the prestressing steel have to be duly accounted for even during stages of construction to ensure effective control of the girder profile.

8.11.4 Connection at Midspan

The choice of the type of interconnection at the mating ends of the two cantilever arms is an important decision affecting the design, construction and performance of the superstructure. The interconnection can be one of the following: (a) Sliding hinge; (b) Suspended span; or (c) Continuous system.

In the case of the sliding hinge, the mating ends of the two cantilever arms are connected through a set of cast steel hinge bearings located in the webs. The central hinge bearing allows transfer of shear forces without structural continuity, and permits free longitudinal displacements of one cantilever girder relative to the other. The central hinge system has a few disadvantages: (a) The hinge is difficult to design and install correctly; (b) The long term performance of the hinge is poor due to wear of the steel plunger and the steel plates below and above the plunger during service; and (c) The riding quality suffers because of progressive settlement of the cantilevered ends during the early part of the service due to shrinkage and creep of concrete and relaxation of steel tendons of the girders. While the other portions of such bridges have performed satisfactorily, the central hinges in some of the cantilever bridges have caused maintenance problems. An alternative design using appropriately designed and housed elastomeric bearings instead of sliding surfaces with steel plates has been proposed to overcome the wear associated with steel hinges [21]. The use of the central hinge is now obsolete.

Instead of a central hinge, an independent suspended span may be used to connect the two cantilevers. The ratio of the suspended span to the main span may vary in a wide range from 0.1 to 0.4. Elastomeric bearings are used at the articulations. The effect of deflection of the cantilever ends is reduced by half in this case compared with the central hinge. This system is sometimes advocated for situations anticipating differential settlement of supports. The bending moment at the supports are reduced due to reduction in the cantilever span for the same main span. However, the riding quality is not significantly improved. The suspended span system is also not favoured in recent designs.

The current trend is to adopt the continuous system by providing a keying segment at the junction of the two cantilevers and prestressing to make the element integral with the cantilever girders. This arrangement is technically superior to the other two arrangements. The deflection at the centre is less and the riding quality is enhanced.

8.12 Stressed Ribbon Bridge

A new type of prestressed concrete bridge construction, called stressed ribbon bridge, has been developed by Finsterwalder[22]. The basic concept of this system is a stressed ribbon of prestressed concrete, hanging in a funicular curve, anchored in the river banks. The bridge is a combination of a suspended concave span with about 2500 m radius and the supported convex part near the piers provided with long cantilevers in order to obtain satisfactory roadway gradients. After the cantilever piers are completed, the tendons are placed piece by piece and stressed to ensure the desired roadway grade at full load. Formwork is hung from the tendons. The stressed ribbon slab is heavily reinforced in the transversal direction, both at the top and bottom, to resist the torsion and bending moments due to traffic loads. The concreting starts in the middle of the freely hanging suspended concave part and continues without interruption to the supports. After the concrete has cured, the formwork is removed. The weight of the steel in the stressed ribbon is very heavy and is around 3.5 kN/m². Stressed ribbon bridge creates substantial horizontal forces, which need to be anchored at the abutments. When soil conditions are poor for cable anchorages, counterweight equal to about twice the horizontal thrust will be needed at the anchorages. A wearing surface of asphaltic concrete is provided for 50 mm thickness.

The first stressed ribbon bridge was built in Switzerland by Rene Walther. Recent applications include the Kent Messenger Millenium bridge designed by Cezary Bednarski of UK and completed in 2001 [23]. For the Kent Messenger bridge, the bridge deck consists of precast concrete segments topped with a cast-in-situ composite deck slab. The segment, suspended on bearing cables, served as falsework and formwork for the in-situ slab. The bearing cables and the tendons were anchored at the abutment blocks. Though they appear to be simple, stressed ribbon bridges are specialist structures, requiring complex computations and competent construction.

8.13 Bonding of Tendons

For pretensioned members, the prestressing tendons are bonded to concrete naturally, and adequate protection against corrosion is ensured. In post-tensioned prestressed concrete members, prestressing tendons are invariably bonded to concrete by grouting of the ducts for the cables. The grout ensures encasement of steel in an alkaline environment for corrosion protection and by filling the duct space, it prevents water collection and freezing. The resulting advantages are: (a) protection of tendons from corrosion; (b) improved ultimate strength; and (c) reduction of crack widths and crack spacing. However, perfect grouting of ducts is a challenging operation, and occasionally voids are formed particularly at inclined anchorages. Such voids aggravate the risk of moisture, chlorides, oxygen and carbon dioxide entering the ducts, resulting in corrosion of tendons.

Guidelines for grouting of post-tensioned cables in prestressed concrete bridges are available in Appendix 5 of IRC: 18[11]. A typical mix for grouting can be 5:2 (Cement: Water) by weight. It is possible to reduce water content by use of water reducing and expansive admixtures. Admixtures for the grout may be used only with caution, as even small quantities of impurities such as chlorides present in the admixture may impair the strength of the stressed high tensile wires[7]. Some admixtures may be incompatible with the cement used causing instability of the grout during setting [24]. The mix to be adopted for any particular application would depend on the length of the cable duct and the grouting equipment. At bridge sites with high ambient temperatures, it would be desirable to use chilled water (10° to 15°C) so as to ensure grout temperature of about 25°C which would result in improved flow conditions for the grout. High strength portland cement from fresh stock should be used for grouting. The compressive strength of 100 mm cubes of the grout should not be less than 17 MPa at 7 days.

Ducts for cables may be of corrugated metal ducts formed by spirally wound sheathing strips (0.3 to 0.4 mm thick) or of plastic tubes of high density polyethylene (HDPE) with minimum thickness of 2 mm. The ducts and their couplers should be such that they do not corrode, they are strong enough to retain their shape during handling and concreting, do not cause chemical attack and are tight enough to prevent entry of cement slurry during concreting. The duct size is determined to ensure adequate space to allow the tendon to be threaded through and to permit free flow of grout around the strand or bars. The area of the duct is usually not less than twice the area of the tendon.

The mixing equipment should ensure grout of uniform colloidal consistency. After mixing, the grout should be kept in continuous movement and should be free of lumps. Hand mixing is not permitted. The grout pump should be electrically operated reciprocating positive-displacement type working at a pressure above 0.3 MPa but below 1 MPa . The capacity of the grout pump should be such as to achieve a forward speed of grout of about 5 to 10 metres per minute. Grouting pressure is preferably kept below 1 MPa, as higher pressures tend to speed up grout flow with possible voids being left.

Grouting of tensioned cables should be performed preferably within 48 hours and not later than one week of tensioning to prevent deleterious consequences like ordinary corrosion and stress corrosion. This is particularly important for bridges in marine environment in tropical areas, where the overall duration from concreting to grouting should be kept within two weeks. Successful grouting demands special knowledge and strict quality control. There have been instances where ducts had not been perfectly filled with grout leading to corrosion of the tensioned steel, attributable to lack of diligence at the execution stage.

Grouting is preferably performed by pumping from the lowest point and venting from the highest point of the duct. Prior to grouting, water is pumped from the lowest point and allowed to flow through the highest points. The water is drained through the lowest point.

Then grouting is performed. When this procedure is inconvenient, grout may be injected from one end anchorage and vented through the other end anchorage, ensuring that the grout of the same consistency emerges from the far end. In this case, the water used for wetting should be expelled with compressed air prior to grouting. After final injection of grout, openings and vents should be hermetically sealed to avoid the ingress of water or other corrosive agents. Grout not used within 30 minutes of mixing should be rejected.

8.14 Precautions to be Observed by the Prestressed Concrete Bridge Engineer

The prestressed concrete bridge engineer should have a thorough understanding of the behaviour of prestressed concrete structures and should be fully familiar with the technique of prestressing, besides properties of the materials used. He would do well to observe the following precautions at the design stage and during construction in order to ensure a satisfactory completed bridge[4].

(a) The designer should familiarize himself with the details and sequence of the construction procedure proposed to be adopted. He should take into account the erection stresses.

(b) The design should provide for shortening of the structure in the direction of prestressing, as prestressing can compress the concrete only when shortening is possible.

(c) Adequate provision should be made in design to cater to radial forces due to change in cross section along the length of the member. Draped cables and splay of cables in plan near supports cause radial forces when the cables are tensioned. Change in the direction of the centroidal axis of the concrete member leads to unbalanced forces which act transverse to the member. The design calculations and structural detailing should take these into account.

(d) The high permissible compressive stresses in concrete can be utilized only if the stiff concrete can be placed and vibrated properly to obtain the designed strength in the field. To ensure proper placement and compaction, special care should be devoted to the choice of the cross sectional dimensions of the concrete and the detailing of the untensioned steel and the tendons.

(e) Tensile stresses should be avoided under dead load. It is prudent in bridge design not to depend on the tensile strength of concrete.

(f) Untensioned steel should be provided in the longitudinal direction to cater to ultimate load conditions, and transverse to the tendons, and specially in the anchorage zones to take care of the concentration of forces.

(g) Prestressing steel should be handled carefully, positioned accurately, and held securely. Prestressing steel is highly sensitive to corrosion, notches, kinks and heat. In thin webs, precise lateral positioning of cables is critical. Lack of precision in the positioning of cables may lead to serious problems due to friction during tensioning.

(h) The design of the formwork and the technique of concreting should be planned with utmost care to ensure adequate vibration of the concrete and to avoid cracking of the young concrete due to deflection of the formwork during concreting. The formwork should be checked for leaks at joints to avoid honeycombing in concrete.

(i) The alignment of ducts should be checked after threading of cables. Excessive wrapping with tape of sheathing at joints should be avoided, particularly in the vicinity of the anchorage.

(j) The sheathing of cables should be leakproof, as otherwise the tensioning of cables will be difficult.

(k) Before commencing the tensioning of cables, it should be checked to see that the structure can move in the direction of tensioning to permit shortening. For reasons of safety, the cable line extended on each side should be kept free of persons.

(l) Prestressing of tendons in long members should be taken up in stages. The first stage should be aimed at providing moderate compression to prevent concrete cracks due to shrinkage and temperature. Full prestress should be applied only when the concrete has attained its designed strength. It is worth remembering that the highest stresses in concrete usually occur during tensioning of the cables.

(m) While tensioning, the cable force should be ascertained from both jack pressure and the cable extension. Records of the tensioning operations should be preserved carefully.

(n) Prior to grouting of the tendons, the ducts are to be ensured to be free from obstructions. The grouting should be done strictly according to the relevant specifications. The rise of grouting pressure should be monitored during grouting. A sudden rise may indicate blockage, whereas a sudden fall may be due to leakage in duct.

(o) When precast girders are used (as in Fig. 8.3), special care should be taken to ensure correct alignment of ducts for transverse prestressing tendons. This would necessitate strict supervision of formwork fabrication and positioning of cable sheathing to close dimensional tolerances.

8.15 External Post-tensioning

External prestressing refers to the method of post-tensioning where prestressing tendons are placed outside the concrete section. The prestressing force is transferred to a structural member through anchorages and deviators. There are three types of applications of external prestressing[25]: (a) The tendons comprise the total prestressing force, e.g. segmental construction of bridge with external cables; (b) Unbonded tendons supply part of the total prestressing force, while the other part is provided by the bonded tendons embedded in concrete, as in repair, rehabilitation or strengthening of existing bridges; and (c) The tendons apply external load to counter subsequent live load, as in strengthening of RCC structures to relieve part of the dead load effects.

In external prestressing, the tendons are unbonded. The tendons may be straight or draped to suit the required combination of axial load and bending. The tendons are outside the concrete section and are attached to it at discrete points of anchorages and deviators. The tendons are straight between points of attachment. Though the tendons and the concrete elements behave as different components of the overall structure, they act in unison by virtue of the connection at anchorages and deviators, thereby contributing to overall strength. External tendons are accessible, easy to inspect and, if necessary, to replace. It is possible to provide reliable corrosion protection to external tendons by covering with grease and encasing in high density polyethylene (HDPE) ducts.

There is greater freedom in positioning the tendons inside the box section, though additional effort is needed to design the deviators to maintain the tendon profile. External

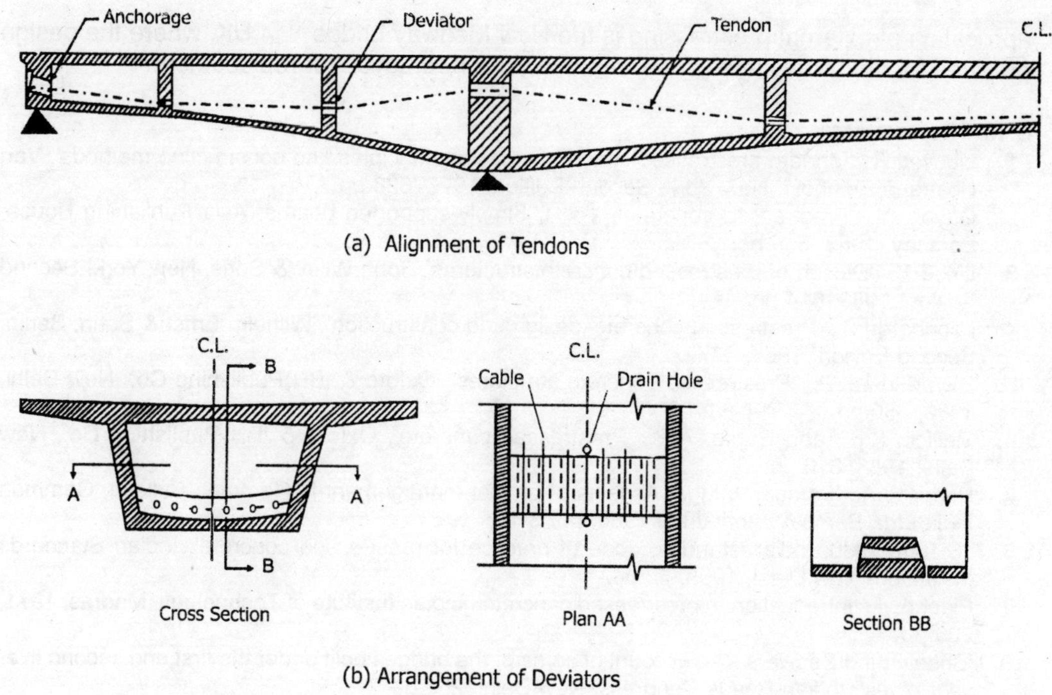

Figure 8.27 Tendon Alignment, Anchorages and Deviators for New Construction.

post-tensioning is applicable to bridges as an aid to reducing construction time. Typical alignment of tendons and arrangement of deviators for new construction are shown in Fig. 8.27. External prestressing is ideal for strengthening existing structures, in which case the deviator may consist of a structural steel bracket or saddle on the soffit of the member or bolted to the stem of the member. Success in external post-tensioning will depend on the design of the anchorage and the method of transfer of the tendon forces to the concrete at the anchorage locations. The tendons may be anchored at existing diaphragms or at the ends of the existing beams. The concrete should be sound and free from chloride contamination.

From the point of view of flexure capacity, externally prestressed bridges are less efficient than internally prestressed ones. This is because the eccentricity is less and due to tendons being unbonded the strain in the tendons does not increase at the same rate as the strain in the adjoining concrete. However, this situation may not be very significant in deep box girders. Though corrosion protection from surrounding concrete is not available, the external tendon facilitates better access for inspection and replacement.

Consequent on the failure of the Mandovi first bridge, the MORTH have stipulated that the design and detailing of prestressed concrete bridges should provide for imparting additional prestressing force to the extent of 20 % of the design prestress in the form of internal or external prestressing at a later date.

Externally prestressed cables are adopted for the continuous box girder deck of the Delhi Noida Toll Bridge, where cables pass through the inside of the box. Another recent application of external prestressing is the New Medway bridge[26] in UK, where the design provided for both limit states even with one tendon removed for replacement.

8.16 References

1. Libby, J.R., 'Modern prestressed concrete – design principles and construction methods', Van Nostrand Reinhold, New York, Second Edition, 1977, 628 pp.
2. Guyon, Y., 'Prestressed concrete', Vol. I, Simply-supported beams, Asia Publishing House, Bombay, 1963, 559 pp.
3. Lin, T.Y., 'Design of prestressed concrete structures', John Wiley & Sons, New York, Second Edition, 1963, 614 pp.
4. Leonhardt, F., 'Prestressed concrete–design and construction', Wilhelm Ernst & Sohn, Berlin, Second Edition, 1964, 677 pp.
5. Dayaratnam, P., 'Prestressed concrete structures', Oxford & IBH Publishing Co., New Delhi, 1982, 680 pp.
6. Mallick, S.K., and Gupta, A.P., 'Prestressed concrete', Oxford & IBH Publishing Co., New Delhi, 1983, 316 pp.
7. Reddi, S.A., 'Pamban bridge: Aspects of project management-I', Gammon Bulletin, Gammon India Ltd., Bombay, April-June 1989, pp. 3-19.
8. 'IS:1343-1980, Indian standard code of practice for prestressed concrete', Indian Standards Institution, New Delhi, 1998, 62 pp.
9. Plaehn, J., 'Introduction to prestressed concrete', Indian Institute of Technology, Madras, 1971, 206 pp.
10. 'Bridging India's rivers - An account of some of the bridges built under the first and second five-year plans', Indian Roads Congress, New Delhi, 256 pp.
11. 'IRC: 18-2000, Design criteria for prestressed concrete road bridges (Post-tensioned concrete)', Indian Roads Congress, New Delhi, 2000, 61 pp.
12. 'Standard plans for highway bridges - PSC girder and RC slab composite superstructure with and without footpaths', Ministry of Surface Transport (Roads Wing), Govt. of India, Published by Indian Roads Congress, New Delhi, Vol. I (1992) and Vol. II (1997).
13. Lee, D.J., 'The selection of box beam arrangements in bridge design', In Rockey, K.C. et al (Ed.), 'Developments in Bridge Design and Construction', Crossby Lockwood and Son Ltd., London, 1971, pp. 400-426.
14. Mathivat, J., 'The cantilever construction of prestressed concrete bridges', John Wiley, 1983, 341 pp.
15. Finsterwalder, U., 'Free-cantilever construction of prestressed concrete bridges and mushroom-shaped bridges', First International Symposium on Concrete Bridge Design, ACI Publication SP-23, American Concrete Institute, Detroit, 1969, pp. 467-494.
16. Subba Rao, T.N., ' Trends in the construction of concrete bridges', Gammon Bulletin, Special Issue – 4 (Technical Series), March 1985, 23 pp.
17. Muller, J., 'Long-span precast prestressed concrete bridges built in cantilever', First International Symposium on Concrete Bridge Design, ACI Publication SP-23, American Concrete Institute, Detroilt, 1969, pp. 705-740.
18. Subba Rao, T.N., and Haridas, G.R., 'Mahatma Gandhi Sethu (the longest river bridge in the world', Gammon Bulletin, Gammon India Ltd., Bombay, Jan.-Mar. 1989, pp. 3-13.
19. Venkatarangaraju, A., 'Construction of a road bridge across Pamban Strait in N.H 49 Madurai Dhanushkodi Road', Jl. of Indian Roads Congress, Vol. 49-3, Paper No. 389, Nov. 1988, pp. 455-516.
20. Woodward, R.J., and Williams, F., 'Collapse of Ynys-y-Gwas bridge, West Glamorgan', Proc. Institution of Civil Engineers, London, Part I, 1988, Vol. 84, pp. 635-669.
21. Chakraborty, S.S. and Ghosh, A.R., 'Rehabilitation of central hinge bearings as replacement to poor performance bearings', International Seminar on Highway Rehabilitation and Maintenance, New Delhi, Nov. 1999, Part II, pp. 201-207.

22. Finsterwalder, U., 'Prestressed concrete bridge construction', Journal of American Concrete Institute, Proc. V. 62, No. 9, Sept. 1965, pp. 1037-1046.
23. Bednarski, C., 'Flamenco in concrete – The Kent Millenium Bridge, Maidstone, UK', In Dhir, R.K. et al, (Eds.), 'Role of concrete in bridges in sustainable development', Thomas Telford, London, 2003, pp. 349-360.
24. Pearson-Kirk, D., 'The performance of post-tensioned bridges', In Dhir, R.K., et al, (Eds.), 'Role of concrete in bridges in sustainable development', Thomas Telford, London, 2003, pp. 129-140.
25. 'IRC:SP:67-2005 Guidelines for use of external and unbonded prestressing tendons in bridge structures', Indian Roads Congress, New Delhi, 2005, 9 pp.
26. Hendy, C.R., and Smith, D.A.., 'The design of the New Medway bridge', In Dhir, R.K. et al, (Eds.), 'Role of concrete in bridges in sustainable development', Thomas Telford, London, 2003, pp. 53-62.

Chapter 9

Steel Bridges

9.1 General

Steel bridges have been adopted in the past for major bridges on the highways and more commonly on the railways. The Howrah bridge at Calcutta is a splendid example of steel construction. In view of the shortage of steel, not many steel bridges have been built in India in the recent past. General concepts pertaining to steel bridges are discussed in this Chapter. For a detailed discussion of design and welded fabrication, the reader may refer to more specialized texts.

Compared to concrete construction, steel superstructure will be of lighter weight, and will facilitate faster construction. Further the construction operations at the bridge site can be reduced with steel superstructure by prefabricating parts of the components at a nearby factory. In the span range of 120 to 140 m, for which prestressed concrete cantilever bridges are being adopted, steel construction can lead to time savings, as in the rail-cum-road bridge across Brahmaputra river at Jogigoppa in Assam. Steel bridges require greater maintenance attention than concrete bridges. For example, the Forth railway bridge needs continuous painting; it takes three years to complete one coat and the process is repeated.

Steel bridges could be a preferred option in Build Operate Transfer (BOT) projects, where speed in construction is crucial. Steel structures may also prove advantageous for urban flyover/elevated road projects as they cause less disturbance to traffic through faster construction and possible prefabrication.

Steel bridges can be classified under the following groups:

(a) Beam bridges;
(b) Plate girder bridges;
(c) Box girder bridges;
(d) Truss bridges;
(e) Arch bridges;
(f) Cantilever bridges;
(g) Cable stayed bridges; and
(h) Suspension bridges.

Beam bridges are used for culverts, using rolled steel joists as the main supporting members. Girder bridges are adopted for simply supported spans less than 50 m and for continuous spans up to 260 m. Truss bridges are suitable for the span range of 40 to 375 m. Arch bridges are competitive for the medium span range of 200 to 500 m. Cantilever bridges have been built with success with main spans of 320 to 549 m. Cable stayed bridges are economical when the span is about 200 to 800 m. For long spans above 800 m, suspension bridges provide the most economical solution.

9.2 Beam Bridges

9.2.1 RAILWAY CULVERT

Steel joists are often employed for superstructure of railway culverts. This type of construction has the advantage of speedy erection, easy fabrication and simplicity of design. For railway loading, the equivalent load tables as given in Chapter 4 are used. The design procedure will be evident from the following example.

9.2.2 EXAMPLE

To design a steel beam culvert with a clear span of 5 m to carry a broad guage single track on main line.

(a) *Dead Load effects*

Assume 2 R.S. joists placed at 2.0 m centres as shown in Fig. 9.1. The dead load due to track can be assumed at 7.5 kN/m and this is equally shared by the two joists. Self weight of the joist may be estimated as (0.2L + 1.0) kN/m, where L is the clear span in m.

$$\text{Total dead load} = \frac{7.5}{2} + (0.2 \times 5 + 1.0) = 5.75 \text{ kN/m}$$

$$\text{Max. B.M.} = 5.75 \times 5^2/8 = 18 \text{ kN.m}$$

$$\text{Max. shear} = \frac{5.75 \times 5}{2} = 14.4 \text{ kN}$$

(b) *Live Load effects*

From Table 4.1,
Equivalent uniform load due to live load for bending moment = 741 kN
Equivalent u.d.l. for shear = 888 kN
Coefficient of dynamic augment (CDA) = 0.877
Impact factor = 1.877

$$\left.\begin{array}{l}\text{B.M. due to L.L.}\\ \text{including impact}\end{array}\right] = \frac{741}{2} \times \frac{5}{8} \times 1.877 = 435 \text{ kN.m}$$

$$\left.\begin{array}{l}\text{Shear due to L.L.}\\ \text{including impact}\end{array}\right] = \frac{888}{2} \times \frac{1}{2} \times 1.877 = 417 \text{ kN}$$

(c) *Design*

Design moment = 18 + 435 = 453 kN.m
Design shear = 14 + 417 = 431 kN
Assume f_y = 250 MPa
From Table 5.12,
Permissible bending stress = 0.62 f_y = 155 MPa

$$\text{Modulus of section required} = \frac{453 \times 1000 \times 100}{155 \times 100} = 2923 \text{ cm}^3$$

Adopt ISWB 600 at 1.31 kN/m
Modulus of section available = 3540 cm^3

Permissible shear stress = 0.38 f_y = 95 MPa

$$\text{Shear stress} = \frac{431 \times 1000}{600 \times 11.2} = 64 \text{ MPa} < 95 \text{ MPa}$$

The procedure for computing the permissible bending compressive stress is given in detail in Clause 508.6.2 of the Code. The details are not shown here. For the present case, the permissible bending stress in compression amounts to 154 MPa, which is more than the actual stress of 128 MPa. Hence the assumed section is adequate.

The bracings shown in Fig. 9.1 will be adequate to take up wind loads.

9.3 Plate Girder Bridges

9.3.1 Components

Since the days of early steel bridge construction, there has been a marked preference for the plate girder bridge system, in view of the elegant aesthetics obtainable with this type and also the convenience in maintenance. Plate girder bridges can be of two types: (a) deck type; and (b) half-through type. Deck type is normally preferred. Half-through type is adopted when the cost of additional embankment to raise the rail level is high. A plate girder highway bridge will consist of the deck slab (normally of reinforced concrete) and stringers running longitudinally and resting on transverse floor beams, which in turn rest on the plate girders.

In the case of a railway bridge, the plate girders carry the wooden sleepers over which the steel rails are fastened. The girder bridges will be braced laterally at the level of the top flange and the bottom flange, besides cross bracings to resist the lateral load due to wind. The cross bracings consist of angles and are provided at the ends and at intervals of about 4 to 5 m.

There is usually a choice available between (a) using two widely spaced longitudinal girders, with the cross girder system supporting the deck, and (b) providing multiple longitudinal girders with small spacing. In the first case, the cross girder system may consist of closely spaced cross girders alone or cross girders supporting a system of longitudinal stringers. The two-girder system necessitates deeper girders and may lead to economy in certain circumstances. For the deck type, the distance between the two girders is kept slightly larger than the gauge of the track to reduce the severity of the impact loads on the girders. In the half-through type of bridge, the railway load is carried at the lower flange.

9.3.2 Example

Typical details of elevation, plan, web splice and cross section for a deck type riveted plate girder railway bridge of effective span 30 m are shown in Figs. 9.2. and 9.3. The bridge is meant for a single track on broad gauge main line.

The steps in the design are briefly outlined here:

Dead load of sleeper, rails
and fittings assumed at 20 kN/m = 600 kN

Dead load of girder assumed at
4 kN/m of span = 120 kN

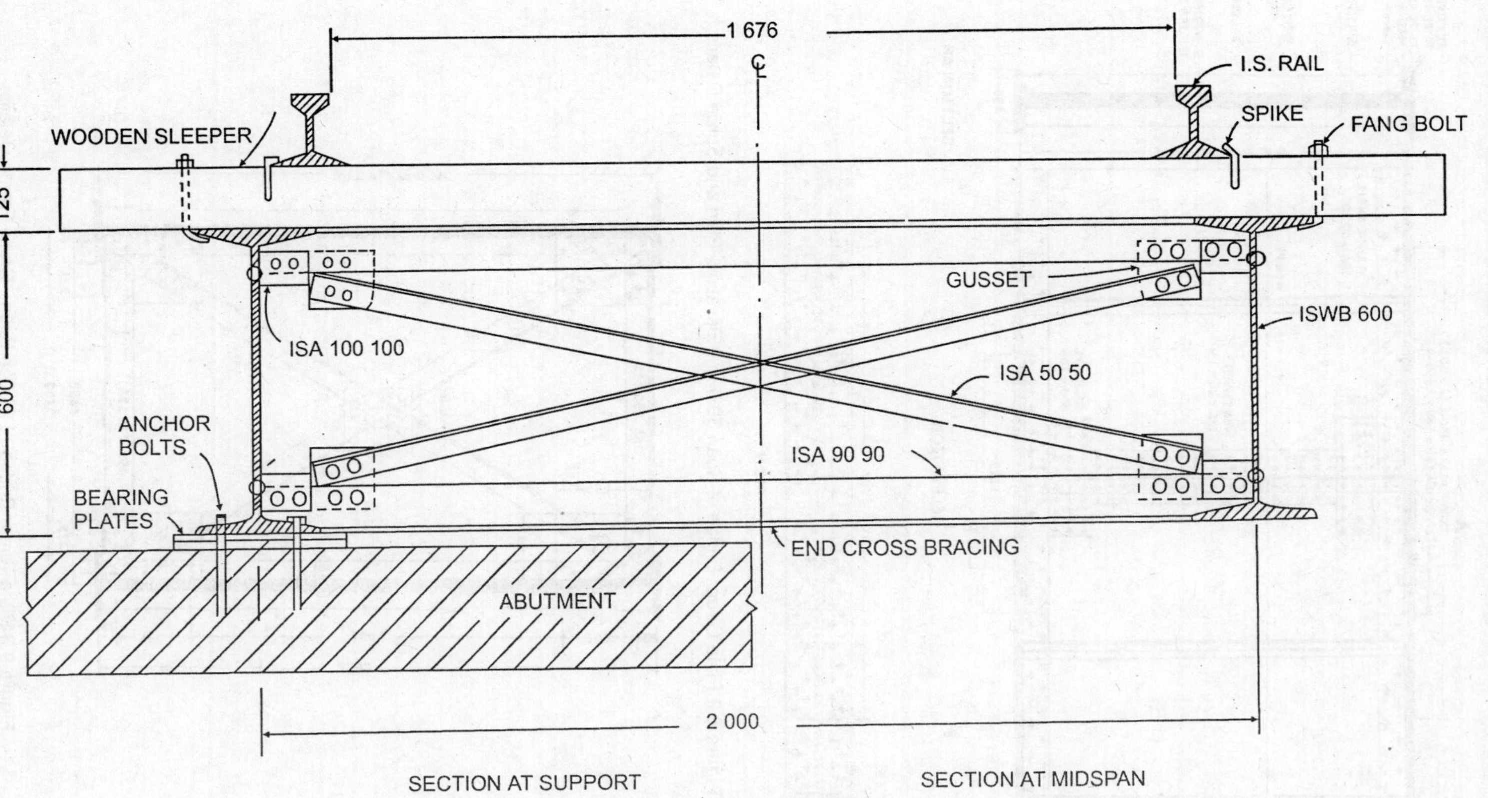

Figure 9.1 Cross Section of a Railway Culvert of Clear Span of 5.0 m.

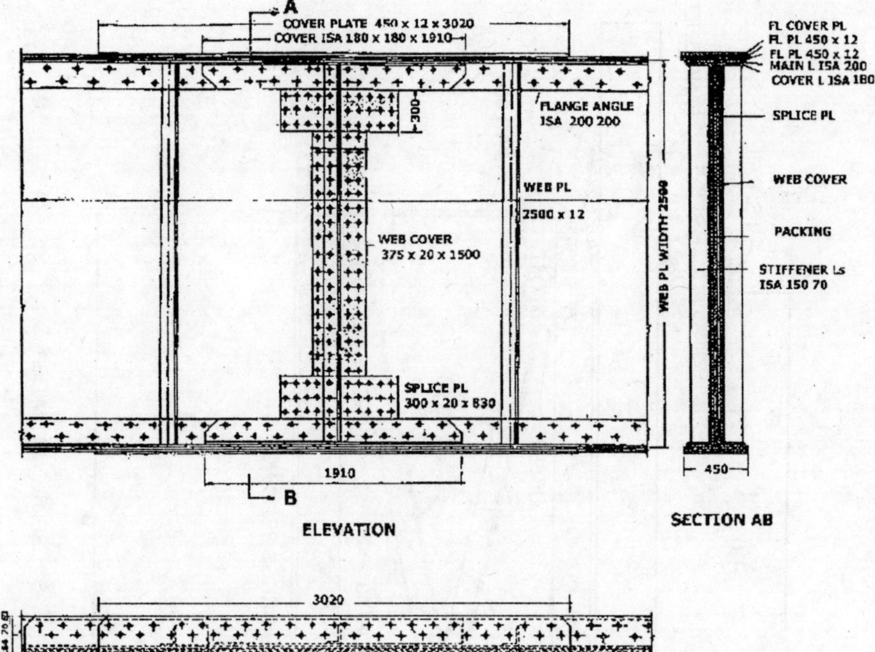

Figure 9.2 Plate Girder Bridge of Span 30.0 m - Elevation, Plan and Splice Details.

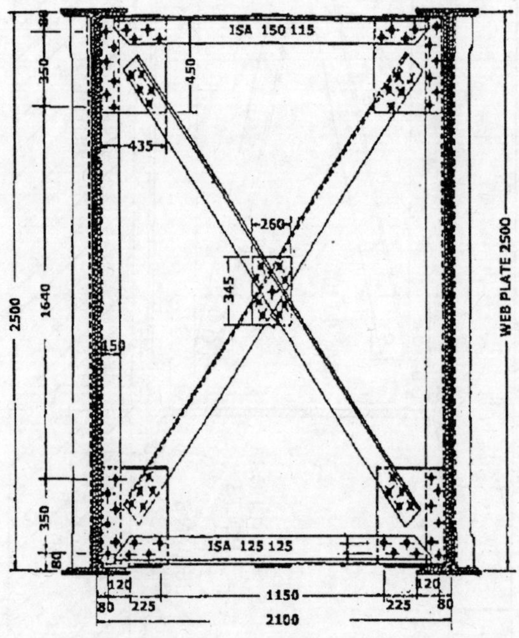

Figure 9.3 Plate Girder Bridge of 30.0 m Span - Cross Frame Details.

Equivalent uniformly distributed
live load for bending moment,
from Table 4.1 = 2727 kN
Impact factor = 1.372
Total U.D. load due to dead load
and live load per truck = 600 + 120 + (2727 x 1.372) = 4461 kN
Design load per girder for B.M. = 2231 kN

By similar procedure,
Design load per girder for shear = 2416 kN
Maximum design bending moment = WL/8 = 8329 kN.m
Economical depth is given by

$$h = k\,[M/f_b]^{0.33} \qquad \text{..... (9.1)}$$

where h = economical depth in mm
k = a constant with value between 5 and 6, taken here as 6
M = bending moment in N.mm
f_b = maximum permissible fibre stress, taken as 141 MPa

Substituting values,

$$h = 6\,[8329 \times 10^6/141]^{0.33} = 2337 \text{ mm}$$

The thickness of the web plate is usually 12 mm. Flange width is usually in the range of L/40 to L/60. The outstand of flange beyond the flange angle should not exceed 20 times the thickness of flange. The cross sectional dimensions are determined by trial method.

Assuming a section consisting of (see Fig. 9.2):
Web plate 2500 × 12
Flange angles 2 – 200 × 200 × 20
Flange plates 2 – 450 × 12

The other calculations are as for a normal plate girder, involving the checking of stresses, and detailing of stiffeners, splices and bracings[1]. Normally, plate bearings will be adequate for plate girder bridges.

Welded plate girder bridges are increasingly used in recent years. Because all the material is effective in resisting loads, a welded plate girder is more efficient than a riveted plate girder. Also the flange plate is welded directly to the web plate, without the need for the use of flange angles. Hence the welded plate girder deck is lighter than the corresponding riveted plate girder deck. A higher level of workmanship is, however, required. The design principles are similar to those for the riveted girder. The behaviour of plates in compression in bridges requires special consideration[2].

9.3.3 Bridges With Multiple Girders

When the number of longitudinal girders is more than two, they are connected with cross beams such that the entire deck acts as a grid. The provision of cross beams permits better transverse distribution of concentrated loads to the longitudinal girders. The number of cross beams is usually three for small spans and five for larger spans.

9.4 Orthrotropic Plate Decks

Modern highway bridges of moderate and long spans increasingly adopt orthotropic plate decks, schematically shown in a simplified form in Fig. 9.4. The deck consists of a

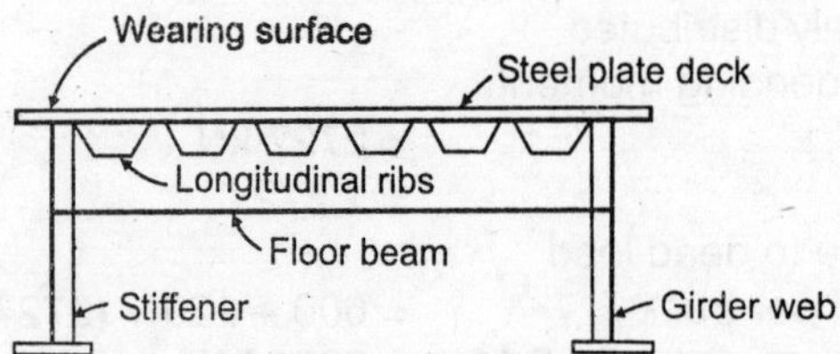

Figure 9.4 Orthotropic Plate Decking for Bridges.

stiffened deck plate over which a thin layer of asphaltic concrete wearing course is directly laid. The steel plate is stiffened in two orthogonal directions: longitudinally by closed rib systems and transversely by the floor beams. Since the stiffness in two orthogonal directions are different, the behaviour of the deck is said to be anisotropic. This type of deck with orthogonally (ortho) placed stiffeners and with anisotropic (tropic) behaviour is known as 'orthotropic plate deck'. Originally developed in Germany in the 1950s, this system has since been used in many bridges worldwide, resulting in substantial savings in materials and cost. The successful application of orthrotropic plate decks is mainly due to advances in mechanised welding.

Essentially, the deck consists of a flat deck plate, stiffened by welded (closed) longitudinal ribs, which span between transverse floor beams, which in turn span between the main girders. The components are interconnected and together form a complex structural system. The deck plate acts as a continuous member supporting the concentrated wheel loads placed between ribs and transfers the reactions to the ribs. The ribs are usually of trapezoidal shape. The deck plate also functions as the top flange of the ribs, the floor beams and the main longitudinal girders. The deck is paved with a wearing course to provide a durable and skid resistant surface for vehicular traffic. Orthotropic plate decks are used for a wide variety of steel bridges such as plate girder bridges, box girder bridges, movable bridges, cable stayed bridges and suspension bridges. The orthotropic plate deck scheme results in reduced weight for the deck, a condition of special importance for long span bridges. It also results in shallower sections and leads to economy in the required length of approaches. Other advantages include faster construction due to lighter components. For a detailed treatment of the design and applications of orthotropic plate decks, the reader may refer to Reference 3.

9.5 Box Girder Bridges

Developments in welding technology and precision gas cutting techniques in the post second world war period facilitated the economical fabrication of monolithic structural steel forms such as steel box girders characterized by the use of thin stiffened plates and the closed form of cross section. A box girder is built up using a deck plate, vertical or inclined webs and a bottom plate. The deck plate carries the heavy traffic loads and so needs stiff stringers and transverse beams to transfer the loads to the box webs by bending. The box webs are subjected to bending and shear stresses. The bottom plate acts as a chord member for bending and also gets axial tension or compression. It should be well stiffened against buckling under axial compression. The box girder deck can have single cell or multiple cells, the latter being uneconomical for short spans.

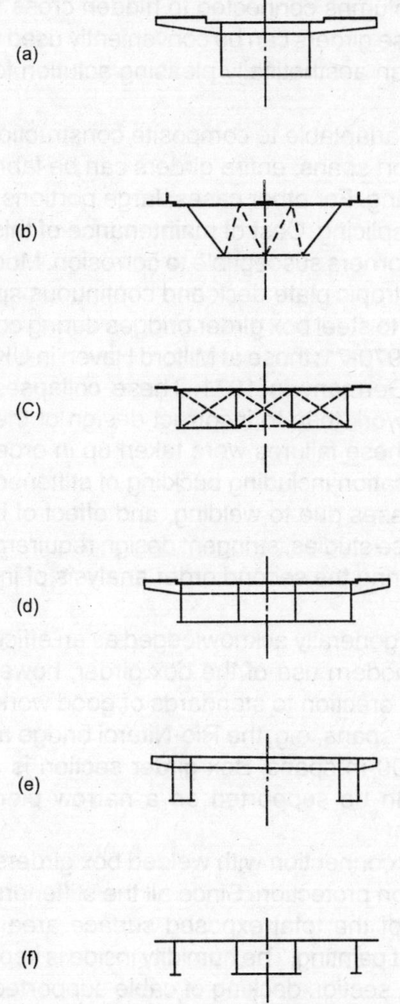

Figure 9.5 Typical Forms of Box Girders.

Typical forms of box girders are shown in Fig. 9.5 and they are detailed below:

(a) Rectangular box with wide cantilevering span on either side;
(b) Trapezoidal box sections;
(c) Two box sections which are connected together by bracing for the integral action of the deck;
(d) One wide box section subdivided into three cells;
(e) Two box sections kept wide apart; an
(f) One middle box section with one longitudinal girder on either side.

Box girder bridges have exceptional torsional rigidity resulting in better transverse load distribution. The depth of superstructure can be shallower with this type of construction, leading to lower gradients on approaches. The intermediate supports of such construction can be individual slender columns connected to hidden cross frames, saving substructure costs and erection time. These girders can be conveniently used for curved and/or continuous bridges, and often provide an aesthetically pleasing solution for urban highway structures like flyovers.

Box girders are easily adaptable to composite construction, for which only narrow top flanges are needed. For short spans, entire girders can be fabricated in the shop, enabling maximum use of shop welding. For other cases, large portions can be shop fabricated and connected together by site splicing. Cost of maintenance of this type of bridge is low, since there are fewer vulnerable corners susceptible to corrosion. Modern steel box girder bridges invariably incorporate orthotropic plate deck and continuous spans.

Three major disasters to steel box girder bridges during construction involving the loss of 51 lives occurred during 1970-71: those at Milford Haven in UK and at Melbourne, Australia in 1970, and at Koblenz, Germany in 1971. These collapses were attributable more to inefficient detailing of steelwork than to incorrect design of the effective section[4]. Several studies into the causes of these failures were taken up in order to clarify many aspects of box girder design and fabrication including buckling of stiffened plates under compression, importance of residual stresses due to welding, and effect of initial imperfections in plates and stiffeners. Based on these studies, stringent design requirements were evolved in Europe for this type of bridge, requiring the second order analysis of instability of thin plates due to geometric imperfections.

The steel box girder is generally acknowledged as an efficient, economical and elegant form of bridge deck. The modern use of the box girder, however, calls for special care in design and fabrication, and erection to standards of good workmanship. Box girders would be economical only for long spans, e.g. the Rio-Niteroi bridge across the Guanabara Bay in Brazil, with 200 – 300 – 200 m spans. Box girder section is also appropriate if only one central girder is required to be supported on a narrow pier for functional or aesthetic considerations.

A recent innovation in connection with welded box girders is to dehumidify the interior of the box girder as corrosion protection. Since all the stiffeners are placed in the interior of the girder, the major part of the total exposed surface area is protected from corrosion without the need for frequent painting. The humidity inside is kept below 40 %. This innovation is being applied also to box section decking of cable supported bridges.

9.6 Truss Bridges

9.6.1 General

Truss bridges have been used economically in the span range of 100 to 200 m. A bridge truss derives its economy from its two major structural advantages: (a) the primary forces in its members are axial forces, and (b) greater overall depths permissible with its open web construction leads to reduced self weight when compared with solid web systems. The erection of a truss bridge is considerably simplified because of the relative lightness of the component members. The aesthetic appearance of a truss bridge is debatable, mainly because of the complexity of the elevation and the different directions of its members.

9.6.2 Types and Components

The major types of bridge trusses are shown in Fig. 9.6. The most common form is the

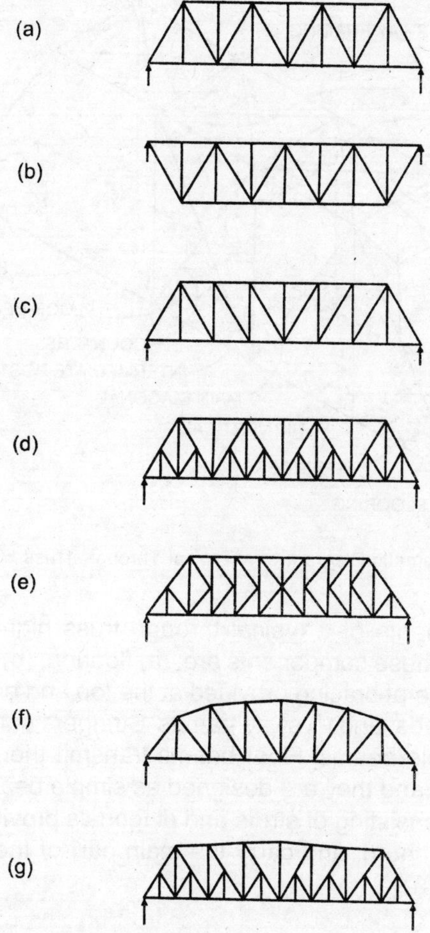

Figure 9.6 Typical Bridge Trusses.

Warren truss, shown in Fig. 9.6 (a) and (b) for the through and deck types, respectively. The Pratt truss shown in Fig. 9.6 (c) is considered to be advantageous in that the longer diagonals are in tension, while the shorter verticals are in compression. Some of the panels towards the middle may be provided with counters if there is a possibility of reversal of stress in the diagonals. The diagonals of the Pratt truss slope downward towards the centre, whereas the diagonals of the Warren truss alternate between downward toward the centre and downward away from the centre. Panels of a Warren truss may be sub-divided as in Fig. 9.6 (d) in order to provide a better support for the deck, the arrangement shown referring to a through truss. Sub-division reduces the unsupported length by half and hence leads to more slender members, especially in compression. However, it also leads to higher unit prices of steel, less attractive final structure and higher secondary stresses in some cases. The K-bracing system shown in Fig. 9.6 (e) is convenient when the depth of a bay is of the order of twice its length. The top chords may be curved in case of longer spans, as in Fig. 9.6 (f) and (g).

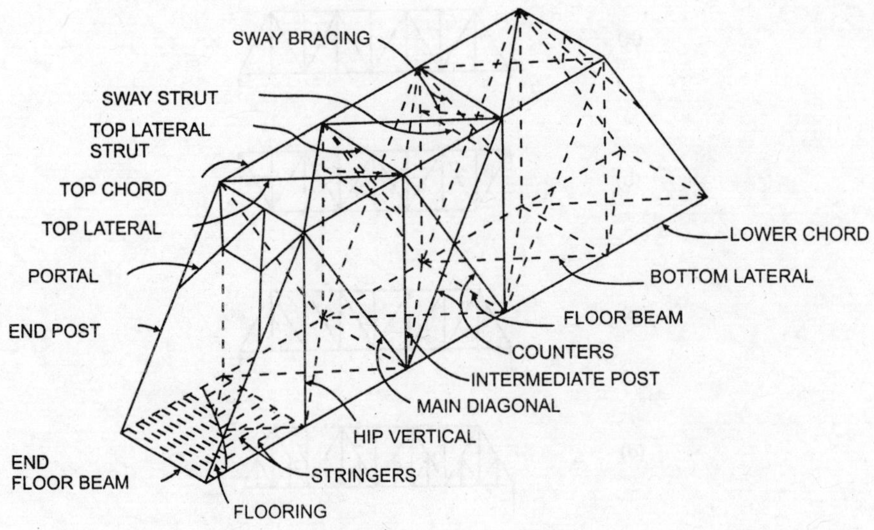

Figure 9.7 Schematic Diagram of a Typical Through Truss Highway Bridge.

The various components of a typical through truss highway bridge are indicated schematically in Fig. 9.7. These components are: (a) flooring; (b) stringers; (c) floor beams; (d)two main trusses; (e) lateral bracing provided at the top and bottom chord levels to cater to horizontal transverse loads; and (f) sway frames. Stringers carry the loads from the floor and are designed as simple beams. Floor beams transmit the load from stringers to the bottom nodes of the truss, and they are designed as simple beams spanning between the trusses. The top laterals consisting of struts and diagonals provide rigidity to the structure, stabilize the compression chord, and carry the main part of the wind loads to the bridge portals.

9.6.3 Design

Practically all triangulated trusses are designed on the assumption that the members are subjected to axial forces only. This assumption is valid when no transverse loads are applied away from the nodal points and when no moments are transmitted at the joints. These direct stresses are called primary stresses, whereas the bending stresses are called secondary stresses.

Secondary stresses are caused by: (a) eccentricity in connections; (b) torsional moments due to floor beams; (c) transverse loads; and (d) rigidity of joints and truss distortion. Usually, these may be neglected in otherwise well proportioned designs, in view of the low permissible stresses normally used in design. For more details on rigorous analysis, including consideration of secondary stresses, truss buckling, influence of gusset plates and deck-truss interaction, the reader is referred to Reference 5.

The main decisions needed in the design of the main trusses of a truss bridge are: (a) truss type; (b) truss depth; and (c) length of deck panel. The depth of truss for a highway bridge can be L/8 to L/20, whereas it would range from L/5 to L/10 for a railway bridge. The type of truss can normally be decided from convenience of detailing and on grounds of aesthetics. The length of the panel should be such that the diagonals should not have a

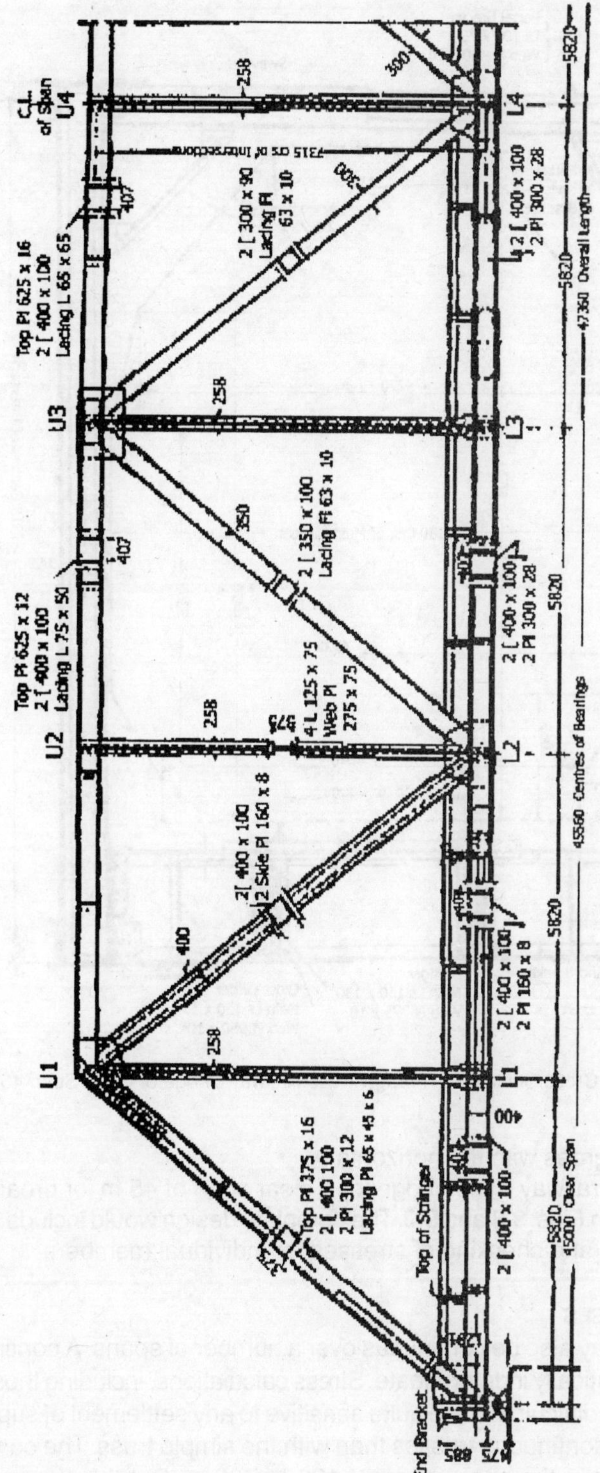

Figure 9.8 Typical Elevation of Through Truss Railway Bridge of Clear Span 45.0 m.

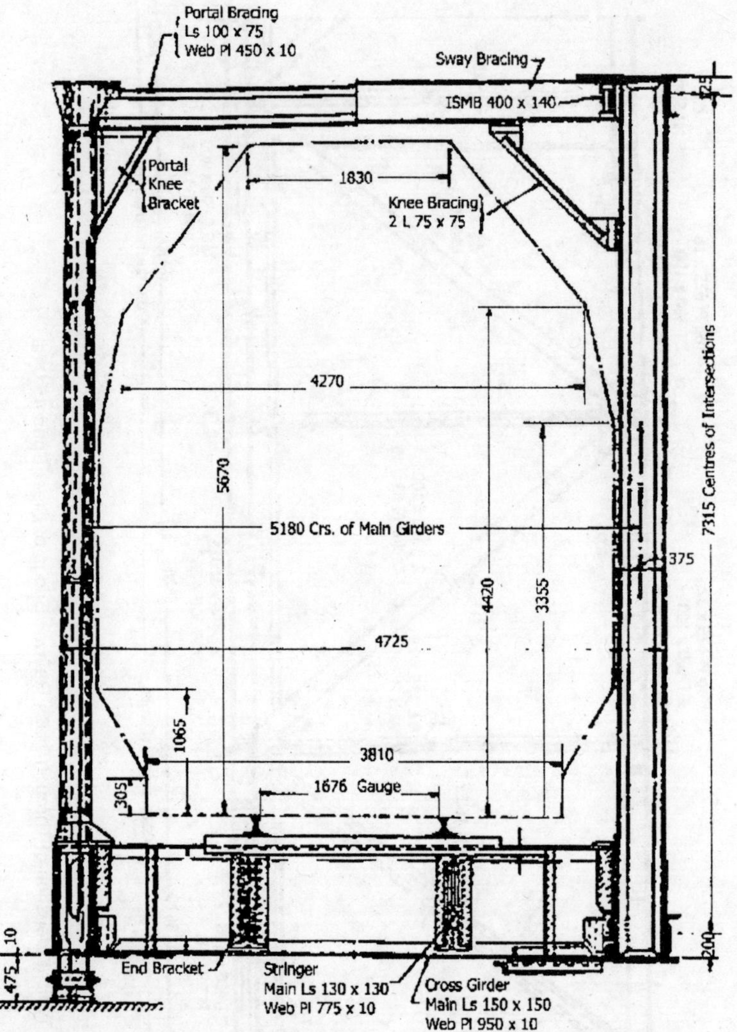

Figure 9.9 Typical Cross Section of Through Truss Railway Bridge of Clear Span 45.0 m.

slope of less than 45 degrees with the horizontal.

Typical details of a railway truss bridge for a clear span of 45 m for broad gauge main line standard are shown in Figs. 9.8 and 9.9. The complete design would include consideration of design of connections and checking of stresses for individual members[1].

9.6.4 Continuous Trusses

The truss bridge may also be continuous over a number of spans. A continuous truss is more rigid but it is also statically indeterminate. Stress calculations, including those for erection stresses, are much more difficult and are quite sensitive to any settlement of supports. Longer spans are possible with continuous trusses than with the simple truss. The continuous truss bridge has been serving in the span range of 180 to 400 m. But this range is now being increasingly taken over by cable-stayed bridges.

9.7 Arch Bridges

The arch form is best suited to deep gorges with steep rocky banks which furnish efficient natural abutment to receive the heavy thrust exerted by the ribs. In the absence of these natural conditions, the arch usually suffers a disadvantage, because the construction of a suitable abutment is expensive and time consuming.

The arch form is aesthetically the most pleasing and has been used in steel bridges in the span range of 100 to 250 m. Typical steel arch bridges are shown in Fig. 9.10. Deck type open spandrel arches can be particularly attractive as in the case of Rainbow bridge across the Niagara river at Niagara Falls. The arch profile is intended to reduce bending moments in the superstructure and will be economical in material when compared with an equivalent straight simply supported girder or truss. The efficiency is made possible by the horizontal reactions provided by the supports and hence the site has to be suitable. The fabrication and erection of an arch bridge would pose more difficult problems than a girder bridge, and

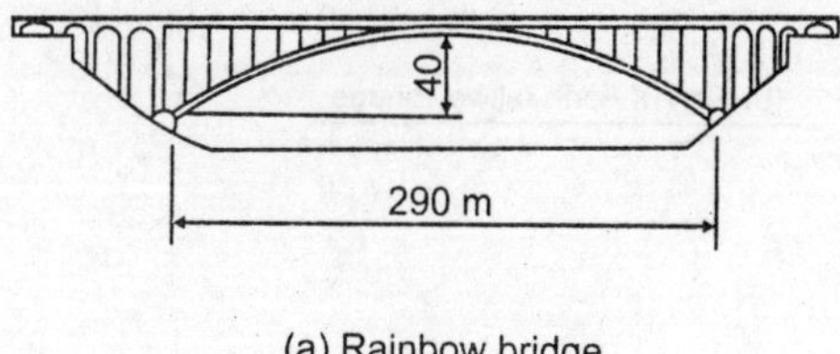

(a) Rainbow bridge

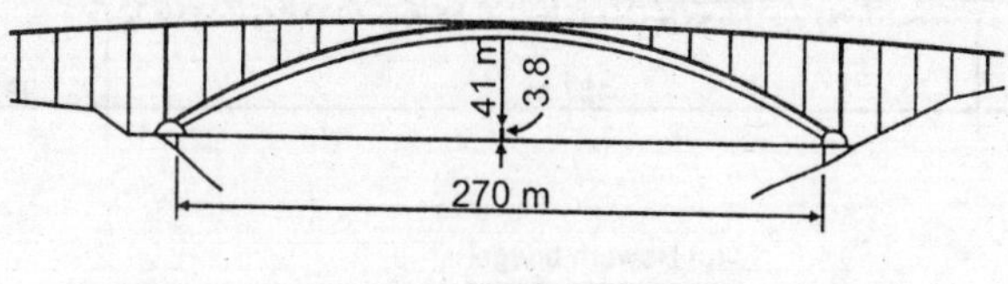

(b) Askerofjord bridge

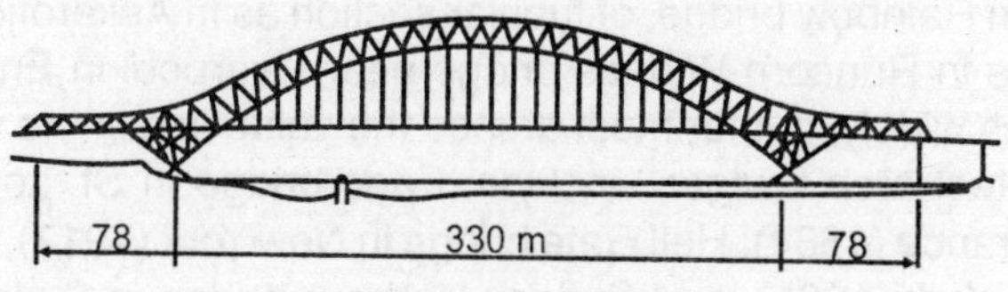

(c) Runcorn-Widnes bridge

Figure 9.10 Typical Steel Arch Bridges.

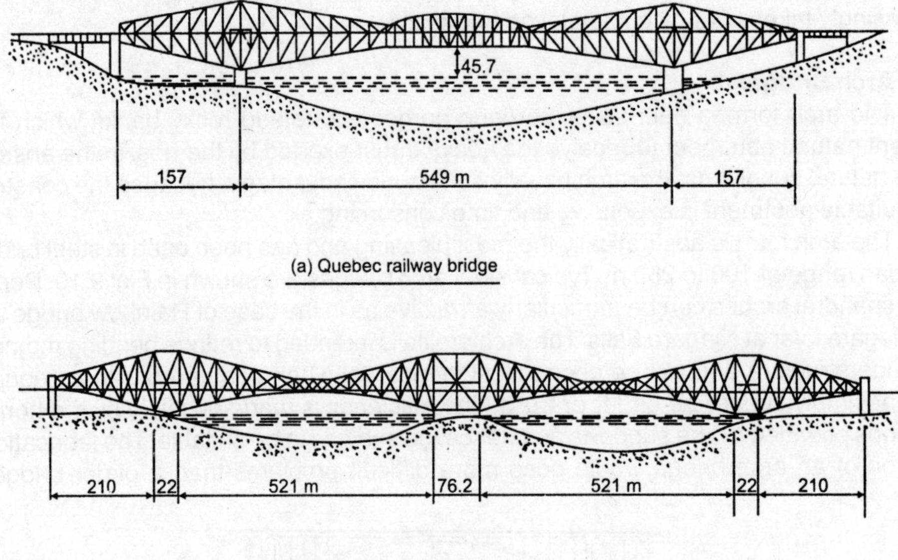

(a) Quebec railway bridge

(b) Firth of Forth railway bridge

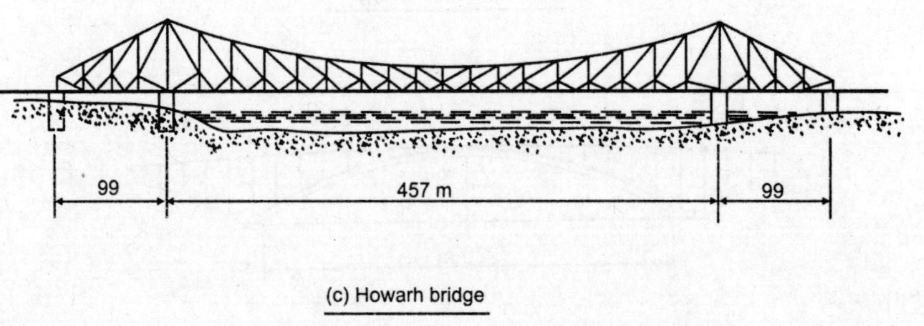

(c) Howarh bridge

Figure 9.11 Typical Cantilever Bridges.

should be properly taken into account by the designer. Arch ribs can be hingeless as in the case of the Rainbow bridge; or may have one, two or three hinges. The arch rib can consist of a box section as in Rainbow bridge, of tubular section as in Askerofjord bridge in Sweden or a trussed form as in Runcorn-Widnes bridge near Liverpool in England. The rise-span ratio of arches varies widely, but for most arches the value lies in the range of 1:4.5 to 1:6.

The famous steel arch bridges[6] include: Eads bridge in St. Louis, Missouri (1874), Garabit Viaduct in France (1884), Hell Gate bridge in New York (1917), Bayonne bridge over Kill Van Kull in New York (1931), and Sydney Harbour bridge in Sydney, Australia (1932). Eads bridge, designed by J.B. Eads, consisted of three spans of 153 m, 158 m and 153 m. This bridge is also historically important because this was the first bridge in America where: (i) steel was used extensively in bridge construction; (ii) steel material was tested for strict conformity to specified strength; (iii) pneumatic caissons were used for foundation work; and (iv) cantilever method of erection was used. Garabit viaduct, designed by Gustav Eiffel is a magnificent two-hinged arch bridge which tapers to the springings, providing increased lateral stiffness. Lindenthal's Hell Gate bridge has a two-hinged arch span of 298 m and is one of the longest steel bridges in the world having an overall length between abutments of 5574 m. It is also the most heavily loaded long span bridge in the world carrying four ballasted railway tracks upon a solid deck floor. Ralph Freeman designed the Sydney Harbour bridge with a two-hinged arch span of 503 m and the bridge was completed in 1932. Bayonne bridge, designed by O.H. Amman, with a two-hinged arch span of 504 m, has been a record bridge till 1977. The New River Gorge bridge in West Virginia completed in 1977 with a span of 519 m using weathering steel has been the world's largest span arch bridge till recently. Currently, the title for the world's longest span arch bridge is held by the Lupu bridge in Shanghai, China, built in 2004 with a main span of 550 m.

The arch form is likely to go out of adoption for new steel bridges in the near future, in favour of the cable stayed form of construction, as the latter tends to be more economica and aesthetically pleasing.

9.8 Cantilever Bridges

A cantilever bridge with a single main span consists of an anchor arm at either end between the abutment and the pier, a cantilever arm from either pier to the end of the suspended span and a suspended span. Such an arrangement permits a long clear span for navigation and also facilitates erection of steel work without the need for supporting centering from below.

Steel cantilever bridges came into general use for long span railway bridges, because of their greater rigidity compared with suspension bridges. Three well known examples are shown in Figure 9.11. The Firth of Forth bridge with two main spans of 521 m each became a milestone in bridge construction on its completion in 1889. The designers, John Fowler and Benjamin Baker, used tubular members of fairly large size with riveted construction for the arch ribs to withstand wind pressures of 2.68 kN/m^2. Though the tubes were large in size, the weight per linear metre of the bridge was still less than that of Quebec bridge.

The design of Quebec bridge was first entrusted to Theodore Cooper, who was then well known for his specifications for railway bridges. The plan envisaged a main span of 549 m with anchor spans of 157 m each, making this bridge the longest span in the world. The first attempt to construct the bridge ended in complete collapse of the south arm killing 75 men (1907). The failure was due to miscalculation of dead load and wrong design of compression members, which errors were not noticed in time. The design was revised by H.A. Voutelet and the structure was reconstructed in 1917 with the same main span.

Howrah bridge with a main span of 457 m was the third longest span cantilever bridge in the world at the time of its construction (1943). The bridge was erected by commencing at the two anchor spans and advancing towards the center with the use of creeper cranes moving along the upper chord. The closure at the middle was obtained by means of sixteen hydraulic jacks of 800 t capacity each. The construction was successfully completed with very close precision.

Osaka Port bridge was completed in 1974 with a clear span of 510 m. The bridge is double decked and is currently the world's third largest span cantilever bridge. The construction has been achieved without accidents and with great precision, testifying to the great advance in technology in bridge construction.

The weight of the structure and the labour involved in the construction of a cantilever bridge are large compared with a cable stayed bridge of the same clear span. Hence the cantilever bridge is not very popular at present.

9.9 Cable Stayed Bridges

9.9.1 General

A cable stayed bridge is a bridge whose deck is suspended by multiple cables that run down to the main girder from one or more towers. The cable stayed bridge is specially suited in the span range of 200 to 900 m and thus provides a transition between the continuous box girder bridge and the stiffened suspension bridge. It was developed in Germany in the postwar years in an effort to save steel which was then in short supply. Since then many cable stayed bridges have been built all over the world, chiefly because they are economical over a wide range of span lengths and they are aesthetically attractive. The wide application of the cable stayed bridge has been greatly facilitated in recent years by the availability of high strength steels, the adoption of orthotropic decks using advanced welding techniques and the use of electronic computers in conjunction with rigorous structural analysis of highly indeterminate structures. The beauty and visibility of a cable stayed bridge at night can be enhanced by innovative lighting schemes. The early cable stayed bridges were mainly constructed using steel for stay cables, deck and towers. In some of the recent constructions, the deck and towers have been constructed in structural concrete or a combination of steel and concrete. Basic concepts of the application and design of the cable stayed bridges are presented here. For a comprehensive treatment of the theory and design, the reader may refer to References 7 to 10.

9.9.2 Types of Cable Stayed Bridges

The main components of a cable stayed bridge are: (i) Inclined cables, (ii) Towers (also referred as Pylons), and (iii) Deck. In a simple form, the cables provided above the deck and connected to the towers would permit elimination of intermediate piers facilitating a larger width for purposes of navigation, as shown in Fig. 9.12 (a).

When the number of stay cables in the main span is between 2 and 6 as in Fig. 9.12 (a and b), the spans between the stay supports tend to be large (between 30 and 60 m) requiring large bending stiffness. The stay forces are large and the anchorages of cables become complicated. The erection of such bridges involves use of auxiliary structures. On the other hand, the use of multiple stay cables as in Fig. 9.12 (c to e) would facilitate smaller distances between points of supports (between 6 and 10 m) for the deck girders, resulting in reduced structural depth and facilitating erection by free cantilever method without auxiliary supports[11]. The multiple stay cable system also permits easy replacement of cables if needed and enhances aerodynamic stability through increased damping capacity.

The deck can be supported by a number of cables in a fan form (meeting in a bunch at the tower) as in Fig. 9.12 (b and c) or in a harp form (joining at different levels on the tower) as in Fig. 9.12 (d). Fig. 9.12 (e) shows a typical fan-shaped cable arrangement with the anchorages at the tower distributed vertically down a certain length (modified fan form). This arrangement facilitates easy replacement of cables at a later date in case of accidents. The fan type configuration results in minimum axial force in deck girders. The harp form requires larger

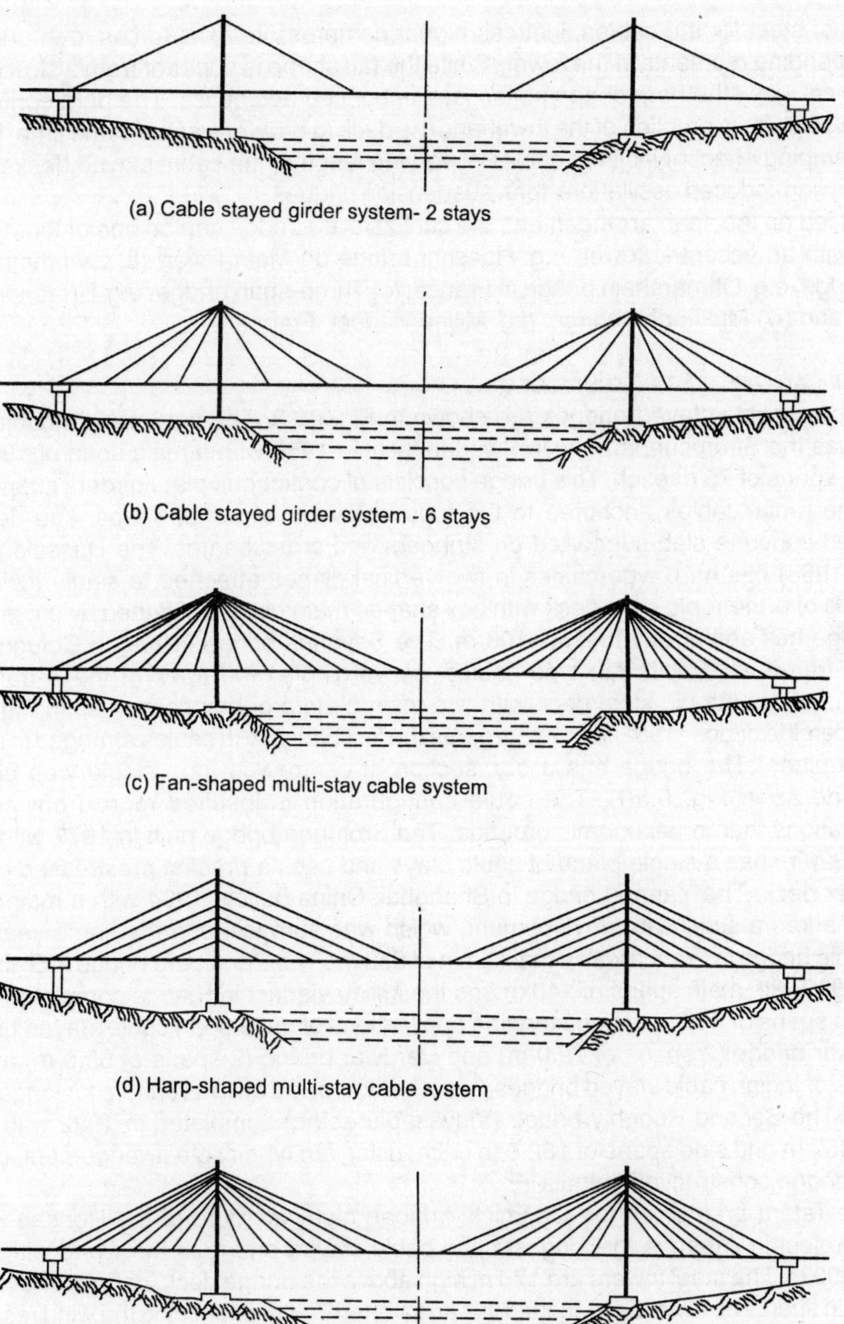

Figure 9.12 Types of Cable Stayed Bridges.

quantity of steel for the cables, induces higher compressive axial forces in the deck, and causes bending moments in the tower. While the fan shape is superior from a structural and economical view, the harp shape possesses enhanced aesthetics. The harp configuration cables also permits erection of the tower and the deck to progress at the same time. Because of the damping effect of inclined cables of varying lengths, the cable stayed decks are less prone to wind-induced oscillations than suspension bridges.

Based on the span arrangement, the cable stayed bridge can be one of four types: (a) Bridge with an eccentric tower, e.g. Hoescht bridge on Main River; (b) Symmetrical two-span bridge, e.g. Ottmarshein bridge in France; (c) Three-span bridge, e.g. Brotonne bridge, France; and (iv) Multi-span bridge, e.g. Millau viaduct, France.

9.9.3 Typical Cable Stayed Bridges

Typical cable stayed bridges are shown in Fig. 9.13. The first modern cable stayed bridge was the Stromsund bridge in Sweden, built in 1956 with a main span of 183 m and two side spans of 75 m each. This bridge consists of continuous plate girders supported by two plane radial cables anchored to the tops of towers of portal shape. The deck is of reinforced concrete slab supported on stringers and cross beams. The Dusseldorf North Bridge (1958) has harp type cables in two vertical planes attached to single towers. The decking is of orthotropic steel deck with box shaped main girders stiffened by cross beams. The bridge has spans of 108–260–108 m. The Severins bridge (1959) in Cologne has a single A-frame tower with fan type cables, converging at the apex of the A-frame. The decking is of orthotropic steel deck with two main girders of box section as in Fig. 9.5(e). The Norderelbe bridge (1962) in Hamburg was the first bridge with cables arranged in star type in single plane. The bridge has a box section at centre with one single web girder on either side as in Fig. 9.5(f). The cable configuration is justified more from aesthetic considerations than on economic grounds. The Brotonne bridge built in 1977 with a main span of 320 m has a single plane of cable stays and uses a precast prestressed concrete box girder deck. The Yangpu bridge in Shanghai, China built in 1994 with a main span of 602 m marked a significant development, which was surpassed in the same year by the Normandie bridge in France with a main span of 856 m . The Sunniberg bridge in Switzerland built in 1999 with main spans of 140 m and the Millau viaduct in France completed in 2005 with main spans of 342 m are outstanding applications of multi-span cable stayed bridges.

Akkar bridge (2 spans of 79.0 m) and Hardwar bridge (2 spans of 65.0 m) are early examples of Indian cable stayed bridges, essentially evolved as forerunners for longer spans to follow. The Second Hooghly bridge (Vidyasagar sethu) completed in 1992 with a main span of 457 m and side spans of 182.9 m each, using fan type cable arrangement, is a land mark of bridge construction in India[12].

The Tatara bridge[9] on the Onomichi-Imabari highway route of the Honshu-Shikoku Bridge Project in Japan is the longest span cable stayed bridge in the world with a main span of 890 m. The steel towers are 176 m high above the bridge deck, corresponding to 0.2 of the main span. The towers are shaped like an inverted Y after examining the wind resistance, structural efficiency and aesthetics. The stay cables have two-plane multi-fan shape. The cables are anchored at spacing of 20 m at deck level and at 3 m spacing at the tower. Based on wind tunnel tests, the surface of the polyethylene cover of the stay cables was provided with indentations, with a view to prevent the turbulence that results from wind blowing on rain water running on the surface of the long stay cables. This innovation provides sufficient

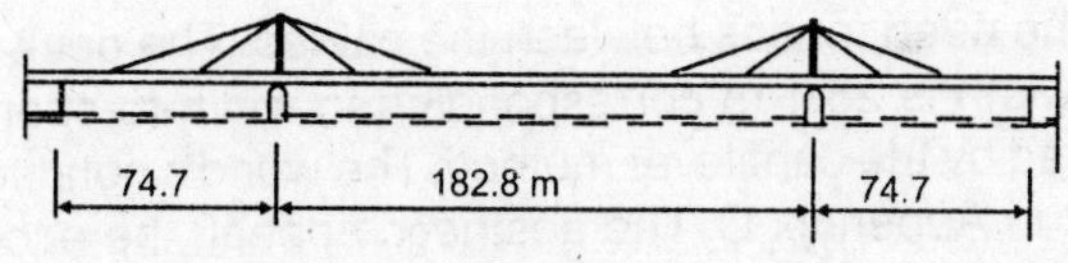

(a) Stromsund bridge

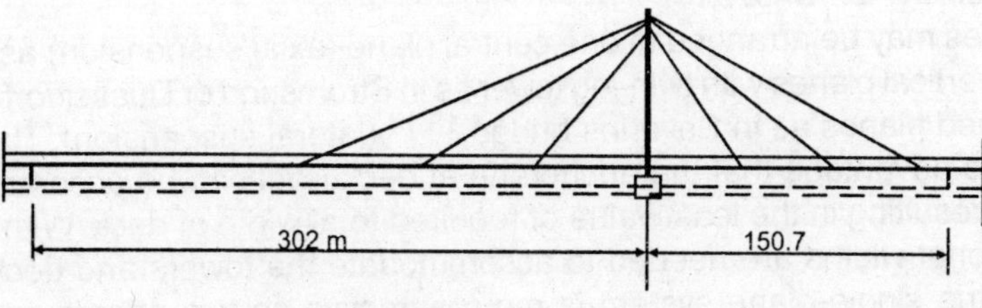

(b) Severins bridge

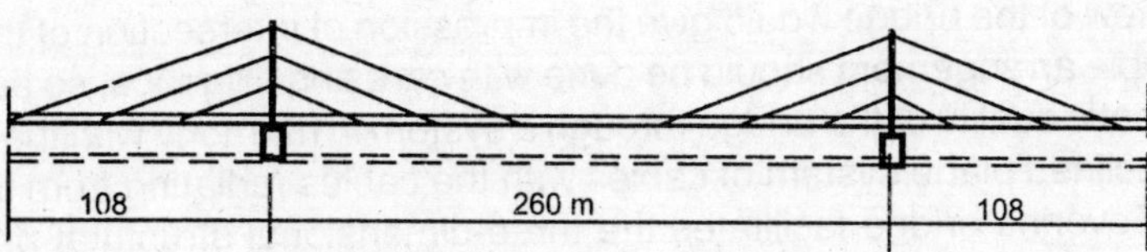

(c) Dusseldorf North bridge

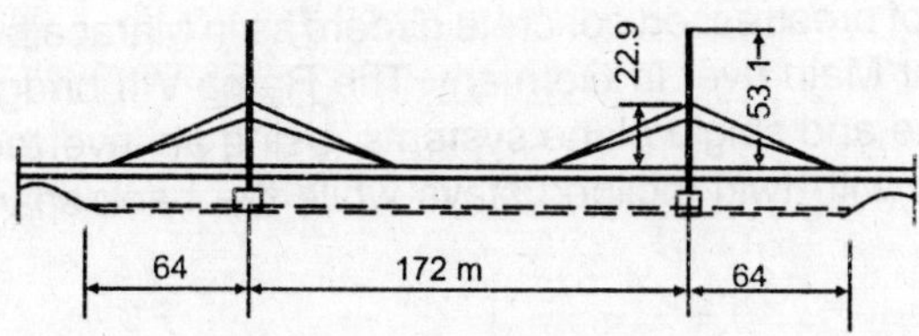

(d) Norderelbe bridge

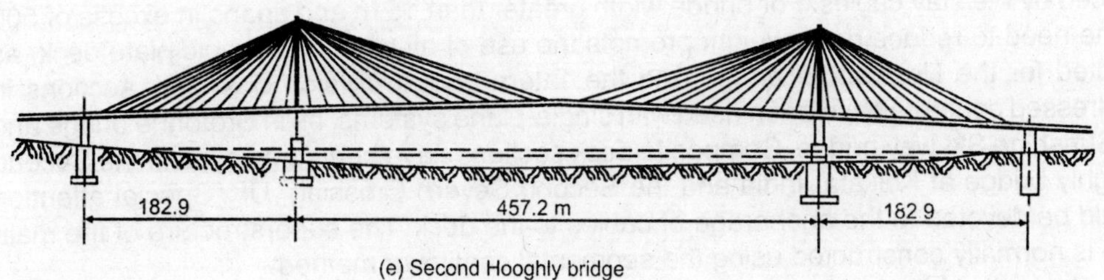

(e) Second Hooghly bridge

Figure 9.13 Typical Cable Stayed Bridges.

damping and avoids the need for ties between the cables. The deck is of streamlined steel box girder. The deck width is 28.1 m corresponding to width-to-span ratio of 1 : 31.7. The centre span was erected by the cantilever method. The world's notable cable stayed bridges are listed in Table C.2 in Appendix C. The aesthetic appeal, the economic advantage and the ease of construction make the cable stayed bridge the preferred option in the span range of 200 to 900 m.

9.9.4 Arrangement of Cables

The cables may be arranged in one central plane (axial suspension) as in Norderelbe bridge, in two vertical planes with twin-leg tower as in Stromsund or Dusseldorf North bridges, or in two inclined planes as in Severins bridge [13,14] (lateral suspension). The single-plane system has the advantage that the anchorage at deck level can be accommodated in the traffic median resulting in the least value of required total width of deck. With the two-plane system, additional widths are needed to accommodate the towers and deck anchorages. Aesthetically, the single-plane system is more attractive as this affords an unobstructed view on one side for the motorist. Other notable examples of single-plane system are the Rama IX bridge (1987) in Bangkok, Thailand, the Sunshine Skyway bridge (1987) in Florida, USA and the Normandie bridge (1994) in France. In the case of a two-plane system of cables, a side view of the bridge would give the impression of intersection of the cables. The choice of the cable arrangement should be done with care and diligence, so as to ensure an enhanced aesthetic quality of the bridge through a system in harmony with the environment.

The two inclined plane system of cables with the cables radiating from the apex of an A-frame as in Severins bridge facilitates the three-dimensional structural performance of the superstructure and reduces the torsional oscillations of the deck due to wind, thus enhancing the aerodynamic stability of the bridge. The torque due to eccentric concentrated loads would necessitate the use of box section orthotropic deck for the single-plane system. The decking is generally of orthotropic plate system with box girders for the two-plane system also, but can be of prestressed concrete girders as in Maracaibo bridge in Venezuela and Hoescht bridge over Main river in Germany. The Rama VIII bridge in Bangkok uses a combination of two-plane and single plane systems. Using an inverted-Y pylon, the 300 m main span is supported with twin inclined stays while the back span has a single plane system of stays.

9.9.5 Deck Structure

While the deck is merely supported by the cables in a suspension bridge, the deck of a cable stayed bridge is an integral part of the structure resisting the axial force and bending induced by the stay cables. For bridge width greater than 15 m and spans in excess of 500 m, the need to reduce dead weight prompts the use of all-steel orthotropic plate deck, as adopted for the Normandie bridge and the Tatara bridge. Torsion box deck sections in prestressed concrete have been used with single-plane systems, as in Brotonne bridge and the Sunshine Skyway bridge. Composite deck sections have been employed in the Second Hooghly bridge at Kolkata, India and the Second Severn Crossing, UK. Special attention should be devoted to the anchorage of cables to the deck. The superstructure of the main span is normally constructed using the segmental cantilever method.

The ratio of the side span (L_s) to the main span (L_m) for the case of a bridge with towers on both sides of the main span usually lies between 0.3 and 0.45. The ratio L_s/ L_m

can be 0.42 for concrete highway bridge decks and not more than 0.34 for railway bridges[15]. This ratio influences the changes in stress in the back stay cables due to variation of live load. It further influences the magnitude of vertical forces at the anchor pier, the anchor force decreasing with increasing L_s/L_m. The choice of L_s/L_m depends also on the local conditions of water depth and foundation.

9.9.6 Towers

Towers carry the forces imposed on the bridge to the ground. They are not replaceable during the life of the bridge. Hence they should be designed to be structurally strong, constructible, durable and economical.

The towers may take any one of the following forms:

(a) Single free standing tower, as in Norderelbe bridge;
(b) Pair of free standing tower shafts, as in Dusseldorf North bridge;
(c) Portal frame, as in Stromsund bridge and Second Hooghly bridge;
(d) A-frame as in Severins bridge or inverted Y-shape as in Yangpu bridge;
(e) Diamond configuration as in Globe Island bridge, Sydney.

When the stay cables are in one plane, a single free standing tower may be adopted. In this case, the pier below the box girder should be sufficiently wide for bearings to resist the torsional moments of the superstructure. For bridges with cables in two planes, the towers can be a free standing pair, or a portal frame with a slender bracing. An additional bracing may be introduced below the deck. The A-shaped tower and the inverted Y-shaped tower have been favoured for long bridges having shallow box girder decks in regions of strong wind forces. The land take at the base can be reduced by adopting a diamond configuration, as used in the Tatara bridge. Typical arrangements of towers are shown in Fig. 9.14.

Since the tower is the most conspicuous component in a cable stayed bridge, besides structural considerations, aesthetics plays a prominent part in the selection of the particular shape of the tower. For example, the proximity of Cologne cathedral influenced the adoption of the A-frame for the Severins bridge. Sometimes, an additional height is provided for the tower above the point of connection of the cable for architectural reasons, as in Norderelbe bridge (in this case, as a tribute to the city fathers). Anchorage of cables at the tower should follow good order. Since the cables at the deck level are anchored along a line along the edges or at the middle of the deck, it is natural that these should end along a vertical line at the tower head. In the case of A-shaped tower, the anchorage line can be parallel to the tower leg. It is not desirable to spread the anchorages transversely in one layer at the tower.

The single tower or towers consisting of a pair of separate columns will be stable in the lateral direction due to the restoring force provided by the cables in case of lateral displacement due to wind forces, as long as the cable anchorages are situated at a level above the base of the tower. The towers may be designed to be hinged or fixed at the base, depending on the magnitude of the vertical loads and distribution of the cable forces. While a tower with a fixed base induces a large moment, the increased rigidity of the total structure resulting from a fixed base at the towers and the relative ease in erection as compared with a hinged base may be advantageous. On the other hand, the hinged base results in reduced bending moments in the towers and may be advantageous with weak soil conditions. The towers should be slender and should have a low bending stiffness in the longitudinal direction so

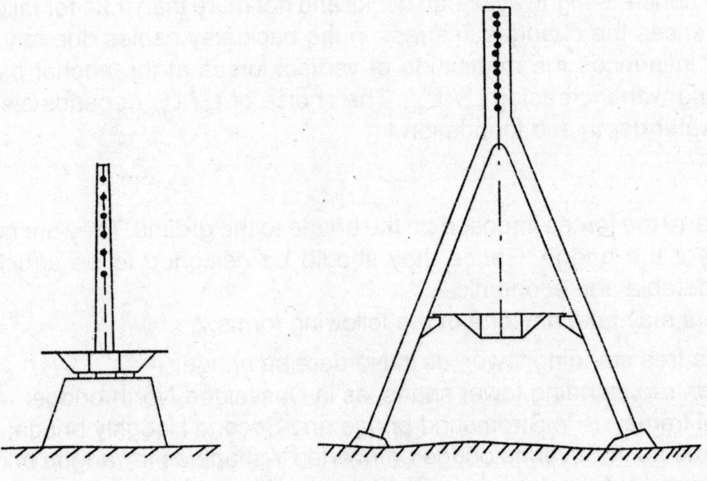

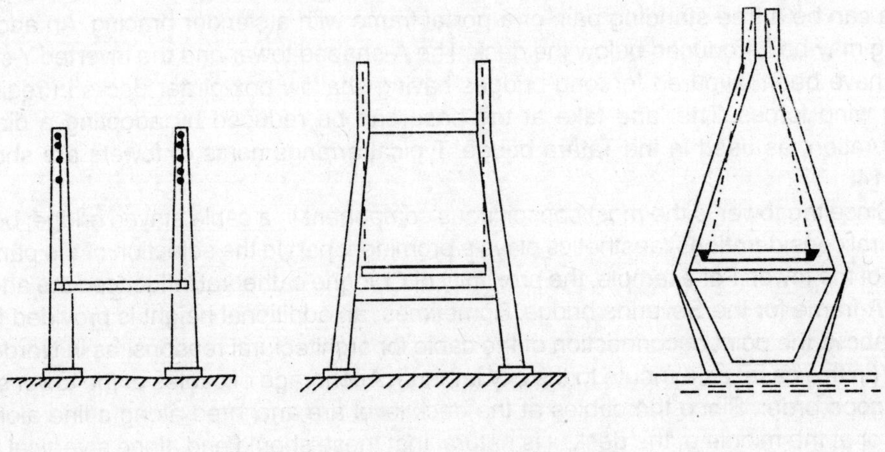

Figure 9.14 Typical Tower Arrangements.

that back stay cables will be functional in partially catering to live loads in the main span. Towers should normally be vertical.

The height of the tower should preferably be in the range of 0.2 to 0.25 L_m. The higher the tower, the smaller will be the quantity of steel required for the cables and the compressive forces[11]. But it is not advantageous to increase the height beyond 0.25 L_m.

9.9.7 Cables

The stay cables constitute critical components of a cable stayed bridge, as they carry the load of the deck and transfer it to the tower and the back stay cable anchorage. So the

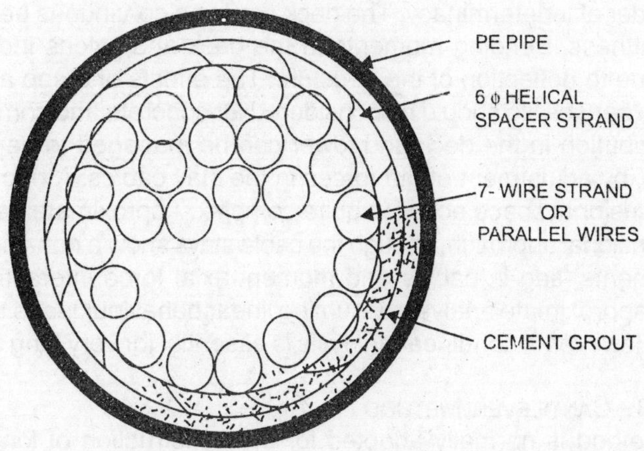

Figure 9.15 Typical Section of Stay Cable.

cables should be selected with utmost care. The main requirements of stay cables are: (a) High load carrying capacity; (b) High and stable Young's modulus of elasticity; (c) Compact cross section; (d) High fatigue resistance; (e) Ease in corrosion protection; (f) Handling convenience; and (g) Low cost. The ultimate tensile strength of wire is of the order of 1600 MPa. A typical section of a stay cable is shown in Fig. 9.15. While locked coil strands have been used in early bridges (e.g. Stromsund bridge), the recent preference is towards the use of cables with bundles of parallel wires or parallel long lay strands. The sizes of cables are selected to facilitate a reasonable spacing at the deck anchorages. Parallel wire cables using 7 mm wires of high tensile steel have been adopted in Second Hooghly bridge. Corrosion protection of the cables is of paramount importance. For this purpose, the steel may be housed inside a polyethylene (PE) tube which is tightly connected to the anchorages. The cables are anchored at the deck and at the tower. The anchorage at the deck is fixed and has a provision for a neoprene pad damper to damp oscillations. The length adjustment is done at the tower end.

The cables are prestressed by introducing additional tensile force in the cables in order to improve the stress in the main girder and tower at the completion stage, to prevent the lowering of rigidity due to sagging of cable, and to optimize the cable condition for the erection. The magnitude of the prestress is determined by taking into consideration the following factors: (i) the horizontal component of each cable tension is balanced such that there is no in-plane bending of the tower due to unbalanced horizontal force due to dead load at the completion stage; and (ii) the net force on the main girder member at the connection of the cable at the completion stage be zero.

Currently the steel used for cables have ultimate tensile strength (UTS) of the order of 1600 MPa. Carbon fibre cables having UTS of about 3300 MPa are under development. The latter cables are claimed to have negligible corrosion and to possess high fatigue resistance. However, carbon fibre cables are presently very expensive.

9.9.8 Analysis

The cable stayed bridge with the multi-stay configuration is a statically indeterminate structure with a high order of indeterminacy. The deck acts as a continuous beam on elastic supports of varying stiffness. Bending moments in the deck and pylons increase due to second order effects due to deflection of the structure. The effects of creep and shrinkage during construction and service life should be considered for concrete and composite decks. The internal force distribution in the deck and tower can be managed to be compression with minimum bending, by adjustment of the forces in the stay cables. A rigorous analysis considering three-dimensional space action is quite complex. Approximate designs can be made using a two-dimensional approach. Though the cable stays show a non-linear behaviour due to large displacements, sag in cables and moment-axial force interactions in stays, girders and towers, an approximate analysis assuming linear behaviour leads to satisfactory results in most cases. However, a non-linear analysis is essential for very long span bridges.

9.9.9 Construction By Cantilever Method

The cantilever method is normally adopted for the construction of long span cable stayed bridges. Here the towers are built first. Each new segment is built at site or installed with precast segment, and then supported by one new cable or a pair of new cables which balances its weight. The stresses in the girder and the towers are related to the cable tensions. Since the geometric profile of the girder or elevation of the bridge segments is mainly controlled by the cable lengths, the cable length should be set appropriately at the erection of each segment. During construction, monitoring and adjustment of the cable tension and geometric profile require special attention.

A notable example of construction of a major cable stayed bridge by cantilever method is the Yangpu bridge in Shanghai, China, built in 1994 with a main span of 602 m. The composite girders of this bridge consisted of prefabricated, wholly welded steel girders and precast reinforced concrete deck slab.

Depending on the bridge site, cable stayed bridges can have any one of four general layout of spans: (a) Cable stayed bridges with one eccentric tower, eccentric with respect to the gap to be bridged, e.g. Severins bridge; (b) Symmetrical two-span cable stayed bridges, e.g. Akkar bridge; (c) Three-span cable stayed bridges, e.g. Second Hooghly bridge, Stromsund bridge; (d) Multi-span cable stayed bridges, e.g. Millau viaduct. Of these the most common type is the three-span cable stayed bridge, consisting of the central main span and the two side spans. Temporary stability during construction is a major problem, particularly just prior to closure at midspan. The structure must be able to withstand the effects due to wind and accidental loads due to mishaps during erection. When intermediate piers are provided in the side spans, the stability is very much enhanced. In this case, the side spans are built first on the intermediate supports, and later the long cantilevers in the main span.

9.10 Suspension Bridges

The suspension bridge is currently the only solution for spans in excess of 900 m, and is regarded as competitive for spans down to 300 m. The discussion in this Section is limited to the three-span configuration shown in Fig. 9.16. The Wheeling suspension bridge across the Ohio river in USA built by Charles Ellet in 1849 with a span of 308 m and rebuilt by John Roebling in 1854 after tornado damage was the first long-span wire-cable suspension bridge in the world. The Brooklyn bridge in New York designed by Roebling was completed in 1886 with a central span of 486 m. This was followed by other notable bridges such as the George

Washington bridge with a main span of 1067 m (1931), the Golden Gate bridge with a central span of 1280 m (1937), the Mackinac bridge with a span of 1158 m (1957), the Verrazano Narrows bridge of span 1298 m (1964), the Severn bridge with a span of 988 m (1966), the Humber bridge of span 1410 m (1981), and the Tsing Ma bridge in Hong Kong (1997) with a span of 1377 m. The Rodenkirschen bridge in Germany, designed and built by Fritz Leonhardt in 1941 with a modest span of 378 m, is an example of structural elegance.

The world's longest span bridge at present is the Akashi Kaikyo bridge across Akashi Straits in Japan with a main span of 1991 m. The second longest span bridge is the East Bridge across the Great Belt Waterway in Denmark with its main span of 1624 m. The Bosporus bridge at Istanbul, Turkey, completed in 1973 with a central span of 1074 m, provided the first permanent highway link between Europe and Asia. A list of record span suspension bridges is given in Table C.1 in Appendix C.

The components of a suspension bridge, as shown in Fig. 9.16, are: (a) flexible main cables, (b) towers, (c) anchorages, (d) hangers, (e) deck, and (f) stiffening systems. The main cables carry the stiffening trusses by hangers and transfer the loads to the towers. The cable normally consists of parallel wires or parallel wire ropes of high tensile steel. The Akashi Kaikyo bridge has two main cables. Each cable, 1122 mm in diameter, consists of 290 parallel wire ropes, each containing 127 high strength (UTS = 1800 MPa) wires of 5.23 mm diameter. Thus each cable contains 36830 parallel wires.

The towers support the main cables and transfer the bridge loads to the foundations. Besides the primary structural function, the towers have a secondary function in giving the entire bridge a robust, graceful and soaring visual image. While earlier bridges had steel towers, concrete towers have been used in the Humber bridge and the Great Belt East bridge. Anchorages are usually massive concrete structures which resist the tension of the main cables. The hangers transfer the load from the deck to the cable. They are made up of high tensile wires. The hangers are usually vertical, as also adopted in Akashi Kaikyo bridge. Only three major suspension bridges, namely Severn, Bosporus and Humber, have inclined hangers.

The deck is usually orthotropic with stiffened steel plate, ribs or troughs and floor beams. The deck may be of strong steel trusses or of streamlined steel box girder. The stiffening system, usually consisting of trusses, pinned at the towers, serves to control aerodynamic movements and to limit the local angle changes in the deck. If the stiffening system is inadequate, torsional oscillations due to wind might result in the collapse of the structure, as illustrated in the tragic failure of the first Tacoma Narrows bridge in 1940. The

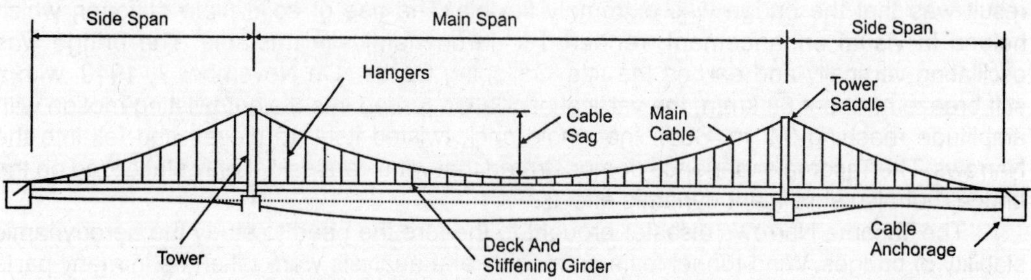

Figure 9.16 Components of a Suspension Bridge.

Akashi Kaikyo bridge and the Verrazano Narrows bridge have steel trusses. Other recent long-span bridges having steel box girder decks include: Great Belt, Bosporus, Severn and Humber.

Decisions have to be taken regarding the following items in the preparation of the design for a suspension bridge:

(a) The main and the side spans;
(b) Layout of the cable including its connections at the towers, to the stiffening girder at midspan and at the anchorages;
(c) Cross section of the deck including traffic arrangement;
(d) Sag of the cable;
(e) Construction of the cable;
(f) Hangers;
(g) Stiffening system; and
(h) Tower arrangement.

As articulated by D.B. Steinman: "A suspension bridge is a naturally artistic composition, with the graceful cable curves and the symmetry of three spans, punctuated by the dominant soaring towers and framed between the two massive, powerful anchorages. There is symmetry about each tower, and over-all symmetry of the three-span ensemble"[16]. The sidespan to main span ratio varies from 0.17 to 0.50. The span to depth ratio for the stiffening truss in existing bridges lies between 85 and 100 for spans up to 1000 m and rises rather steeply to 177 for 1300 m span. The adoption of aerodynamically shaped box cross sections in recent bridges (e.g. the Severn bridge and the Great Belt East bridge) permits impressive lightness of deck. The width of deck between cables is also to be detailed carefully and should not be made too narrow. The ratio of span to width of deck for existing bridges ranges from 20 to 56. The sag is normally adopted as 0.1 of the main span. The towers can be of concrete or steel. While concrete pylons cost less, the steel pylons facilitate faster construction, an important factor because the tower construction is situated in the critical path in the construction schedule. Since the main span and the tower height are usually large, it is necessary to take into account the curvature of the earth. For example, the towers of the Humber bridge are 36 mm further apart at the top than at the bottom.

A brief discussion of the salient features of three historic suspension bridges may be interesting. The Tacoma Narrows bridge – I designed by L. Moissieff was opened to traffic on July 1, 1940 with a span of 853 m, the world's then third longest span. In an attempt to slenderize the bridge, the designer provided shallow plate girders for stiffening the deck. The span to depth ratio was 350, where the earlier bridges of this span range provided around 90. The span to width of deck ratio was 72 as against the normal value of 35. The result was that the bridge was extremely flexible. The use of solid plate stiffener, which helped in visual enhancement, rendered it aerodynamically unstable. The bridge was oscillating vertically and earned the title 'Galloping Gertie'. On November 7, 1940, with a stiff breeze of about 68 km/h, the vertical oscillation turned into violent twisting motion with amplitude reaching 9.2 m. Soon the whole deck twisted itself to pieces and fell into the Narrows. The Tacoma experience demonstrated that wind causes not only static load on the bridge, but also significant dynamic effects.

The Tacoma Narrows disaster brought to the fore the need to study the aerodynamic stability of bridges. Wind tunnel tests on models and analysis were taken up in many parts of the world. D. B. Steinman, who took a prominent part in the aerodynamic investigation, indicated that the responsibility for the Tacoma failure was to be shared by the profession, since 'the profession had neglected to combine, and apply in time, the knowledge of aerodynamics and of dynamic vibrations with its rapidly advancing knowledge of structural engineering'.

A cable supported bridge subjected to wind may suffer wind-induced drag (the static component), flutter (aerodynamic instability) and buffeting ('shaking' due to gusts). Drag is the governing consideration for moderate spans. Flutter becomes critical when the wind acting on the structure reaches a critical velocity that triggers a self-excited unstable condition. Buffeting affects the fatigue of the bridge materials as well as the users' comfort. Based on investigations, it is now known that winds blowing slightly upwards under the deck of a bridge cause the worst disturbance. In order to mitigate such adverse effects, provisions are incorporated into the designs to set up eddies or turbulence in the wind and to increase the rigidity of the structure. The first object is achieved by leaving open in the deck a few slots and by providing cantilevered footpaths to act as horizontal fins. To achieve the second object, the dimensions of the deck are increased and diagonal suspenders may be used instead of vertical hangers and artificial damping devices may be incorporated. Such provisions were applied to many existing suspension bridges in the interest of safety, in view of the Tacoma failure. The bridge was redesigned with increased depth of stiffening trusses and wider deck and the construction was completed in October, 1950.

The Verrazano Narrows bridge, designed by O.H. Amman with a clear span of 1298 m, was opened to traffic in 1964. The bridge caters to twelve lanes of highway traffic and is two-decked, as shown in Fig. 9.17. Its towers are 210 m high and weighing 274 MN and are of cellular construction. The deck rises 69.5 m above high tide level, adequate to pass any ocean linear below. Each of the four main cables is 910 mm diameter, made up of 26 108 galvanized, parallel cold-drawn wires of 5 mm diameter. According to one estimate, the overall length of wire used in this bridge is 228 000 km, adequate to encircle the globe five times. The stiffening trusses each 7.3 m deep, spaced at 31.4 m apart, interconnect the two

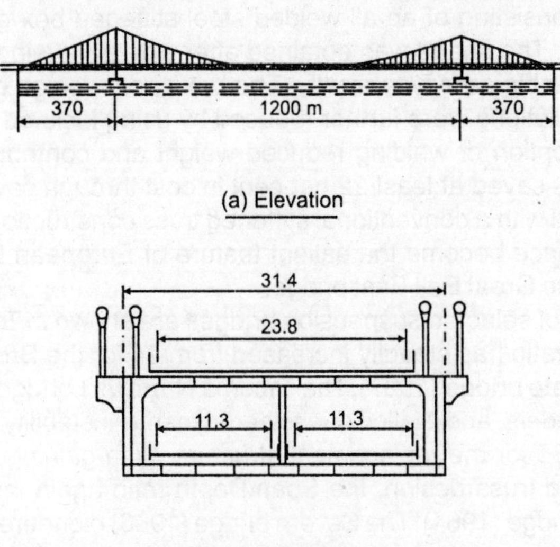

Figure 9.17 Verrazano Narrows Bridge.

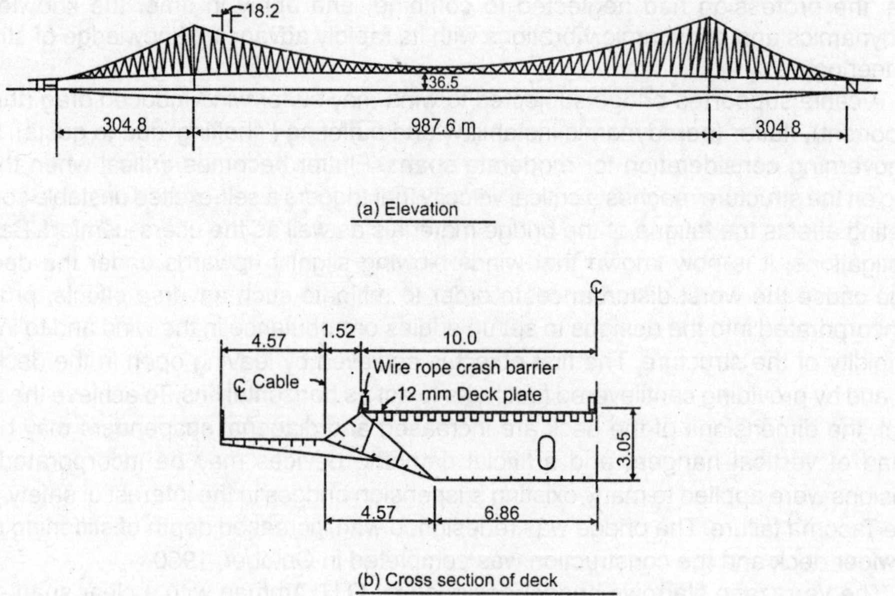

Figure 9.18 Severn Bridge.

decks. The resulting tubular framework for the stiffening member facilitated a superior technical solution even with a slender appearance. Special care was taken to provide light but strong roadway.

The Severn bridge (Fig. 9.18) completed in 1966 became a landmark in bridge construction. It was designed by G. Roberts of Freeman, Fox and Partners as a suspension bridge with a deck consisting of an all-welded steel stiffened box girder, streamlined and tapered at the edges. The shape was obtained after extensive wind-tunnel tests, and was found to be entirely stable aerodynamically. The box girder design gives torsional stability. Possible vertical oscillations were further reduced by using inclined suspenders instead of vertical hangers. Adoption of welding reduced weight and contributed to economy. This innovative design has saved at least 22 per cent in cost through saving in quantity of steel used, when compared with a conventional stiffened truss construction. The closed steel box girder concept has since become the salient feature of European bridge design, e.g. the Humber bridge and the Great Belt East bridge.

Some statistics of selected suspension bridges are shown in Table 9.1. It can be seen that the Span/Depth ratio has steadily increased from 94 for the Brooklyn bridge (1863) to 168 for the Golden Gate bridge (1937). The Tacoma Narrows I bridge adopted a ratio of 350 with stiffening plate girders, and it failed due to aerodynamic instability. Conservatism dictated the use of a ratio of 85 for the replacement structure. With gaining of confidence with the conventional stiffening truss design, the Span/Depth ratio again increased to 177 for the Verrazano Narrows bridge (1964). The Severn bridge (1966) pioneered the all-welded closed box deck with inclined suspenders, and the innovative design achieved a Span/Depth ratio of 324. This was also followed in the Great Belt East bridge with a ratio of 406. The Akashi Kaikyo bridge, which has the longest span of 1991 m, has adopted the conventional design for stiffening trusses and thus maintained a ratio of 142. The aerodynamic stability will have to be investigated thoroughly by detailed analysis as well as wind tunnel tests on models. A detailed treatment of the design of suspension bridges is outside the scope of this book and the interested reader may see the References 17 and 18 for additional details.

9.11 Cable Vibration

Cables of long span suspension or cable stayed bridges are vulnerable to wind induced vibration. Some of the mitigating measures adopted in the recent bridges include: (a) Increase damping, usually at the cable ends; (b) Alter the natural frequency of the cables by reducing cable length with the use of spacers or cross cables; and (c) Change cable characteristics by increasing surface roughness. Reliable prediction of cable vibration is not yet clear.

Table 9.1 Some Statistics on Selected Suspension Bridges.

Name of Bridge	Year of Completion	Main Span m	Depth of Stiffening Truss/Girder m	Width between cables m	Span/ Depth	Span/ Width
Brooklyn	1863	486	5.2	25.9	94	19
Ambassador	1929	564	6.7	18.1	84	31
George Washington	1931	1067	9.1	32.3	117	33
Golden Gate	1937	1280	7.6	27.5	168	47
Tacoma Narrows – I	1940	853	2.4	11.9	350	72
Tacoma Narrows – II	1950	853	10.1	18.3	85	47
Mackinac	1957	1158	11.6	20.7	100	56
Verrazano Narrows	1964	1298	7.3	31.4	177	41
Forth	1964	1006	8.4	23.8	120	42
Severn	1966	988	3.0	22.9	324	43
Humber	1981	1410	4.5	22.0	313	64
Akashi Kaikyo	1998	1991	14.0	35.5	142	56
Great Belt East	1998	1624	4.0	31.0	406	52

9.12 References

1. Arya, A.S., and Ajmani, J.L., "Design of steel structures', Nem Chand Bros., Roorkee, Fifth Edition, 1996, 901 pp.
2. Chatterjee, S., 'The design of modern steel bridges', Blackwell Science, Oxford, 2003, 207 pp.
3. Mangus, A.R., and Sun, S., 'Orthotropic deck bridges', In Chen, W-F., and Duan, L., (Eds.), "Bridge Engineering Handbook', CRC Press, Boca Raton, Florida, 2000, pp. 14-1 to 46.
4. 'Steel box girder bridges', Institution of Civil Engineers, London, 1973, 315 pp.
5. O'connor, C., 'Design of bridge superstructures', Wiley-Interscience, New York, 1971, 522 pp.
6. Shirley-Smith, H., 'The world's great bridges', English Language Book Society, London, 1964, 250 pp.
7. Troitsky, M.S., 'Cable stayed bridges - Theory and design', Crosby Lockwood Staples, London, 1977, 385 pp.
8. Podolny, W., and Scalzi, J.B., 'Construction and design of cable-stayed bridges', John Wiley, New York, Second Edition, 1986, 336 pp.

9. Ito, M., Fujino, Y., Miyata, T., and Narita, N., (Ed.), "Cable stayed bridges – Recent developments and their future", Elsevier, Tokyo, 1991, 438.
10. Mathivat, J., 'The cantilever construction of prestressed concrete bridges', John Wiley, 1983, 341 pp.
11. Leonhardt, F., 'Certain aspects of bridge design', In Yilmaz, C., and Wasti, S.T., (Ed.), 'Analysis and design of bridges', Martinus Nijhoff Publishers, The Hague, 1984, pp. 23-45.
12. Subba Rao, T.N., 'Foundations and piers of the Second Hooghly bridge', ING/IABSE Seminar on Cable Stayed Bridges, Bangalore, Preliminary Publication, Oct. 1988, pp. V-63 to V-85.
13. Beckett, D., 'Great buildings of the world - Bridges', Paul Hamlyn, London, 1969, 191 pp.
14. 'Stahlerne strassenbrucken', Beratungsstelle fur stahlverwendung, Dusseldorf, Merkblatt. 380, 1965, 50 pp.
15. Leonhardt, F., and Zellner, W., "Past, present and future of cable stayed bridges", In Reference 9, pp. 1-33..
16. Steinman, D.B., 'The design of the Mackinac Bridge', In Rubin, L.A., 'Mighty Mac', Wayne State University Press, Detroit, 1958, pp. 14-19.
17. Steinman, D.B., 'A practical treatise on suspension bridges', John Wiley, New York, Second Edition, 1949.
18. Pugsley, A., 'The theory of suspension bridges', Edward Arnold, London, 1957.
19. Hammond, R., 'Civil engineering today', Oxford University Press, New York, 1960, 229 pp.
20. Plowden, D., 'BRIDGES - The Spans of North America', The Viking Press, New York, 1974, 328 pp.

Chapter 10

Masonry and Composite Bridges

10.1 Masonry Arches

10.1.1 General

Stone masonry bridges are by nature strong and they require very little maintenance. Their inherent reserve strength is evident from the many stone arch bridges existing in service now, carrying many times their originally designed load. An example of the longevity of stone masonry bridges is the Ponte Milvio across the Tiber river in Rome, built in 109 BC and still standing[1]. The disadvantage of stone masonry is the slow pace of construction; 'it takes time' to build them, piece by piece; each stone requires to be quarried, dressed and individually matched with the surrounding stones. Hence in affluent countries with high labour costs, stone masonry bridges are not in favour. Developing countries seem to follow suit. Stone masonry bridge construction, being labour intensive, deserves much better adoption and attention in developing countries than at present.

Except for very primitive slab bridges (also known as clapper bridges), most masonry bridges have been arch bridges. They imply permanence and have a graceful appearance. The arches require unyielding abutments and piers. Many fine river bridges with masonry arches have been built in the eighteenth and nineteenth centuries, e.g. those by Smeaton, Telford and Harrison. It is interesting to note that elegant stone masonry arch bridges are still being built in China, e.g. Wuchao river bridge (1990) with a main span of 120 m (currently the world record for this type of bridge) [2].

The components of a masonry arch are shown in Fig. 10.1. The arch consists of voussoirs (wedge shaped stones) rising between the skew back at the abutment level and the keystone at the crown. The exterior and interior faces of the arch are called the extrados and intrados, respectively. The space between the extrados and the top of the keystone is called the spandrel. To serve as a cushion, an earth fill is provided above the extrados up to road or sleeper level, usually for a minimum depth of 900 mm. Spandrel walls are provided along the edges of the arch vault to retain the earth fill.

Masonry arches may be of the following types: (a) Semi-circular, (b) Segmental, (c) Pointed, (d) Semi-elliptical, (e) Parabolic, and (f) Multi-centred. The horizontal thrust at the abutments will be least when the arch is semi-circular, and small when a semi-elliptical profile is used. Where sufficient height is available, pointed or semi-circular arches can be used. Segmental or parabolic profile will be quite attractive for a medium span length. Multi-centred arches are convenient only for long spans. The rise of an arch should not ordinarily be less than one-third of span. For spans up to 12 m clear, a simple segmental or circular curve may be adopted. For spans over 12 m, the shape of the arch axis should be designed such that the axis conforms as nearly as possible to the equilibrium polygon for dead load

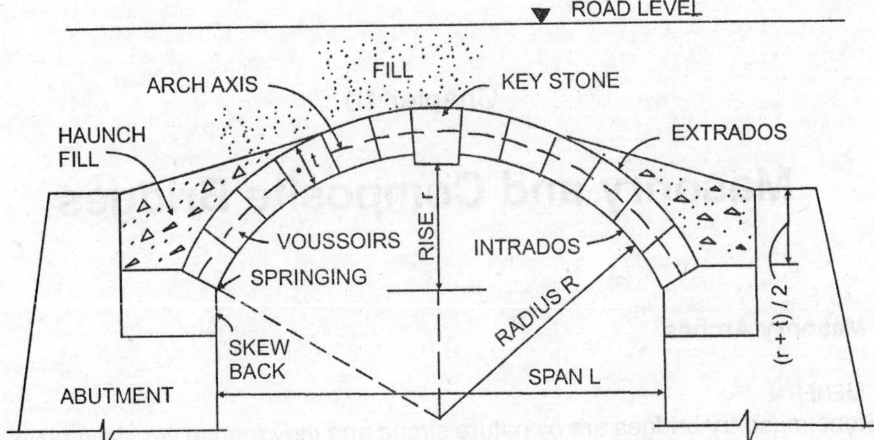

Figure 10.1 Components of a Masonry Arch.

plus 50 per cent of the live load taken as equivalent uniformly distributed load covering the entire span.

10.1.2 Design Principles

The dimensions of the arch ring are first assumed based on experience and then checked for adequacy. For economy in foundations, the rise of the arch should be at least one-fourth of the span; otherwise the horizontal thrust will be of high magnitude. However, shallower arches have been adopted for masonry bridges in China, e.g. 1:5 for Wuchao river bridge.

The radius of a segmental arch is computed from Equation (10.1).

$$R = [(L^2 + 4r^2)/8r] \quad \text{... (10.1)}$$

where R = radius of segmental arch
r = rise of arch
L = span of arch.

The determination of the appropriate arch thickness to accommodate the range of eccentricities of the thrust line is the challenge to the bridge engineer. Based on past experience and observations of successful structures, several empirical relationships between the arch thickness 't' and the span have been developed. One of these, according to Trautwyne, for first class cutstone masonry, adapted to metric units, is given by Equation (10.2).

$$t = 0.138\sqrt{R + 0.5L} + 0.06 \quad \text{... (10.2)}$$

where L, R and t are expressed in metres. For arches in brick masonry or plain concrete, the value of 't' is taken as about 33% more than that for cutstone masonry.

For spans upto 12 m, the arch ring can be of uniform thickness. For larger spans, the thickness may be kept uniform for the middle third of the span and increased symmetrically on either side gradually with due regard for conditions of stress, economy and appearance. The depth of haunch filling is taken as 0.5 (r + t) as shown in Fig. 10.1.

10.1.3 Stresses in Arch Ring

The following conditions are stipulated by the Arch Bridge Code[3] of Indian Railways.

For spans less than 12 m with uniform thickness of arch ring, the loading condition with live load for the end 5/8 of the span should be investigated.

For spans less than 12 m with variable thickness of arch ring, stresses are to be investigated at three critical sections with the positions of uniform load given as:

Section	Max. positive moment	Max. negative moment
Crown	Middle 1/4 of span	End 3/8 of span (both ends)
Quarter Points	End 3/8 of span (adjacent end)	End 5/8 of span (opposite end)
Springing	End 5/8 of span (opposite end)	End 3/8 of span (adjacent end)

For spans exceeding 12 m, the stresses should be worked out throughout the arch, the number of sections to be investigated depending on the span of the arch.

The line of pressure under the above conditions of loading should lie within the middle half of the arch ring if the line of pressure is determined by elastic theory, or within the middle third of the arch ring if graphical or other approximate methods are used.

The maximum pressure anywhere in the arch section should not exceed the permissible working stress of the arch material. The inclination of the thrust line with the normal to a section should not be more than the angle of friction at the joint, which is usually about 35 degrees.

The working stresses for masonry may be taken as below, when the crushing strength is determined by actual tests.

Compressive stress = 0.125 crushing strength
Tensile or shear stress = 0.025 crushing strength

Where tests are not made, the crushing strength may be assumed as below:

Cutstone masonry in cement mortar	14 MPa
Brickwork in cement mortar	7 MPa
Brickwork in lime mortar	4 MPa

For road bridges, the specifications given in IRC : 40[4] will govern the design.

10.2 Analysis of Arches

10.2.1 Two methods of analysis are usually available as below:

(i) Elastic method, and
(ii) Graphical method.

In either method, preliminary dimensions should be first assumed and then the stresses should be checked for the prescribed combined loading conditions.

10.2.2 Elastic Method

Masonry arches are generally treated as fixed arches and the analysis is based on the consideration of the following three conditions:

(i) The length of span remains unchanged.
(ii) Continuity of the arch axis is maintained, and one end does not move vertically with respect to the other end.
(iii) The inclination of the arch axis at each abutment remains unchanged. It is evident therefore that temperature changes, shrinkage and rib shortening, which alter the span length, will introduce additional stresses. Also settlement, spread or rotation at supports will modify the computed stresses.

The formulae for horizontal thrust, bending moment at crown and the vertical components of the reaction are given below[5].

$$H_c = \frac{\Sigma_0^{L/2}\frac{S}{I}.\Sigma_0^{L/2}M_R.Y.\frac{S}{I} - \Sigma_0^{L/2}Y.\frac{S}{I}.\Sigma_0^{L/2}M_R.\frac{S}{I}}{2.\Sigma_0^{L/2}\frac{S}{I}.\Sigma_0^{L/2}Y^2.\frac{S}{I} - 2.\left[\Sigma_0^{L/2}Y.\frac{S}{I}\right]^2} \times \left\{1 - \frac{\Sigma_0^{L/2}\frac{S}{A}.\Sigma_0^{L/2}\frac{S}{I}}{\Sigma_0^{L/2}\frac{S}{I}.\Sigma_0^{L/2}Y^2.\frac{S}{I} - \left[\Sigma_0^{L/2}Y.\frac{S}{I}\right]^2}\right\} \quad \text{... (10.3)}$$

$$M_c = \frac{\Sigma_0^{L/2}M_R.\frac{S}{I} - 2.H_c.\Sigma_0^{L/2}Y.\frac{S}{I}}{2.\Sigma_0^{L/2}\frac{S}{I}} \quad \text{... (10.4)}$$

$$V_c = \frac{\Sigma_0^{L/2}M_R.x.\frac{S}{I}}{2.\Sigma_0^{L/2}x^2.\frac{S}{I}} \quad \text{... (10.5)}$$

where H_c = horizontal thrust
L = span length
S = length of one segment
I = moment of inertia of average section of the segment
M_R = moment due to any applied force P about centre of gravity of the segment
x = horizontal distance from crown to the centre of the segment
Y = vertical distance between the crown and the centre of the segment
A = area of cross section of the segment
M_c = moment at crown
V_c = vertical reaction at a cut section at the crown.

The latter half of the expression for H_c is the factor accounting for rib shortening. It can be seen that the denominator in the rib shortening factor is similar to that of the primary term for H_c.

The effects due to temperature and shrinkage may usually be neglected in the design of masonry arches, when the rise-span ratio is not less than 0.3. In the case of reinforced concrete arches, the thrust or pull caused by a rise or fall of temperature can be computed from Equation (10.6).

$$H_c = +\frac{T \cdot t_c \cdot L \cdot E_c}{2.\sum_0^{L/2} Y^2 .\frac{S}{I} - \frac{2\left(\sum_0^{L/2} Y.\frac{S}{I}\right)^2}{\sum_0^{L/2} \frac{S}{I}}} \quad \text{... (10.6)}$$

where T = range of temperature variation in degree C

t_c = coefficient of linear expansion of concrete per degree

E_c = modulus of elasticity of concrete.

Moment at crown due to temperature is computed from Equation (10.7).

$$M_c = \frac{H_c \sum_0^{L/2} Y.\frac{S}{I}}{\sum_0^{L/2} \frac{S}{I}} \quad \text{... (10.7)}$$

(The effect of shrinkage is similar to that of fall of temperature. Under normal conditions, it would suffice to take the effects due to shrinkage as half of those due to temperature).

In order to evaluate the summations involved in the above equations, the arch is divided into a number of equal segments, and computations are made in a tabular form to evaluate the summations required in the above equations. The number of segments for one-half span may be five when the span is less than 20 m and may be increased to ten for a span of 40 m. Applying unit loads at the crown and other points at the ends of segments, the influence lines for horizontal thrust and moment at the crown can be constructed. The dead load and live load effects can be computed using these influence lines. By a similar procedure, the effects at other sections can also be calculated. It may be verified that the net stresses at every cross section of the arch are within permissible limits. If not, the sections originally assumed would be modified and the calculations repeated.

10.2.3 Graphical Method

In the graphical method, the funicular polygon of the given system is checked to see if the line of thrust could be accommodated within the middle third zone at every cross section along the entire arch length.

For this purpose, the arch profile is drawn to scale and is divided into a number of equal segments. The dead load due to arch ring and filling is plotted as ordinates representing equivalent surcharge, by dividing the dead load by the unit weight of masonry. Similarly, the live load is also plotted over the dead load ordinates as equivalent surcharge. For any one condition, the funicular polygon is drawn passing through three arbitrarily chosen points, one each at the two springings and one at the crown. If the line of thrust (funicular polygon) passes within the middle third band throughout, the arch is stable. Otherwise, the chosen points may be modified or the section may need revision.

10.3 Composite Bridges

10 3.1 TYPES

Composite construction in the broad sense would embrace any construction involving the combination into one structural element of two dissimilar elements. Composite bridge deck using a reinforced concrete slab on top of two or more steel girders is popular for highway bridges.

Steel girders may be of rolled steel joists for short spans up to about 10 m or can be of plate girders for longer spans. The girders may be spaced about 3 m. In case of spans over 50 m, cross beams may be introduced to reduce the depth of slab and also to reduce the number of longitudinal girders. Box girders may also be used instead of plate girders. Other variants of composite bridge deck construction may include precast reinforced concrete or precast prestressed concrete girder topped by a cast-in-place reinforced concrete slab.

A composite girder consists of the following components: (i) precast or prefabricated girder, which may be of precast reinforced concrete girder or precast prestressed concrete girder, or a rolled beam, rolled beam with cover plates or built up girder; (ii) cast-in-place reinforced concrete slab; and (iii) shear connectors. Composite construction with prestressed concrete girders has been discussed in Chapter 8. In this chapter, discussion is mainly directed to the steel-concrete composite construction.

Composite construction combines the advantages of prefabricated construction, namely, speedy erection, better quality control, and reduced cost of formwork, at the same time minimising the cost of transport of finished components inherent in precast construction. Economy in the quantity of steel to the extent of 8 to 60 per cent against pure steel beam has been reported[6]. The overall construction depth (top of slab to underside of beam) is of the order of L/18 to L/20 for simple spans and L/20 to L/30 for continuous spans, L being the span. The possible saving in the overall depth of the beam for a composite superstructure leads to savings in lengths of approaches and, in the case of embankments, leads to additional savings in foundations for abutments. The flexural stiffness of a steel-concrete composite beam will be about 2 to 4 times that for a corresponding steel beam; and this property results in reduced deflections and vibrations.

The effective width of flange for design of a composite beam may be computed as for a normal reinforced concrete beam as in Section 5.9. For calculation of deflection, the full width of slab may be considered effective.

10.3.2 METHODS OF CONSTRUCTION

Composite beams may be constructed with or without temporary supports (shoring). When shores are used till the attainment of the required strength by the concrete, it would be permissible to assume that the composite section carries all the loads. If temporary supports are not used, the beam should be designed to carry its own weight, besides the weight of the formwork and in-situ concrete for the slab during casting and curing. Thus shored construction requires less steel than unshored construction. However, this economy is often offset by the additional cost of shoring. The construction sequence should be carefully examined prior to detailed design.

10.4 Analysis of Composite Section

When a reinforced concrete slab is just supported over a steel I-beam, the two components have equal deflection; but their deformation and hence the stress pattern are

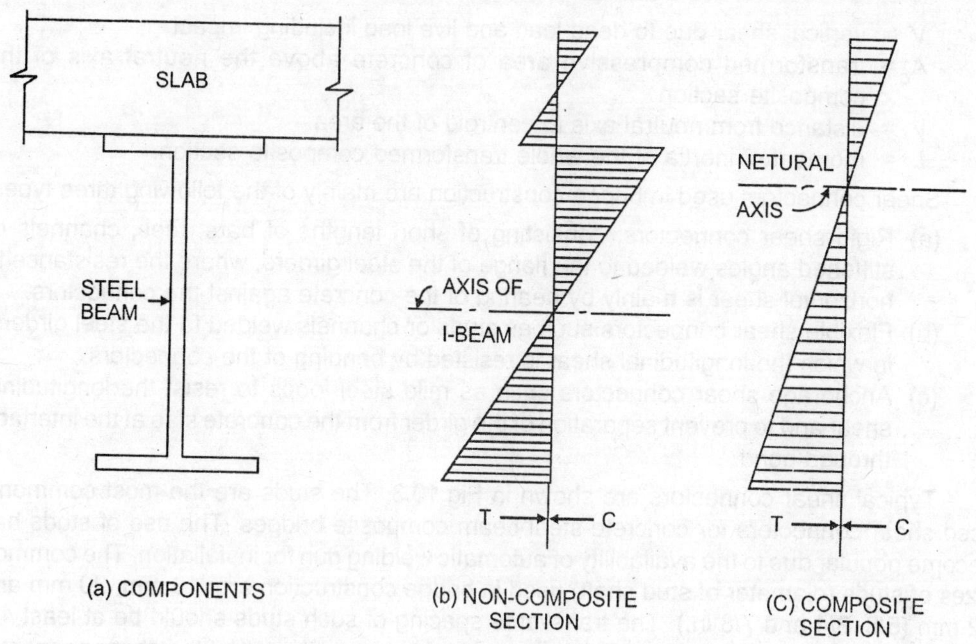

Figure 10.2 Stress Block for Composite Section.

different as shown in Fig. 10.2(b). The bottom of the slab will be in tension, while the top of the steel beam will be in compression. The two components act in a non-composite manner.

In the case of a composite section, where the total longitudinal shear is fully transferred at the junction of the steel beam and the in-situ concrete slab by means of shear connectors, the deformations of the slab and the steel beam at the junction are the same, and the stress pattern will be as shown in Fig. 10.2(c). The tensile stress at the bottom of the steel beam is now smaller than for the non-composite steel beam, because the section modulus for the composite section is larger than for the non-composite section. The stress at the bottom of the slab is usually compressive. The deflection of the composite section is much less than that for the corresponding non-composite section due to increased moment of inertia.

10.5 Shear Connectors

In order to prevent separation between the prefabricated steel beam and the in-situ concrete slab, and to facilitate the full transfer of longitudinal shear along the contact surface without slip, shear connectors are provided as specified in IRC:22[7]. These connectors are designed for the effects of dead load applied after the concrete has been cast and for live loads including impact. The effects of creep and shrinkage may be neglected in normal designs.

The longitudinal shear per unit length at the interface of the prefabricated unit and the in-situ unit is computed from Equation (10.8).

$$V_L = V.\,A_c\,.y\,/\,I \qquad \text{... (10.8)}$$

where V_L = longitudinal shear per unit length at the contact surface at the point in the span under consideration

V = vertical shear due to dead load and live load including impact
A_c = transformed compressive area of concrete above the neutral axis of the composite section
y = distance from neutral axis to centroid of the area
I = moment of inertia of the whole transformed composite section.

Shear connectors used in bridge construction are mainly of the following three types:

(a) Rigid shear connectors, consisting of short lengths of bars, tees, channels or stiffened angles welded to the flange of the steel girders, where the resistance to horizontal shear is mainly by bearing of the concrete against the connectors.
(b) Flexible shear connectors such as studs or channels welded to the steel girders, in which the longitudinal shear is resisted by bending of the connectors.
(c) Anchorage shear connectors such as mild steel loops to resist the longitudinal shear and to prevent separation of the girder from the concrete slab at the interface through bond.

Typical shear connectors are shown in Fig.10.3. The studs are the most commonly used shear connectors for concrete-steel beam composite bridges. The use of studs has become popular due to the availability of automatic welding gun for installation. The common sizes of studs (diameter of stud shaft) used in bridge construction are: 16 mm, 20 mm and 22 mm (5/8, 3/4 and 7/8 in.). The transverse spacing of such studs should be at least 4d, where d is the diameter of the stud. The concrete cover over the top of the shear connector should be at least 50 mm. Also the embedment of the stud within concrete slab should be more than 50 mm. Tee connectors have been used in the Rajendra Bridge over the Ganga at Mokameh. Anchorage type is used for composite section with precast reinforced or prestressed concrete girder and reinforced concrete cast-in-situ slab.

For mild steel shear connectors, the safe shear for each shear connector, is computed from Equations (10.9) to (10.11) for the respective type of connector[7].

For channel, Tee or angle shear connector, made of mild steel with minimum ultimate strength of 420 MPa, yield strength of 230 MPa and elongation of 21%,

$$Q = 3.32\,(h_f + 0.5t)\,L \,.\, \sqrt{f_{ck}} \quad \text{... (10.9)}$$

For welded stud connector of steel with minimum ultimate strength (f_u) of 460 MPa, yield strength (f_y) of 350 MPa and elongation of 20%, when h/d is less than 4.2,

$$Q = 1.49\,h.d.\,\sqrt{f_{ck}} \quad \text{... (10.10)}$$

For welded stud connector of steel as above, when h/d is equal to or greater than 4.2,

$$Q = 6.08\,d^2\,\sqrt{f_{ck}} \quad \text{... (10.11)}$$

where Q = safe shear resistance in Newton of one shear connector
h_f = maximum thickness of flange measured at the faces of the web in mm
t = thickness of web of shear connector in mm
L = length of shear connector in mm
f_{ck} = characteristic compressive strength of concrete in MPa
h = height of stud in mm
d = diameter of stud in mm

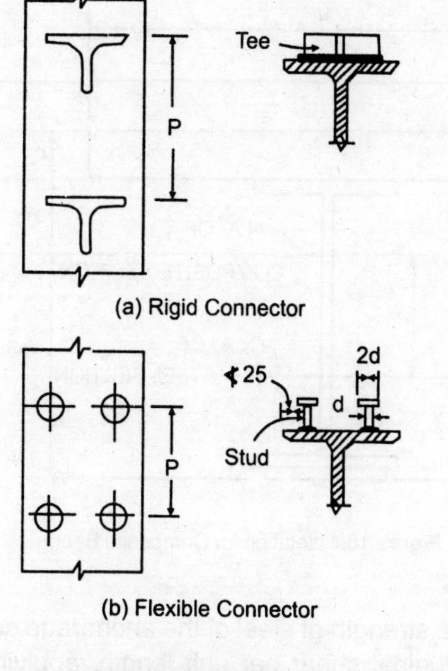

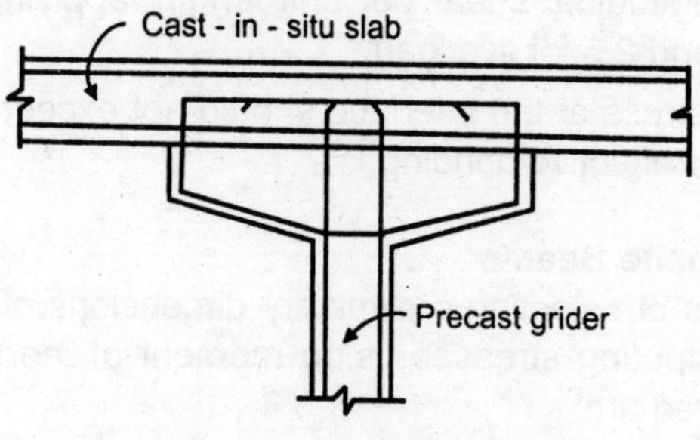

(C) Anchorage connector

Figure 10.3 Typical Shear Connectors.

The pitch (spacing) p of the shear connector is given by the Equation (10.12).

$$p = (\Sigma Q) / V_L \qquad \text{... (10.12)}$$

where V_L = longitudinal shear per unit length from Equation (10.8)

ΣQ = total shear resistance of all connectors at one transverse cross section of the girder.

When anchorage shear connectors are used, the ultimate shear resistance of one connector and the spacing are obtained from Equations (10.13) and (10.14), respectively.

$$Q_u = A_s \cdot f_u \times 10^{-3} \qquad \text{... (10.13)}$$

$$p = (\Sigma Q_u) / V_{Lu} \qquad \text{... (10.14)}$$

where Q_u = ultimate shear resistance of each connector in kN

A_s = cross sectional area of the connector in mm²

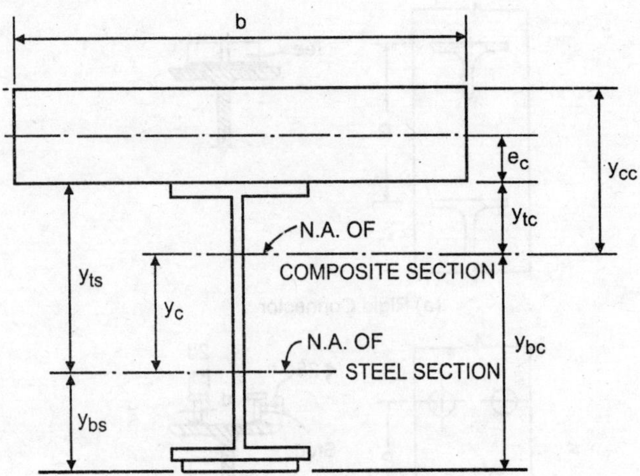

Figure 10.4 Notation for Composite Beam.

f_u = ultimate tensile strength of steel of the anchorage connector in MPa
V_{Lu} = ultimate longitudinal shear per unit length, applying load factors of 1.5 for dead load and 2.5 for live load.

The ultimate bond stress at the interface should not exceed 2.1 MPa. The interface should be made rough for effective bonding.

10.6 Design of Composite Beams

The design consists of selecting preliminary dimensions of the steel beam and the deck slab and then computing stresses using moment of inertia method. The loading conditions to be considered are:

(i) live load and dead load including creep on composite beam, and
(ii) dead load caried by steel beam alone.

The net stresses should not exceed the permissible values.

Stresses in the composite beam due to the moment M_c can be computed by the use of the following formulae (see Fig. 10.4 for notation):

m = modular ratio E_s/E_c
k = factor depending on type of loading taken as 1 for live load and 3 for dead load

$$A_c = \frac{bt}{km} \quad \text{... (10.15)}$$

$$K_c = A_c/(A_c + A_s) \quad \text{... (10.16)}$$

$$\bar{y}_c = (y_{ts} + e_c)K_c \quad \text{... (10.17)}$$

$$I_c = (y_{ts} + e_c)\,\bar{y}_c . A_s + I_s + A_c . \frac{t^2}{12} \quad \text{... (10.18)}$$

$$\overline{y}_{tc} = \overline{y}_{ts} - \overline{y}_c \quad \text{... (10.19)}$$

$$\overline{y}_{bc} = \overline{y}_{bs} - \overline{y}_c \quad \text{... (10.20)}$$

$$y_{cc} = y_{tc} + e_c + \frac{t}{2} \quad \text{... (10.21)}$$

f_t = stress at top of steel beam

$$= \frac{M_c . y_{tc}}{I_c} \quad \text{... (10.22)}$$

f_b = stress at bottom of steel beam

$$= \frac{M_c . y_{bc}}{I_c} \quad \text{... (10.23)}$$

f_c = stress at top of concrete

$$= \frac{M_c . y_{cc}}{I_c} \quad \text{... (10.24)}$$

The above formulae apply for the condition when the neutral axis of the composite section is located below the slab. However, the use of the above formulae leads to adequately accurate results if d/t is less than or equal to $A_c/3A_s$, which condition is invariably satisfied in bridge design.

The vertical shear is assumed to be taken entirely by the web of the steel section.

Shrinkage stresses are computed as the stresses in an eccentrically loaded column with a load of 0.0002 $E_s.m.A_c$ applied at the centroid of the slab. Since $m = E_s/E_c$, the stresses can be expressed as below:

$$f_t = 0.0002 E_s \left(K_c + \frac{y_{cc} - 0.5t}{S_{tc} / A_c} \right) \text{ compression} \quad \text{...(10.25)}$$

$$f_b = 0.0002 E_c \left(-K_c + \frac{y_{cc} - 0.5t}{S_{bc} / A_c} \right) \text{ tension} \quad \text{...(10.26)}$$

When shrinkage is considered, the permissible stresses can be exceeded by about 25 per cent. Hence, ordinarily shrinkage stresses can be neglected in design.

For a more comprehensive coverage of the design of bridges with composite beams, the reader is referred to References 6, and 8 to 10.

10.7 References

1. Melaragno, M., 'Preliminary design of bridges for architects and engineers', Marcell Bekkar, New York, 1998, 534 pp.
2. Li, G., and Xiao, R., 'Bridge design practice in China', In Chen, W-F., and Duan, L., 'Bridge Engineering Handbook', CRC Press, Boca Raton, Florida, 2000, pp. 63-1 to 37.
3. 'Indian railways standard code of practice for the design and construction of masonry and plain concrete arch bridges (Arch Bridge Code)', Govt. of India, Ministry of Railways, 1965, 8 pp.

4. 'IRC: 40-2002, Standard specifications and code of practice for road bridges, Section IV (Brick, Stone and Cement Concrete Block Masonry)',Indian Roads Congress, New Delhi, 2002, 40 pp.
5. Chettoe, C.S., and Adams,H.C., 'Reinforced concrete bridge design', Chapman & Hall, London, 1952, 416 pp.
6. Yan, H.T., 'Composite construction in steel and concrete', Orient Longmans, Calcutta, 1965, 210 pp.
7. 'IRC: 22-1986, Standard specifications and code of practice for road bridges, Section VI–Composite construction', Indian Roads Congress, New Delhi, 1991, 32 pp.
8. Antia, K.F., 'Composite construction with steel beams', Jl. Institution of Engineers (India), Vol. 40, No. 11, Pt. 1, July 1960, pp. 681-701.
9. Antia, K.F., 'Composite construction with prestressed concrete', Jl. Institution of Engineers (India), Vol. 40, No. 11, Pt. 1, March 1960, pp. 421 - 450.
10. Viest, I.M., Fountain, R.S. and Singleton, R.C., 'Composite construction in steel and concrete for bridges and buildings', McGraw-Hill Book Co., New York, 1958, 176 pp.

Chapter 11

Temporary and Movable Bridges

11.1 Temporary Bridges

Temporary bridges are those which are required as temporary expedients to cross obstacles in cases where, due to lack of resources or time, permanent works could not be put up speedily. Military operations need temporary bridges to be put up at short notice. Due to susceptibility to damage by fire or rot, all timber bridges are classified by Indian Roads Congress as temporary structures. Temporary causeways are often needed on diversions during the reconstruction of an existing permanent bridge. The life span of a temporary bridge is approximately ten years.

11.2 Timber Bridges

11.2.1 Timber Construction

Timber has been used in the past for short and medium span bridges. In view of its cheaper first cost than other types of construction, timber bridges can be considered in forest areas with abundant timber and cheap skilled labour. With the present day traffic and regulations regarding lane widths, timber bridges, particularly truss types, are practically obsolete. Under favourable conditions, timber bridges of small spans can be built, maintained and replaced every 15 years economically.

Hence, whenever a timber bridge is proposed, very careful study should be directed to its applicability for a given site.

The following points are to be considered:

(a) Materials-species of timber and kinds of materials to be used for joints;
(b) Length of life desired;
(c) Loads to be carried;
(d) Availability of skilled labour;
(e) Erection methods; and
(f) Environmental conditions at site.

Timber used in bridge construction should be well seasoned. Green timber will shrink, lose weight, crack, warp and check due to seasoning. These should be taken into account in design. The available size and length of individual pieces would affect spans for trestle bridges and panel lengths for trusses. Good site inspection and control of workmanship are essential.

11.2.2 Working Stresses

Typical working stresses for some Indian timbers used in bridge construction are given

Table 11.1 Safe Working Stresses of Some Indian Timbers.

Sl. No.	Species	Weight per m^3 at 12% moisture content kN	Working stresses for structural grade MPa				
			Modulus of elasticity GPa	Bending and tension along grain	Shear (Horizontal)	Compression: Parallel to grain	Compression: Perpendicular to grain
1.	Andaman paduk	7.2	11.2	14.4	1.02	10.5	4.3
2.	Bishop wood,	7.7	5.6	6.3	0.46	3.9	1.1
3.	Bullet wood, bukol	8.8	12.4	14.4	1.27	9.8	4.4
4.	Chilauni	6.9	9.6	9.1	0.88	6.0	1.8
5.	Deodar	5.6	9.5	8.8	0.70	7.0	2.1
6.	East Indian satin wood	10.2	11.6	14.8	1.37	9.9	5.3
7.	Oak	8.6	12.5	12.3	1.20	8.1	3.5
8.	Teak	6.7	11.2	13.7	0.98	9.1	3.4

Note: 1 GPa = 1000 MPa

in Table 11.1[1]. It may be noted that the strength along the grain is higher than that perpendicular to the grain. As far as possible, timber must be so placed that its greatest stress is with the grain. When the load is applied at an angle to the grain, the corresponding bearing value can be estimated from Hankinson formula[2] given in Equation (11.1).

$$N = \frac{PQ}{P \sin^2\theta + Q \cos^2\theta} \quad \text{... (11.1)}$$

where N = the allowable unit stress on the inclined surface
P = unit stress in compression parallel to the grain
Q = unit stress in compression perpendicular to the grain
θ = the angle between the direction of the load and the direction of the grain.

11.2.3. Hardware

The strength of connections usually constitutes the "weakest link in the chain" in timber construction. Hence in a majority of cases, the design effort is directed more towards the proper detailing of joints of members than in their design for strength.

The hardware used in timber bridge construction includes the following:

Bolts and nuts with washers,
Drift bolts,
Coach screws,
Spikes,
Patented connectors, like split rings, spike grids, etc.

Bolts are provided with at least two steel washers of 2.5 times the diameter of the bolt. Drift bolts are round steel pieces with heads or points. These are driven like spikes into holes bored into timber with diameter equal to 0.8 diameter of drift bolts. To join two pieces of timber with spikes, auger holes slightly smaller than the size of spikes are made in the first piece. The spikes are driven so as to force their way into the connecting piece of timber. Patented connectors of different designs are being used increasingly in heavy construction. The split-ring connector is of steel, rectangular in cross section. It fits into precut grooves in the timber faces with half the depth of the ring in the groove of each member. Spike grids are embedded into the wood surfaces by pressure. In view of the liability for corrosion, the minimum size of bolts should be at least 16 mm for normal exterior exposures and at least 25 mm for marine environment. Where mild steel plates are used at joints, they should be at least 6 mm and 12 mm thick for the above two conditions.

11.2.4 Timber Road Bridges

IRC Bridge Code classifies timber bridges as temporary structures and hence requires these to be designed for Class B loading. No impact allowance need to be considered. Timber bridges are not suitable for railways under current loading conditions.

Timber bridges may be constructed with stringers (also known as road bearers) spanning between abutments in the case of small spans up to 6 m or with trusses for spans beyond 6 m. Typical details of a timber truss bridge for a span of 12 m on a forest road are shown in Fig. 11.1.

The flooring consists of planks 300 mm wide resting on stringers. These planks can be designed as continuous beams. A wearing deck of 300 mm wide planks laid perpendicular to the main deck takes the wear due to traffic. The stringers are designed as simply supported between transoms. The entire deck should be so proportioned that the deflection of the deck is kept within 1/300 of span. Ample margin should be allowed for wear in the planking. The exact amount of extra thickness needed cannot be determined by calculations, but should be decided by the engineer by a close study of the behaviour of similar wooden bridges under traffic.

The truss is analysed for stresses due to different loading conditions and the severest effects are tabulated. Once the member forces are known, the problem of design of the chord is a combination of column and beam design and the use of judgment about the amount of material that can be cut for making the connections. Since it is difficult to vary the size of any chord from panel to panel, it is customary to provide the same size throughout one chord. Thus it is necessary only to calculate the severest force in the centre panel and to provide about one-third more than required so as to allow for connections.

Web members are arranged in such a manner as to have wooden compression diagonals and steel rod tension verticals (Howe truss). Counterbracing is provided in the centre panels to take care of stress reversals.

Timber bridges of spans up to 64 m span have been built in USA and Canada, but spans over 45 m are rare, especially with modern loadings. The long span bridges have invariably used split ring connectors for joints. For conditions obtaining in India, timber bridges are not generally recommended on important roads.

11.3 Temporary Causeways

When a temporary causeway is needed on a diversion road or on a road formed for an emergency military operation, a causeway may be formed either by using timber stringers and planking over cribs used as piers or by constructing a culvert using pipes. Typical

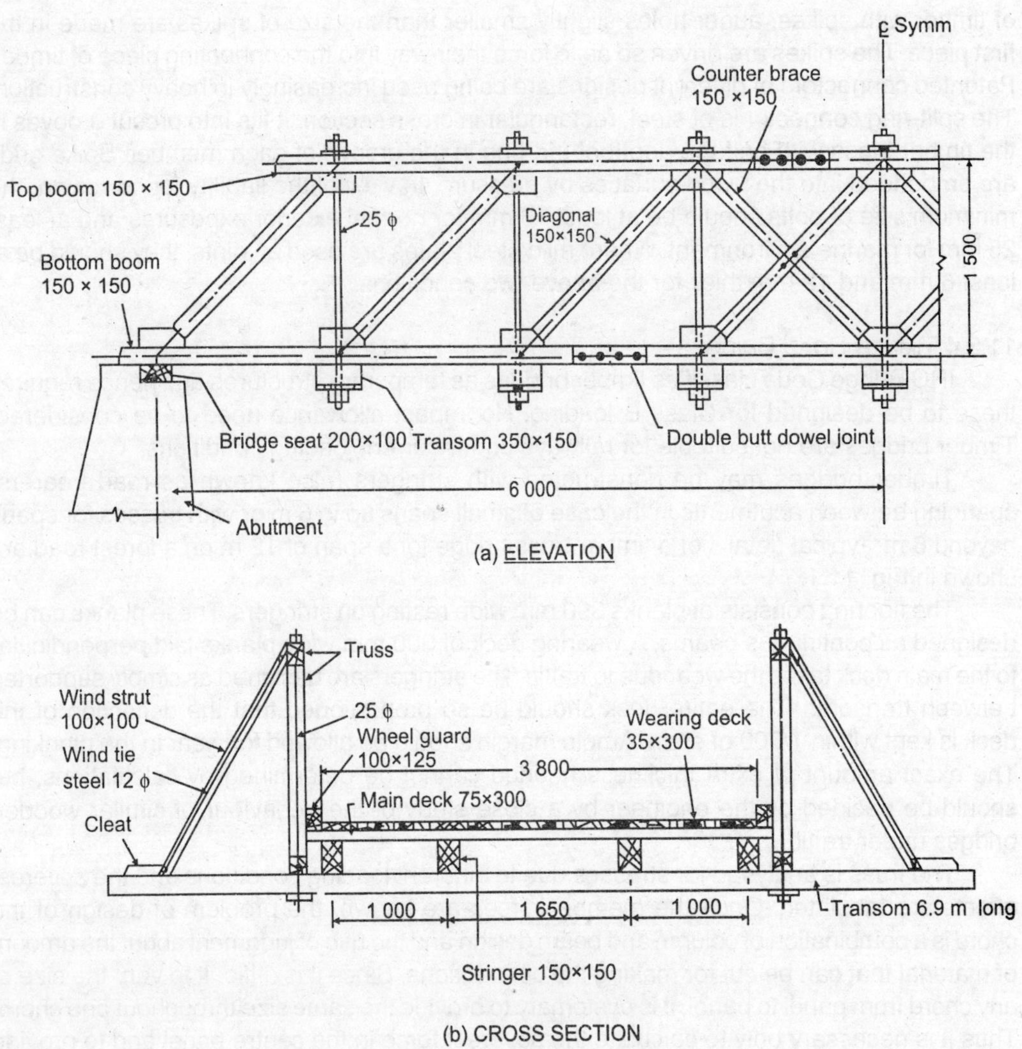

Figure 11.1 Typical Details of a Howe Truss Bridge of 12 m Span.

details of a temporary pipe culvert are shown in Fig. 11.2. In this case, an earth cushion of at least 400 mm is provided.

The pipes may be concrete pipes, armco pipes, earthenware pipes or even drums, depending on availability of materials and urgency of operations.

11.4 Military Bridges

During field operations in case of a war, the army engineers have to put up bridges across rivers and dry gaps to afford crossings where no bridges exist or where existing bridges have been demolished by the enemy.

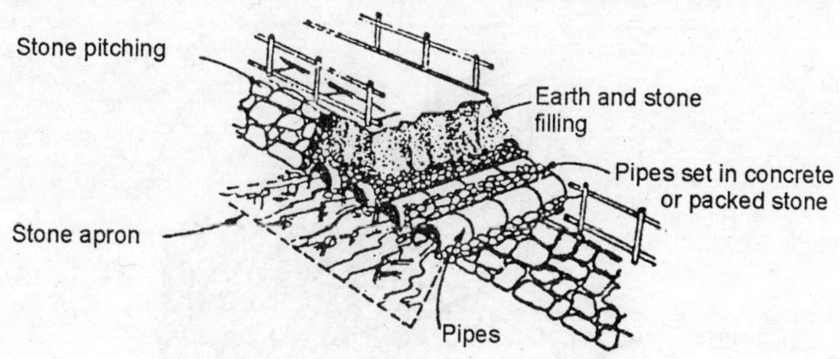

Figure 11.2 Temporary Causeway.

Military bridges can be broadly classified into two types: (i) Fixed bridges, normally used across narrow streams or dry gaps, and (ii) Floating bridges, also known as pontoon bridges, where a roadway is laid on floating supports. From the point of view of tactical use, these bridges may be classified into two groups: (a) Assault bridges for use by frontline infantry troops and by supporting troops with heavy equipment, and (b) Transport bridges of more or less permanent construction in rear areas.

Military bridges differ from civilian bridges in that they have to be designed for speedy construction with minimum labour and also for speedy dismantling, transporting and storing. For this purpose, the final design should be of reasonably small and light modular prefabricated units, which are standardised and held in stock. One such design used extensively during World War II is the Bailey bridge [3,4], designed by Sir Donald Bailey. This type of bridge has since been used in many civilian applications also for speedy construction of temporary road bridges. Other designs of similar concept are also available, e.g., D-brucke, a bridge with triangular modular units.

The Bailey bridge was designed essentially as a through-type fixed bridge built up of standard truss panels, bracing frames, transoms, stringers, bearings, links and pins. Typical details of a Bailey bridge are shown in Fig. 11.3. All the components were designed to be handled and erected by manpower. By combining the panels suitably, bridges can be built to carry up to 100 t over spans of 9.15 to 61 m (30 to 200 ft.). Bridges of any length are possible by incorporating intermediate supporting piers. The roadway consists of wooden chesses laid over stringers, which rest on transoms supported on trusses on the two sides of the roadway. Each truss is composed of a number of panels, each 3.05 m or 10 ft long, pinned together. Depending on the span length and the load class, the Bailey bridge may have one or more rows of trusses and one or more storeys of panels. Thus a bridge having single-truss and single-story is referred as single-single (SS); a bridge having double-truss and single-storey is called double-single (DS), and so on till a maximum of triple-triple (TT).

The most common erection procedure for a Bailey bridge is by cantilever launching with the aid of a launching nose. A skeleton framework of panels and transom is erected on rollers on the near bank for adequate length of the nose. Then the first bay of the bridge is added. With the completion of additional bays, the entire bridge is pushed forward on the rollers across the gap. When the forward end of the nose reaches the far bank, it is guided on the rollers on the far bank. After the first bay of the bridge comes on the far end rollers, the launching nose is dismantled. End posts are then attached. The complete bridge is

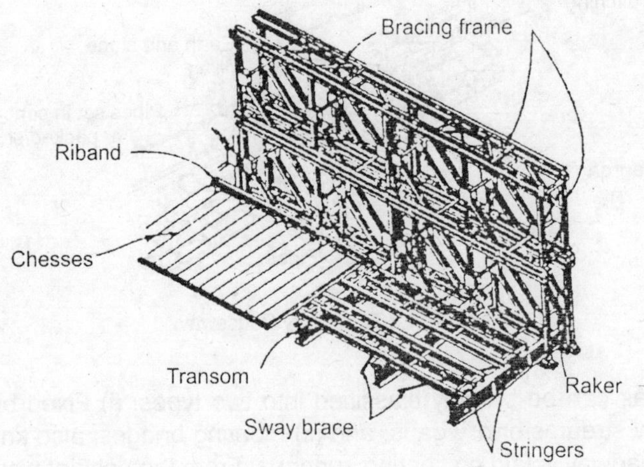

(a) Cut-away view showing components

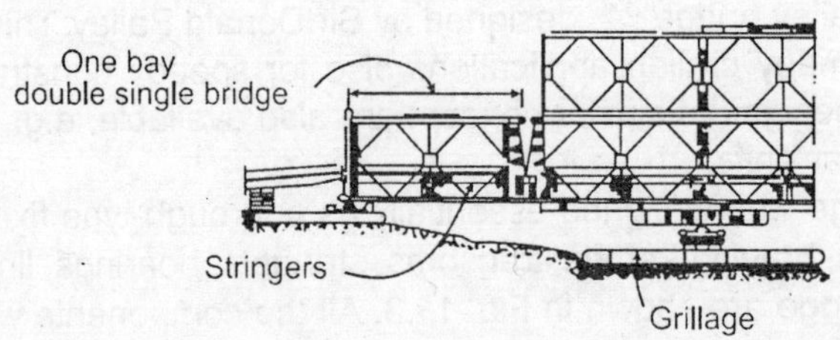

(b) Elevation at bank end

Figure 11.3 Typical Details of Bailey Bridge.

jacked off the rollers, one end at a time, and lowered on to firm bearing. End ramps are then provided to prepare the bridge for traffic.

The Bailey bridge is considered to be a major breakthrough in military engineering. The bridge was originally intended as a fixed bridge, with the piers, if needed, being constructed of cribs. With certain modifications in articulations, this bridge has also been adapted as floating bridge. The life of a Bailey bridge is about 20 years with proper maintenance. Since all parts are modular and interchangeable, only damaged parts need be replaced without disturbing the full bridge.

A semi-permanent Bailey bridge of span 27.4 m founded on ice body at Khardungla in Ladakh valley in Kashmir at an altitude of 5603 m is an example of the resourcefulness of the Indian military engineers[5]. The bridge has the distinction of being the highest altitude motorable bridge in the world. Another innovative application of the Bailey bridge is the construction of the 122 m (400 ft.) span Shatrujeet Bailey suspension bridge at Wangtu, Himachal Pradesh across River Sutlej[6] on NH-22 by the Indian Army engineers in June 1998. The bridge was constructed in 21 days surmounting difficult site constraints regarding available space around the towers on the banks.

The load class for a military bridge is denoted by a number such as 3, 5, ... 9, 30, 40, 50, 60, and 70, the class number referring to the approximate load in tonnes of the hypothetical standard vehicles appropriate to the class. While a jeep can go over a bridge of Class 3, and trucks correspond to Class 9, a Centurion tank moving on its own power will need Class 50 and the same tank when carried over a trailer will require a bridge of Class 70. The classifications in frequent use are classes 3, 9 and 50. The approximate equivalents to IRC Class AA, A and B loadings are military Classes 70, 40 and 24, respectively.

The present trend is to use aluminium alloys for military bridges because of the more favourable strength-to-weight ratio, the most common combination being Al-Zn-Mg alloy, with additives like Zr, Ti, Cr and Mn. New designs have been developed by American and European armies with aluminium alloys.

11.5 Floating/Pontoon Bridges

From time immemorial, floating or pontoon bridges have been used for the temporary passage across river—both during war and peace, but more during war. An early example is a pontoon bridge built by the Persian ruler Xerxes across the Hellespont at about 480 BC for military use. Floating bridges provide cost-effective solutions for crossing large bodies of water with large depth and soft bottom soil where conventional piers are uneconomical. An interesting and unusual example of a modern pontoon bridge is the 2018 m long Murrow bridge (1940) in Seattle, Washington, USA. This bridge is supported by hollow concrete pontoons, floating in Lake Washington, which is over 1610 m wide and 61 m deep with another 61 m of soft bottom.

The concept of a floating bridge is based on the buoyancy of water to support the dead load and the live loads of the bridge. While conventional piers and foundations are not required, an anchoring system is needed to maintain the longitudinal and transverse alignments of the bridge. The major advantages of floating bridges over fixed bridges are: (i) they are quicker to construct; (ii) they can be built where the river bed is unsuitable for improvised or crib piers; (iii) depth of water is not critical; and (iv) these can be assembled and used during hours of darkness and dismantled before sunrise, making them less vulnerable to enemy attack than fixed bridges[7].

The general requirements of a floating bridge are: (i) the bridge should have a minimum number of different types of components; (ii) the components should be interchangeable; (iii) each component should be of such size and weight as to enable easy manual handling and easy transport; (iv) the time required for construction and dismantling should be small; (v) the material used should be strong, durable and non-corrosive; (vi) the pontoons used should be of multi-cellular construction such that if one cell gets damaged during small arms fire, the pontoon would still have sufficient buoyancy to keep the bridge afloat; (vii) the two ends of the bridge should have properly designed ramps which could be adjusted to connect with the banks even with the rise and fall of the water level; (viii) the completed bridge should have low elevation over the water level, so that it can be easily camouflaged and concealed; and (ix) replacement of damaged parts should be easy and fast. Special provisions are to be incorporated in the design if navigational openings for the passage of boats and vessels are needed.

A modern floating bridge consists essentially of two parts: (i) the floating supports or pontoons; and (ii) superstructure. The pontoons are located at regular intervals along the bridge with their main axis parallel to the current of the river. These provide the buoyancy necessary to prevent the structure from sinking under live load on the bridge. The pontoons may be of rigid type, e.g., flat-bottomed boat or decked barge, or of pneumatic type. The pneumatic pontoon, which is being used increasingly in preference to the rigid type has three rubberized fabric main tubes which can be inflated prior to use. A saddle is placed on top of the float and also provides for the connection with the superstructure. The float has numerous bulkheads to form a number of separate air compartments, so that the flotation of the pontoon will not suffer much due to damage at one or a few of the air pockets.

The superstructure provides the roadway and is designed to act as one continuous member when all parts are connected. Thus a vehicle load coming on any one span will be carried by a number of pontoons on either side of that span. This facilitates the design of pontoons for smaller reactions than would be necessary if the superstructure were on simple spans. In modern floating bridges, the superstructure consists of aluminium alloy girders of box section which are placed alongside each other and supported on inflatable pontoons and connected at midway between pontoons to secure continuity.

The spacing of pontoons is a very important consideration. Too close a spacing will tend to obstruct floating debris in the river and result in damage to the bridge. With wider spacing, the strength of superstructure required and hence the sizes of superstructure members and the pontoons will increase, posing difficulties in handling, transport and erection Thus the design is a compromise to produce a bridge achieving maximum capacity with minimum weight, requiring minimum transportation and facilitating rapid assembly with a reasonable number of soldiers. Adequate anchorage of the pontoons both on the upstream and the downstream sides of the bridge is to be provided to maintain the alignment of the bridge against river currents. Particular attention should be paid to the connections at the two banks to properly take care of the rise and fall of the water level. In recent applications, inflatable rubber pontoons are used in place of wooden or metal barges. It is possible to use the pontoons for substructure and to use Bailey components for superstructure.

Depending on the equipment in use and the site conditions, there are four methods of erection of a pontoon bridge: (i) Successive pontoons, (ii) Successive bridge sections, (iii) Booming out, and (iv) Swinging full bridge. The first method is the simplest and is useful for narrow streams or shallow streams. The pontoons are added successively at the far end of the bridge and the roadway added on top of them. In the second method, complete sections of the bridge, usually with two pontoons, are built in convenient positions by the near bank, floated and towed by motorised boats to position and joined together. This method is usually adopted in wider and deeper streams and it facilitates faster assembly. In the booming out method, pontoons are added successively at the shore end and the bridge portion built till then is pushed out. Though this method provides saving in transport of roadway materials to the far end of the bridge, it is nowadays adopted only for light bridges. By the fourth method, the entire bridge is built alongside the near bank and swung into position and anchored.

Under favourable site conditions, this method may lead to very rapid erection. However, the operation is risky in fast streams and may result in severe damage to the bridge. This method is hence not favoured.

Every floating bridge presents its own set of technical, social, environmental and economic issues to be resolved. Diligently designed, competently constructed and properly maintained floating bridges can provide safe and efficient means of transportation over a long service life of 75 to 100 years with low life-cycle costs.

11.6 Movable Bridges

Movable bridges are used over a canal or waterway, when the requirements of navigation necessitate a high vertical clearance between the water level and the bottom of deck, but it is inconvenient to raise the road level to that extent from the point of view of cost and road traffic. Most movable bridges are railway structures in flat terrain, where construction of high-level bridges with vertical clearance necessary for navigation would involve very long and costly approaches to cater to the easy gradients required for the trains. The decision to opt for a movable bridge depends on the relative amounts of vehicle and waterway traffic and the required vertical clearance for the waterway. Movable bridges are also found in cities where long approach arrangements are inconvenient. Chicago in USA has the maximum number of movable bridges in the world.

The movable bridges are generally of the following types: (a) Swing bridge, (b) Bascule bridge, (c) Lift bridge, and (d) Transporter bridge. A brief description of these types is given below. Besides carrying land traffic over waterways, movable bridges perform like huge complex machines[8]. The structural steel members are vulnerable to fatigue due to large variation in stresses during opening and closing movements. The movable bridges are specialized structures and, while designing these structures, special attention must be paid to the machinery involved and their housing, operation and maintenance. If the bridge can be constructed with aluminium alloy, the saving in weight will be substantial, resulting in considerable saving in cost of power required for operation.

The swing bridge consists of a two-span continuous truss which is supported at the centre on a horizontally rotating gear arrangement which in turn is mounted on a pier located in the middle of the channel (Fig. 11.4). The truss may be of the deck type or the through type. Usually the two spans will be equal, and if they are unequal the shorter span will have to be counterweighted to balance the longer span. The swing bridge reduces the navigational capacity of the waterway by dividing the wider waterway into two narrower halves and also by taking away space for the swing deck. The pier also obstructs the flow of water and deflects the currents towards the banks. The bridge normally has to swing by ninety degrees to allow passage even to a small vessel, thus requiring a long time for each operation. The land traffic gets no protection at the water edge when the swing span is rotated. The machinery required for the operation of a swing bridge are more complicated than for a bascule or a vertical lift bridge. In view of these disadvantages, the swing bridge is currently not in favour. The largest swing bridge in the world is the El-Ferdan railway bridge across the Suez Canal in Egypt built in 2001 with a main span of 340 m.

A single-span opening can be conveniently managed by a bascule bridge, which opens up to an angle of about 80° by rotating in the vertical plane (Fig. 11.5). If the bridge is in two parts, each part can open by rotating about its support. In the former case, it is called a single-bascule while the latter is called a double-bascule. The bascule leaves are precisely

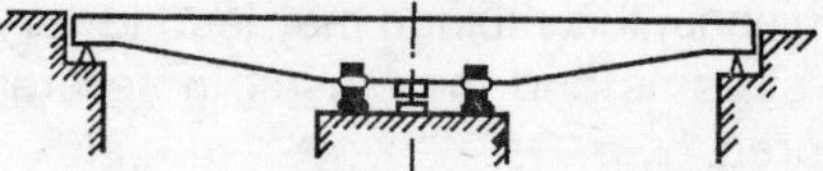

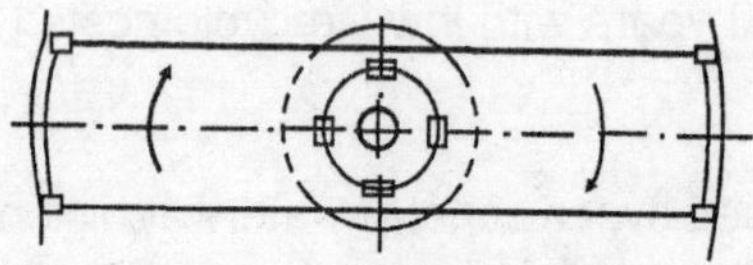

Figure 11.4 Swing Bridge.

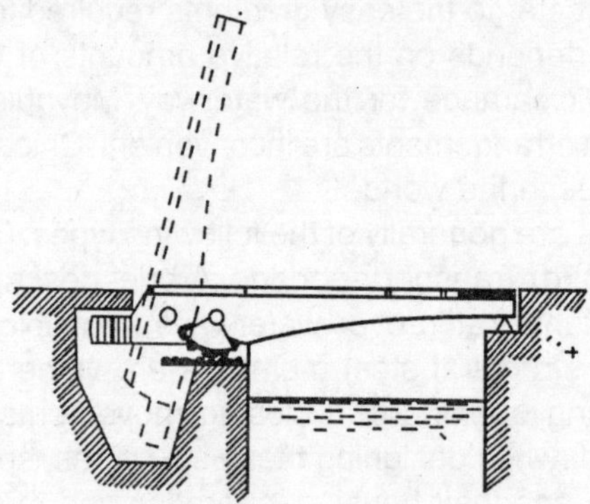

Figure 11.5 Bascule Bridge.

balanced by substantial counterweights at the bank end of the bascule, so that relatively low motor power is needed for their operation - usually just adequate to overcome inertia, frictional resistance and wind loads.

Different arrangements using cables or rack-and-pinion are possible to achieve the necessary movements. The bridge can be of either the deck type or the through type. The obvious advantage of the double-bascule is that the two smaller leaves can be raised faster than a single larger one, and requires smaller counterweights and moving parts. The bridge leaves are mostly made of steel in order to permit easier configuration of the moving elements. Modern bascule bridges have the counterweights aesthetically shaped and integrated in the bank-side cantilever. The current designs incorporate electro-hydraulic drive with hydraulic cylinders for opening and closing of the bridge, and bronze bearings for the turning point. The required locking devices at the centre and at the ends of the bridge are also driven hydraulically. The realisation of proper balance in weights of the span components and the counterweights in the erected structure requires ingenuity in design and expertise in erection. Examples of bascule bridges include the railway bridge at Mandapam near Rameswaram in S. India, London Tower bridge and Chicago's Van Buren Street bridge. There are several advantages to the bascule over the swing bridge. The operation is faster. The opening can be either partial to allow small crafts or full to allow large vessels. In case of expansion of service, at a later stage, it is easy to build another bascule adjacent to the existing one.

A vertical lift bridge is used when it is desired to provide a clearance for navigation across a channel (Fig. 11.6). The movable span is lifted by raising the span on a vertical plane while the span remains horizontal. Two towers are provided, one on each bank. The road or rail span is usually of a truss, which may be of deck or through type. The truss is connected at the ends by cables which pass over pulleys mounted at the top of the towers and are connected to counterweights at the other end. Regular inspection and maintenance of the cables are essential to ensure safe and reliable operation of the lift bridge. The truss is lifted up vertically to allow navigation in the channel. A light overhead truss is usually provided at the top connecting the two towers resulting in better bracing and also providing support to a walkway for maintenance and for cable ducts, pipe lines, etc. A notable example of lift bridge is the 166 m span Cape Cod Canal bridge at Buzzards Bay, Massachusetts. When completed in 1935, it was the longest lift span in the world. It was also the first to employ roller bearings on its counterweight sheaves, which innovation resulted in considerable saving in operation costs over the years. The towers of the bridge were given architectural treatment to resemble a lighthouse. Currently, the record for the longest lift span bridge in the world is held by Staten Island Rapid Transit's Arthur Kill bridge, which was built in 1959 with a span of 170 m. The lift bridge compares very favourably with the bascule bridge in operation and is cheaper for long spans. The aesthetics of vertical lift bridges is debatable, particularly for spans with trusses.

A transporter bridge (Fig. 11.7) is used within a harbour area to provide for transfer of men and materials across a channel. Here two towers are provided which are connected by a truss. A moving cage is hung by wire ropes which terminate on a roller arrangement moving along the bottom chord of the overhead truss. This works more as a ferry than a bridge for vehicular transport. In view of the limited capacity and the high cost, transporter bridges are not adopted anymore.

Movable bridges need greater maintenance attention than fixed bridges, and call for a broader range of skills and knowledge. Inspection of movable bridges involves activities similar to those required for fixed bridges, besides additional requirements for the machinery and moving components. The main activities for maintenance include ensuring proper balance in moving parts and greasing of cables at regular intervals to minimize friction and wear. Despite significant technical progress regarding machinery and maintenance, movable bridge design still lacks in aesthetics.

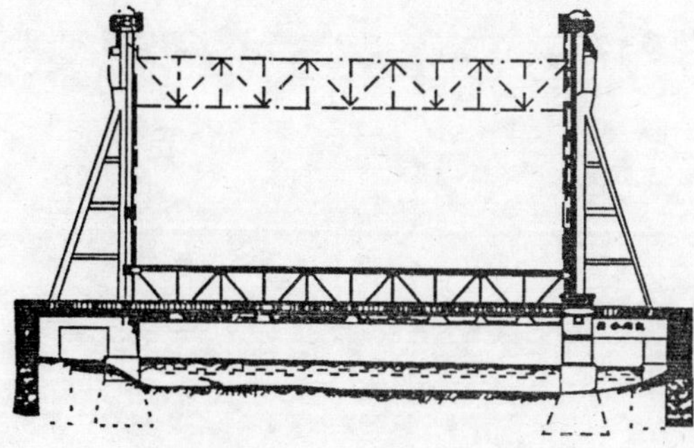

Figure 11.6 Typical Lift Bridge.

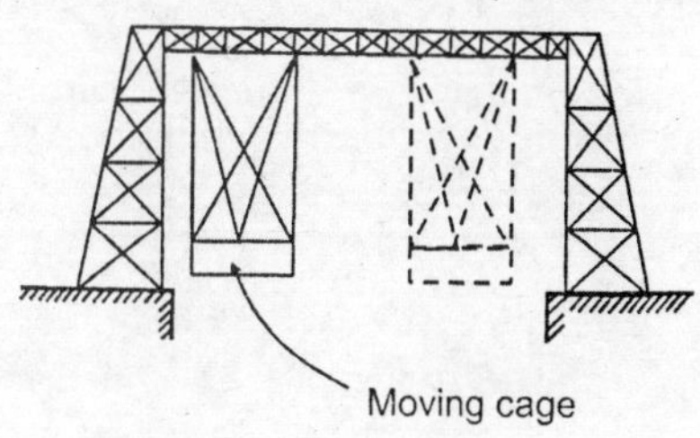

Figure 11.7 Transporter Bridge.

11.7 References

1. Masani, N.J., 'Lecture notes on simple calculations in the design of forest bridges of stock spans of 15, 20, 30 and 40 feet', Manager of Publications, Delhi, 1952, 171 pp.
2. Hool, G.A, and Kinne, W.S., 'Steel and timber structures', McGraw-Hill Book Co New York, 1942, 733 pp.
3. 'Stahlbau - Ein handbuch fur studium and praxis', Deutschen Stahlbau Verband, Koln, Vol. 2, 1964, 840 pp.
4. 'Encyclopaedia Brittanica', Vol. 4, Encyclopaedia Brittanica Ltd., London, 1958, 990 pp.
5. Vombatkere, S.G., 'Bridge resting on an ice body at high altitude', Journal of Construction Engineering and Management, American Society of Civil Engineers, Vol. 5, No. 6, 1986, pp. 429-442.
6. Kumar, M.J., 'Construction of Bailey Suspension Bridge at Wangtu on River Sutlej', In Rajagopalan, N. (Ed.), 'Advances and Innovations in Bridge Engineering', Allied Publishers, Chennai, 1999, pp. 356-372.
7. Goel, I.B., Thomas, A, Wele, S.M., 'Analysis of superstructure of floating bridge', Proceedings of Seminar on Design, Development and Production of Light Weight Bridges, Jan. 1975, Paper No. 15.
8. Koglin, T.L., "Movable Bridge Engineering", John Wiley, 2003, 691 pp.

Chapter 12

Substructure

12.1 Definition

The portion of the bridge structure below the level of the bearing and above the foundation is generally referred to as substructure. Thus for a river bridge with well foundation, the substructure will consist of the piers, the abutments and wing walls, the pier caps and the abutment caps.

12.2 Pier and Abutment Caps

The pier cap or abutment cap (also known as bed block or bridge seat) is the block resting over the top of the pier or the abutment. It provides the immediate bearing surface for the support of the superstructure at the pier or abutment location, and disperses the strip loads from the bearings to the substructure more evenly. The pier cap should cover the entire area of the top of the pier and should project 75 mm beyond the pier dimensions. This offset prevents rain water from dripping down the sides and ends of the pier and also improves the appearance of the pier. The top of the pier cap except at bearings should have suitable slope towards the outside.

The cap should be of M20 concrete with a minimum thickness of 225 mm up to a span of 25 m and 300 mm for longer spans[1]. The thickness is reduced at the end over the cutwaters. The cap is provided with nominal reinforcement of not less than 1 per cent steel distributed equally at top and bottom and provided in two directions both at top and bottom. The reinforcement along the shorter side is in the form of hoops, extending for the full width of the pier cap. The reinforcement along the length of the pier should extend from end to end of the pier cap. In addition, provision should be made for local strengthening of the cap with two layers of mesh reinforcement one at 20 mm from top and the other at 100 mm from top of pedestal or pier cap each consisting of 6 mm bars at 75 mm centres in both directions placed directly under the bearings. Typical details of a pier cap over a solid masonry pier are shown in Fig. 12.1.

12.3 Materials for Piers and Abutments

Piers and abutments may be constructed with masonry, mass concrete or reinforced concrete. Masonry piers or abutments may use stone masonry (granite) in cement mortar, or composite construction with stone masonry facing and mass concrete hearting. Concrete construction will be economical in situations where good stones suitable for masonry and skilled stone masons are not available locally.

Stone masonry used for the pier construction should be of coursed rubble masonry, first sort, in cement mortar 1:4. In the past mass concrete was adopted in some cases, using 1:3:6 mix by volume with 38 mm size aggregate. It was then permissible to add

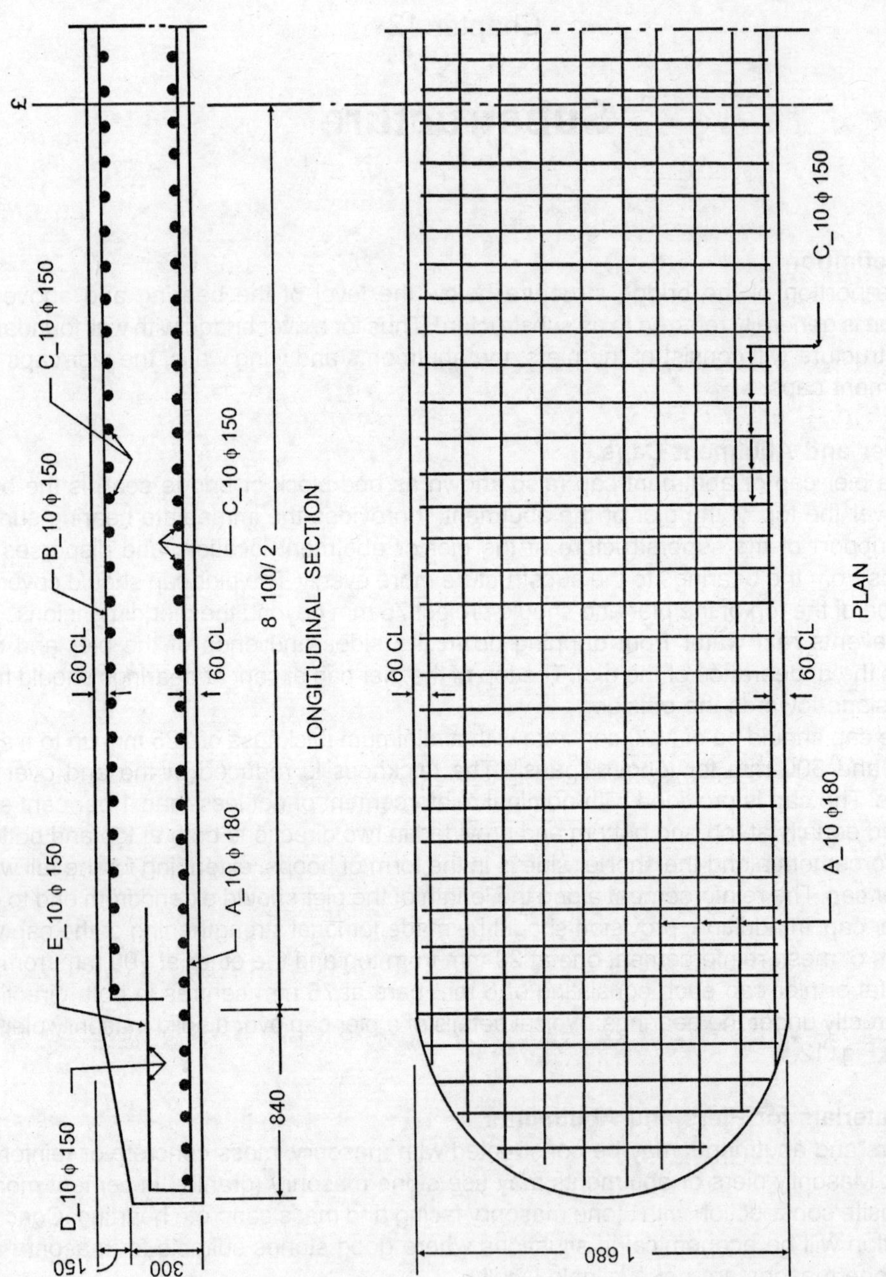

Figure 12.1 Typical Details of Pier Cap.

'plums', i.e., stones of 100 to 150 mm size, up to a volume of about 20% of the mass concrete in order to save cost. Such stones had to be placed by hand not closer than 300 mm centres. However, in recent practice, concrete of grade M20 with nominal surface reinforcement is being adopted. For reinforced concrete piers, especially single column piers, the concrete grade used may correspond to M25 to M35. Typical values of permissible stresses for mass concrete and masonry are given in Table 12.1.

Table 12.1 Permissible Stresses for Substructure.

S.No.	Material	Maximum Compressive Stress MPa	Maximum Tensile Stress in Bending MPa
1.	Mass concrete 1:3:6 mix by volume	2.7	0.28
2.	Plain concrete M 20	5.0	0.50
3.	Coursed rubble granite in cement mortar	1.5	0.10
4.	Sound brick in cement mortar	1.0	0.10
5.	Sound brick in lime mortar	0.6	0.12

Note: 1 MPa = 1000 kN/m^2

12.4 Piers

Piers are structures located at the ends of bridge spans at intermediate points between the abutments. The function of the piers is two-fold: to transfer the vertical loads to the foundation, and to resist all horizontal forces and transverse forces acting on the bridge. Being one of the most visible components of a bridge, the piers contribute to the aesthetic appearance of the structure.

The general shape and features of a pier depend to a large extent on the type, size and dimensions of the superstructure and also on the environment in which the pier is located. Piers can be solid, cellular, trestle or hammerhead types (Fig.12.2). Solid and cellular piers for river bridges should be provided with semicircular cutwaters to facilitate streamlined flow and to reduce scour. Other designs such as reinforced concrete framed type as shown in Fig.12.3 have also been used. Solid piers can be of mass concrete or of masonry for heights up to about 6 m and spans up to about 20 m. It is permissible to use stone masonry for the exposed portions and to fill the interior with lean concrete. This would save expenses on shuttering and would also enhance appearance. The stone layers should be properly bonded with the interior with bond stones.

Single column piers are increasingly used in urban elevated highway applications, and also for river crossings with a skew alignment. In an urban setting, single column piers provide an open and free-flowing perception to the motorists using the road below. Such piers when used for a skew bridge across a river results in least obstruction to passage of flood below the bridge.

Cellular, trestle, hammerhead and single column types use reinforced concrete and are suitable for heights above 6 m and spans over 20 m. The cellular type permits saving in the quantity of concrete, but usually requires difficult shuttering and additional labour in placing reinforcements. The thickness of the walls should not be less than 300 mm. The lateral reinforcement of the walls should be more than 0.3 per cent of the sectional area of

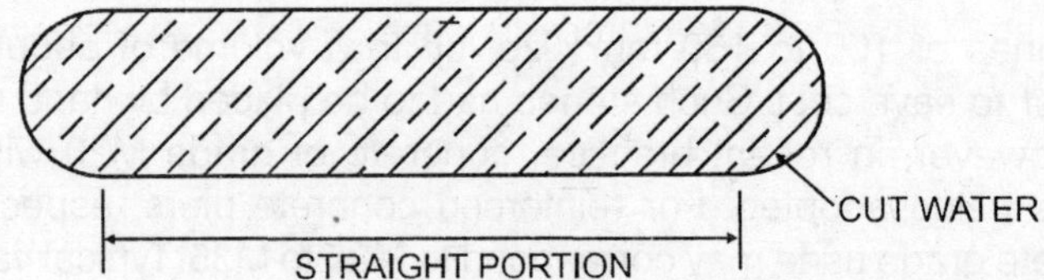

(a) SOLID PIER OF MASONRY

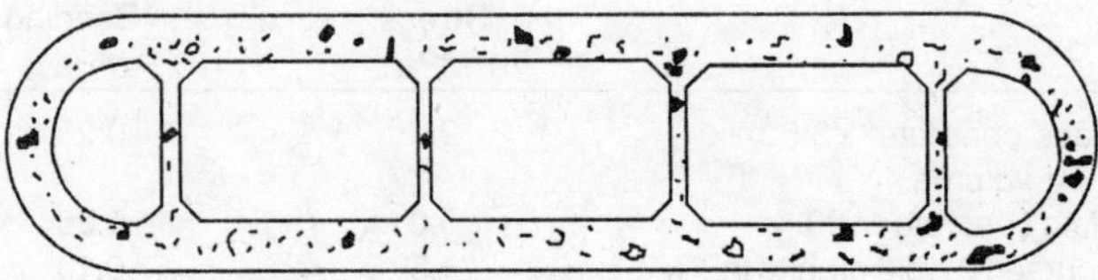

(b) CELLULAR R.C. PIER

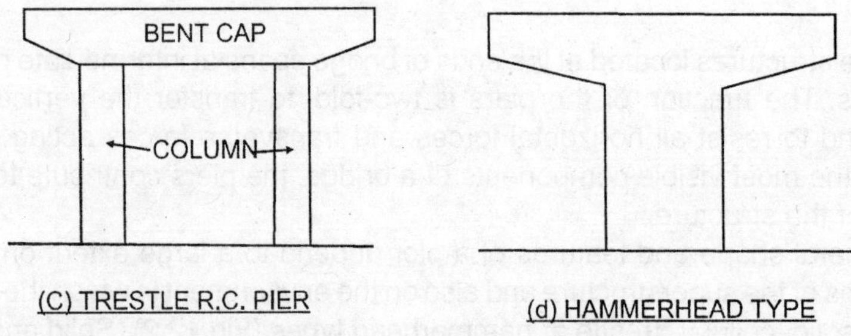

(C) TRESTLE R.C. PIER

(d) HAMMERHEAD TYPE

Figure 12.2 Typical Shapes of Piers.

the wall of the pier, and the quantity should be distributed as 60 per cent on the outer face and 40 per cent on the inner face.

The trestle type consists of columns (usually circular or octogonal) with a bent cap at the top. In some recent designs, concrete hinges have been introduced between the top of the column and the bent cap in order to avoid moment being transferred from deck to the columns. For tall trestles, as in flyovers and elevated roads, connecting diaphragms between the columns may also be provided. The hammerhead type provides slender substructure and is normally suitable for the elevated roadways. When used for a river bridge, e.g., Jawahar Setu across Sone river at Dehri[2], this design leads to minimum restriction of the waterway. The construction procedure should be arranged such that the construction joints are minimized, by adopting continuous concreting or by use of slip form technique to the extent possible. Simple geometry of the pier leads to reduced construction costs.

The top width of the pier depends on the size of the bearing plates on which the superstructure rests. It is usually kept at a minimum of 600 mm more than the out-to-out dimension of the bearing plates, measured along the longitudinal axis of the superstructure.

The length of the pier at the top should not be not less than 1.2 m in excess of the out-to-out dimension of the bearing plates measured perpendicular to the axis of the superstructure. The bearing plates are so dimensioned that the bearing stress due to dead and live loads does not exceed 4.2 MPa.

When the length of a pier is narrower in plan than the width of bridge deck carried so that the deck cantilevers beyond the pier edges, the pier is called an inboard pier. Such piers are used in a multi-span urban interchange featuring a flyover and ground level slip roads, wherein the adoption of inboard piers offers considerable saving in the use of costly urban land that would have been required when full-width piers are adopted. In addition, inboard piers for urban interchanges facilitate improved sight lines for vehicles passing the piers and enhance the overall appearance. Other innovative designs for piers to suit urban site requirements include H-shaped piers flaring at the top, which provide wider base at the top of the pier for stability of the deck and limited use of space at the base of the pier at the ground level.

The bottom width of pier is usually larger than the top width so as to restrict the net stresses within the permissible values. It is normally sufficient to provide a batter of 1 in 25 on all sides for the portion of the pier between the bottom of the bed block and the top of the well or pile cap or foundation footing, as the case may be.

In the case of river bridges, the portion of the pier located 'between wind and water', that is, the portion of the masonry surface which lies between the extreme high and extreme low water, is particularly vulnerable to deterioration and hence needs special attention. This surface is subject to damage due to the impact of floating debris, the erosive action of the current, waves, and in the case of sea water or chemical environment to chemical attack.

Reinforced concrete framed type of piers as shown in Fig. 12.3 have been used in recent years. The main advantage in their use is due to reduced effective span lengths for

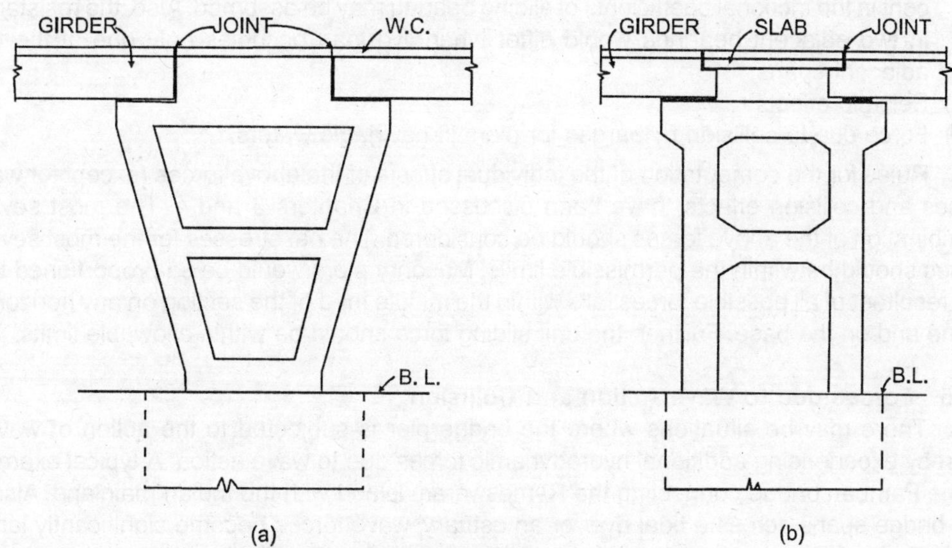

Figure 12.3 Typical Framed Piers.

girders on either side of the centre line of the pier leading to economy in the cost of superstructure. However, the author would suggest caution in their wide adoption. Firstly, such framework would be conductive to accumulation of debris and especially floating trees if used in rivers subjected to sudden floods near hills and forests. Secondly, such designs call for two expansion joints at close intervals of about 1 to 2 m on each pier, resulting in riding discomfort besides maintenance problems. If this type is to be adopted, the author would suggest that the ends of the decking on either side of the pier centre line be cantilevered beyond the bearing so that one expansion joint would be adequate.

Reinforced concrete framed piers of 'V' shape supporting a short length of reinforced concrete decking have been used successfully in conjunction with suspended spans of prestressed concrete for bridges in hilly areas, e.g., Gambhirkhad bridge[3]. In such cases, the horizontal member between the top and base of the pier is avoided to enhance aesthetics.

The load and forces to be considered in the design of piers are as below:

(1) Dead load of superstructure and the pier itself
(2) Live load of traffic passing over the bridge. The effect of eccentric loading due to the live load occurring on one span only should be considered.
(3) Impact effect for the top 3 m of the pier only
(4) Buoyancy of submerged part of substructure. If the pier is anchored to rock by dowels, it is permissible to neglect the effect of buoyancy.
(5) Effect of wind on moving loads and on the superstructure
(6) Force due to water current
(7) Force due to water action, if applicable
(8) Longitudinal force due to tractive effort of vehicles
(9) Longitudinal force due to braking of vehicles
(10) Longitudinal force due to resistance in bearings. In order to reduce the net longitudinal force in bearing, it is usual to make bearings of two spans located on a pier to be of the same type, i.e., expansion bearings or fixed bearings. Still a variation of about 10 per cent in the frictional coefficients of sliding bearing may be assumed. Also, the resistance in two adjacent bearings would differ when live load occupies only one of the two adjacent spans.
(11) Seismic effects
(12) Force due to collision by barges for piers in navigable waters.

Rules for the computation of the individual effects of the above forces (except for wave forces and collision effects) have been discussed in Chapters 3 and 4. The most severe combination of the above forces should be considered. The net stresses for the most severe cases should be within the permissible limits. Masonry piers would be so proportioned that the resultant of all possible forces falls within the middle third of the section on any horizontal plane and on the base. Further, the unit sliding force should be within allowable limits.

12.5 Forces due to Wave Action and Collision

There may be situations where the bridge pier is subjected to the action of waves, thereby experiencing additional hydrodynamic forces due to wave action. A typical example is the Pamban bridge connecting the Rameswaram island with the Indian mainland. Also, if the bridge spans across a tidal river or an estuary, wave forces become significantly large.

The wave motion is essentially an unsteady fluid flow. The fluid particles are subjected not only to a velocity in the horizontal and vertical directions, but also to accelerations in these directions. Since the piers are normally rigidly fixed at the bottom, the horizontal forces only are of interest to the designer. The fluid particle velocity and acceleration induce a horizontal drag force and an inertial force which taken together may be considerably large.

In the case of major bridges in wide navigable waters, it is necessary to provide for the possibility of boats, barges and vessels colliding with the piers during storms and foggy weather. For example, Mandovi and Zuari bridges on National Highway No. 17 in Goa have to allow very heavy barge traffic in the rivers to transport iron ore for export. Collision of the barges with the piers should be prevented to the extent possible, to avoid damage to both the bridge and the barge. The best protection from the point of view of safety to the barge is a fender of wooden piles around the pier. However, such wooden piles have not been effective in the case of Mandovi first bridge. Hence, more complicated concrete fenders were adopted for the Zuari bridge. The determination of the magnitude and direction of the collision force to be provided for in the design is difficult, requiring considerable engineering judgment, and can best be done after conducting model tests and theoretical assessment of probability of occurrence of collision. When barge impact is to be provided for, only solid type piers are to be adopted.

12.6 Example of Design of Pier

(i) *Data*

Superstructure : Simply supported T-beam of 21.3 m span
Foundation : Well foundation
Dimensions : As in Fig. 12.4
Dead load from each span = 2250 kN
Reaction due to live load on one span = 900 kN
Maximum mean velocity of current = 3.6 m/sec
Material for pier: Cement concrete M20 grade
Live load: IRC Class AA or Class A whichever produces severer effect
Only the straight portion of the pier will be considered in design here.
It is required to check the adequacy of the dimensions.

(ii) *Stresses due to dead load and self weight*

D.L. from superstructure = 2 x 2250 = 4500 kN

Self weight of pier $= 8.2\times\left(\frac{1.8+2.7}{2}\right)\times 9\times 22$

$= 3653$ kN

Total direct load $= 8153$ kN

Stress at base of pier $= \frac{8153}{8.2\times 2.7} = 368\ \text{kN/m}^2$

(iii) *Effect of buoyancy*

Width of pier at H.F.L = 1.89 m

Submerged volume of pier $= 8.2\times\left(\frac{1.89+2.7}{2}\right)\times 8.1 = 152.4\ \text{m}^3$

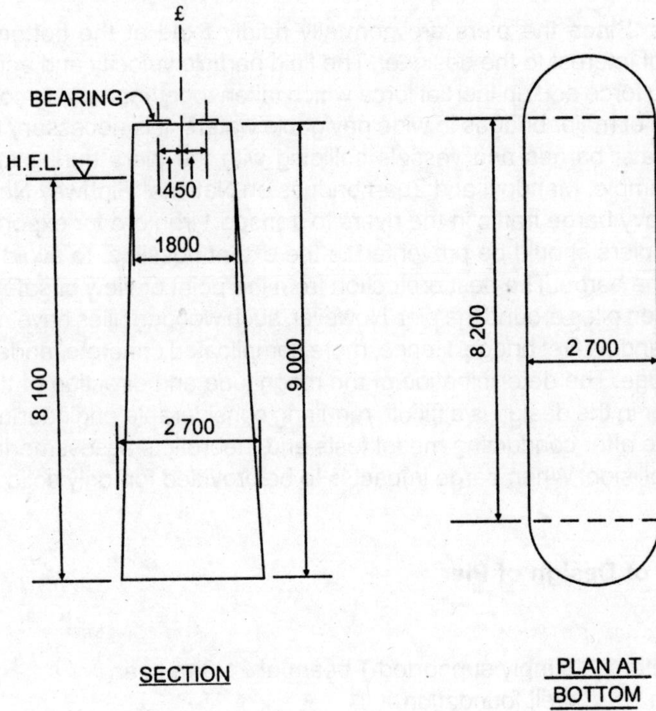

Figure 12.4 Dimensions of Pier for Example.

Reduction in weight of pier due to buoyancy = 1524 kN

$$\text{Stress at base due to buoyancy} = -\frac{1524}{8.2 \times 2.7} = -68.8 \text{ kN/m}^2$$

(iv) *Stresses due to eccentricity of live load*

Reaction due to live load from one span is 900 kN and acts at 0.45 m from the centre line of pier.

Moment due to ecentricity = 900 x 0.45 = 405 kN.m

Modulus of section of pier at base about transverse axis (axis at right angle to the direction of traffic)

$$Z_t = \frac{8.2 \times 2.7^2}{6} = 9.96 \text{ m}^3$$

Stress at base due to eccentric live load

$$= \frac{900}{8.2 \times 2.7} \pm \frac{405}{9.96} = +81.3 \text{ or } -0.1 \text{ kN/m}^2$$

(v) *Stresses due to longitudinal forces*

(a) Due to tractive effort or braking forces:

Longitudinal force for Class AA load = 0.2 x 700 = 140 kN

Effect due to Class A loading will be less.
Moment at base = 140 x 9 = 1260 kN.m

$$\text{Stress at base due to tractive effort} = \pm\frac{M}{Z_t} = \pm\frac{1260}{9.96}$$

$$= \pm 126.5 \text{ kN/m}^2$$

(b) Due to resistance in bearings to movement due to temperature:

It is possible that the frictional coefficients of the two bearings on the pier may happen to be different due to unequal efficiency of the bearings. For the severest effect, we shall assume the live load to be on left span and the frictional coefficients of bearings to be 0.25 and 0.225 on the left and right sides, respectively.

Total resistance by left side bearings = 0.25 (2250 + 900) = 787 kN
Total resistance by right side bearings = 0.225 x 2250 = 506 kN
Unbalanced force at bearing = 281 kN
Moment at base = 281 x 9 = 2529 kN.m

$$\text{Stress at base} = \pm\frac{2529}{9.96} = \pm 253.9 \text{ kN/m}^2$$

(vi) *Stresses due to wind load*

(a) Area of structure seen in elevation due to deck and handrails (to be computed from dimensions of superstructure) = 71.7 m^2
Assuming the average height of exposed surface above the bed level to be 10 m, the intensity of wind load is taken as 0.91 kN/m^2 from Table 3.2.
Total wind force = 71.7 x 0.91 = 65.2 kN

(b) Wind force against moving load, considering Class A train = 20.4 x 3 = 61.2 kN

(c) Total wind force as in (a) and (b) above= 126.4 kN

(d) Minimum limiting force on deck at 4.5 kN/m = 23.1 x 4.5 = 104 kN

(e) Minimum limiting force at 2.4 kN/m^2 on exposed surface = 71.7 x 2.4 = 172 kN

Since the force in (e) is the maximum, this will be adopted. This force will be assumed to act at the bearing level for the purpose of calculating the moment at the base of the pier.

Moment at base = 172 x 9 = 1548 kN.m
Modulus of section of the pier at base about the longitudinal axis

$$Z_1 = \frac{2.7 \times 8.2^2}{6} = 30.2 \text{ m}^3$$

$$\text{Stress at base} = \pm\frac{1548}{30.2} = \pm 51.2 \text{ kN/m}^2$$

(vii) *Stresses due to water current*

Intensity of pressure according to Eq. (3.4)

$= 0.5\,KV^2 = 0.5 \times 0.66 \times 3.6^2 = 4.3 \text{ kN/m}^2$

$$\text{Force due to water current} = \left(\frac{18 + 2.7}{2}\right) \times 9 \times 4.3 = 87 \text{ kN}$$

This force acts at a height of $2/3 \times 8.1$ or 5.4 m above the base.
Moment at base = $87 \times 5.4 = 470$ kN.m

Stress at base $= \pm \dfrac{470}{30.2} = \pm 15.6$ kN/m^2

If the current direction varies by 20 degrees,
pressure parallel to pier = 4.3 cos 20° = 4.0 kN/m^2
pressure perpendicular to pier = 4.3 sin 20° = 1.5 kN/m^2
Stress at base due to component parallel to pier

$$= \pm 15.6 \times \frac{4.0}{4.3} = \pm 14.5 \text{ kN/m}^2$$

Force perpendicular to pier = $1.5 \times 8.2 \times 8.1 = 99.6$ kN
Moment at base = $99.6 \times 5.4 = 538$ kN.m
Stress at base due to component perpendicular to pier

$$= \pm \frac{538}{9.96} = \pm 54.0 \text{ kN/m}^2$$

Maximum stresses due to water current = ± 69.6 kN/m^2

(viii) *Summary of stresses*

The net stresses should be considered for the cases when the river is dry and under floods. The summary of stresses is given in Table 12.2

It is seen from Table 12.2 that the extreme compressive and tensile stresses are well within the permissible limits (Table 12.1). The calculations shown are typical and simplified. In any practical case, minor modifications and additional consideration of seismic effects may be necessary.

Table 12.2 Summary of Stresses on Pier.

Sl. No.	Loads	Stresses, kN/m^2 When dry	Stresses, kN/m^2 During floods
1.	Dead load and self weight	+ 368.0	+ 368.0
2.	Buoyancy	—	– 68.8
3.	Eccentric live load	+ 81.3 or 0.1	+ 81.3 or 0.1
4.	Longitudinal forces		
	(a) Tractive effort	± 126.5	± 126.5
	(b) Bearing resistance	± 253.9	± 253.9
5.	Wind load	± 51.2	± 51.2
6.	Water current	—	± 69.6
	Net Maximum	+880.9	+881.7
	Net Minimum	– 63.5	– 201.9

12.7 Abutments

An abutment is the substructure which supports one terminus of the superstructure of a bridge and, at the same time, laterally supports the embankment which serves as an approach to the bridge. For a river bridge, the abutment also protects the embankment from scour of the stream. Bridge abutments can be made of masonry, plain concrete or reinforced concrete.

An abutment generally consists of the following three distinct structural elements: (i) the breast wall which directly supports the dead and live loads of the superstructure, and retains the filling of the embankment in its rear; (ii) the wing walls, which act as extensions of the breast wall in retaining the fill though not taking any loads from superstructure; and (iii) the back wall (also known as dirt wall), which is a small retaining wall just behind the bridge seat, preventing the flow of material from the fill on to the bridge seat.

In abutment design, the forces to be considered are :

(1) Dead load due to superstructure
(2) Live load on the superstructure
(3) Self weight of the abutment
(4) Longitudinal forces due to tractive effort and braking and due to temperature variation and concrete shrinkage
(5) Thrust on the abutment due to retained earth and effect of live loads on the fill at the rear of the abutment. The latter effect is considered in design as an equivalent surcharge. The Bridge Code (Clause 714.4) requires all abutments to be designed for a live load surcharge of 1.2 m height of earth fill.

Of the above forces, the earth pressure is the most difficult to compute correctly. The magnitude of earth pressure varies with the character of the material used for back fill and the moisture content. The earth pressure may be computed as indicated in Section 3.16. It is important in abutment construction to place the fill material carefully and to arrange for its proper drainage. A good drainage system may be secured by placing rock fill immediately behind the abutment.

The braking force is usually larger than the tractive effort and is taken as 0.2 of the weight of the design vehicle. The other longitudinal forces due to temperature variation and concrete shrinkage at the bearing level may be conservatively assumed as 10 % of the dead load from superstructure[4]. However, the more elaborate procedure to compute the latter component as per IRC Code[1] is indicated in the illustrative example in Section 12.8.

The design of an abutment is performed by assuming preliminary dimensions of the abutment section depending on the type of superstructure, substructure and foundations, and checking for stability against overturning, base pressures and sliding. The factor of safety against overturning should be greater than 2.0. Further, the eccentricity of the resultant of all forces on the abutment should lie within one sixth of the base width so that there is no tension at the base. The maximum stress should be less than the safe bearing capacity of the soil. The factor of safety against sliding should be more than 1.5. The required calculations are indicated in detail in the example.

For masonry abutments, it is usual to provide a batter of about 1 in 25 to 1 in 12 for the front face of the breast wall. The rear batter is adjusted to get the width required to restrict the net pressures within the prescribed limits. When reinforced concrete abutments are adopted, it would be permissible to have vertical faces both in front and at the rear faces of the breast wall. The toe and the heel portions of the base slab are so proportioned that the eccentricity of the resultant is limited to on sixth of the base width.

Wing walls will normally have sections similar to those shown in Fig. 6.3. A wing wall can be cast monolithically with the abutment breast wall to form a single monolithic structure. It is often desirable to provide a construction joint between the abutment and wing walls when these are of stone masonry or mass concrete, especially if the levels of foundation are different. Wings can be splayed or made perpendicular to the breast wall depending on the site conditions.

Typical forms of reinforced concrete abutments are shown in Fig. 12.5. The wing walls have been cantilevered without extending the base of breast wall for support, as would have been necessary for masonry abutments. The length of the cantilever returns where adopted may be restricted to 4.0 m. The slope of the bottom edge of the wing should be such as to have this edge below the level of the revetment of the embankment. A gravity type breast wall is used in (a), whereas the wall at (b) is of the counterfort type. Designs (c) to (f) attempt to reduce the overturning moment due to earth pressure. For overpasses over expressways, the scheme at (g) will enhance aesthetics. The buried abutments shown in (h) and (i) are more adaptable for overpasses with side spans and sloping cuts, than for river bridges where the embankments may be vulnerable to attack by floods.

A bridge abutment may fail in several ways as below, and the final design should be checked to avoid these failures. The breast wall may fail by tensile cracks, crushing or shear. The wall may tilt forward due to excessive overturning moment due to earth pressure. The

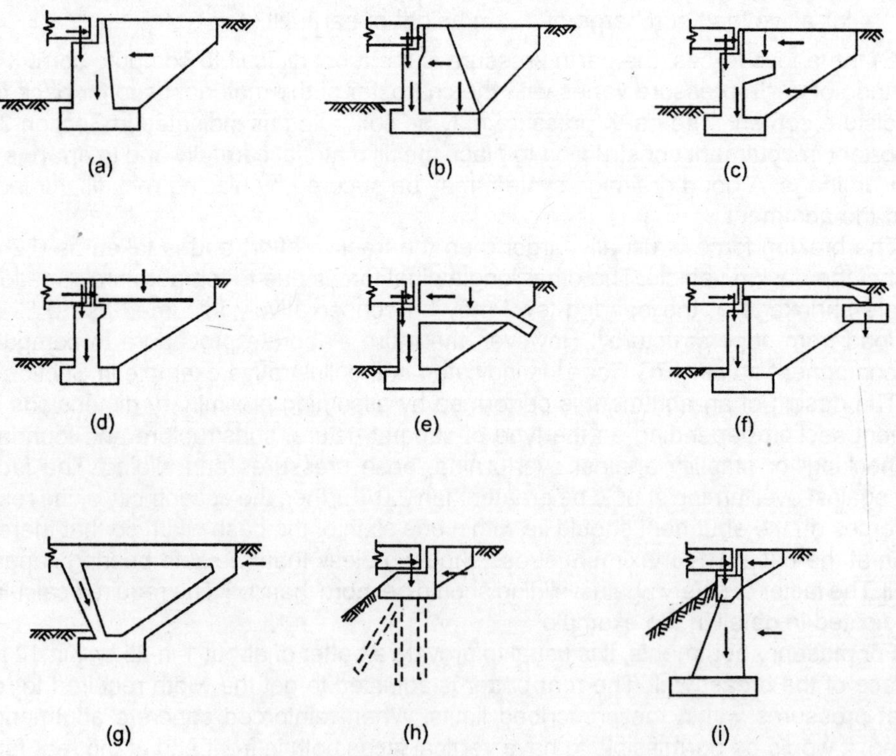

Figure 12.5 Typical Reinforced Concrete Abutments.

wall may slide forward due to earth pressure if the vertical forces are inadequate. Though the wall may be structurally strong, failures may occur along a curved surface by rupture of the soil due to inadequate shear resistance.

12.8 Example of Design of Abutment

(a) *Data*

Preliminary dimensions : Assumed as in Fig. 12.6
Superstructure : T-beam two- lane bridge of effective span 16.1 m
Overall length = 17.26 m
Type of abutment : Reinforced concrete
Loading : As for National Highway
Back fill : Gravel with angle of repose $\phi = 35°$
Unit weight of back fill, w = 18 kN/m^3

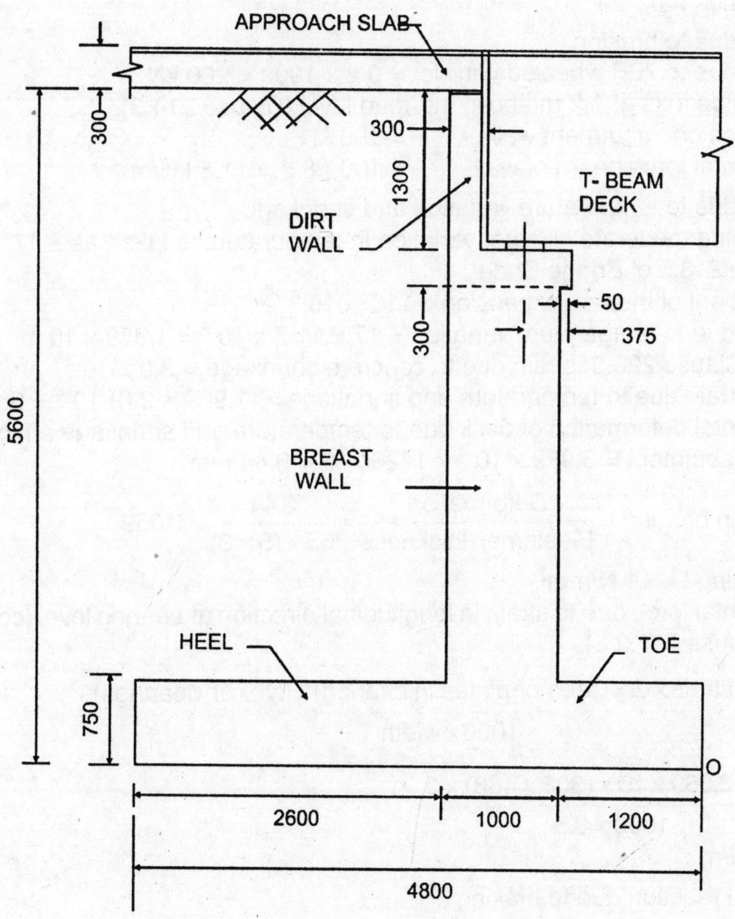

Figure 12.6 Preliminary Dimensions of Abutment.

Angle of internal friction of soil on wall, z = 17.5°
Approach slab: R.C. slab 300 mm thick, adequately reinforced
Load from superstructure per running foot of abutment wall:

Dead load = 119 kN/m
Live load = 85 kN/m

(The above two values are to be obtained from the calculations for superstructure, and are taken to act over a width of 8.5 m).

Bearings: Neoprene pads of overall size 320 x 500 x 65 mm, embedding 5 plates of 3 mm thickness and 6 mm clearance in plan. G = 1 kN/mm².

It is required to check the adequacy of the assumed section. The reinforcement details are not computed here.

(c) *Self weight of abutment*

Treating the section as composed of 4 elements as shown in Fig. 12.7, the weight of each element and moment about the point O on the front toe are computed as in Table 12.3.

(d) *Longitudinal forces*

(i) Force due to braking
Force due to 70R wheeled vehicle = 0.2 x 1000 = 200 kN
This force acts at 1.2 m above the road level (Clause 214.3).
Force on one abutment wall = 100 kN
Horizontal force per m of wall = 100 / 8.5 = 11.8 kN/m

(ii) Force due to temperature variation and shrinkage
Assuming moderate climate, variation in temperature is taken as ± 17° C as per Clause 218.5 of Bridge Code.
Coefficient of thermal expansion = 11.7×10^{-6} / °C
Strain due to temperature variation = $17 \times 11.7 \times 10^{-6} = 1.989 \times 10^{-4}$
From Clause 220.3, strain due to concrete shrinkage = 2.0×10^{-4}
Total strain due to temperature and shrinkage = $(1.989 + 2.0)\ 10^{-4} = 3.989 \times 10^{-4}$
Horizontal deformation of deck due to temperature and shrinkage affecting one abutment = $3.989 \times 10^{-4} \times 17260 / 2 = 3.44$ mm

$$\text{Strain in bearing} = \frac{\text{Deformation}}{\text{Elastomer thickness}} = \frac{3.44}{65-(5\times 3)} = 0.069$$

Assuming G = 1 N/mm²
Horizontal force due to strain in longitudinal direction at bearing level (computed as per Clause 707.2.3)

$$= \frac{1.10 \times \text{strain} \times G \times (\text{area of plates in bearing}) \times (\text{No. of bearings})}{1000 \times \text{width}}$$

$$= \frac{1.10 \times 0.069 \times 1.0 \times (308 \times 488) \times 3}{1000 \times 8.5}$$

= 4.0 kN/m

(iii) Vertical reaction due to braking

$$\text{Vertical reaction at one abutment} = \frac{200\ (1.2 + 1.6)}{16.10 \times 8.5} = 4.1 \text{ kN/m}$$

(d) *Earth pressure*

Active earth pressure P = 0.5 wh² . K_a
where K_a is obtained from Equation (3.5)
Here θ = 90°, φ = 35°, z = 17.5°, δ = 0°
Substituting values in Equation (3.5), we get K_a = 0.496
Height of backfill below approach slab = 5.6 m
Active earth pressure = 0.5 x 18 x 5.6² x 0.496 = 140.0 kN/m
Height above base of centre of pressure = 0.42 x 5.6 = 2.35 m
Passive pressure in front of toe slab is neglected.

(e) *Live load surcharge and approach slab*

Equivalent height of earth for live load surcharge as per Clause 714.4 is 1.20 m.
Horizontal force due to L.L. surcharge = 1.2 x 18 x 0.496 x 5.6 = 60.0 kN/m
Horizontal force due to approach slab = 0.3 x 24 x 0.496 x 5.6 = 20.0 kN/m
The above two forces act at 2.8 m above the base.
Vertical load due to L.L. surcharge and approach slab
= (1.2 x 18 + 0.3 x 24) 2.6 = 74.9 kN/m

(f) *Weight of earth on heel slab*

Vertical load = 18 (5.6 – 0.75) 2.6 = 227 kN/m

(g) *Check for stability - overturning*

The forces and their positions are as shown in Fig. 12.7.
The forces and moments about the point O at toe on the base are tabulated as in Table 12.3. Two cases of loading condition are examined: (i) Span loaded condition; and (ii) Span unloaded condition.

Case (i) Span loaded condition
See Row 15 of Table 12.3
Overturning moment about toe = 623.1 kN.m
Restoring moment about toe = 1740.9 kN.m
Factor of Safety against overturning = 1740.9 / 623.1 = 2.8 > 2.0 Safe.
Location of Resultant from O
$x_0 = (M_V - M_H) / V$ = (1740.9 - 623.1) / 691.4 = 1.62 m
Eccentricity of resultant
e_{max} = B/6 = 0.80 m.
e = (B/2) - x_0 = 0.78 m < 0.80 m .

Case (ii) Span unloaded condition
See Row 11 of Table 12.3
Overturning moment about toe = 572.4 kN.m
Restoring moment about toe = 1607.2 kN.m
Factor of safety against overturning = 1607.2/572.4 = 2.8 > 2.0 Safe.
x_0 = 1.72 m
e = 0.68 m < 0.80 m.

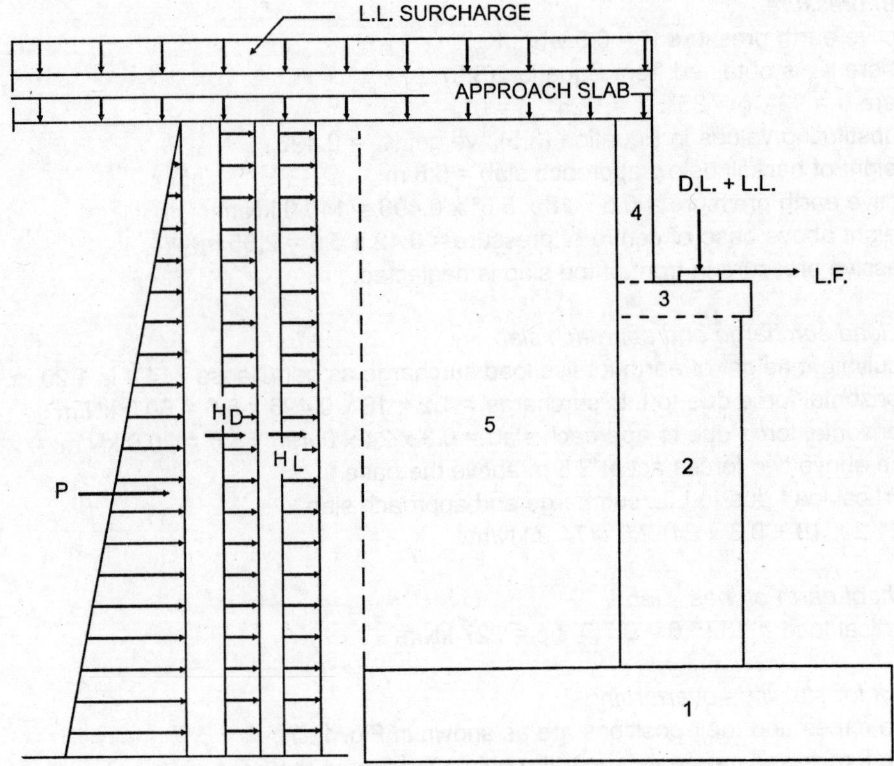

Figure 12.7 Loads and Forces on the Abutment.

(h) *Check for stresses at base*

For Span loaded condition

Total downward forces = 691.4 kN

$$\text{Extreme stresses at base} = \frac{6914}{4.8 \times 1.0}\left(1.0 \pm \frac{6 \times 0.78}{4.8}\right)$$

$$= 284.5 \text{ or } 3.6 \text{ kN/m}^2$$

Maximum pressure = 284.5 kN/m² < 350 kN/m² permissible

Minimum pressure = 3.6 kN/m² > 0 (No tension)

Hence safe.

(i) *Check for sliding*

See Row 15 of Table 12.3

Sliding force = 235.8 kN

Force resisting sliding = 0.6 x 691.4 = 414.8 kN

Factor of Safety against sliding = 414.8 / 235.8 = 1.76 > 1.5. Safe.

(j) *Summary*

The assumed section of the abutment is adequate.

Table 12.3 Forces and Moments About Base for Abutment.

Sl. No.	Details	Force, kN V	Force, kN H	Arm m	Moment about O, kN.m M_V	Moment about O, kN.m M_H
1.	D.L. from superstructure	119.0	—	1.50	178.5	—
2.	Horizontal force due to temperature and shrinkage	—	4.0	4.85	—	19.4
3.	Active earth pressure	—	140.0	2.35	—	329.0
4.	Horizontal force due to L.L. surcharge and approach slab	—	80.0	2.80	—	224.0
5.	Vertical load due to L.L. surcharge and approach slab	74.9	—	3.50	262.2	—
6.	Self weight – part 1 4.8 × 0.75 × 24	86.4	—	2.40	207.4	—
7.	Self weight – part 2 3.25 × 1.0 × 24	78.0	—	1.70	132.6	—
8.	Self weight – part 3 0.3 × 1.05 × 24	7.6	—	1.68	12.8	—
9.	Self weight – part 4 0.3 × 1.3 × 24	9.4	—	2.05	19.2	—
10.	Weight of earth on heel slab	227.0	—	3.50	794.5	—
11.	Σ Items 1 to 10 (Span unloaded condition)	602.3	—		1607.2	572.4
12.	L.L. from superstructure Class 70 R wheeled vehicle	85.0	—	1.50	127.5	—
13.	Vertical force due to braking	4.1	—	1.50	6.2	—
14.	Horizontal force due to braking	—	11.8	4.30	—	50.7
15.	Σ Items 11 to 14 (Span loaded condition)	691.4	235.8		1740.9	623.1

12.9 Backfill Behind Abutment

The design and construction of the backfill and drainage behind the abutment should be carefully attended to. A layer of filter material well packed to a thickness of 600 mm should be provided over the entire surface behind the abutment, with smaller size towards the soil and the larger size towards the wall. Adequate number of weep holes should be provided to prevent any accumulation of water and building up of hydrostatic pressure behind the walls. The weep holes may be of 100 mm diameter with 1 in 20 slope placed at about 1.0 m spacing in both directions above the low water level.

The backfill should be of clean broken stone, gravel, sand or any other pervious material of adequate length to form a wedge of cohesionless backfill. The fill should be compacted in layers. Cohesive backfill should be compacted in layers by rollers to maximum dry density at optimum moisture content. The sequence of filling behind the abutment should be controlled to conform to the assumptions made in the design. For example, if the earth pressure in front of the abutment (normally ignored) has been assumed in design, the front filling should be done along with the filling behind the abutment layer by layer. Similarly, if the design assumes that the dead load of the superstructure exists when the earth pressure due to embankment is applied, the filling behind the abutment should be deferred till the placement of the superstructure.

12.10 References

1. 'IRC: 78- 2000 Standard specifications and code of practice for road bridges: Section VII–Foundations and substructure', Indian Roads Congress, New Delhi, 2000, 97 pp.
2. 'Jawahar Setu', Pamphlet by Gammon India Ltd., Bombay, 1985, 18 pp.
3. 'Long span bridges and long span roof structures', Gammon India Limited, Mumbai.
4. Xanthakos, P.P., 'Bridge substructure and foundation design', Prentice Hall PTR, New Jersey, 1995, 844 pp.
5. Dunham, D.W., 'Foundations of structures', McGraw-Hill Book Co., New York, International Student Edition, 1962, 722 pp.
6. Andersen, P., 'Substructure analysis and design', Ronald Press, New York, 1956.

Chapter 13

Foundations

13.1 General

The design of foundations is an important part of the overall design for a bridge and affects to a considerable extent the aesthetics, the safety and the economy of the bridge. The purpose of any foundation is to transfer the load from the superstructure to the earth in such a manner that the stresses on the soil are not excessive and the resulting deformations are within acceptable limits. The design demands a detailed knowledge of hydraulics, soil mechanics and structural analysis. In this chapter, the main considerations in bridge foundations will be reviewed. For a more exhaustive treatment of the subject, reference may be made to textbooks on foundation engineering.[1-3] Guidelines for the design of bridge foundations are available in IRC Bridge Code - Section VII[4].

Foundation engineering is as much an art as a science. The engineer will have to gather the data on soil profile at the site and on the waterway characteristics. Since the soil at any place is not of one uniform type, the design of a suitable foundation would involve the exercise of considerable judgment. A young engineer would do well to study critically many case studies from the past practice and try to develop a feel for the subject. The details furnished in this chapter should be taken only as a preliminary guide.

In order to design the foundation for a bridge, the designer must determine the following reasonably and accurately:

(i) the maximum likely scour depth,
(ii) the minimum grip length required,
(iii) the soil pressures at the base, and
(iv) the stresses in the structure constituting the foundation.

The foundation should be taken to a depth which is safe from scour, and is adequate from considerations of bearing capacity, settlement stability and suitability of strata at the founding level.

13.2 Scour at Abutments and Piers

Scour denotes the washing away of the bed material in a stream due to the passage of a flood. The pattern of scour occurring at a bridge across a river depends on many factors including discharge, bed slope, bed material, direction of flow, and alignment of piers, their shape and their size. Hence the prediction of scour depth is difficult. The normal depth of scour according to the formula developed by Lacey for alluvial bed and regime channel is given by Equation (2.8). The maximum depth of scour is taken as a multiple of the normal scour depth as indicated in Section 2.11. It has been noticed that the maximum scour in the case of abutments occurs at the upstream corner, while in the case of a pier, it occurs at the downstream end[5]. The scour will be further aggravated if the pier is not aligned in the direction of flow.

13.3 Grip Length

Unless the foundations are rested on rock, adequate grip length (embedment) below the maximum scour level should be provided. The minimum required grip length is specified as one-third the maximum scour depth for road bridges and one-half of the maximum scour depth for railway bridges. The purpose of the grip length is to ensure stability under heavy flood conditions and to facilitate mobilisation of passive pressure against horizontal forces. In case of open foundations resting on rock, the minimum embedment should be 0.6 m in hard rocks with ultimate crushing strength of 10 MPa, 1.5 m in rocks with ultimate crushing strength of 2 MPa and a suitably increased depth for other types of rocks.

13.4 Types of Foundations

The foundations used in bridge structures may be broadly classified as:

(i) Shallow foundations, and
(ii) Deep foundations.

A shallow foundation is sometimes defined as one whose depth is smaller than its width. For the purpose of discussion in this chapter, a shallow foundation is taken as one which can be prepared by open excavation, and a deep foundation would refer to one which cannot be prepared by open excavation. Footings and raft foundations are examples of shallow foundations. Shallow foundations transfer the load to the ground by bearing at the bottom of the foundation. In the case of a deep foundation, the load transfer is partly by point bearing at the bottom of foundation and partly by skin friction with the soil around the foundation along its embedment in the soil.

Deep foundations are further classified as:

(a) Pile foundations, and
(b) Caisson foundations.

A pile is defined as a column-support type of foundation which may be precast or formed at site. Caisson (well) foundation is a structure with a hollow portion, which is generally built in parts and sunk through the ground to the prescribed depth and which subsequently becomes an integral part of the permanent foundation.

Caisson foundations are of two types:

(a) Open caissons (also known as well foundations in India) ; and
(b) Pneumatic caissons.

An open caisson is one that has no top or bottom cover during its sinking. It is more popularly known as well foundation. A pneumatic caisson is a caisson with a permanent or temporary roof near the bottom so arranged that men can work in the compressed air trapped under it. Pneumatic caisson can be used for a depth of about 30 m below water level, beyond which pile foundations would have to be resorted to.

The selection of the foundation system for a particular site depends on many considerations, including the nature of subsoil, the presence or otherwise in the subsoil of boulders, buried tree trunks, etc., and the availability of expertise and equipment with the contractors operating in the region where the bridge work is located. Constructibility is of primary importance in selecting the method of construction and deciding on the details of the foundation components. This is particularly important for over water bridge piers. Generally, piles would be suitable when a thick stratum of soft soil overlays a hard soil. Caissons are preferred in sandy soils. It is not uncommon to come across cases where a contractor, owning pile driving machinery, would quote lower for precast piling, even when the client has specified open caissons.

13.5 Shallow Foundations

Shallow foundations can be laid using open excavation by allowing natural slopes on all sides. This is normally convenient above the water table and is practicable up to a depth of about 5 m. For larger depths and for work under water, it would be necessary to use shoring with sheet piles or to resort to the provision of cofferdams. The purpose of shoring and cofferdams is to permit excavation with minimum extra width over the foundation width and to facilitate working on the foundation "in the dry", using suitable water pumping arrangements. In case of shoring, sheathing with timber planks supported by wales and struts is provided as the excavation proceeds. The size of the excavation at the bottom should be sufficiently large to permit adequate space for fixing formwork around the footing and to leave a working space of about 300 mm all around. The limiting depth of cofferdams is normally about 10 m. When the excavation reaches the foundation level, the exposed area of the bottom of the pit is leveled and compacted by ramming. In case pumping of water is necessary, a sump is provided to drain the water. A leveling course of about 150 mm thickness with lean concrete (1:3:6 or 1:4:8 by volume) is laid. The plan of the pier is marked on the top of the leveling course, and construction is commenced.

A shallow foundation usually consists of spread footings in concrete or coursed rubble masonry. The bottom most footing over the leveling course is of plain concrete 1:3:6 mix or of reinforced concrete of suitable thickness. The depth of foundation should be such that the foundation rests on soil with adequate bearing capacity. The maximum pressure on the foundation should be checked to ensure adequate factors of safety for different combinations of loads as specified in the IRC Code[4]. The area around the footing due to excavation should be carefully backfilled with riprap stone and not with erodible soil.

13.6 Pile Foundations

13.6.1 General

Pile foundations may be considered appropriate for bridges in the following situations: (a) when the founding strata underlies deep standing water and soft soil; (b) when the foundation level is more than 30 m below the water level, so that pneumatic sinking of wells is difficult; (c) when suitable founding strata is available below a deep layer of soft soil; and (d) in conditions where pile foundations are more economical than wells.

Pile foundations may be divided into two groups:

(i) Foundations with friction piles, and
(ii) Foundations with point bearing piles.

As the name suggests, a friction pile develops bearing capacity to a major extent by skin friction, i.e., it transfers the load to the adjacent soil by friction along the embedded length of the pile. Friction piles are driven in ground whose strength does not increase appreciably with depth. A point bearing pile transfers practically all of its load by end bearing to a hard stratum on which it rests. It should be ensured that the strata beneath the bearing layer are not too weak to carry the additional loads. The load carrying capacity of the pile is the lesser of the values of its structural capacity and the capacity of the supporting soil to carry the load. For an economical design, the dimensions of the pile should be so chosen that the structural capacity of the pile is nearly equal to the estimated supporting capacity of the soil for the pile. The pile tip should be rested on hard strata having a thickness of about twelve times the diameter or side of the pile. Since the soil conditions are generally non-uniform even at one site of a major bridge, considerable study and exercise of engineering judgment are necessary to design a proper pile foundation. The minimum spacing of piles should be 2.5 to 3.0 times the diameter of the larger pile.

Piles can be of timber, steel, reinforced concrete or prestressed concrete. Timber piles are not used for bridges nowadays, firstly due to lack of suitable logs of long length, and secondly due to susceptibility to damage by rot, borers, etc. H-section steel piles can be used, but they are not common in India due to shortage of steel. Reinforced concrete piles are in general use. Prestressed concrete tubular piles of diameter 1.5 to 6.0 m have been used in USA and Japan, in view of better ductility and high axial compressive capacity.

Concrete piles can be precast or cast-in-situ. Each type has advantages as well as disadvantages. Since bridge structures involve foundations under water or in soil with a high water table, precast piles are often preferred. Precast piles can be made to high quality with respect to dimensions, reinforcement disposition and concrete strength, and hence the structural capacity of these piles can be relied upon. However, the length of pile is limited to about 20 m depending on the driving equipment. The main advantage of an in-situ pile is that there is no wastage of concrete and no chance of damage to pile during driving exists. Also, the soil occurring at the foundation level is seen prior to placing the concrete for the pile. A serious disadvantage is the possibility of inadequate curing due to chemical attack in aggressive subsoil water, which condition is likely to go undetected during construction. Further, the occurrence of necking of concrete during removal of the forming shell and mixing of the concrete with the surrounding soil should be carefully guarded against.

A hybrid procedure is sometimes used combining a few features of precast and cast-in-situ piles. In this case, a steel shell is driven into the ground to the required depth and the earth inside the shell is removed. A precast pile which has a central duct throughout its length is lowered into the shell. Cement mortar grout is pumped through the central duct and the outer shell is also removed simultaneously. The grout fills the space between the pile and the surrounding earth ensuring good friction. This procedure is specially desirable in locations with aggressive subsoil water.

13.6.2 Load Carrying Capacity Of Pile

The load carrying capacity of a single pile can be found best by conducting a load test at the site. It can also be assessed by use of any one of the available empirical formulae[6,7]. Such a formula is an attempt to estimate the static longitudinal load that a pile can safely carry as a part of a permanent structure from an evaluation of the resistance of the pile to the dynamic forces applied on it during driving. The above assessment can only be taken as a guide for cautious interpretation by an experienced engineer.

The ultimate load carrying capacity (R_u) of a pile may be computed from Equation (13.1).

$$R_u = R_b + R_f \qquad \text{... (13.1)}$$

where R_b = ultimate base resistance

$= A_b . f_{bu}$

A_b = plan area of base of pile
f_{bu} = ultimate bearing capacity of the soil at the pile base
R_f = ultimate side frictional resistance

$$= \sum_{i=1,n} A_{si} \cdot f_{si}$$

n = number of layers of soil
A_{si} = surface area of pile in a particular strata
f_{si} = average skin friction per unit area of the pile for that particular stratum

The safe load carrying capacity of the pile is obtained by applying a factor of safety of 2.5.

13.6.3 Load Carrying Capacity Of Pile Group

For point bearing piles, the safe load per pile when used in a group can be taken nearly the same as that obtained from a load test. In the case of friction piles, the bearing capacity of an individual pile in a group of piles is less than the individual capacity under identical conditions of the soil and depth of embedment. In other words, the efficiency of the pile in a group is less than 100 per cent. This is because most of the soil which assists in supporting a single pile by radial distribution of the load through vertical shear will also be required to assist in the support of other piles driven nearby. Increase in spacing of piles will tend to improve the efficiency, but will also necessitate larger size of pile cap. The minimum spacing on centres is generally taken as three times the size of the pile or 1 m, whichever is greater.

The efficiency factor F, by which the bearing capacity of a single pile is to be multiplied to arrive at the capacity of a friction pile in pile group, is frequently obtained from Converse-Labarre formula, given in Equation (13.2).

$$F = 1 - \theta \frac{(n-1)\,m + (m-1)n}{90\,mn} \qquad \dots (13.2)$$

where n = number of piles per row
m = number of rows
θ = arc tan (d/s), in degrees
d = diameter of pile
s = spacing of pile centres.

Vertical piles are generally adopted. Batter piles (also known as raking piles) are more efficient to carry horizontal forces occurring at abutments and piers. However, batter piles are difficult to drive with accuracy on the slope. These also need special equipment and expertise. The maximum rake permitted for precast driven piles is 1 in 4. When using a combination of vertical and batter piles, the analysis should consider the total system of the pile group with the pile cap, as these act together.

13.6.4 Pile Cap

A rigid pile cap in reinforced concrete should be provided to transfer the load from the pier to the piles as uniformly as possible under normal vertical loads. The plan dimensions of the pile cap should be such that there is a minimum offset of 150 mm beyond the outer faces of the outermost pile in the group. The cap thickness is usually kept at least one-half of the pile spacing. The reinforcement of the pile cap should consider the shape of the cap and the disposition of the piles. The top of the piles should be stripped to an adequate length and the exposed longitudinal reinforcement of the piles should be fully embedded into the pile cap. The top of the pile cap is generally specified to be at the bed level in case of rivers with seasonal flow or at the low water level (LWL) in case of perennial rivers. In some cases, the bottom of the pile cap is kept at about 150 mm above the LWL. While the former approach leads to better aesthetics but with difficulties in construction, the latter method facilitates faster construction and inspection of the underside of the pile cap during service.

13.6.5 Precast Concrete Piles

Precast concrete piles can be round, square or octagonal in section. A circular pile is difficult to make due to difficult formwork, whereas a square pile is easy to cast; and octagonal pile is a compromise between the two. The minimum size of a precast pile is 350 mm square; the maximum size is seldom above 600 mm square, with the practical maximum around 450 mm square. The pile is generally cast to its full length. The maximum length of a driven pile is usually about 36 m from transportation consideration. Field splicing, though possible, involves delay and additional cost. The heavy weight of the pile is a disadvantage. The bearing capacity of the individual pile normally ranges from 300 to 1500 kN.

The concrete used for the pile should be M35 with cement content not less than 400 kg/m³ of concrete. The steel reinforcement provided in a pile is mainly intended to take care of the handling and driving stresses rather than to provide column strength in its final position. Pile handling requires two operations: (i) lifting with two slings, and (ii) hoisting with one sling prior to driving. In order to minimize the bending moment due to handling, toggle holes

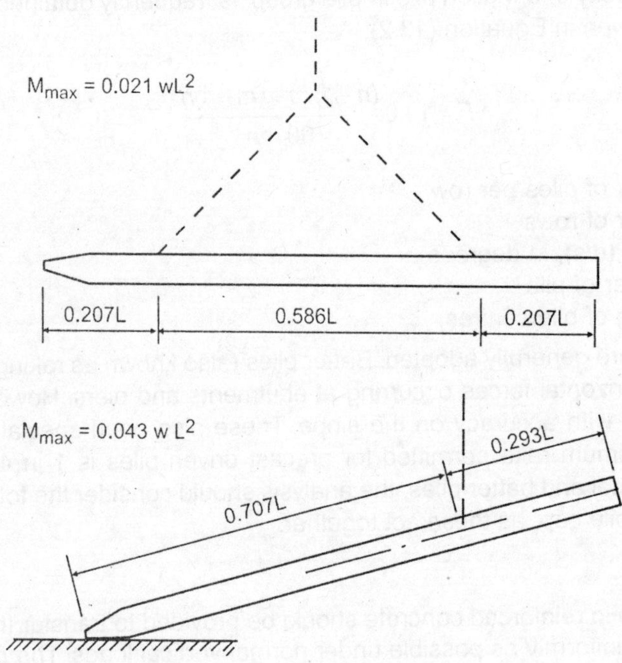

Figure 13.1 Pile Handling.

(with sleeves) are usually provided at 0.207 L from either end for lifting; and at 0.293 L from the top end for hoisting. The arrangements are shown schematically in Fig.13.1, where the maximum moments are also indicated in terms of the pile length L and weight of pile per unit length w.

The longitudinal reinforcement is usually heavy, ranging from 1.25 to 4 per cent of the cross-sectional area of the pile. The minimum amount of longitudinal steel in terms of cross-sectional area of the pile will be 1.25, 1.5 and 2.0 per cent, for L/D ratio of less than 30, between 30 and 40, and over 40, respectively, where D is the least lateral dimension or diameter of the pile.

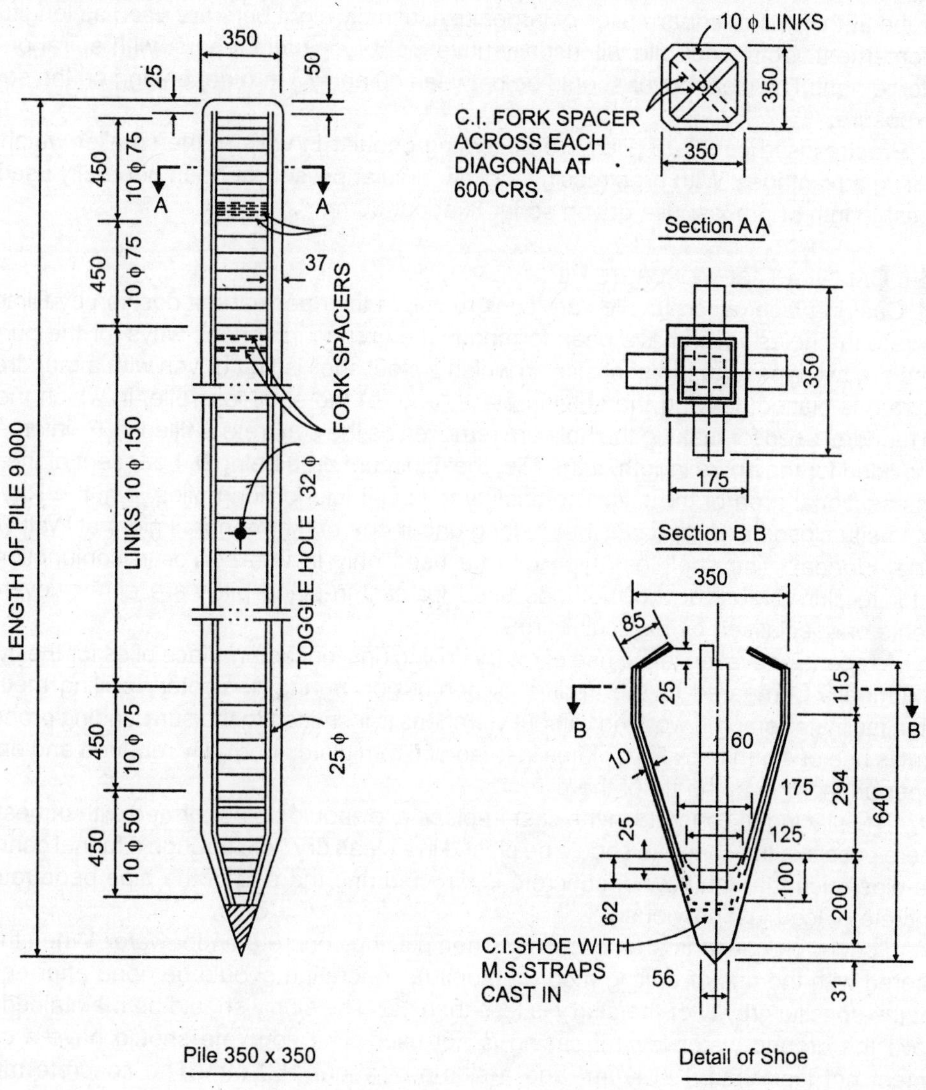

Figure 13.2 Typical Precast R.C. Pile.

Lateral reinforcement consisting of hoops of minimum 6 mm diameter are provided to resist driving stresses besides the usual function as ties in a column. The amount of lateral reinforcement should be not less than 0.6 per cent of the gross volume for a distance 3D from either end of the pile, and should be not less than 0.2 per cent of the gross volume in the body of the pile.

Typical reinforcement details for a precast pile of size 350 mm square are shown in Fig. 13.2. The longitudinal steel consists of 4-25φ bars. Links with 6 mm rod are provided with varied spacing, the spacing being closer towards the ends to provide for the driving stresses. The concrete in the piles should be of M35 grade. The normal curing operation would involve striking end shutters at 1 day, curing till 7th day, lifting at 10th day and driving after the 28th day. For square piles of larger size, usually eight bars are used as longitudinal reinforcement. Octagonal pile will usually have eight longitudinal bars, with spiral or hoop reinforcement. The clear cover should be between 40 and 75 mm depending on the severity of exposure.

Prestresssed concrete piles are becoming popular in view of the smaller weight and handling advantages. With prestressing, hollow circular piles have been generally used. The longest length of precast pile driven so far is about 60 m.

13.6.6 Cast-in-Place Concrete Piles

Cast-in-place concrete piles are constructed in their permanent position by filling with concrete the holes which have been formed in the ground in various ways for the purpose. There are two types: (a) the shell pile, in which a steel shell is first driven with a mandrel and concrete is placed, leaving the shell in place, and (b) the shell-less pile, in which the pipe and mandrel used for making the hole are removed as the concrete is filled in. Reinforcement is provided for the entire length of the pile, the minimum area being 0.4 per cent of the gross cross-sectional area of the pile. The shell type is suitable for long piles, e.g. the Raymond cast-in-situ closed end pipe pile 114 m long under one of the monorail piers at Walt Disney World, Florida[8]. The shell-less type can be used only in firm soil or in conjunction with bentonite slurry. Most of the methods used for cast-in-place piles are either covered by patents or specialised by particular firms.

A recent innovation is the use of rotary drilling rigs for cast-in-place piles for the flyovers at Mumbai[9]. Compared to the traditional percussion boring, the rotary drilling method is faster, facilitates smooth working without vibrations transmitted to the surrounding properties, permits use of 4 mm thick liner plates instead of 6 mm plates normally required and ensures proper anchorage at the tip of the pile.

The placing of concrete in the cast-in-place pile should be performed with utmost care. Where possible, the concrete should be placed in a clean dry hole through a funnel connected to a pipe such that the flow of concrete is directed and the end of the pipe penetrates the concrete placed just previously.

Special precautions are necessary when placing concrete under water. When the hole is bored with the use of drilling mud (bentonite), concreting should be done after ensuring that the specific gravity of the slurry is less than 1.2. The slurry should be maintained 1.5 m above the ground water level if casing is not used. The concrete should have a cement content not less than 400 kg/m³ and a slump of about 150 mm. The concrete must be placed by the tremie method, and must be completed in one continuous operation. The tremie pipe, should have a minimum diameter of 150 mm for 20 mm aggregate. The pipe should always penetrate well into the placed concrete. The top of the concrete in a pile should be brought above the cut-off level so that the pile will have good concrete after removal of laitance and weak concrete before pile cap is laid.

13.6.7 Typical Pile Foundation for Bridge Pier

A typical pile foundation for a bridge pier is shown in Fig. 13.3, wherein sixteen precast R.C. piles as detailed in Fig. 13.2 have been used. The foundation relates to a pier for a bridge with 18 m clear spans and caters to a load of 5000 kN, occuring along with a moment of 1400 kN.m at pile cap level. The piles are of point bearing type resting on firm stratum.

The total reaction on any pile due to vertical reaction and moment is computed from Equation (13.1).

$$P = \frac{\Sigma V}{n} \pm \frac{\Sigma M_1 d_1}{\Sigma d_1^2} \pm \frac{\Sigma M_2 d_2}{\Sigma d_2^2} \qquad \text{... (13.1)}$$

where P = total pile reaction resulting from moment and direct load

ΣV = sum of the vertical loads acting on the foundations

ΣM = sum of moments about the centre of gravity of the group, the subscript denoting the centroidal axis about which the moment acts

n = number of piles in the group

d = distance from the centre of the gravity of the group to the pile in question, the subscript indicating the centroidal axis about which the moment acts

Σd^2 = sum of the squares of distances to each pile from the centre of gravity of the group, the subscript denoting the corresponding centroidal axis.

The individual piles are checked for the maximum and minimum pile reaction. Handling stresses are to be computed and checked for two cases: (a) hoisting from ground by one-point slinging, the slinging point being at 0.293 L from the top end, and (b) lifting by two-point slinging, the support points being located at 0.207 L from either end, L denoting the overall length of the pile.

If the pile is to function as a bearing pile, the column strength should also be investigated. The volume of lateral ties used in the body of the pile will be about 0.4% of the gross volume of the pile. At the two ends, the spacing will be reduced to half the normal spacing for a length of about two times the side or diameter of the pile.

A pile cap is needed to distribute the load from the pier to the piles. Normally, a thickness of about 0.9 m will be adequate. The pile head will be stripped and the longitudinal bars embedded inside the cap for a length of about 0.6 m. As the base of the pier covers all the pile heads, the cap need not be checked for bending and shear. However, nominal reinforcement of 18 mm diameter bars at 300 mm centres will be provided as shown in Fig. 13.3.

13.6.8 Pile Splicing

Splicing of a precast concrete pile at site has always been a time-consuming, costly and inconvenient operation. If a precast concrete pile is required to be extended in length at site for reasons such as non-availability of satisfactory soil stratum at the expected depth, the conventional splicing procedure requires stripping the top of pile to expose reinforcement for sufficient lap length (lap length or minimum 200 mm if full penetration butt welding is

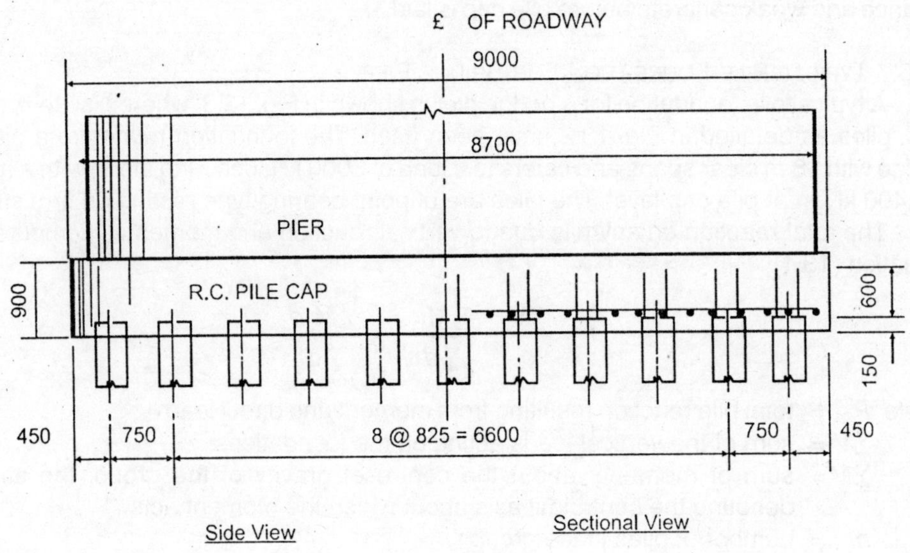

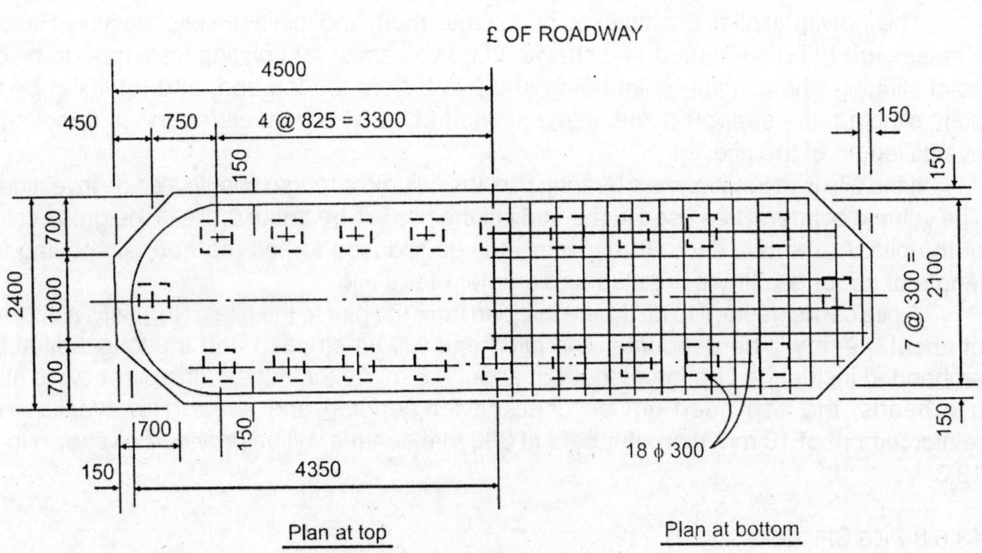

Figure 13.3 Typical Pile Foundation.

adopted), erection of shutters, adding on the reinforcement cage and concreting at site for the additional length. In order to do this, the pile driving rig has also to be shifted and to be re-erected for further driving after the curing period. In a typical major bridge involving deep piling in large numbers, adoption of such procedure can only be inefficient and wasteful in time, effort, material and ultimately cost. Many methods[10] have been proposed for pile splicing, but these methods are either covered by foreign patents or are too cumbersome requiring high levels of field precision and complicated prior preparation.

The author[11] has developed a simple method of splicing precast concrete piles as shown in Fig. 13.4. Two alternative designs of the basic method are indicated in the above figure, the first design for normal use and the second for use in chemically aggressive environment. The details of the splicing method are described below

Let A and B be two pile units. One end of the pile unit A is immovably attached to one side of a metallic jointing member C, which is provided with two or more recesses G on its other side. So also one end of pile unit B is immovably attached to one side of a metallic jointing member F, with recesses G matching with those of the jointing member C. The jointing members may be immovably attached to the respective pile units by welding to dowel bars D to be lapped with the longitudinal reinforcement prior to casting. The recesses G may usually be circular to suit cylindrical plugs which can be conveniently placed by hand into the recesses. The height of the plug is such as to fit loosely the total recess space when jointing members C and F are aligned with F over C.

Assume that pile unit A is the driven pile and pile unit B is to be spliced to pile unit A. The plugs are now placed in the recesses G on jointing member C. The pile unit B is then lowered on to pile unit A and aligned so as to engage the plugs in the recesses in jointing member F. Once so aligned, the members C and F are welded together along the chamfered periphery H. The pile thus spliced may then be driven into the ground. This operation can be repeated to secure the desired total length of the pile. In the embodiment shown in Fig. 13.4(a) the lateral periphery of each of the pile units at the end to be spliced is surrounded by a steel casing E flush with the bounding edge of the jointing member. The casing is intended for preventing spalling of concrete while the piles are driven.

In the case of piles located in chemically aggressive environment,, the embodiment shown in Fig.13.4 (b) is proposed. Here the bounding edges of C and F are disposed in relief adjacent to the bounding edges of the concrete portion of A and B. The casing E is omitted. After the pile unit B is spliced to pile unit A in the manner described already, the gap at K is filled with a bounding material such as epoxy mortar to serve as a protective cover for the members C and F.

The thickness of the plates and weld sizes can be designed to provide for bending, pulling out of pile and driving stresses. To ensure speedy splicing by proper matching while aligning, the splicing plates C and F along with the recesses G and plugs may be fabricated in a factory in matched sets prior to casting of the pile units. The splicing can be completed within a few hours, as against the long curing period and cumbersome site work associated with the conventional method.

The proposed splicing procedure can be used in any situation where precast concrete piles of over 25 m are involved. Specifically, the procedure is advantageous in bridge construction in backwater area. For example, piling for a major bridge around Cochin in Kerala would require a pile length of about 45 m, with over 10 m of standing water. This is an ideal situation to use the proposed splice, as the conventional method splicing is not suitable

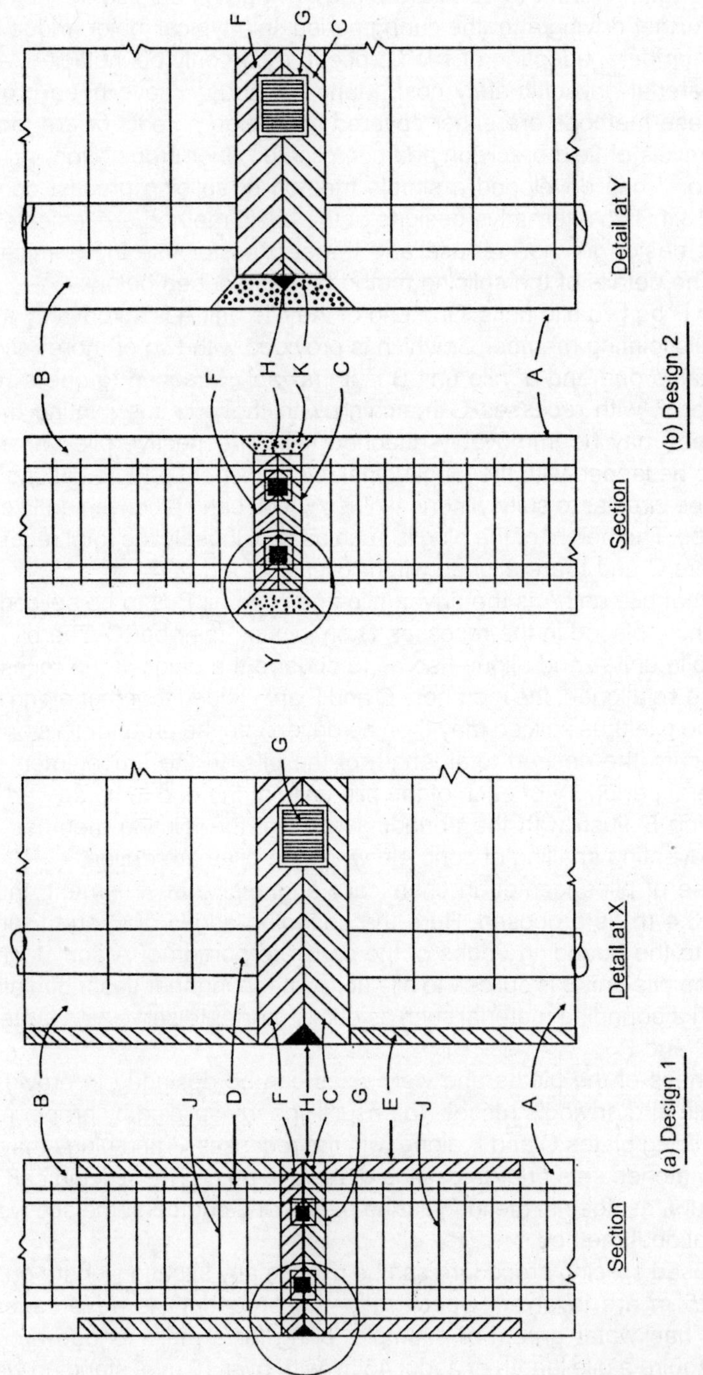

Figure 13.4 Splicing of Precast Concrete Piles.

for economic and efficient working. If a pile is used in three sections of about 15 m each involving two joints, considerable economy can be achieved in time, effort, materials and cost. The splicing can be done in the field within a few hours as against the normal 28-day curing period and the need for shifting the driving rig associated with the conventional method. Since handling here requires only shorter lengths of pile units, the pile section and reinforcement can be reduced significantly. The saving in steel and concrete and in time will offset by a wide margin the additional cost of splicing by this method.

The procedure can also be used advantageously in the case of hybrid piling, where a precast pile is lowered into a driven casing tube and the space between the pile and ground is grouted with simultaneous withdrawal of the tube. The shorter lengths of individual units would reduce the pile section and reinforcements. The cracking associated with handling long piles can be eliminated. The jointing plates C and F of Fig. 13.4 may be provided with a central hole to facilitate grouting. The dowel bars, jointing plates and peripheral weld will need special attention in design to transfer by tension the load due to self weight of pile units below a joint during insertion into the casing tube.

13.7 Well Foundations

13.7.1 General

Well foundation (open caisson) is the most commonly adopted foundation for major bridges in India. This type evolved in India, and has been adopted for the Taj Mahal. Since then, many major bridges across wide rivers have been founded on wells. Well foundation is preferable to pile foundation when the foundation has to resist large lateral forces, the river bed is prone to heavy scour, heavy floating debris are expected during floods and when boulders are embedded in the substrata.

The basic shapes adopted for wells are shown in Fig. 13.5. The foundation may consist of a single large diameter well or of a group of smaller wells of circular or other shapes. The shape of wells may be circular, double-D, square, rectangular, dumb-bell, etc. The circular well has the merit of simplicity for construction and sinking. Double-D, rectangular and dumb-bell shapes are used when the bridge has multi-lane carriageway. For piers and abutments of very large size used in cantilever, cable stayed or suspension bridges, large rectangular wells with multiple dredge holes of square shape may be used. The size of the dredge hole is decided so as to facilitate ease of construction and inspection of the foundation.

While deciding the bottom level of a well foundation, the following considerations may be kept in view:

(a) Normally, a sandy strata with adequate bearing capacity is preferred to a clayey strata.

(b) A thin stratum of clay occurring between two layers of sand is not relied upon but pierced through.

(c) If foundation has to be laid on a clayey layer, a well is rested on stiff clay only.

The size of the well for a particular foundation would depend on a large number of factors. The minimum size of a dredge hole is 2.5 m. The thickness of steining should be adequate to permit sinking of the well without excessive kentledge. From considerations of sinking effort and erection of shuttering, a single large diameter well is nowadays preferred to a group of small sized wells. From practical considerations, the maximum size of a single circular well should be limited to about 12 m diameter for a concrete steining and 6 m diameter for a brick masonry steining. If the wells are to rest on sloping rock, smaller wells

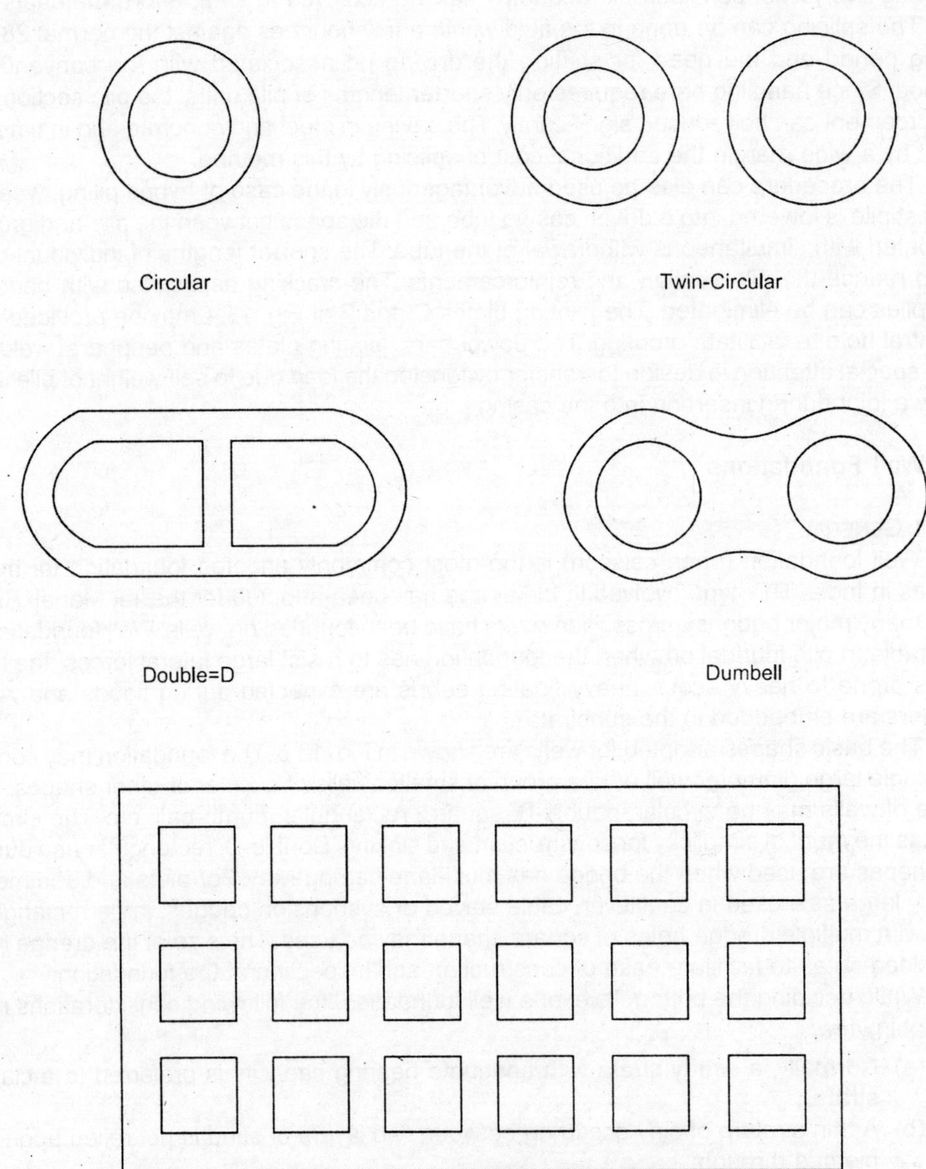

Figure 13.5 Shapes of Wells in Plan.

may be advantageous for ensuring even seating of the wells on the rock. However, in the case of the Joghigoppa bridge across Brahmaputra river, an innovative design of 18 m diameter well with steining thickness of 3.5 m has been adopted for two piers, incorporating other special features in view of steeply sloping bed rock.

Twin-circular wells or three circular wells with a common capping slab have been used in earlier constructions for wide piers and abutments. The wells in these cases had small diameter. Currently, this type is not popular. Double-D and Dumbell shaped wells are suitable for deep foundations and double line railways. Multi-cellular wells are adopted for deep foundations for long span bridges, as in the Howrah bridge and the Second Hooghly bridge.

13.7.2 Components

A well foundation consists of the following components:

(a) Steining
(b) Well curb
(c) Bottom plug
(d) Sand filling
(e) Top plug
(f) Well cap.

Typical details of the above components are shown in Fig.13.6.

13.7.3 Steining

The steining is normally of reinforced concrete. The concrete used for steining should be M15 for normal exposure. In areas of marine or adverse exposure, the concrete for steining should be at least M20 with cement content not less than 310 kg/m³ of concrete with water/cement ratio not more than 0.45.

The following factors are to be considered while determining the thickness of the steining:

(a) It should be possible to sink the well without excessive kentledge.
(b) The wells should not get damaged during sinking.
(c) If the well develops tilts or shifts during sinking, it should be possible to rectify the tilts and shifts without damaging the well.
(d) The well should be able to resist safely the earth pressure developing during a sand blow that may occur during sinking.
(e) At any level of the steining, the stresses under all conditions of loading that may occur during sinking or during service should be within permissible limits.

The thickness of concrete steining should not be less than 500 mm nor less than that given by Equation (13.2).

$$t = KD\sqrt{L} \qquad \text{... (13.2)}$$

where t = minimum thickness of concrete steining in m

D = external diameter of circular well or dumb bell shaped well or smaller plan dimension of twin D well in m

L = depth of well in m below L.W.L. or top of well cap whichever is greater

K = a constant depending on the nature of subsoil and steining material (taken as 0.030 for circular well and 0.039 for twin-D well for concrete steining in sandy strata and 10% more than the corresponding value in the case of clayey soil).

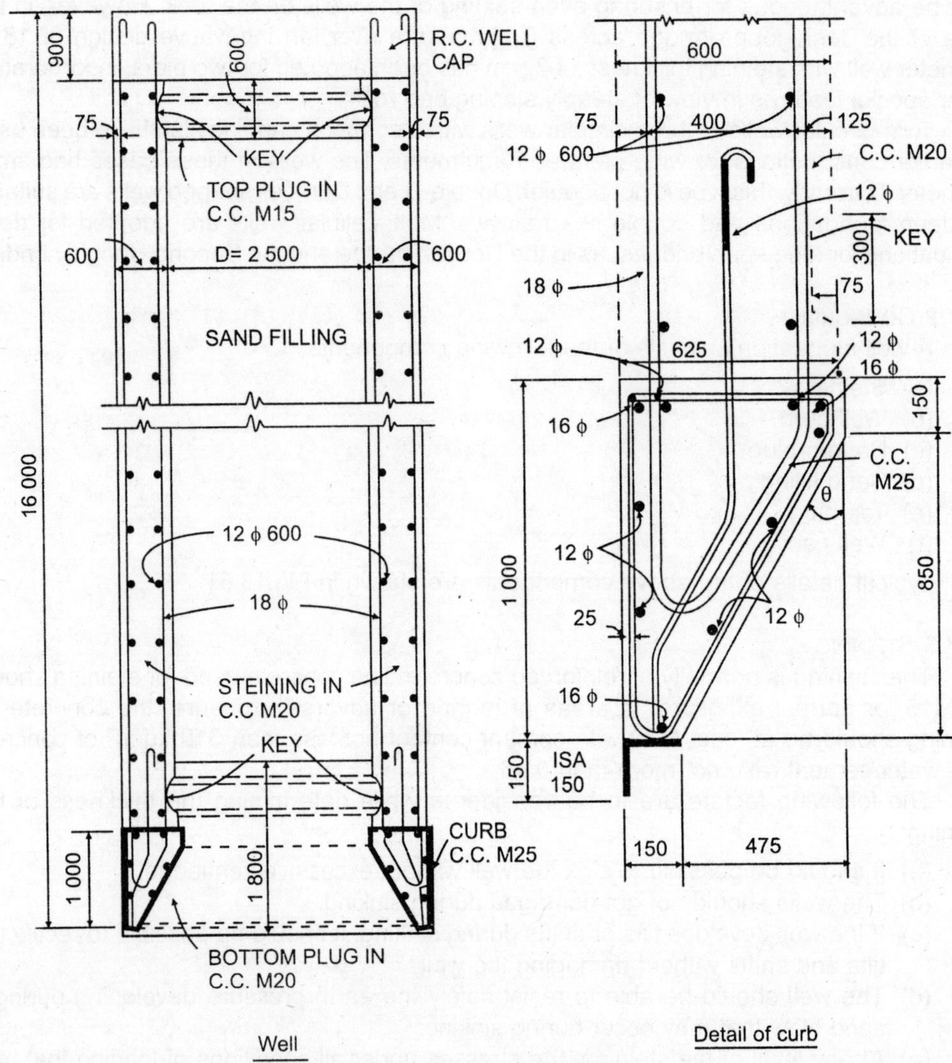

Figure 13.6 Typical Foundation Well.

The concrete well steining is reinforced with longitudinal bond rods and hoop rods on both faces of the well. The minimum vertical reinforcement (either mild steel or deformed bars) in the steining will be 0.12 per cent of gross cross sectional area distributed equally on both faces of the steining. The hoop steel shall not be less than 0.04 per cent of the volume per unit length of the steining. The minimum cover to bond rods is 75 mm. For wells of over 8 m diameter, the thickness of steining should be at least 1 m. In recent constructions, the jack-down technique has been employed to assist in sinking the well, leading to a reduction in the thickness of the steining, e.g. the Second Narmada bridge on NH8 in Gujarat, and the New Nizamuddin bridge on NH24 in New Delhi.

13.7.4 Well Curb

The well curb carries the cutting edge for the well and is made up of reinforced concrete using controlled concrete of grade M25. The cutting edge usually consists of a mild steel equal angle of side 150 mm. The angle will have one side projecting downward from the curb as shown in Fig. 13.6 for soils where boulders are not expected. In soils mixed with boulders, the angle will have the vertical leg embedded in the steining in such a manner that the horizontal leg of the angle is flush with the bottom of the curb. The reinforcement in a well curb will be arranged as shown in Fig. 13.6, assuring a minimum quantity of steel of 72 kg/m^3 of concrete. The angle θ as shown in the figure will be normally about 30 degrees and will be less for stiff strata. In case blasting is anticipated, the outer face of the well curb should be protected with 6 mm thick steel plate for half the height of the curb and the inner face should have 10 mm plate up to the top of the curb and 6 mm plate further up to a height of 3 m above the top of the curb. The steel plates are to be properly anchored to the concrete. In such a case, the curb and the steining for a height of 3 m above the top of curb should have additional hoop steel.

13.7.5 Bottom Plug, Filling and Top Plug

A bottom plug is essential to transfer the load from the well steining to the base soil. It is usually provided for a thickness of about half the diameter of the dredge hole. In practice the bottom plug is provided up to a height of 0.3 m above the top of well curb. The concrete used is of M20 grade, the richness of the mix being necessitated by the possibility of loss of part of the cement due to under-water placing of the concrete. The concrete is usually placed in one continuous operation by the tremie method, i.e., pouring through a funnel and tube whose bottom end is kept immersed in concrete in plastic state placed just earlier or very near the bottom at start. When wells are founded on bed rock, it is normal to provide anchor bars of 32 mm diameter to anchor the bottom plug into the rock by about 2.0 m depth.

A top plug of M15 concrete is generally provided for a thickness of about 0.6 m beneath the well cap and on top of the compacted sand filling.

The space inside the well between the bottom of the top plug and the top of the bottom plug is usually filled with clean sand, so that the stability of the well against overturning is increased. While this practice is good in case of wells resting on sand or rock, the desirability of sand filling for wells resting on clayey strata is doubtful, as this increases the load on the foundation and may lead to greater settlement[12]. In the latter case, the sand filling is done only for the part of the well up to scour level and the remaining portion is left free.

13.7.6 Well Cap

A well cap of reinforced concrete with M20 concrete is needed to transfer the loads and moments from the pier to the well or wells below. The shape of the well cap is normally kept the same as the well with a possible overhang all round of about 150 mm. If two or three wells of small diameter are needed to support the sub-structure, the well cap will be extended to cover the wells. Typical details of a well cap over three wells for the foundation of a pier are shown in Fig. 13.7. The well cap is designed as a slab resting over the well or wells. It is admissible to consider partial fixity at the edges of the wells. The top of the well cap is usually kept at the bed level in case of rivers with seasonal flow or at about the low water level in case of perennial rivers. In the latter case, the construction of a suitable

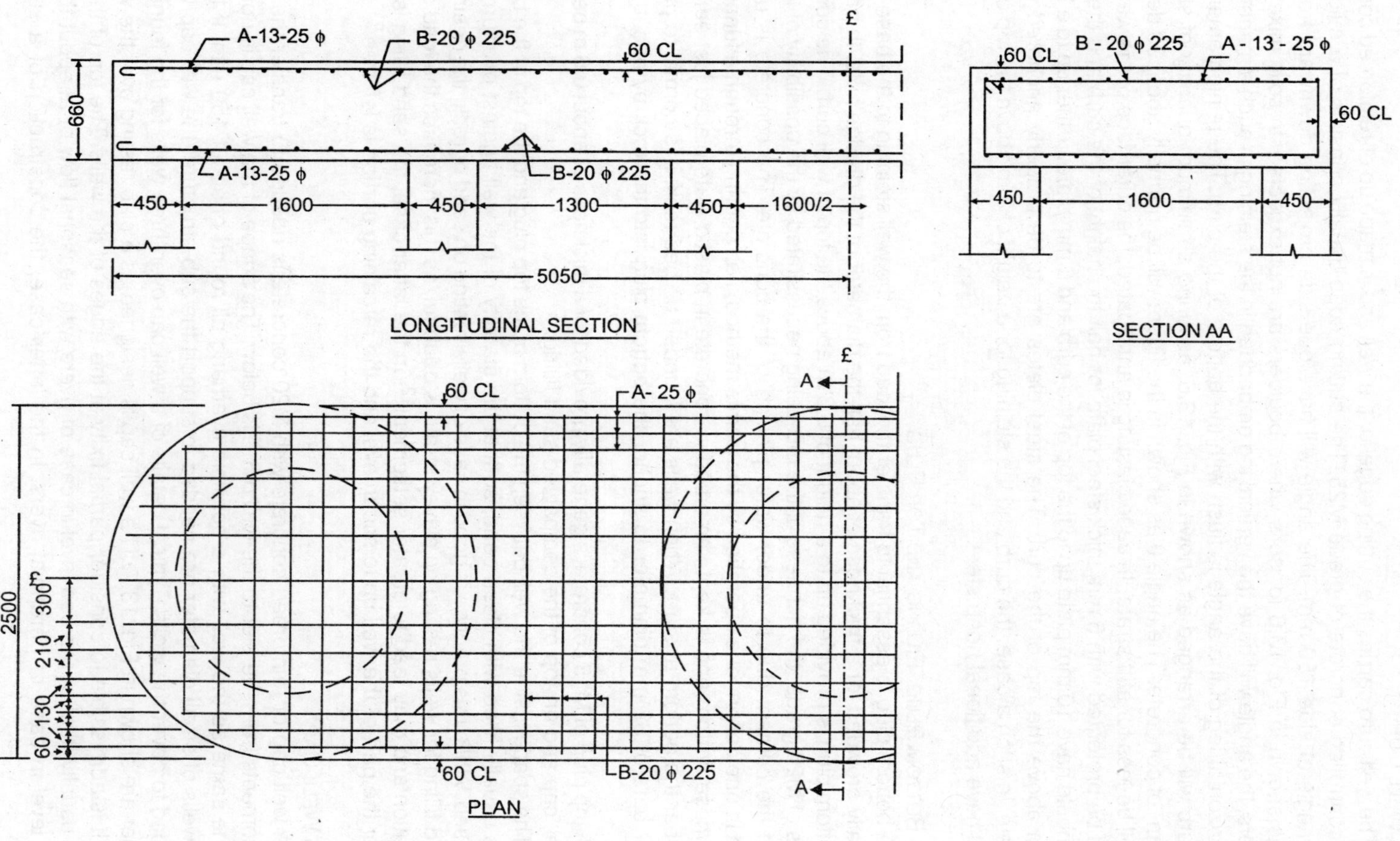

Figure 13.7 Typical Details for Well Cap for Pier.

cofferdam and dewatering will be required during concreting of the well cap. In some cases, the well cap has been located such that the bottom of the cap is about 150 mm above the low water level. While this arrangement is convenient for fast construction, the aesthetics of the bridge will be affected to some extent.

13.7.7 Design Procedure

The design procedure for a single well sunk in sandy strata by elastic theory is outlined as below. The depth of foundation is decided in such a manner as to provide adequate grip length below the maximum scour level and to rest on a suitable bearing strata. This is checked for the following two conditions of stability, namely: (i) the maximum soil reaction from the sides cannot exceed the maximum passive pressure at any depth, and (ii) the soil pressure at the base should be compressive throughout and the maximum pressure should not exceed the allowable pressure on the soil at the base.

The first condition will be satisfied if

M/I is not greater than $w(k_p - k_a)$

where M = total applied moment about base of well

w = density of soil, submerged density to be taken when under water or below water table

k_p = coefficient of passive earth pressure

$$= \frac{\cos^2\phi}{\left[1 - \sqrt{\dfrac{\sin(\phi + z)\sin\phi}{\cos z}}\right]^2}$$

k_a = coefficient of active earth pressure

$$= \frac{\cos^2\phi}{\left[1 + \sqrt{\dfrac{\sin(\phi + z)\sin\phi}{\cos z}}\right]^2}$$

ϕ = angle of internal friction of soil

z = angle of friction between well and soil, to be taken as 0.67 ϕ, limited to 22.5 degrees

$I = I_b + I_v(1 + 2\alpha \,.\, \tan z)$

I_b = moment of inertia of base about the axis normal to direction of horizontal forces passing through the centre of gravity of the base

I_v = moment of inertia of the projected area in elevation of the soil mass offering resistance

$$= \frac{LD^3}{12}$$

L = projected width of soil mass offering resistance multiplied by the appropriate value of shape factor, which is 0.9 for circular well and 1.0 for rectangular well

D = depth of well below scour level

αD = 0.5 base width for rectangular wells or 0.318 diameter for circular wells.

If the above condition is not satisfied, the grip length required is determined by putting $M/I = w(k_p - k_a)$ and the revised value is adopted.

For the second condition, the maximum and minimum soil pressures at base are computed from Equation (13.3).

$$\left.\begin{matrix}\sigma_1\\ \sigma_2\end{matrix}\right\} = \frac{W}{A} \pm \frac{MB}{2I} \qquad \text{... (13.3)}$$

where σ_1, σ_2 = maximum and minimum base pressures, respectively

W = net downward load acting at the base of well, including the self weight of well and the upward vertical force equal to P. tan z due to wall friction and the total horizontal soil reaction P

A = area of the base of well

B = width of base of well in the direction of forces and moments.

If the above condition is not satisfied, the well will have to be redesigned suitably. The reader may see Reference 13 for detailed derivation of the above procedure.

13.7.8 Sinking of Wells

Well sinking is a specialized operation requiring considerable skill. When a concrete well is to be sunk on shore or with shallow water depth, it is usual to sink the well up to about 6 m by excavating the soil in the dredge hole by employing skilled divers, and after dredging, pumping out water from the sump to induce sinking. A tripod and a mechanical grab operated by a power winch may also be used. The sinking of the well through the soil is resisted by skin friction along the external surface of the well and by bearing on the cutting edge at the bottom. These resistances are overcome by the dead weight of the well steining reduced by the buoyancy of the submerged portion. If the side friction is so great as to retard the sinking under its own weight, additional load (known as kentledge) is added at the top of the steining to aid sinking. When addition of kentledge is inadequate, a practical remedy is to suspend dredging for a brief period to allow the water in the dredge hole to reach its normal level, and to pump out the water under observation so that the resulting differential head reduces the skin friction leading to better sinking (as done in the Vasista bridge in Andhra Pradesh). The average rate of sinking per day for a medium sized well will be about 900 mm in sandy strata and about 500 mm in clayey strata. For large sized wells, the rate of sinking per day may be about 600 mm in sandy strata and about 400 mm in clayey strata.

For water depth less than 7 m and the stream velocity less than 2 m/s, an artificial temporary island may be built at the pier location and the curb of the caisson can be built on this island. For depths of water 7 to 10 m and velocity in excess of 2 m/s, a cofferdam made from wooden poles, bamboo matting and clay filling is used to protect the sand island. This method would facilitate accurate location of the curb of the well for sinking. Typical details of sand bar-cum-island for well foundation using local materials for piling as used for a major bridge in India are indicated in Fig. 13.8.

Tilting of wells accompanied by shift is a common hazard in well sinking and special care should be taken to keep it to a minimum and to rectify it as and when it occurs. Tilting may be caused by non-uniform excavation in the dredge hole, by the presence of non-homogeneous soil around the cutting edge or due to sudden relative variation in the skin friction developed around the well. It is usually possible to rectify a tilt by one of many methods, including: (i) use of eccentric kentledge, (ii) use of water jets on the outer periphery

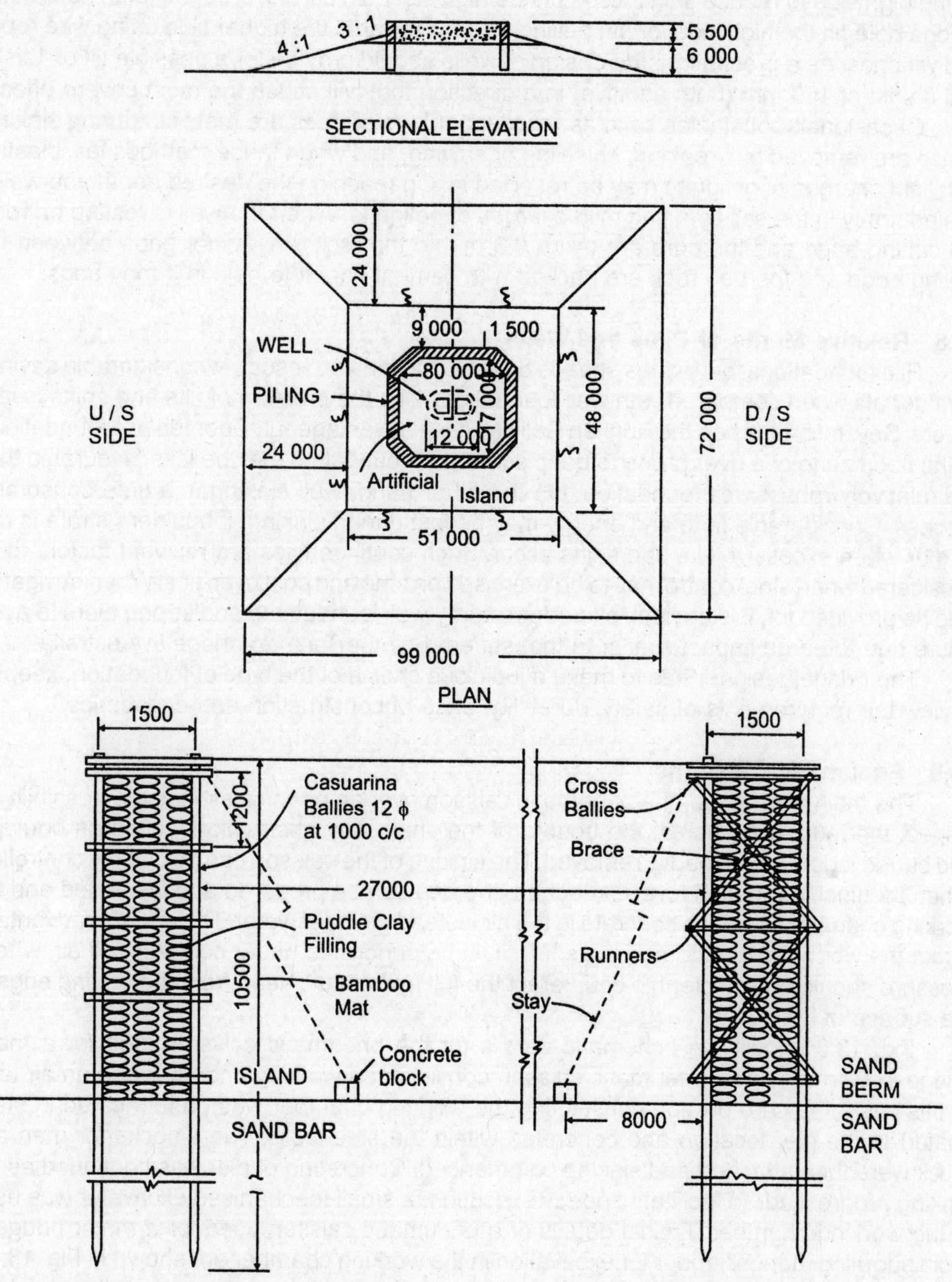

Figure 13.8 Sand Bar–cum–Island for Well Foundation.

on the high side to reduce skin friction on the high side, (iii) careful dredging of the soil in the dredge hole on the high side, or (iv) pulling the well towards the higher side using wire ropes and winches. As a precaution, the design of wells should provide for a possible tilt of 1 in 80 and a shift of 150 mm (both additive) in a direction that will cause the most severe effect.

Occasionally, obstacles such as boulders or logs of trees are met with during sinking. These are removed by breaking, chiseling or sawing, and when these methods fail, blasting with light charges of gelignite may be resorted to. On reaching the desired depth, the well is seated firmly in the soil by using mild charges of gelignite. When the well is resting on rock, the cutting edge and the curb are taken 0.3 m into the rock. Any minor gaps between the cutting edge and the bed rock are packed with cement concrete held in gunny bags.

13.8 Relative Merits of Piles and Wells

Pile foundations for bridges speeds up construction and results in considerable savings in materials when compared with well foundation. Also the problem of tilts and shifts is less severe. Several bridges on the Konkan Railway have advantageouly adopted pile foundations. In the flood zone of a river prone to deep scour, pile foundation may be less favourable than the relatively robust well foundation. On the other hand, well sinking is a time-consuming process. Considerable time and energy may be lost during sinking, if bouldery strata is met with or when excessive tilts and shifts occur. Such contingencies are relevant factors to be considered when strict control has to be exercised on time and cost overruns. When navigation is to be provided for, it is advisable to adopt sturdy well foundations and strong piers to avoid failure due to barge impact similar to that suffered by the Tasman bridge in Australia.

The bridge designer has to make a judicious choice of the type of foundation, keeping in view the requirements of safety, durability, ease of construction and economics.

13.9 Pneumatic Caissons

The main advantage of a pneumatic caisson over an open caisson is the possibility to inspect the foundation soil at the bottom of the shaft. Any obstructions, such as boulders and buried logs, can be readily removed. The sinking of the caisson can be closely controlled. When the final foundation level is reached, the bed can be prepared and inspected and the working chamber filled with concrete in the air instead of under water. The main disadvantage is that the work at the bottom of the shaft must be performed under compressed air, whose pressure should be sufficient to counteract the full hydrostatic head from the cutting edge to the surface of the water.

Fig. 13.9 shows the schematic details for the pneumatic caisson used for a major bridge in Germany. The pneumatic caisson consists of a working chamber and an air lock. In this case, the curb portion constituting the working chamber was prefabricated in steel, floated to the pier location and concreted within the steel cells. The airlocks for men and muck were then attached and sinking commenced. Concreting of pier was continued as the sinking progressed. To facilitate speedy execution, a small mechanised excavator was used in the working chamber. Typical details of a pneumatic caisson used for a major bridge in India adopting manual labour for excavation in the working chamber are shown in Fig. 13.10.

The roof of the working chamber must be sufficiently strong to support the weight of the pier and should also be airtight. The airlock is made of steel and should have two doors, one of them to be closed tight while the other is opened for use. Provision is made for gradually increasing or decreasing the pressure inside the airlock. Usually two airlocks are provided, one for men and another for muck.

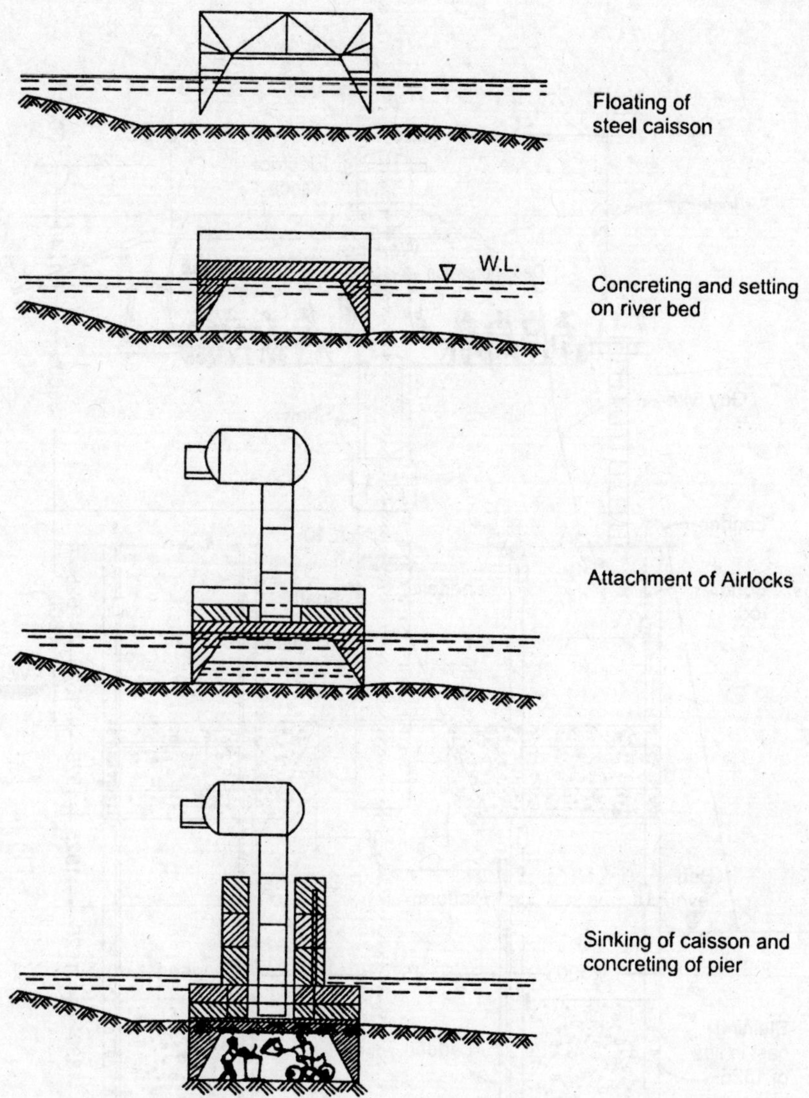

Figure 13.9 Pneumatic Caisson with Steel Curb.

A person working in an enclosure of pressure different from the atmospheric pressure is liable to suffer 'caisson sickness' due to accumulation of gases in the blood at the joints or other critical parts of the body at the time of decompression prior to leaving the airlock. An accumulation of gases at the joints cause the 'bends' while that in the spinal cord could result in paralysis. The maximum air pressure which a healthy person can stand is about 3.5 atmospheres above normal. This pressure will be reached at a depth of 40 m. Hence it is not convenient to go below this depth with pneumatic caissons. At the air pressure of 3.5 atmospheres above normal, some men cannot work at all and only some can work for about half an hour at a time. Also, the workmen should spend considerable time in the

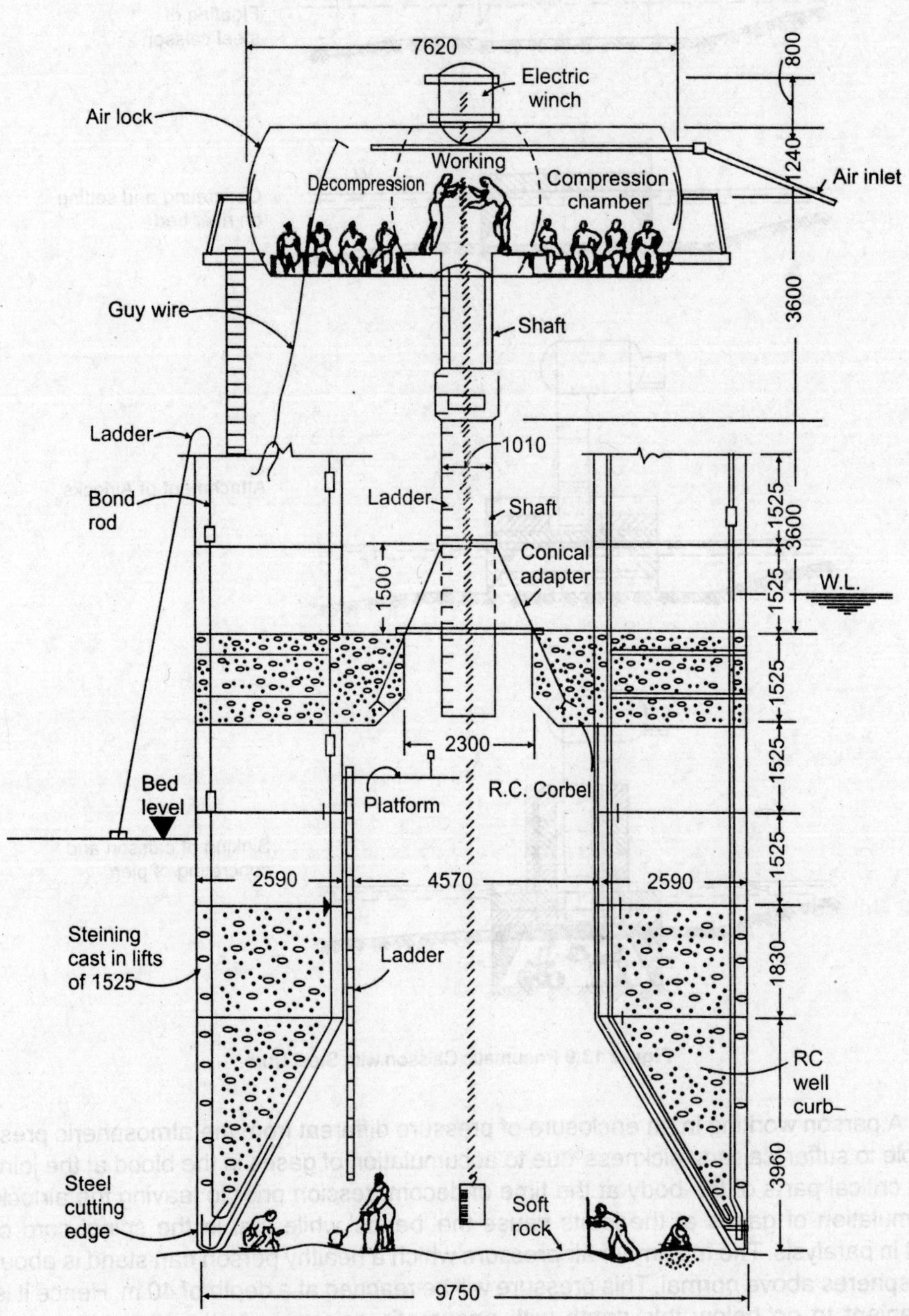

Figure 13.10 Pneumatic Caisson with R.C. Curb.

airlock for gradual change of pressure from and to this level. Hence pneumatic caisson sinking beyond 30 m depth below the low water level is not advisable. The air pressure required can be reduced if the water table at the location can be significantly lowered by dewatering.

After the caisson has been sunk to its final level, the working chamber is filled with concrete. The concrete is transported through the airlock and the air pressure is maintained until the concrete has attained its design strength.

13.10 Cofferdams for Bridge Piers

Cofferdam refers to construction in a water body to provide an enclosed site, which can be dewatered to facilitate construction of a bridge pier in almost dry condition. Medium sized cofferdams of 10 to 50 m diameter have been used for construction of bridge piers in water up to about 20 m depth. Sheet piles have been used for such cofferdams with internal bracing against external pressures. For very large bridge piers and suspension bridge anchorage, cellular cofferdams with soil filling in cells can also be used. The cofferdams must be designed to resist the lateral force due to water currents, waves and unbalanced soil loads. A concrete seal is usually provided at the bottom of the cofferdam by tremie method to facilitate dewatering. This seal should have emergency relief pipes to prevent structural failure of the seal in case of failure of the dewatering system.

13.11 Box Caissons

A recent development in overwater pier foundation is the box caisson, which is essentially a variation of the open caisson. The box caisson is usually a structural shell of steel or concrete section, often with cellular cross section, with sizes ranging from hundred to many thousands of tonnes. The box caisson is fabricated on shore, towed to the pier location using tugs and barge mounted cranes, set on prepared bed and filled with concrete placed by the tremie method. The pier bases are then built on these caissons. Examples of application of box caissons include the Akashi Kaikyo bridge, the Great Belt East bridge and the Second Severn bridge at Bristol.

13.12 References

1. Varghese, P.C., 'Foundation Engineering,' Prentice-Hall of India, New Delhi, 2005, 570 pp.
2. Hool, G.A., and Kinne, W.S., 'Foundations, abutments and footings', McGraw-Hill Book Co., New York, 1943, 417 pp.
3. Peck, R.B., Hanson, W.E., and Thornburn, T.H., 'Foundation engineering', Asia Publishing House, Bombay, First Indian Edition, 1959, 410 pp.
4. 'IRC: 78-2000 Standard specifications and code of practice for road bridges, Section: VII, Foundations and substructure', Indian Roads Congress, 2000, 97 pp.
5. Balwant Rao, B., and Muthuswamy, C., 'Considerations in the design and sinking of well foundations for bridge piers', Jl. Indian Roads Congress, Vol. XXVII-3, August 1963, pp. 1-82.
6. Dunham, C.W., 'Foundations of structures', McGraw-Hill Book Co., New York, 1962, 722 pp.
7. Anderson, P., 'Substructure analysis and design', Ronald Press Co., New York, 1956, 336 pp.
8. 'World's longest pile?', Foundation Facts, Raymond International Inc., Vol. VII, No.1, 1971, p. 20.
9. Tambe, S.R., Bongirwar, P.L., and Kanhere, D.K., 'Critical appraisal of Mumbai flyover project', Jl. Indian Roads Congress, New Delhi Vol. 60-2, Oct. 1999, pp. 235-263.

10. Bruce, R.N., and Hebert, D.C., 'Splicing of precast prestressed concrete piles: Parts 1 and 2', Journal of Prestressed Concrete Institute, Vol. 19, No.5, Sept. Oct. 1974, pp. 70-97, and No. 6, Nov.-Dec., 1974, pp 40-66.
11. Victor, D.J., 'A new method for splicing concrete piles for bridge foundation', Jl. Indian Roads Congres, Vol. 38-2, Oct. 1977, pp. 305-320.
12. Saxena, R.K., 'Well foundations for road bridges', Jl. Indian Roads Congress, Vol. 34, No.2, Nov. 1971, pp. 391-435.
13. 'IRC: 45-1972 Recommendations for estimating the resistance of soil below the maximum scour level in the design of well foundations of bridges', Indian Roads Congress, New Delhi, 1996, 27 pp.
14. Lee, D.H., 'An introduction to deep foundations and sheet piling', Concrete Publications Ltd., London, 1961, 264 pp.
15. Ponnuswamy, S., 'Bridge Engineering', Tata McGraw-Hill Publishing Co., New Delhi, 1986, 544 pp.

Chapter 14

Bearings, Joints and Appurtenances

14.1 Importance of Bearings

Bearings are provided in bridges to transmit the load from the superstructure to the substructure in such a manner that the bearing stresses induced in the substructure are within permissible limits. They should also accommodate certain relative movements between the superstructure and the substructure. The latter are usually due to one or more of the following:

(a) longitudinal movement due to temperature variation;
(b) rotation due to deflection of the girders; and
(c) vertical movement due to sinking of the supports.

In addition, there can be movements due to shrinkage, prestressing or creep. The movements and rotations may be reversible or irreversible. The reversible effects are usually cyclic and are due to temperature changes and live loads. Effects due to settlement of supports, prestressing, creep or shrinkage are irreversible.

The magnitude of thermal movement depends on the coefficient of linear expansion and the temperature range to which the member is subjected. For the purpose of preliminary estimates, the maximum range of movement due to all causes, expressed as a proportion of the expansion length, may be assumed as below[1]:

In-situ reinforced concrete	0.0009
Precast reinforced concrete	0.0007
In-situ prestressed concrete	0.0016
Precast prestressed concrete	0.0011
Steel	0.0009
Composite steel and concrete	0.0008

The end rotation of a beam of uniform section may be estimated for initial assessments as four times the maximum permissible midspan deflection divided by the span.

Bearings are of two categories: fixed bearing and expansion bearing. Fixed bearings allow rotation but restrict translation, while expansion bearings allow both rotation and longitudinal translation. Different designs are possible for each of these two categories. The particular design of bearing to be adopted for any given bridge depends upon the type of superstructure, type of supports and also on the length of the span. A simply supported span requires a fixed bearing at one end and an expansion bearing at the other. A continuous girder would require expansion bearings on all supports except at one. Thus a two-span girder would have a fixed bearing at the central support and expansion bearings at the two abutments.

Bearings form an important part of a bridge, and are included among the most highly stressed elements in a bridge. They call for care in design, skill in execution and regular attention in maintenance. Faulty design or malfunctioning of the bearings have often contributed to the collapse of bridges. On certain major bridges, the cost of bearings accounts for as much as 10 to 15% of the total cost of the bridge. Hence there is scope for economy by careful design of bearings. Guidelines for the design of bearings for road bridges are available in IRC: 83 [2,3]. The bearing should be placed on a raised plinth with self-draining arrangement on the bearing shelf. Since the useful life of a bearing is generally less than the service life of the bridge, specific provision should be made in the design of every new bridge for access to the bearing and for temporary jacking of the deck to permit replacement of the bearing if such a need should arise.

14.2 Bearings for Slab Bridges

For high-level bridges with slab spans, no special bearings are usually provided. A thick layer of tarfelt or kraft paper is inserted between the slab and the substructure at the supports. This arrangement is sufficient to allow for the small longitudinal movements. To take care of the rotating movement, each bearing area should be beveled or rounded at the edge.

For submersible bridges, which are generally of slab spans, uplift under submersion condition should be adequately provided for. A typical detail at the support as proposed by the author in an earlier paper[4] is shown in Fig. 14.1.

14.3 Bearings for Girder Bridges

Girder bridges should be provided with fixed and expansion joints. Fixed bearings should have provision for rotation and expansion bearings should be able to allow longitudinal movements as well as rotation. During construction, expansion bearings should be properly aligned to correct for the temperature prevailing at the time of erection. For bridges in gradient, the bearing plates are to be placed in a horizontal plane.

In seismic areas, suitable guides should be incorporated in the bearings to prevent the roller and rocker from being displaced during earthquakes.

For skew bridges with skew angle less than 20 degrees, the metallic bearings to be

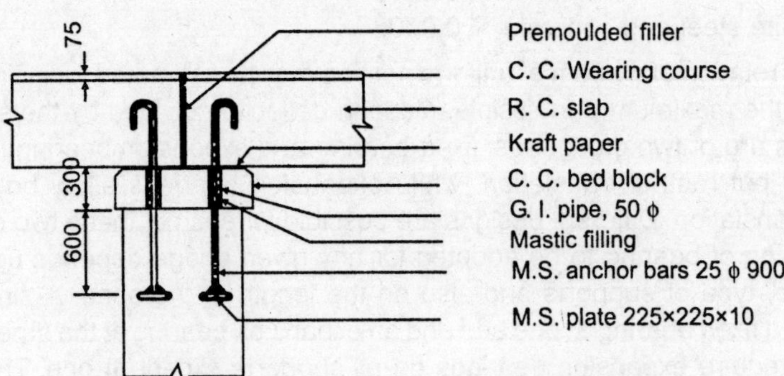

Figure 14.1 Bearings for Submersible Bridges.

provided are placed at right angles to the longitudinal axis of the bridge. When the skew angle is more than 20 degrees and the span length along the longitudinal axis is less than 10 m, arrangement for sliding should be made at both supports. If the span length exceeds 10 m, the fixed bearing is provided at the obtuse corner of the bridge. Since the axis of rotation and the direction of longitudinal movement are not perpendicular, the fixed bearing should be capable of allowing rotation in any direction, and the free bearing should allow movement and rotation in any direction. In the case of curved bridges, bearings which allow movement and rotation in any direction are used.

Expansion bearings for girder bridges are of the following types:

(a) Sliding plate bearing
(b) Sliding-cum-rocker bearing
(c) Steel roller-cum-rocker bearing
(d) R.C. rocker expansion bearing
(e) Elastomeric bearing

Fixed bearings can be classified as below:

(a) Rocker bearing
(b) Steel hinge
(c) Steel rocker bearing
(d) R.C. rocker fixed bearing

Salient features of the major types of expansion bearings and fixed bearings are discussed briefly in Sections 14.4 and 14.5. In view of their increasingly extensive use in modern bridges, elastomeric bearings are dealt with in detail in Section 14.6.

14.4 Expansion Bearings

14.4.1 Sliding Plate Bearing

The sliding plate bearing is the simplest form of expansion bearings. This type is useful for girder bridges of spans up to 15 m.

Past practice has been to provide sliding plate bearings with the contact surfaces flat. Such plate bearings have a tendency to 'freeze', i.e. to restrict movement, and the frictional resistance is sometimes enough to pull the abutment away from the approach fill during cold weather and to force the deck beyond its normal travel during the following warm weather. Dillon and Edwards[5] have found during an investigation of the condition of eight major bridges in the city of London, Ontario, Canada, that for three bridges expansion bearings were 'frozen' and two of these bridges had abutments rendered unstable due to increased horizontal forces developed by the bearings during expansion and contraction of the deck. When the contact surfaces are flat, it is preferable to use teflon coatings for the contact surfaces in order to reduce friction.

Fig. 14.2 (a) shows typical details of a sliding plate bearing as per current practice. Here the contact surface of the top plate is given a curved shape. The bearing has a line contact and it permits rotation and horizontal movement in the direction of the span.

14.4.2 Sliding-Cum-Rocker Bearing

In the case of modern urban bridges with curved alignment, it is often necessary to provide bearings permitting sliding movement and rotation in all directions. A suitable design

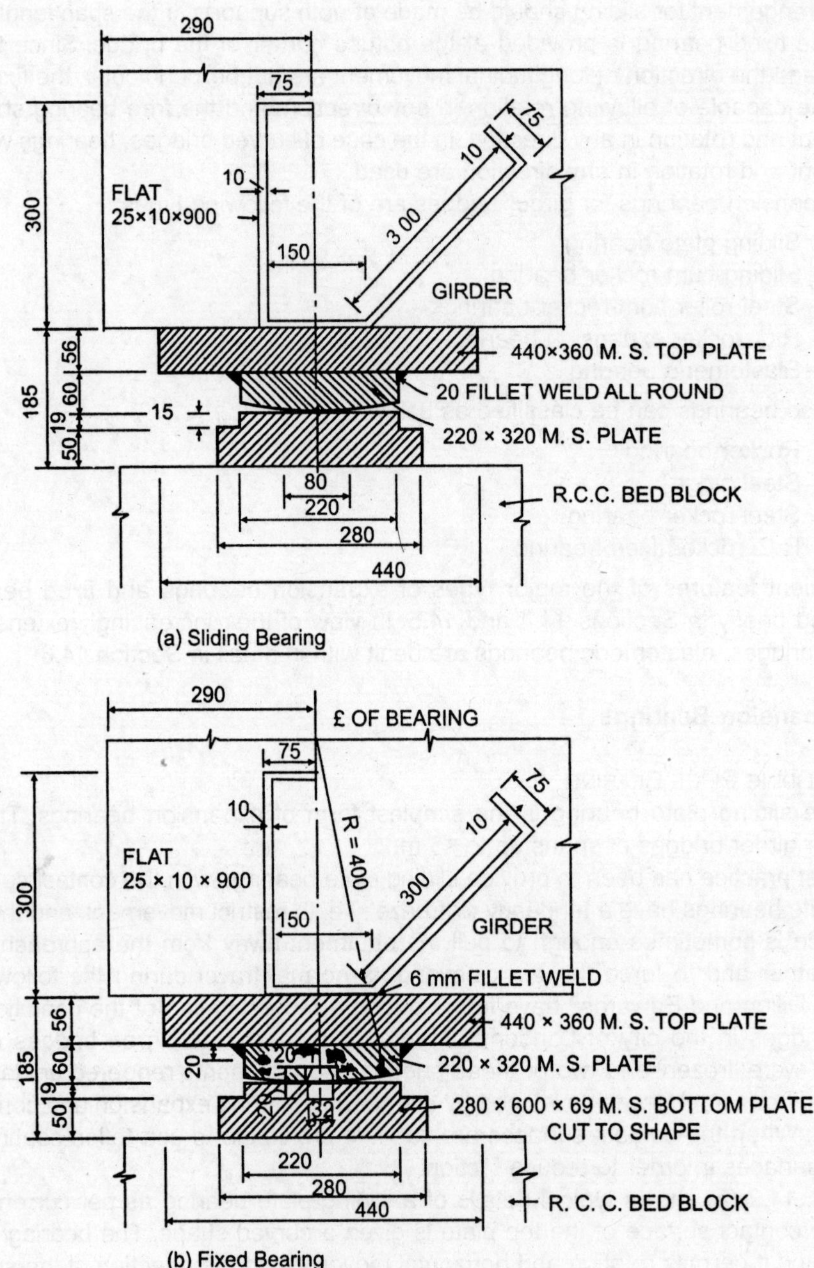

Figure 14.2 Typical Plate Bearing.

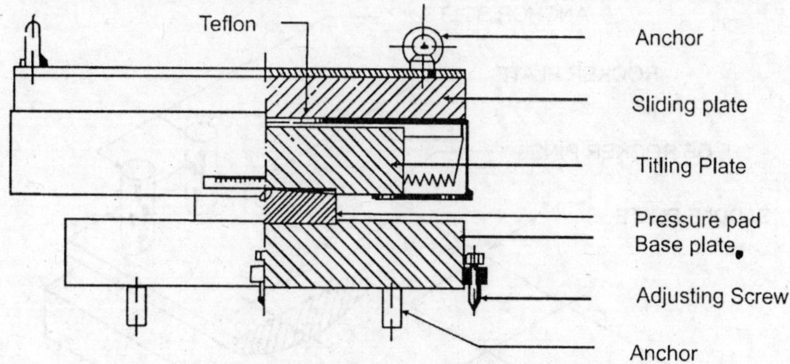

Figure 14.3 Sliding-cum-Rocker Bearing.

for such a bearing known as sliding-cum-rocker bearing is shown in Fig. 14.3. The bearing consists of a sliding plate, a tilting plate, a pressure pad and a base plate. The pressure pad facilities a point contact and permits rotation in any direction. Teflon coated contact surfaces ensure least friction.

14.4.3 Steel Roller-cum-Rocker Bearing

Steel roller-cum-rocker bearing permits longitudinal movement by rolling and simultaneously allows rotational movement. Cast steel roller bearings have been in general use for major bridges with span above 20 m until recently. The bearing may consist of single roller, two rollers or a nest of rollers. Where a nest of rollers is used, the rollers are cylinders of 100 mm to 150 mm diameter. Nests of small rollers have been found to be difficult to keep clean from dirt and dust. They tend to 'freeze' in time and fail to function freely. Further, they often get misaligned and suffer damage due to temperature movement. Single large diameter roller has been reported to be functioning satisfactorily[6]. When two large diameter rollers are used, sometimes only the central segment for a certain width (about half the diameter) is used instead of the full cylinder. Fig. 14.4 shows typical details of a roller-cum-rocker bearing with two segmental rollers. A preferred alternative design with two full cylinder rollers with a capacity of 1250 kN is shown in Fig. 14.5. When roller bearings are provided over abutments, the base plate should have sufficient length to cater for the large movements over the abutments.

14.4.4 Reinforced Concrete Rocker Bearing

Cast steel roller bearings being expensive, cheaper reinforced concrete rocker expansion bearings are sometimes used for concrete bridges. R.C. rocker expansion bearing is sometimes referred as R.C. "roller" bearing in some references. When properly designed, these are as effective as steel roller bearings. The cost of maintenance of these bearings is negligible compared with that for steel bearings. The bearing consists of a R.C. rocker of special quality concrete and adequately reinforced. The top and bottom of the rocker are made level. Lead sheets are placed between the contact surfaces of the rocker with the pier and the superstructure. The area of contact of the lead sheet is determined by dividing the load to be carried by the allowable unit bearing pressure on concrete. The rocker is designed as a short column, being built with high grade concrete and reinforced with vertical bars and hoops. The amount of horizontal reinforcement in the rocker longitudinally with the bridge is

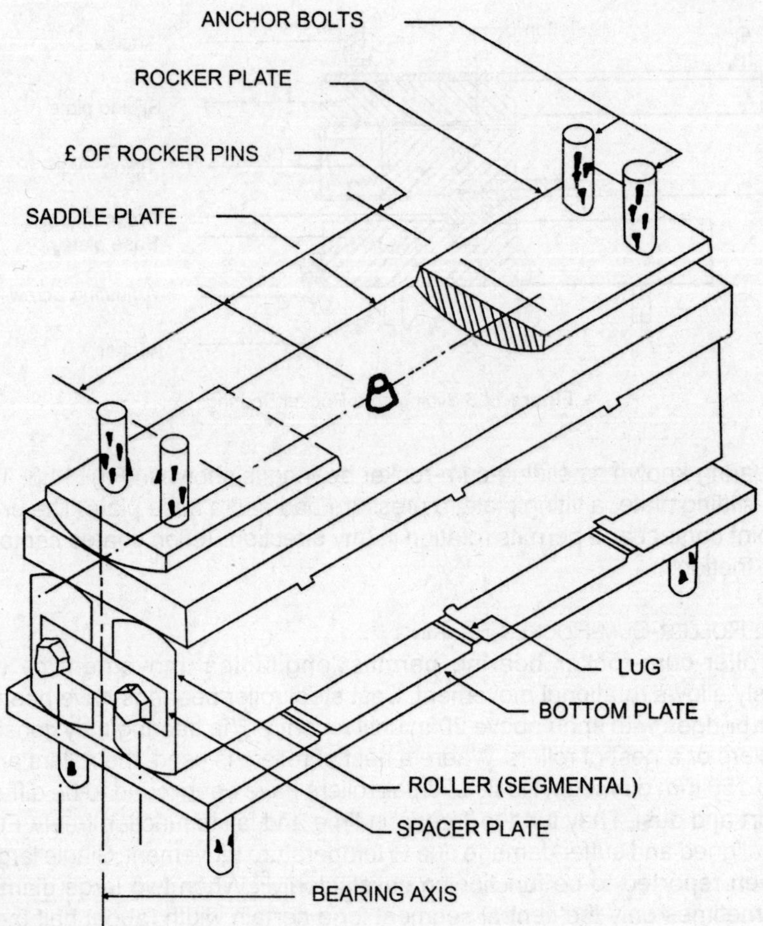

Figure 14.4 Roller-Rocker Bearing (with Segmental Roller).

determined on the assumption that the concentrated reaction develops in the block a tension equal in magnitude to one third of the total reaction acting on the rocker. The reinforcement is placed as hoops or spirals spaced throughout the height of the block and uniformly distributed along the block[7]. R.C. rocker bearing used in Dimni bridge across Kunwary river in U.P. is shown in Fig. 14.6.

The rotation and longitudinal motion of the bearing are accomplished as follows. For rotation, the girder compresses the lead sheet along the inner edge, and the underside of the girder assumes the desired inclination. For longitudinal movement, the girder causes the block to rotate upon the lead sheets. Thus, to expand, the lower lead sheet is compressed along the outside which makes the block tilt outside. Simultaneously the upper lead sheet is compressed along the inner edge. It could be seen that the larger the height of the concrete rocker block the more effective is its action because a smaller angular compression of lead will suffice to get the desired longitudinal motion. This type of bearing is currently not in popular use.

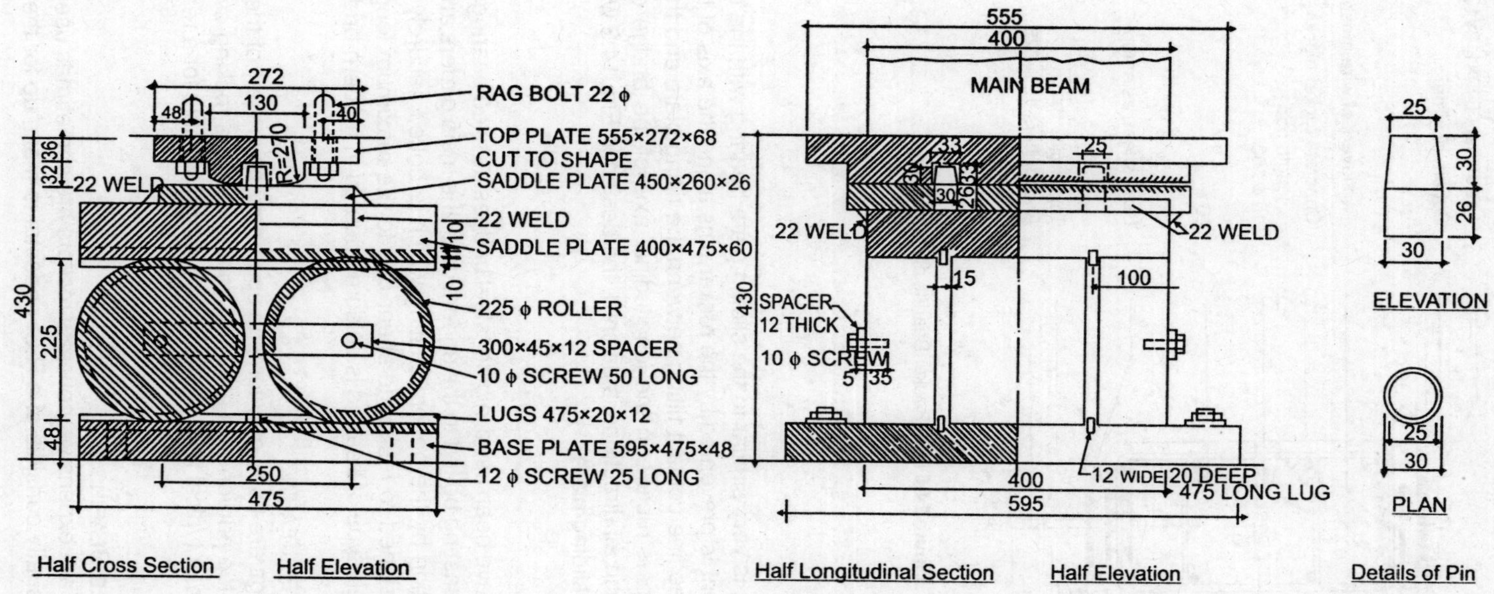

Figure 14.5 Mild Steel Roller-cum-Rocker Bearing of 1250 kN Capacity.

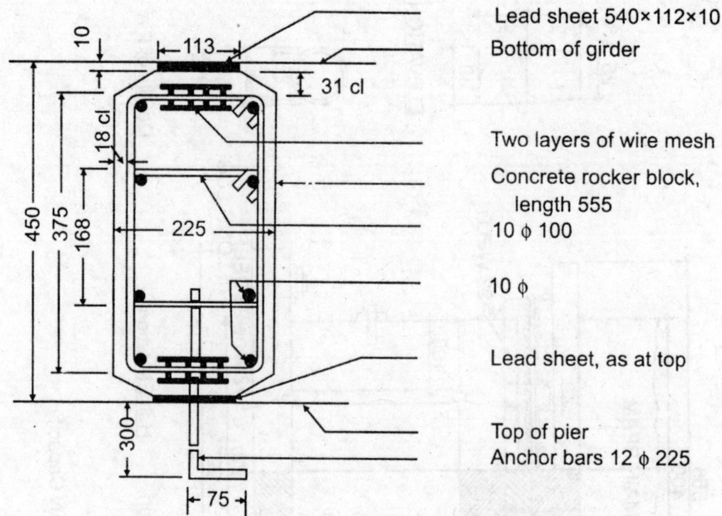

Figure 14.6 R.C. Rocker Bearing of Dimni Bridge.

14.5 Fixed Bearings

14.5.1 Rocker Bearing

The rocker bearing is very similar to the sliding plate bearing with the difference that the longitudinal movement is prevented by the rocker pins along the axis of the bearing as shown in Fig. 14.2(b). The line contact tilting between the tilting plate and the base plates permits tilting of the superstructure in the longitudinal axis of the bridge only. If a fixed bearing allowing rotation in all directions is desired, the design in Fig. 14.3 will be modified with the omission of the sliding plate.

14.5.2 Cast Steel Hinge

Cast steel hinges have been used in older steel bridges as fixed bearings. The bearing consists of a top saddle casting bolted to the underside of the main girders and resting on a knuckle pin held in position by the bottom casting which is bolted securely to the pier or abutment. The pin is designed to resist the shear due to the maximum longitudinal force acting on the bearing. This type of bearing is seldom adopted in modern bridges.

14.5.3 Mild Steel Rocker Bearing

Steel rocker bearings are used only for long span bridges in view of their cost. These consist of two parts: the top portion with a curved contact surface rocking over the bottom with a flat contact surface. A typical mild steel rocker bearing suitable for a reaction of 1250 kN is shown in Fig. 14.7.

14.5.4 R.C. Rocker Fixed Bearing

Since steel rockers are expensive, concrete rockers are sometimes used for concrete bridges. A R.C. rocker usually consists of a R.C. pedestal to make up for the height above

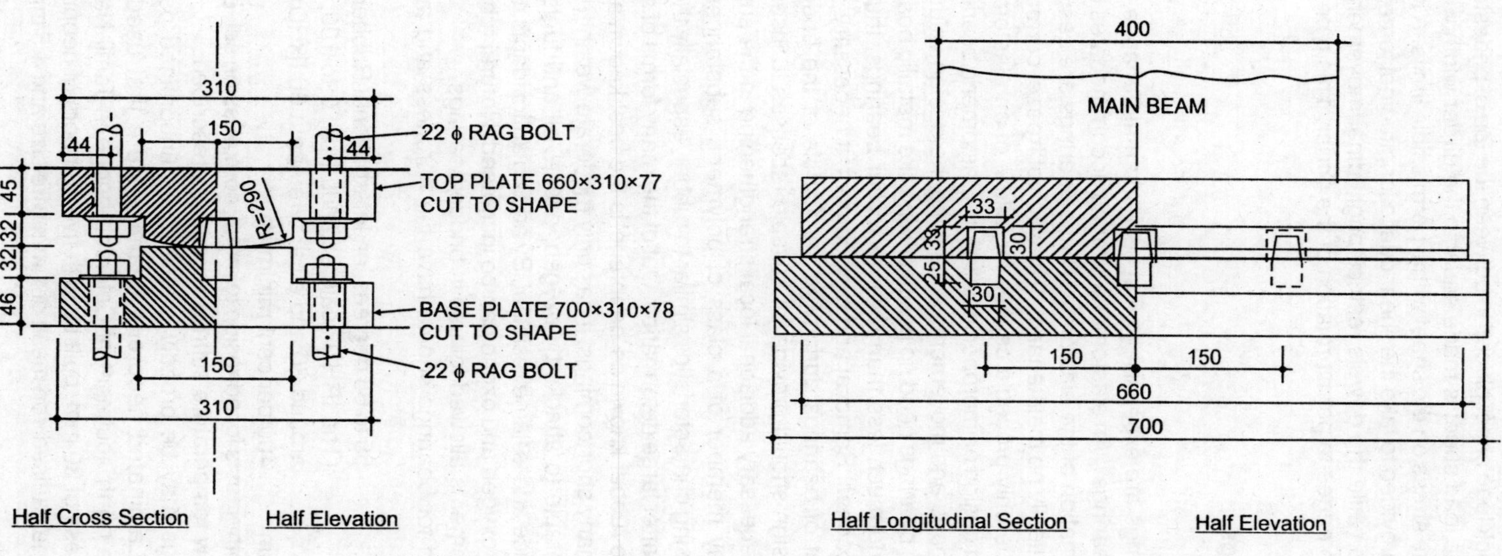

Figure 14.7 Mild Steel Rocker Bearing of 1250 kN Capacity.

bed block and a 10 mm thick lead sheet placed in between the top of pedestal and the bottom of girder. The length of the lead sheet is made equal to the girder width while its breadth is made sufficient to bring the stress on the sheet within permissible limits. A number of 12 mm dia. M.S. dowel rods are provided to take the shear due to longitudinal forces. The lead sheet allows the girder to rotate while the dowels restrict longitudinal movement. The top of the pedestal is beveled at the edges to permit rotation of the girder. This type of bearing is not popular now.

14.6 Elastomeric Bearings

14.6.1 General

Since metallic bearings are expensive in first cost and maintenance, the recent trend is to favour elastomeric bearings[8]. An elastomeric bearing accommodates both rotation and translation through deformation of the elastomer. These bearings are easy to install, low in first cost and require practically no maintenance. They do not freeze, corrode or deteriorate. Barring an earthquake, the only probable causes for failure of an elastomeric bearing are inferior materials, incorrect design or improper installation. Elastomeric bearings are 'forgiving' in that they can tolerate loads and movements exceeding the design values.

Further, elastomeric bearings need no positive fixing like metallic bearings. The height of bearing is minimum and much less than roller or rocker bearings, thus contributing to reduction in cost of approaches. Removal and replacement, if necessary, can be achieved easily. Since replacement of bearings during the service life of the bridge may become necessary, the bridge design should provide for suitable recesses to insert jacks for lifting the deck along with the necessary additional local strengthening in the superstructure.

An elastomer is any member of a class of polymeric substances obtained after vulcanization and possessing characteristics similar to rubber, especially the ability to regain shape almost completely after large deformation. Out of the many forms of synthetic rubbers available, polychloroprene rubber known as 'neoprene' is the best known and best tested in use. Natural rubber has many shortcomings. It has only moderate weathering resistance, is inflammable and is vulnerable to attacks by oxygen, ozone, oil and fuels. Elastomer has better weathering resistance and is flame resistant. By adding antioxidants and antiozonants, its resistance to attack by oxygen and ozone can be increased. Natural rubber bearings are not permitted. Only elastomer is allowed for use in bridge bearings.

The elastomer used for bearings should have the hardness and tensile elongation properties as below:

Hardness (IRHD) : 55 to 65 degrees, on International Rubber Hardness scale (The IRHD scale extends from 0 to 100, an eraser being around 30 and a car tyre about 60 IRHD).

Ultimate tensile strain : 400 per cent minimum.

Besides the above, other tests for adhesion to metal, compression set, ozone resistance, ageing resistance, and low temperature stiffness are also prescribed.

Elastomeric bearings may be broadly classified as unconfined pad bearings and confined pot bearings. Laminated neoprene bearings were first used in India in the articulations and over the north abutment of Coleroon bridge in Tamil Nadu State[9]. Each bearing had a total thickness of 30 mm, built up with five numbers of neoprene pads of 300 x 500 mm reinforced with galvanised iron mesh of 1 mm wire size and 4.5 mm mesh spacing.

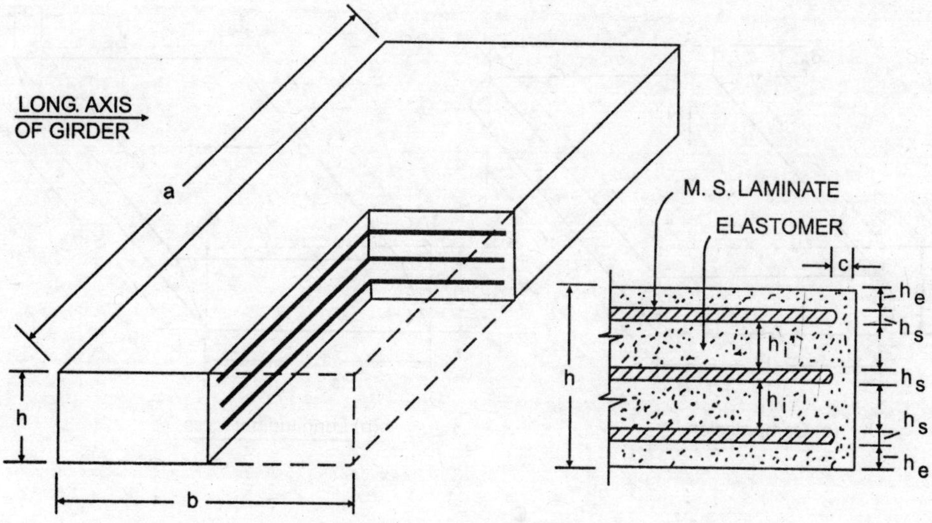

Figure 14.8 Components of Elastomeric Pad Bearing.

Since then elastomeric bearings have been used in many bridges in India. The present trend is to use reinforced moulded pad bearings as per IRC:83 (Part II).

14.6.2 Moulded Pad Elastomeric Bearing

The free elastomeric bearing for girder bridges consists of one or more internal layers of elastomer bonded to internal steel laminates of rectangular shape by the process of vulcanization. The internal layers of elastomer shall be of equal thickness, having elastomer cover on top, bottom and sides. The bearing cast as a single unit in a mould should be vulcanized under uniform heat and pressure. It should then be cured properly and uniformly to ensure balanced properties for the internal elastomer. The components of a typical elastomeric pad bearing are shown in Fig. 14.8. The top and bottom elastomer surfaces bear directly on the structure surfaces and any displacement of the bearing is safeguarded by only intersurface friction, without the aid of adhesive or other anchoring devices. The bearing caters for translation and/or rotation of the superstructure by elastic deformation of the elastomer.

14.6.3 Design of Elastomeric Pad Bearing

(a) *General*

The basic deformations of an elastomeric pad bearing under load are shown in Fig. 14.9. The bearing allows horizontal movement by shear deflection, and permits rotation by angular deformation. The design values of normal load N, horizontal load H, imposed translation u and rotation α shall be determined by suitable analysis of the structure under any critical combination of loads as specified in IRC: 6. The basic criteria for the design are that the elastomeric bearings accommodate the horizontal movement by shearing and the rotation by non-uniform linearly varying compressive deformation. The vertical stiffness of

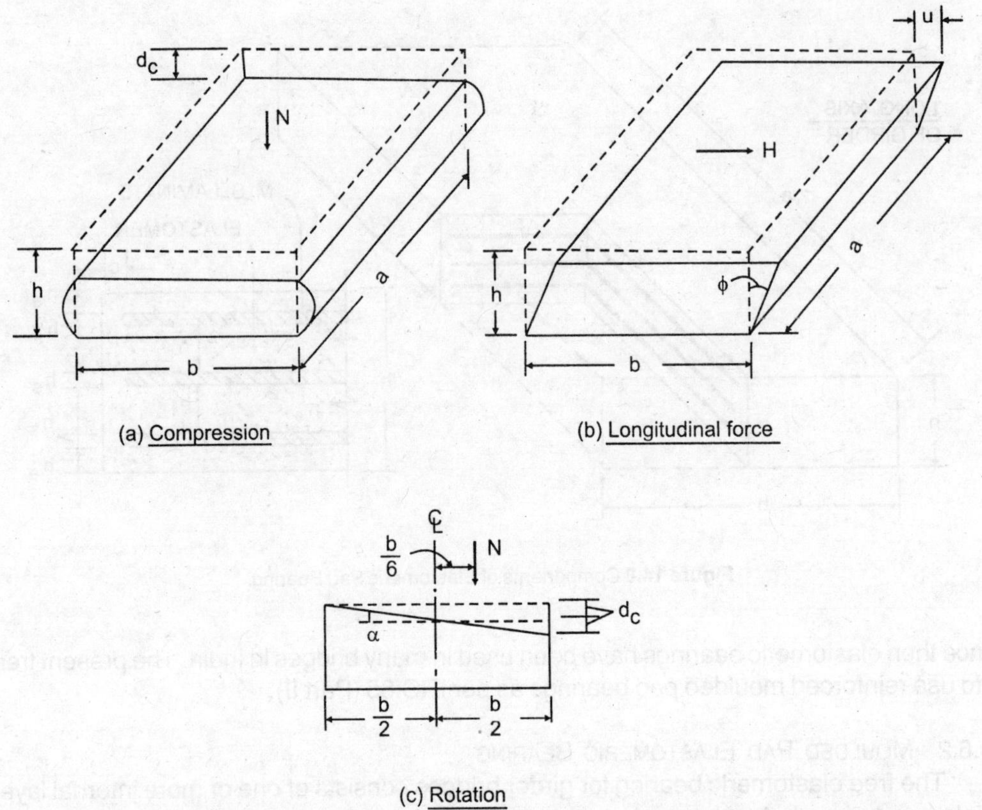

Figure 14.9 Basic Deformations of Elastomeric Bearings Under Load.

the bearing should be adequate to avoid significant changes in height by bulging at the sides under vertical compression; and this is ensured by restricting the value of the 'shape factor' S, which is defined as the ratio of one loaded surface area to the surface area free to bulge for an internal layer of elastomer excluding side cover.

The design requires the determination of the plan dimensions of breadth (along span) and length (perpendicular to span) of the pad bearing, besides detailing the thicknesses of internal layers of elastomer and the steel laminates. The elastomer cover on the sides and at top and bottom are also to be specified.

(b) *Design Rules*

The dimensioning of the bearing and the number of internal layers of elastomer chosen are arranged in such a manner that the following design criteria are satisfied:

(i) The plan dimensions shall conform to 'preferred numbers' R'20 series of IS: 1076[10], e.g., 10, 12, 14, 16, 18, 20, 22, 25, 28, 32, 36, 40, 45, 50, 56, 63, 71, 80, 90, 100 cm

(ii) The effective area of the bearing (equal to plan area of the laminate) should be adequate such that the average normal stress is less than the permissible contact pressure for the concrete structure.

(iii) The ratio of overall length to breadth is equal to or less than 2.

(iv) The total elastomer thickness is between one-fifth and one-tenth of the overall breadth of the bearing.

(v) Translation: The thickness of the elastomer in the bearing should be adequate to restrict the shear strain due to horizontal load and horizontal movement due to creep, shrinkage and temperature to a value less than 0.7. In the absence of more accurate analysis, the longitudinal translation due to creep, shrinkage and temperature can be computed assuming a total longitudinal strain of 5 x 10^{-4} for common R.C. bridge decks. The shear modulus of the elastomer is assumed as 1 N/mm². [IRC: 83 permits the value of shear modulus to be between 0.8 N/mm² and 1.2 N/mm².]

(vi) The thickness of an internal layer of elastomer h_i, the thickness of a laminate h_s, and the elastomer cover at top and at bottom h_e are related as below :

h_i, mm	8	10	12	16
h_s, mm	3	3	4	6
h_e, mm	4	5	6	6

(vii) The side cover of elastomer for the steel laminates is 6 mm on all sides.

(viii) The shape factor S is between 6 and 12.

For the notation indicated in Fig. 14.8, the shape factor may be computed from Equation (14.1).

$$S = \frac{(a-2c)(b-2c)}{2(a+b-4c)h_i} \quad \text{... (14.1)}$$

(ix) Rotation:

The number of elastomer layers provided shall satisfy the relation

$$\alpha \leq \beta . n . \alpha_{bi,max} \quad \text{... (14.2)}$$

where α = angle of rotation, which may be taken as 400. $M_{max} \cdot L/(EI) . 10^{-3}$

M_{max} = maximum midspan bending moment in superstructure

L = effective span of superstructure

I = moment of inertia of superstructure section

E = elastic modulus of concrete in superstructure (which may be taken as half the normal elastic modulus to cater for creep effects due to permanent loads)

n = number of elastomer layers

β = $\sigma_m / \sigma_{m, max}$

$\alpha_{bi, max}$ = $0.5\, \sigma_m \cdot h_i / (b.S^2)$

For calculating $\alpha_{bi,max}$, σ_m may be taken as 10 MPa.

$\alpha_{m, max}$ = 10 MPa

(x) Friction: Under any critical loading, the following limit shall be satisfied, to ensure adequate friction.

$$\text{Shear strain} < 0.2 + 0.1\, \sigma_m \quad \text{... (14.3)}$$

$$10 \text{ MPa} > \sigma_m > 2 \text{ MPa} \quad \text{... (14.4)}$$

(xi) Total shear stress:

The total shear stress due to normal load, horizontal load and rotation should be less than 5 MPa.

Shear stress due to normal load $= 1.5\, \sigma_m / S$ MPa ... (14.5)

Shear stress due to horizontal load assuming shear modulus as 1 MPa

= (Shear strain) MPa ... (14.6)

Shear stress due to rotation $= 0.5\,(b/h_i)^2\, \alpha_{bi}$ MPa ... (14.7)

(xii) **Standard plan dimensions and design data as given for guidance in IRC: 83 are shown in Table 14.1.**

14.6.4 EXAMPLE

Problem: It is required to design an elastomeric pad bearing for a two-lane R.C. T-beam bridge of 15.0 m clear span with the following data:

Maximum dead load reaction per bearing = 280 kN
Maximum live load reaction per bearing = 520 kN
Vertical reaction induced by longitudinal forces per bearing = 12 kN
Longitudinal force per bearing = 33 kN
Concrete for T-beam and bed block over pier : M20

$A_1/A_2 > 2$

Rotation at bearing of superstructure due to D.L. and L.L. = 0.0025 radian

Solution:

Effective span = 15.7 m

$N_{max} = 280 + 520 + 12 = 812$ kN

$N_{min} = 280$ kN

Try plan dimensions 250 x 500 mm and thickness 39 mm as per standard size index No. 6 of Table 14.1.

Loaded area $A_2 = 11.60 \times 10^4$ mm²

From Clause 307.1 of IRC: 21

Allowable contact pressure = $0.25\, f_c \sqrt{A_1 / A_2}$

A_1 / A_2 is limited to 2

Allowable contact pressure $= 0.25 \times 20 \times \sqrt{2} = 7.07$ MPa

Effective area of bearing required = 812 x 1000 / 7.07 = 11.5×10^4 mm²

σ_m = 812 x 1000 / 11.6 x 10 = 7.0 MPa

Thickness of individual elastomer layer h_i = 10 mm

Thickness of outer layer h_e = 5 mm

Thickness of steel laminate h_s = 3 mm

Adopt 2 internal layers and 3 laminates.

Overall thickness of bearing = 39 mm

Total thickness of elastomer in bearing = 39 – (3 x 3) = 30 mm

Side cover c = 6 mm

Translation

Shear modulus assumed = 1.0 N / mm²

Shear strain due to creep, shrinkage and temperature is assumed as 5×10^{-4} and this is distributed to two bearings.

Table 14.1 Standard Plan Dimensions for Elastomeric Bearings (According to IRC: 83 – Part II).

Size Index No.	b mm	a mm	$A.10^{-4}$ mm^2	N_{max} kN	N_{min} kN	h_i mm	n_{max}	n_{min}	h_{max} mm	h_{min} mm	$\alpha_{bi,\,max}$ $\times 10^{-3}$
1.	160	250	3.5	350	70	8	3	1	32	16	8
2.	160	320	4.6	460	90	8	3	1	32	16	7
3.	200	320	5.8	580	120	8	4	2	40	24	4
4.	200	400	7.3	730	150	8	4	2	40	24	3.5
5.	250	400	9.2	920	180	10	4	2	50	30	4
						12	3	1	48	24	6.5
6.	250	500	11.6	1160	230	10	4	2	50	30	3
						12	3	1	48	24	5.5
7.	320	500	15.0	1500	300	10	5	2	60	30	2
						12	4	2	60	36	3
8.	320	630	19.5	1900	380	10	5	2	60	30	1.5
						12	4	2	60	36	2.5
9.	400	630	23.9	2400	480	12	6	3	84	48	1.5
10.	400	800	30.6	3100	600	12	6	3	84	48	1.3

Note :
1. See Figs. 14.8 and 14.9 and section 14.6.3 for notation.
2. Marginal increase in N_{max} not exceeding 10 percent over the specified value may be permitted.
3. Where two values of h_i are given, the higher value may be adopted only when the lower value cannot cater for $\alpha_{bi,\,max}$ specified.

Shear strain per bearing due to creep, shrinkage and temperature

$$= \frac{5\times10^{-4}\times15.7\times10^{3}}{2\times30} = 0.13$$

Shear strain due to longitudinal force $= \dfrac{33.3\times10^{3}}{116\times10^{4}} = 0.29$

Shear strain due to translation $= 0.13 + 0.29 = 0.42 < 0.7$

Rotation

$$\alpha_{bi,\,max} = 0.5\ \sigma_m.\ h_i\,/(b.S^2)$$

$$S = \frac{116\times10^{4}}{2(238+488)10} = 7.99 > 6 < 12$$

Assuming $\sigma_{m,\,max}$ = 10 MPa

$$\alpha_{bi,\,max} = \frac{0.5\times10\times10}{238\times7.99^{2}} = 0.0033 \text{ radian}$$

$$\beta = \sigma_m/10 = 7/10 = 0.7$$

Permissible rotation $= \beta.n.\ \alpha_{bi,\,max}$

$= 0.7 \times 2 \times 0.0033 = 0.0046 > 0.0025$ actual

Friction

Shear strain = 0.42 as calculated

$0.2 + 0.1\,\sigma_m = 0.2 + 0.1\times7 = 0.9 > 0.42$

Also σ_m satisfies 10 MPa > σ_m > 2 MPa

Total shear stress

Shear stress due to compression $= 1.5\,\sigma_m / S$

$= 1.5 \times 7 / 7.99 = 1.31$ MPa

Shear stress due to horizontal deformation = 0.42 x 1.0 = 0.42 MPa

Shear stress due to rotation $= 0.5\,(b/h_i)^2 \alpha_{bi}$

$$= 0.5 \times \left[\frac{238}{10}\right]^2 \times 0.0025$$

$= 0.71$ MPa

Total shear stress $= 1.31 + 0.42 + 0.71 = 2.44$ MPa < 5 MPa

Summary

The elastomeric pad bearing has the characteristics:

Plan dimensions = 250 x 500 mm

Overall thickness = 39 mm

Thickness of individual elastomer layer = 10 mm

No. of internal elastomer layers = 2

No. of laminates = 3

Thickness of each laminate = 3 mm

Thickness of top or bottom cover = 5 mm

Conforms to Index No. 6 of Appendix I of IS: 83 (Part II).

14.6.5 Lead Rubber Bearing

Lead rubber bearing (LRB) is a variant of the laminated elastomeric bearing in that the bearing has a central cylinder of lead extending for the full depth. Such bearings are used as base isolation in bridges located in seismic regions. The size of the lead core is determined such that it remains elastic under service loads but deforms plastically in the event of an earthquake, providing a high level of energy dissipation through hysteresis damping. These bearings are characterized by high horizontal flexibility, increased vertical stiffness and capability to return to original alignment besides enhanced damping.

14.6.6 Installation of Elastomeric Pad Bearings

For in-situ construction, the bearing pad should be placed at the correct location with proper alignment. The bearing should be protected to avoid grout or concrete encasing or damaging the sides of the bearing. This can be achieved by surrounding the bearing with expanded polystyrene and taping adequately between the top surface of the bearing and the polystyrene. After the structure has been cast, the polystyrene should be carefully removed.

When precast concrete girders are seated on the pad bearings, the bearings should be first placed at the correct location with proper alignment. The lowering of the precast girders and seating should be done gradually without any jerk. Movement of the bearing during seating of the girder should be carefully prevented. This can be done by using epoxy-based resin bedding mortar below the bearing to provide sufficient bond between the bearing and the pedestal. In addition, a skim coating layer of mortar can be placed above the bearing prior to beam seating to allow for minor irregularities between the two surfaces. Though the design permits the bearing to be placed without any mortar, it will be prudent to apply the epoxy mortar as above. All bearings installed along a single line of support should be of identical dimensions.

The author has come across situations where the construction inspectors have shown alarm even with moderate bulging of the elastomer between steel plates of the pad bearing. Such moderate bulging is to be expected and allowed. The bulging becomes unacceptable if it is excessive, uneven among layers and accompanied by cracks of the elastomer.

14.7 Elastomeric Pot Bearings

When the vertical forces and the rotation at a bearing are large, unconfined elastomeric pad bearings may be inadequate. In such cases, confined elastomeric bearings known as pot bearings are used. The pot bearing consists of a circular unreinforced neoprene pad of relatively thin section, which is totally enclosed in a steel cylinder (pot). A steel piston fits inside the cylinder and bears on the neoprene. The confined rubber cannot bulge and is found to behave similar to a fluid under high pressure. Rotations up to 1 in 50 can be managed. By incorporating teflon layers, it is also possible to provide for horizontal movement besides rotation. This type of bearing is particularly suited for bridges on curves and for continuous spans of box section.

Schematic details of elastomeric pot bearings are shown in Fig. 14.10. The bearing at (a) permits longitudinal movement in one direction, while that at (b) permits movement in any direction. No longitudinal movement is permitted in (c). All the three types will permit rotation in any direction. Pot bearings have proved to be compact, efficient and economical. They are now widely used.

14.8 Bearings for Skew Bridges

For skew bridges it is preferable to use solid slab type of deck. Since the axis of rotation and the direction of longitudinal movement are not perpendicular, fixed bearings for skew slab bridges should have point contact allowing rotation in any direction, e.g. elastomeric pot bearing. Movable bearings in skew bridges must allow for rotation and movement in any direction, e.g. elastomeric pot bearing with sliding surface made of teflon. Wide skew bridges need special attention in design. For skew bridges with T-beam type of construction and with girders parallel to the longitudinal axis of the bridge, the bearings should be placed at right angles to the longitudinal axis of the girders. In order to avoid torsional fixity to the end cross girder, the end cross girder along the support line should be made slender. Load distribution could be ensured by stiff cross girder at midspan[11].

14.9 Joints

The joint is the weakest and most vulnerable area in bridge design. Unless properly designed, the distress at bridge joints will lead to many maintenance problems, ranging from spalling of concrete edges at the joint to deterioration of pier caps. With the extremely high density of traffic occurring on most major bridges, maintenance work on the bridge should be restricted to a minimum length of time. Hence the joints on a bridge should be so designed as to perform satisfactorily for a long time without requiring repair or replacement.

Three types of joints occur on a bridge structure: (a) construction joint, (b) expansion joint, and (c) contraction joint. Construction joint is necessary whenever the placement of

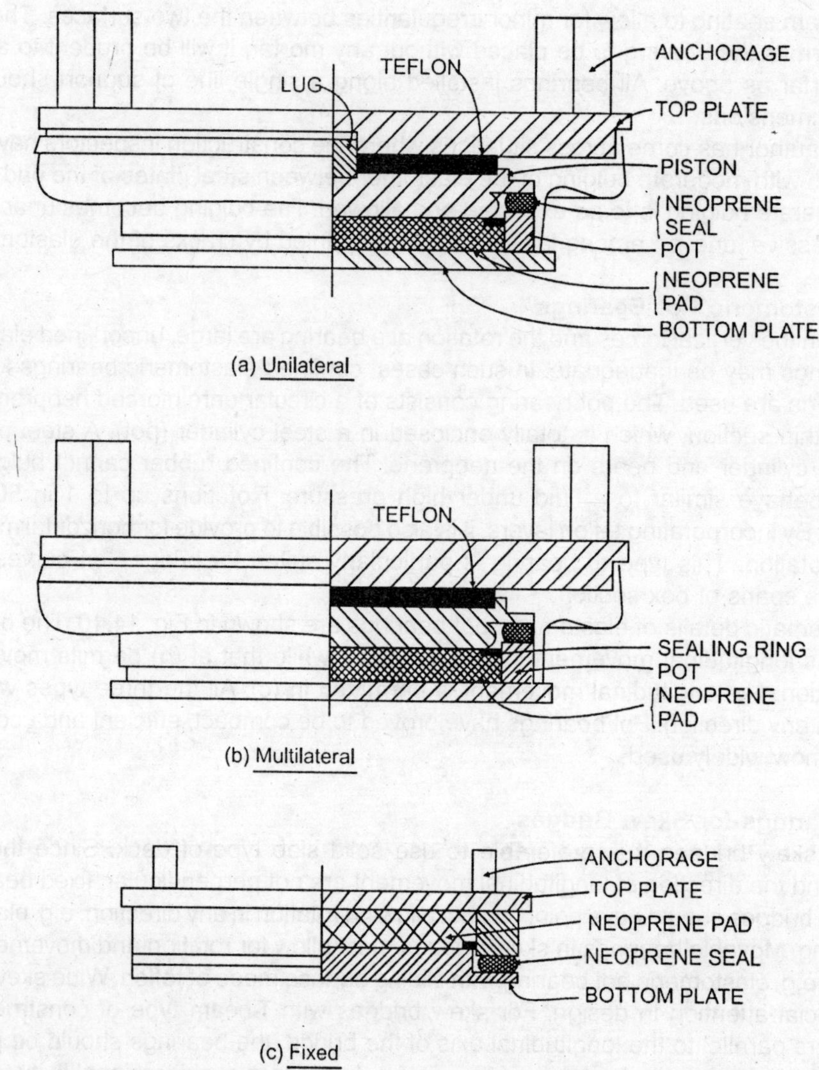

Figure 14.10 Elastomeric Pot Bearings.

concrete has to be stopped temporarily before the completion of the entire monolithic portion under construction. Such temporary suspension of concrete placement may sometimes be unexpected, if it is due to failure of machinery such as concrete mixer, vibrator, etc. But often, it may be scheduled to facilitate addition of reinforcements for a top portion, as in the case of the stem of a retaining wall. When foundations of adjacent parts of the structure are at different levels, as in the case of the junction between the abutment and the wing wall, a construction joint should be provided. Construction joints should be positioned to minimize the effects of discontinuity on the durability, structural integrity and appearance of the structure. Joints should be located away from regions of maximum stress caused by loading, particularly where shear and bond stresses are high.

Expansion joints and contraction joints are provided to take care of deformations due to change in temperature. The difference between the two types is in the depth of the joint and also in the width. Contraction joints, where provided, will be only for a part of the depth of the slab and will often be of smaller width. Expansion joints will be for the full depth of the member.

14.10 Expansion Joints

Expansion joints are important structural elements in any type of road bridge. They accommodate the relative movements of the bridge elements, especially those due to concrete shrinkage, change in temperature and long term creep. Located at the road level, the expansion joints are subject to the impact and vibration due to wheel loads and are exposed to the effects of water, dust, petroleum derivatives and salt solutions. By virtue of their function, expansion joints form a source of weakness in the performance of the bridge. Leakage of water from the deck at the joints have in the past led to deterioration of bearings and substructure. The expansion joints should be robust, durable, watertight and replaceable, besides facilitating good riding quality. Satisfactory long-term performance and durability of expansion joint systems require diligent design, quality fabrication, competent construction, adequate inspection and meticulous maintenance. However, a fully satisfactory expansion joint has not been evolved yet.

The expansion joint and the bearings should be designed together to be compatible. The author has noticed that designers have the tendency to attempt sophisticated techniques of structural analysis and design of the superstructure while they would be quite oblivious of the seriousness of the correct design of expansion joints. Thus even in prestressed concrete bridges of advanced design, expansion joints have been unsatisfactory. Another aspect of bridge design on which very little attention has been devoted by designers in the past is the joint seal. The surface of the deck expansion joint should be made as watertight as possible. The sealing system is particularly important in case of medium and long span bridges. Trouble-free lifetime performance and durability of expansion joint systems depend on the application of appropriate design, high-quality fabrication, correct construction practices, systematic inspection and proper maintenance.

The different types of expansion joints and their suitability for adoption are listed in Table 14.2. Some of these joints and seals are briefly discussed below. Since many of the available joint systems are proprietary, the shape, dimensions and orientation of the various components shown in the figures in this text are to be taken as indicative for educational purposes. Guidelines and specifications for expansion joints are available in IRC:SP:69-2005[12].

A simple expansion joint, known as buried joint, may be used to cater for small movements associated with simply supported spans up to 10 m. Here the width of the joint gap is kept at 20 mm. A steel plate 200 mm wide and 12 mm thick of weldable structural steel as per IS:2062 is placed symmetrical to the centerline of the joint to bridge the gap resting freely over the top surface of the deck concrete. Bituminous/asphaltic wearing course is laid continuously over the steel plate. The emphasis is on securing a good riding quality over the joint.

For highway bridges of spans up to about 10 m, the expansion joint known as the filler joint may be used. This joint is 20 mm wide, with details as shown Fig. 14.11. A copper sheet 2 mm thick is bent to form a bulb in the middle with a plan width of 220 mm and laid

Table 14.2 Suitability Criteria for Different Types of Expansion Joints.

Sl. No.	Type of Joint	Suitability	Expected Service Life, Years	Special Considerations
1.	Buried Joint	Simply supported spans with movement up to 10 mm	10	Only for decks with bituminous/asphaltic wearing course
2.	Filler Joint	Simply supported spans with movement up to 10 mm	10	Joint filler may need replacement if found damaged
3.	Single Strip Seal Joint	Movement up to 80 mm	25	Elastomeric seal may need replacement during service life
4.	Modular Strip/Box Seal Joint	Movement over 80 mm	25	Elastomeric seal may need replacement during service life
5.	Finger Joints	Movement between 80 mm and 200 mm	25	Not suitable for joints involving differential vertical movements. Joint requires sound anchorage with deck.
6.	Special Joints	Movement between 80 mm and 200 mm	25	Joints in high seismic zones. Special design to accommodate large longitudinal movements

integral with the deck slab and the adjoining deck slab/approach slab as in Fig. 14.11. The gap above this sheet is filled with a premoulded resilient joint filler. This arrangement would allow a movement up to 10 mm.

For medium spans, when the movement is of the order of 40 to 50 mm, an expansion joint consisting of mild steel angles and plates as shown in Fig. 14.12 has been adopted. The details shown are suitable for composite girder bridge of about 27 m span. In tropical climates, this type of joint seldom functions properly due to corrosion. It is also difficult to seat the sliding plate accurately on the mating plate throughout the width of the carriageway. Based on observation of performance in the field, the author would recommend that this type should not be used for midspan joint of a prestressed concrete cantilever bridge, and that its use even in T-beam bridges in coastal areas should be with caution. A modern solution for expansion joints for movement up to 80 mm, known as the single strip seal joint, is shown in Fig. 14.13. The joint device consists of continuous, flexible, extruded neoprene seal held in place by steel extrusions which are anchored either to the steel supports or directly to the concrete deck. Satisfactory performance of the joint is governed by the correct setting of the expansion gap with reference to the theoretical requirement at a specified reference temperature, corrected for the actual temperature prevailing at the time of concrete placement. Further, the installation

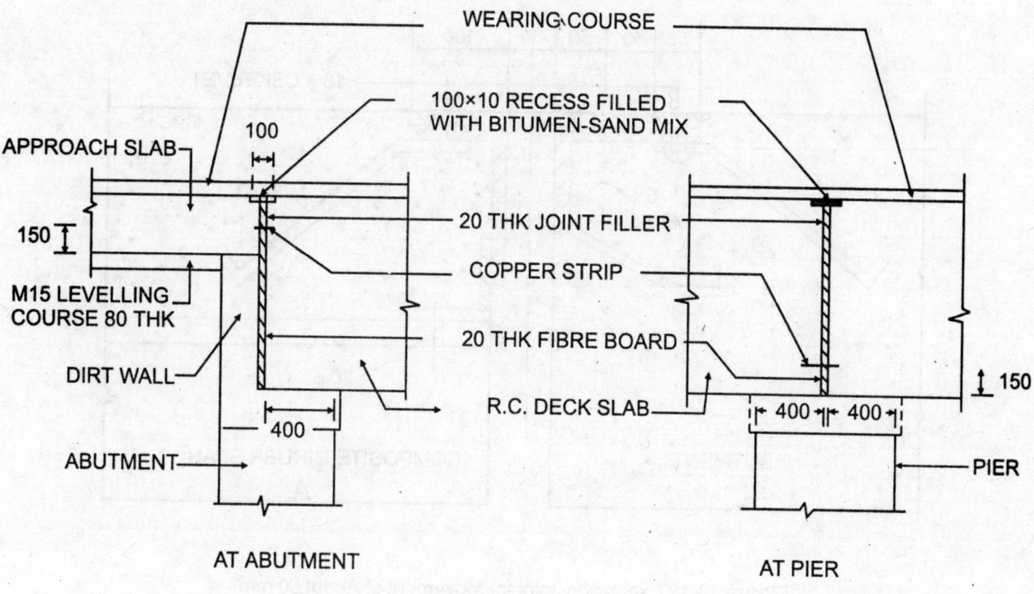

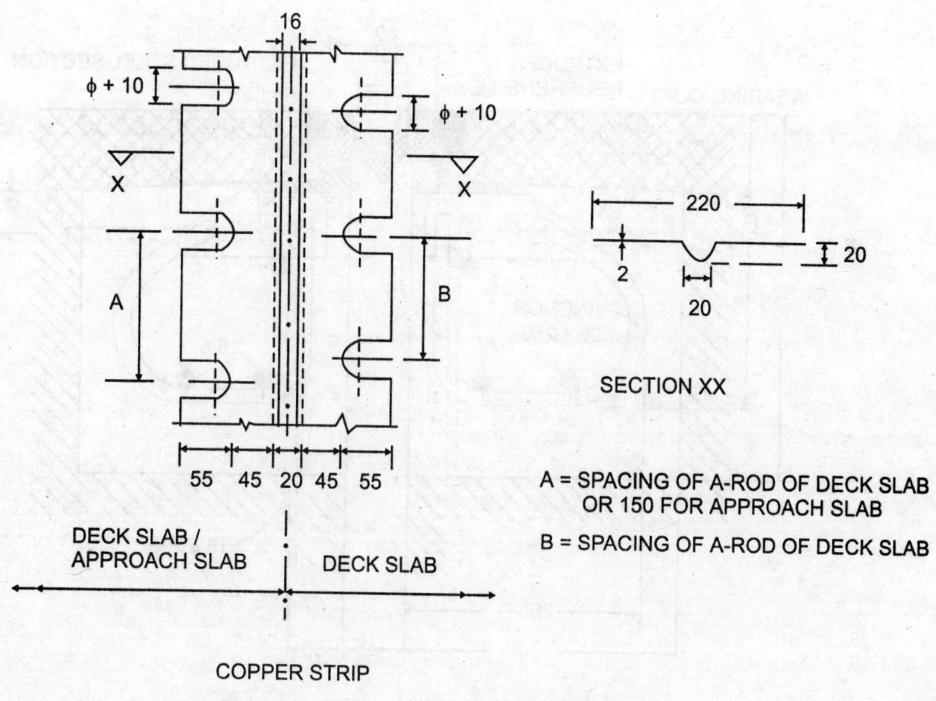

Figure 14.11 Expansion Joint for Slab Bridges.

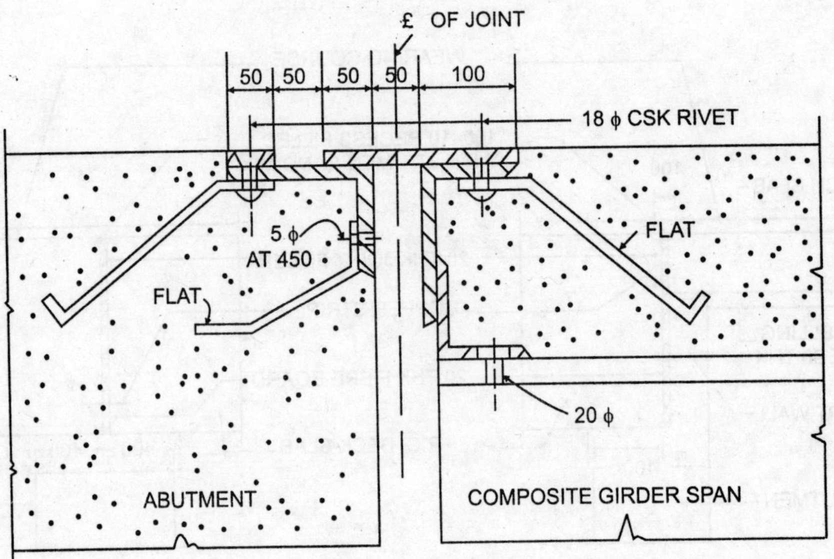

Figure 14.12 Expansion Joint for Movement of About 50 mm.

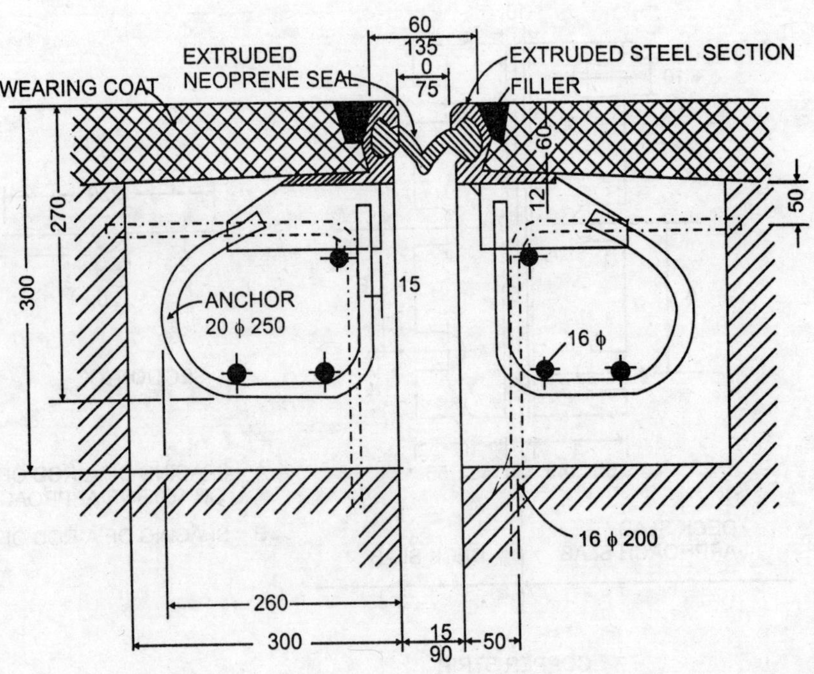

Figure 14.13 Typical Expansion Joint for Movement up to 75 mm.

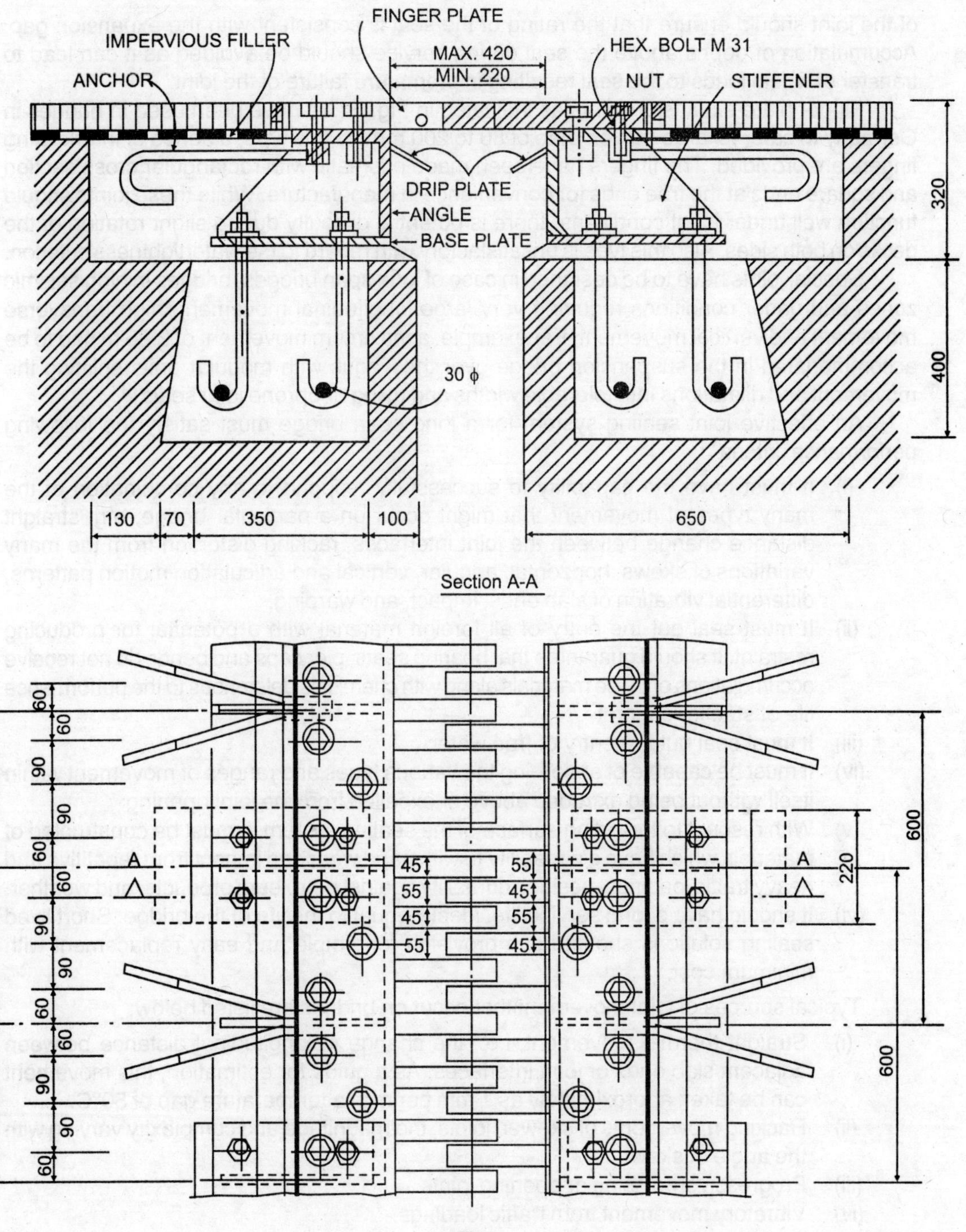

Figure 14.14 Typical Finger Plate Expansion Joint for Movement of 200 mm.

of the joint should ensure that the rating of the seal is consistent with the expansion gap. Accumulation of debris above the seal during service should be avoided as it can lead to transfer of wheel loads to the seal resulting in premature failure of the joint.

Finger plate expansion joints as indicated in Fig. 14.14 have been used in bridges in Germany to cater to large movements of 80 to 200 mm. In this type, a series of interlocking fingers are provided. The fingers have been made prismatic with rectangular cross section and square ends at the free ends for convenience in manufacture. While these joints should function well under ideal conditions, there is potential difficulty due to slight rotation of the decks on both sides. Also this type is unsatisfactory with regard to the watertightness criterion.

Special joints have to be designed in case of long span bridges, bridges in high seismic zones and under conditions requiring very large longitudinal movement, high transverse movement and vertical movement. For example, a maximum movement of 2.85 m had to be accommodated in the suspension bridge over the Tagus with modular units splitting the movement and dilatations into tolerable widths and using neoprene joint seals[13].

An effective joint sealing system for a long span bridge must satisfy the following performance criteria[14]:

(i) It must have the capability to successfully respond to any combination of the many types of movement that might occur on a particular bridge, e.g. straight distance change between the joint interfaces, racking distortion from the many variations of skews, horizontal, angular, vertical and articulation motion patterns, differential vibration of slab ends, impact, and warping.

(ii) It must seal out the entry of all foreign material with a potential for producing restraint. It should guarantee that bearing seats, pier caps and bends do not receive accumulations of these materials along with chemicals deleterious to the performance life of steel or concrete.

(iii) It must seal out the entry of free water.

(iv) It must be capable of absorbing the various types and ranges of movement within itself without being extruded above or expelled from the joint opening.

(v) With respect to the riding surface of the sealing system, it must be constructed of materials which have a capability to withstand wear and impact from repetitive and heavy traffic loadings, besides durability against petroleum products, and weather.

(vi) It should have a long service life, ideally equal to the life of the bridge. Short lived sealing solutions should have provision for simple and easy replacement with minimum cost.

Typical sources of joint movement that occur on bridges are listed below:

(i) Straight thermal movement, i.e., the change in longitudinal distance between adjacent slab ends or joint interfaces. As a guide for estimation, this movement can be taken approximately as 1 mm per m in a temperature gap of 50°C.

(ii) Racking movements of skewed joints, the magnitude and complexity varying with the angle of skew.

(iii) Progressively closing or opening joints.

(iv) Vibratory movement from traffic loadings.

(v) Slab end rotation, which may be temporary due to heavy traffic loading at midspan or permanent due to progressive increase in dead load deflection.

(vi) Articulating movements.

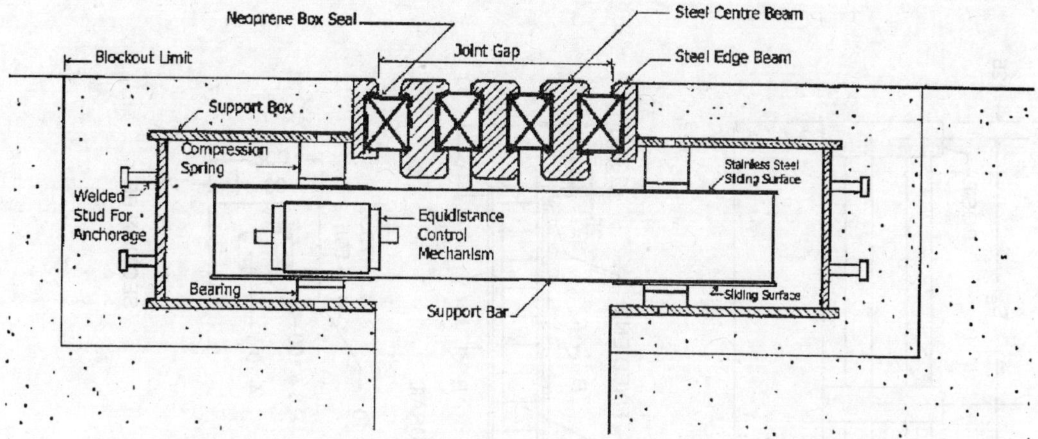

Figure 14.15 Typical Modular Expansion Joint for Large Movement.

Whether the movement at a joint is small as 25 mm in the case of a simply supported T-beam bridge of 20 m span or large as 1.8 m as predicted for the Forth road bridge in Scotland, the movement must be accounted for in the design of the jointing and sealing system with an additional provision for some margin of safety.

A recent innovation in the design of expansion joints is the prefabricated modular compression sealing system to cater to horizontal movement over 80 mm. The modular expansion joints are complex structural systems which provide watertight wheel load transfer across expansion joint openings. Several systems of modular joints are available[15,16]. A typical four-module expansion joint is shown schematically in Fig. 14.15. The modular joint consists of four main components: two edge beams, mechanically locked elastomeric seals, centre beams (for seals), and support bars (for centre beams) The seals can be of strip type or extruded webbed box type. The centre beams are extruded metal shapes to facilitate assembly of the seals in series, and these are supported by independent multiple support bars, which are welded to the centre beams. The support bars are suspended over the joint openings by sliding elastomeric bearings mounted within support boxes, which rest on cast-in-place concrete installed into a preformed blockout. Resilient or shock absorbing support system for the centre beams and the support bars facilitate damping the dynamic loading and also serve to accommodate vertical and transverse movements apart from longitudinal movement. An equidistance control system incorporated in the design ensures equal distribution of the movement among the various seals. The completed expansion joint should be continuous across the full width of the bridge. The joint for a bridge is prefabricated at a factory with strict quality control. The dimensional details of the expansion joint for any particular bridge are to be obtained from the specialist manufacturers of the joint.

In Europe and in North America, leakage of deicing salt through the deck expansion joints causes spalling of concrete surrounding the joint and corrosion of reinforcing bars and prestressing tendons in the deck, besides deterioration of the underlying piers. Modern designs of multi-span bridges in these countries tend to minimize the number of joints by favouring continuity and to discourage suspended decks of balanced cantilever bridges.

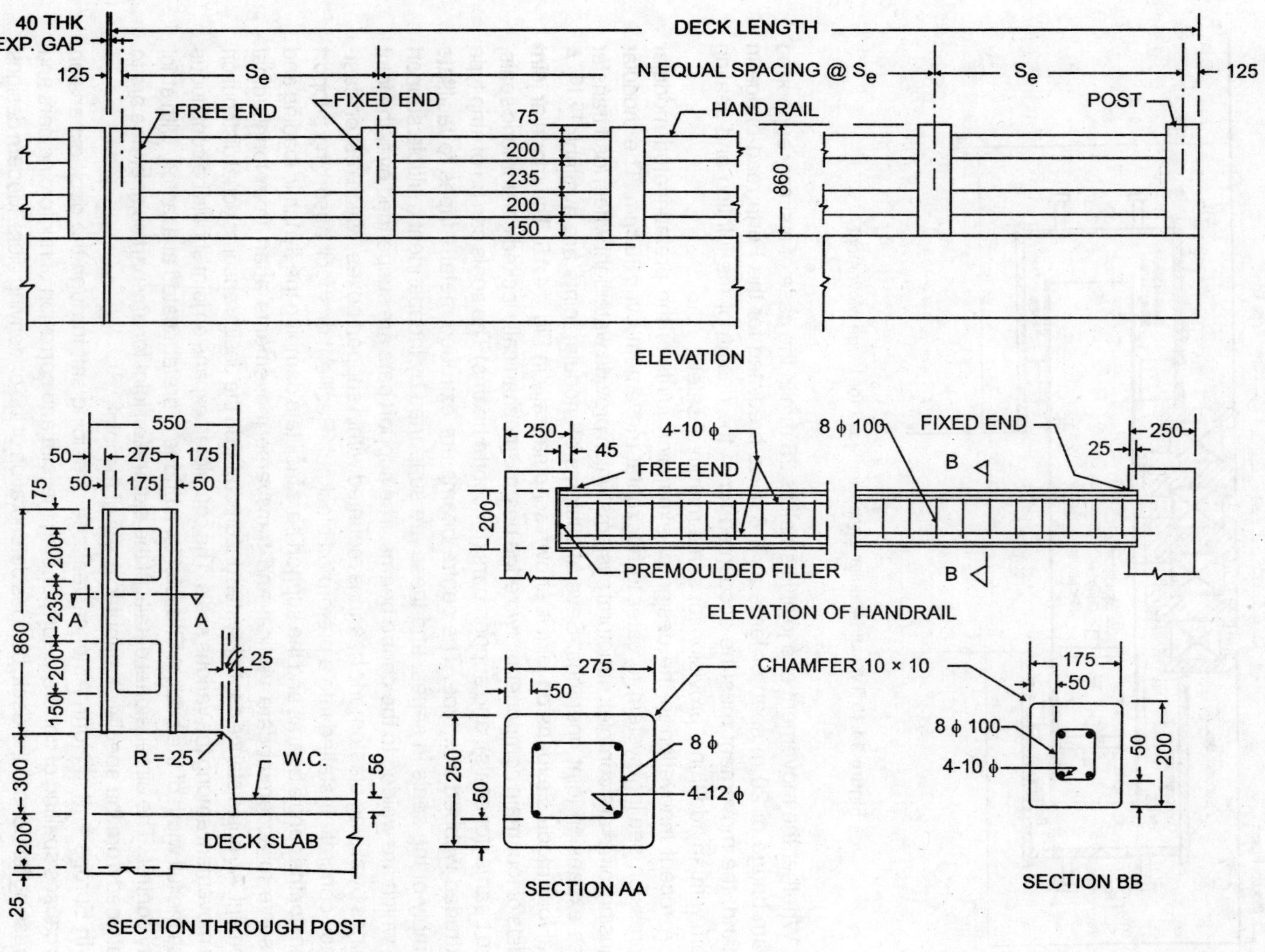

Figure 14.16 Typical Details of Handrails for Road Bridges.

14.11 Appurtenances

An appurtenance, in the context of a bridge, is any part of a bridge or bridge site which is not a major structural component, but still performs a useful function towards the performance of the bridge. The major appurtenances include: handrails, parapets, footpaths on bridges, drainage spouts, wearing course over the deck, river training works, embankments for approaches, approach slab and crash barriers.

14.12 Handrails

Parapets and handrails are provided on bridges as a protective measure to keep bridge users from falling to the depth over the sides of the bridge. While parapet walls would be adequate for culverts and small bridges, handrails are adopted for longer bridges. Since these are the parts prominently visible in the neighbourhood, the design should be such as to enhance the aesthetics of the bridge.

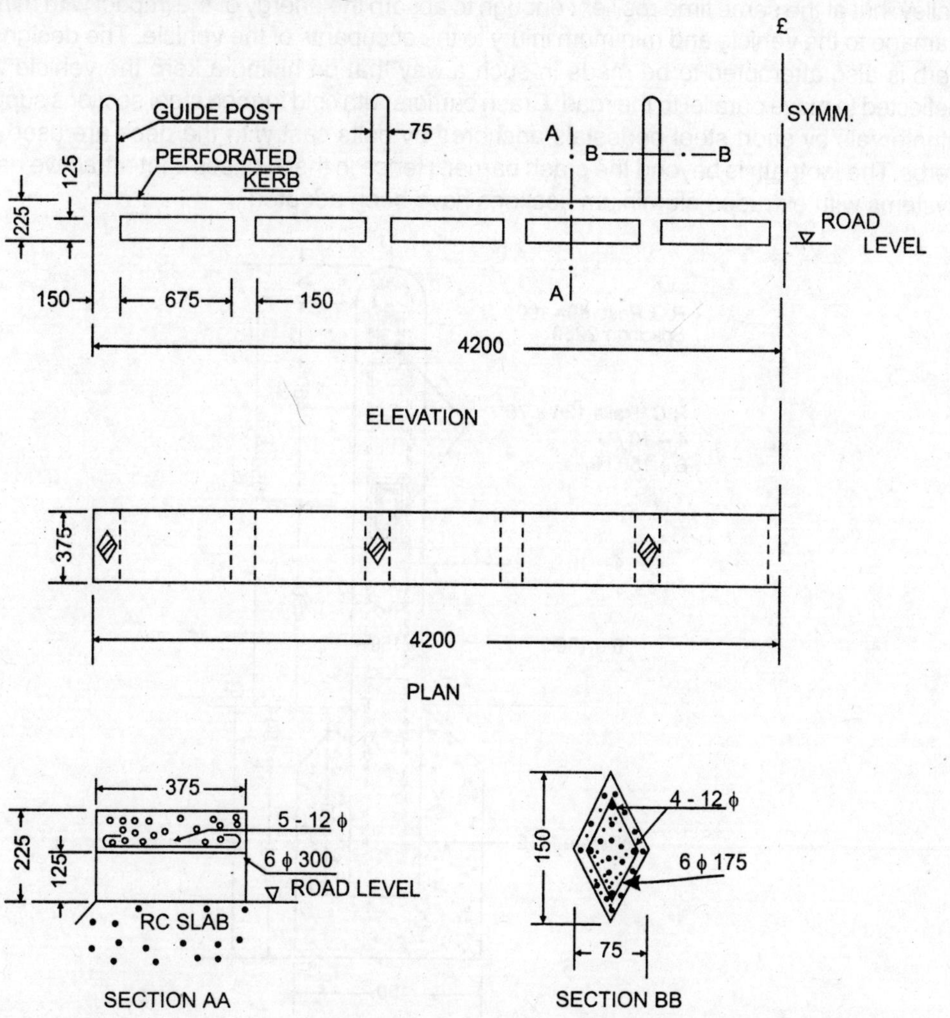

Figure 14.17 Typical R.C. Perforated Kerbs and Guideposts for Submersible Bridges.

Depending on the situations, the handrails can be either of the post and rail system or panel slab and post system. The post and rail system can be of steel or reinforced concrete. Different designs are possible. Fig. 14.16 gives typical details of post and rail system of handrails in reinforced concrete for a highway bridge. Here the posts are spaced at suitable intervals of less than 1.8 m and cast monolithic with the kerb. The rails are of rectangular shape and are precast. R.C. railings are vulnerable for corrosion damages. Hence their construction should be done with utmost care. Reinforcements of the posts should be suitably anchored in the deck slab. The post is cast in a single placement operation after accurately positioning the precast handrail. One end of the handrail is fixed and the other end is freely supported as shown in Fig. 14.16.

Considerable research has been undertaken in developed countries to evolve a handrail system which would be strong enough to prevent a colliding vehicle from falling over to the valley, but at the same time resilient enough to absorb the energy of the impact with minimum damage to the vehicle and minimum injury to the occupants of the vehicle. The design of the kerb is also attempted to be made in such a way that on hitting a kerb the vehicle will be deflected to move parallel to the road. Crash barriers with cold formed steel sections supported at intervals by short steel pedestals anchored by bolts cast with the deck are used at the kerbs. The footpath is beyond the crash barrier. Hence in these cases light attractive handrail systems with extruded aluminium sections have been adopted.

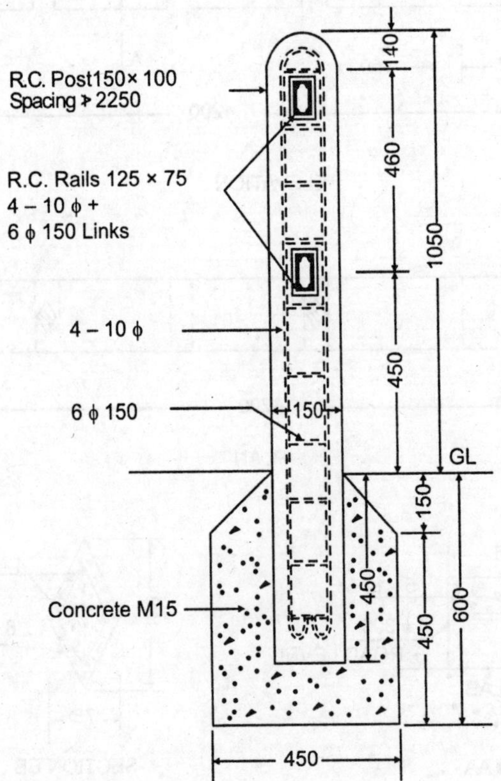

Figure 14.18 Typical Handrails for Approach Roads.

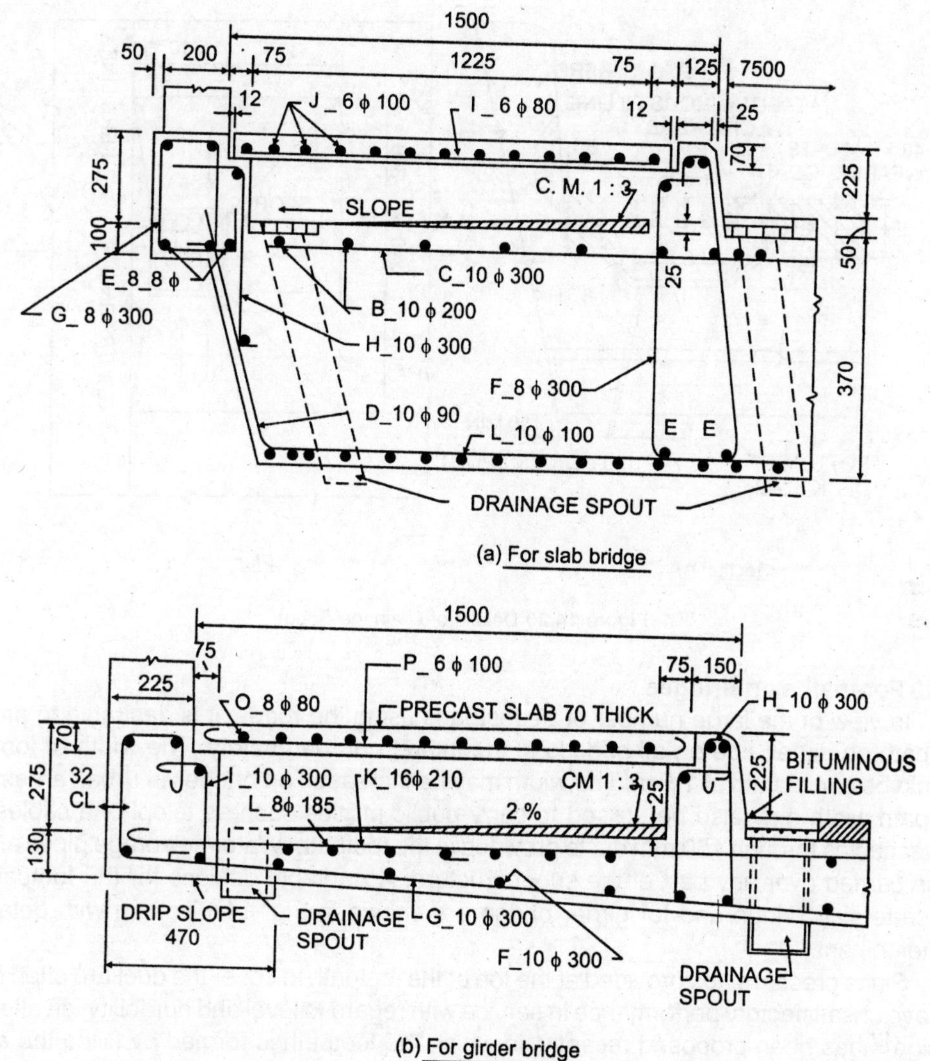

Figure 14.19 Typical Details of Footpath on Concrete Bridges.

In the case of submersible bridges, handrails, if provided, should be collapsible during floods, so as to minimise obstruction to flow of water and passage of floating debris. It is, however, preferable to provide perforated kerbs along with diamond shaped guide posts[4] as shown in Fig. 14.17.

It would be often desirable to provide handrails for approaches to a bridge, especially if the approaches are on high embankments or on a curve. The design, layout and materials chosen for the rails should blend with the surroundings. The posts may be erected on lean concrete blocks cast at site. Typical details of such an arrangement are shown in Fig. 14.18. If reflectors are fixed to the posts, the safety for the vehicles on the approaches will be considerably enhanced and the rails shown in the figure may be omitted.

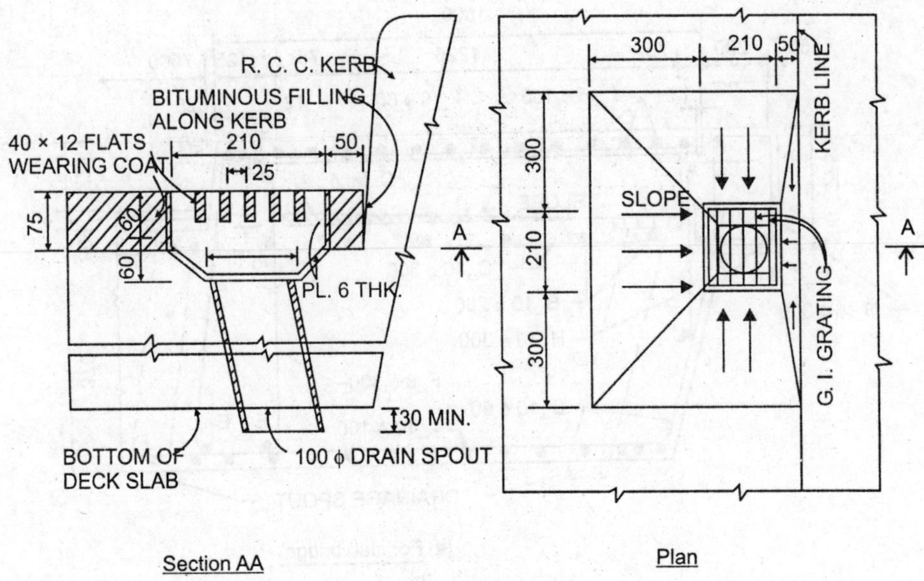

Figure 14.20 Details of Drainage Spout.

14.13 Footpaths on Bridges

In view of the large number of pedestrians using the roads, it is desirable to provide footpath on either side on all bridges in the interest of road safety. The width of footpath should be a minimum of 1.5 m. The width may be increased on bridges in urban areas. The footpath width may also be utilised to carry public utilities such as telephone cables and power cables through 150 mm ducts provided in the footpath. Water/sewerage pipes are not to be carried over any part of the superstructure. Acceptable designs for the footpath for concrete slab bridge and for girder bridge are shown in Fig. 14.19 along with details of reinforcement.

Since precast slabs provided at the top of the footpath to cover the duct are often found to have unsatisfactory performance in service with regard to level and durability, an alternate design[17] has been proposed recently, in which the footpath is formed by filling the space between the road edge kerb and the bridge kerb with concrete. Three 150 mm diameter pipes are embedded in the concrete to convey the utility cables. Inspection chambers are provided at suitable intervals. This design permits the provision of chequered tiles on the footpath in urban areas.

14.14 Drainage Arrangements

Adequate arrangements should be provided for proper drainage of rain water from the bridge deck in order to ensure good durability. For this purpose, drainage spout with grating as indicated in Fig. 14.20 should be provided at spacing not more than 10 m apart. The drainage spout should be galvanized after welding the plates and flats together. Such spouts should be extended well below the soffit of the deck structure to avoid damage to concrete due to spray of salt-laden water. Care should be taken in design to ensure that drains do not discharge on the bridge elements or on traffic passing below the structure in case of a road overbridge. External piping down the sides of the deck to the piers may look ugly. It may be desirable to use concealed internal lateral connecting drains from kerb gullies to drains in concealed slots in piers. By effective maintenance inspection, the drains should be kept free of debris so that the drains function properly. Inadequate deck drainage may lead to ponding of water and hydroplaning of vehicles, leading to traffic safety hazards besides deterioration of the deck surface.

A drip groove 25 mm deep is located at the bottom of the deck slab at about 160 mm from the fascia parallel to the roadway to prevent rain water from flowing down from fascia to the beams. This groove should be stopped 900 mm from the face of the abutment to prevent moisture accumulating on the bed block over the abutment.

14.15 Wearing Course

A wearing course (sometimes referred as wearing coat) is provided over concrete bridge decks to protect the structural concrete from the direct wearing effects of traffic and also to provide the cross camber required for surface drainage. The wearing course may be of asphaltic concrete or cement concrete. Asphaltic concrete wearing course is currently the preferred option as this permits the use of buried expansion joint for short spans facilitating a smooth transition between the bridge and the approaches for the riding surface. The thickness of the wearing course is kept uniform and the top of the deck slab is adjusted to facilitate the cross camber for surface drainage.

(a) Asphaltic Concrete Wearing Course

Asphaltic wearing course of 56 mm uniform thickness is desirable when the road pavement on the approach on either side of the bridge is of asphaltic concrete. The wearing course consists of the following: (i) A coat of mastic asphalt 6 mm thick with a prime coat over the deck slab; and (ii) 50 mm thick asphaltic concrete wearing course in two layers of 25 mm each.

(b) Cement Concrete Wearing Course

Cement concrete wearing course of 75 mm uniform thickness in M30 concrete over concrete deck slab may be adopted in case of isolated bridges where use of asphaltic concrete is inconvenient. The wearing course should be reinforced with 6 ϕ 200 in both directions where the deck slab is in compression and with 6 ϕ 100 in both directions where the deck slab is in tension. The reinforcement is placed at the middle of the wearing course. The free ends of the reinforcement at panel joints should be bent down to protect the ends of the joints.

The cement concrete wearing course should be laid in two longitudinal strips with casting of alternate panels of equal length in each strip. The joints of the panels in the two strips shall be staggered. While concreting the left out panels, bituminous papers will be placed at the joints with the previously placed panels in order to get a separation between the panels. Shuttering will have to be provided at the free ends for ensuring vertical face and also to attain good compaction.

14.16 Approach Slab

A reinforced concrete approach slab is usually provided on either side of a concrete bridge to function as a smooth transition between the paved roadway and the riding surface of the bridge. The slab serves to minimize bumps to traffic and the resulting impact to abutment due to potential differential settlement between the approach embankment and the abutment. The slab should cover the full width of the roadway and should extend for a length of not less

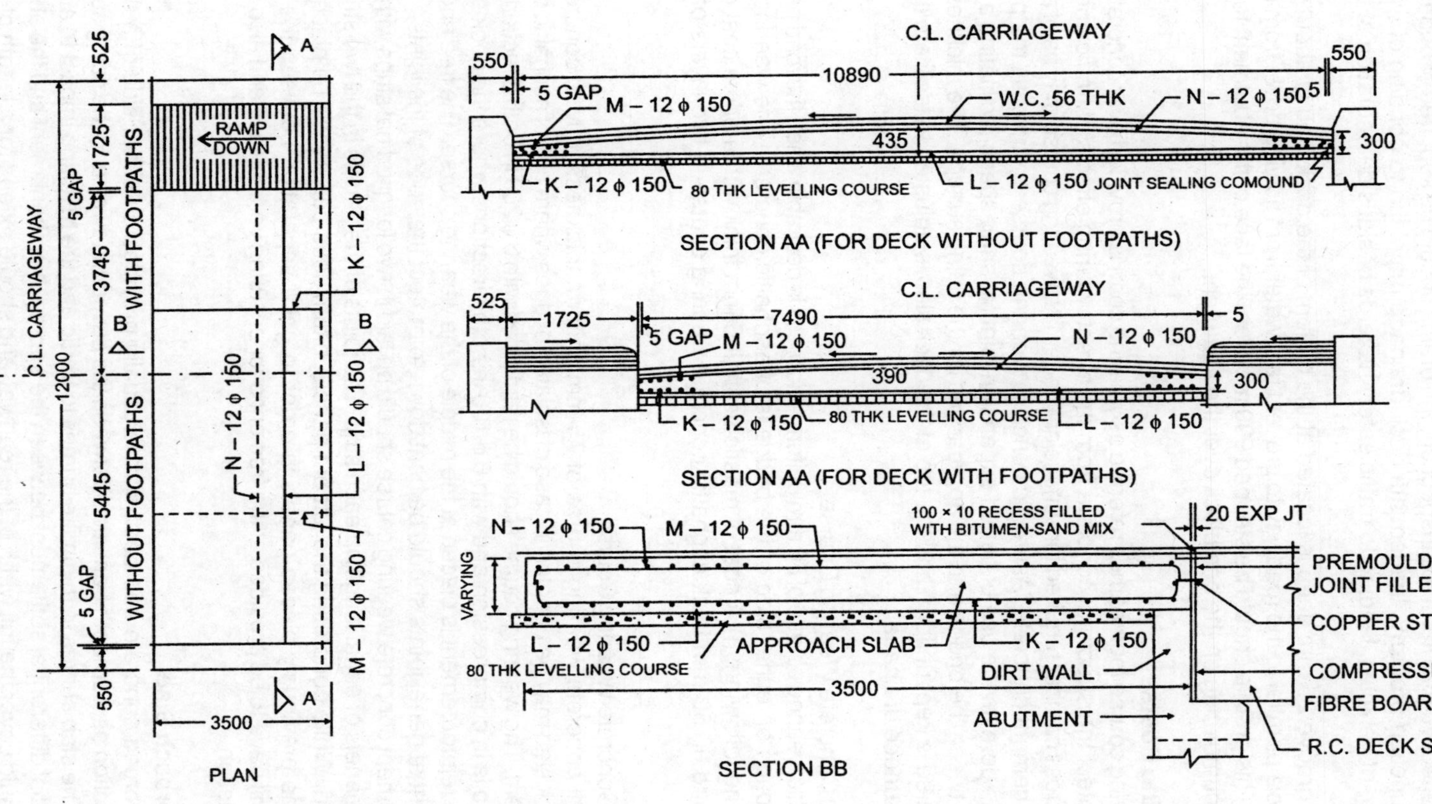

Figure 14.21 Details of Approach Slabs.

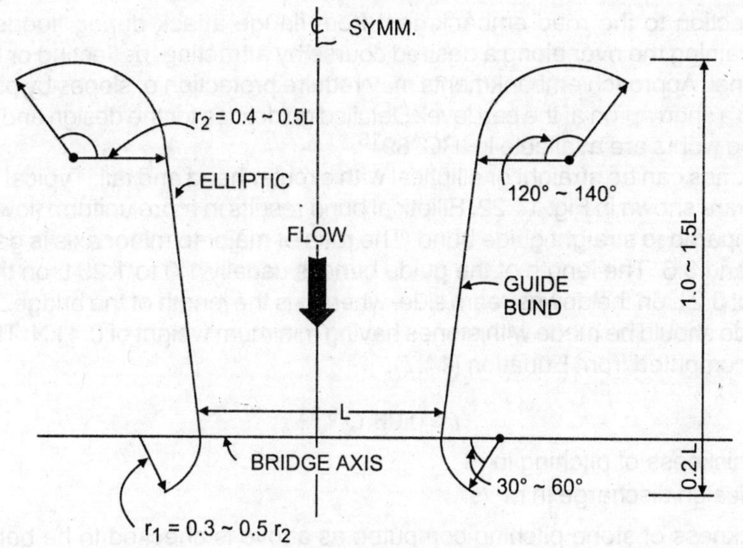

(a) Geometrical shape

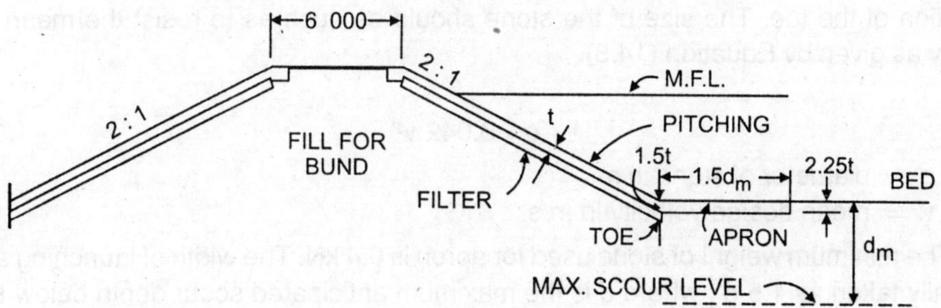

(b) Cross section of bund

Figure 14.22 Details of Guide Bund.

than 3.5 m into the approach. The top of the approach slab should conform to the cross profile of the top of the deck slab. The slab has a minimum thickness of 300 mm at the ends with the maximum thickness adjusted to suit the cross camber. Typical details of the approach slabs for the case of bridge without footpaths and for the case of bridge with footpaths are shown in Fig. 14.21[17]. It may be noted that the approach slab is a heavily reinforced slab having reinforcement 12 ϕ 150 both ways at top and at bottom.

14.17 River Training Works

In case of major bridges across wide rivers, river training works such as guide bunds, spurs and approach road protection works may sometimes be required. Guide bunds are provided to channel the flow of flood waters in the river towards the ventway of the bridge and to afford protection to the road embankment from flange attack during floods. Spurs are provided for training the river along a desired course by attracting, deflecting or repelling the flow of a channel. Approach embankments may require protection of slopes by pitching along the slopes and a short apron at the bed level. Detailed guidelines for the design and construction of river training works are available in IRC: 89[18].

Guide bunds can be straight or elliptical with circular head and tail. Typical details of an elliptical bund are shown in Fig. 14.22. Elliptical bund results in more uniform flow through the bridge as compared to straight guide bund. The ratio of major to minor axis is generally kept between 2.0 and 3.5. The length of the guide bund is usually 1.0 to 1.25 L on the upstream side and about 0.2 L on the downstream side, where L is the length of the bridge. The pitching on the river side should be made with stones having minimum weight of 0 4 kN. The thickness of pitching is computed from Equation (14.7).

$$t = 0.06\, Q^{0.33} \qquad \text{... (14.7)}$$

where t = thickness of pitching in m

Q = design discharge in m^3/s.

The thickness of stone pitching computed as above is checked to be between 0.3 m and 1.0 m. A filter is provided under the slope pitching to prevent the escape of embankment material through the voids in the pitching.

An apron (known as launching apron) is provided at the toe of the river side slope for the protection of the toe. The size of the stone should be such as to resist the mean design velocity as given by Equation (14.8).

$$d = 0.042\, v^2 \qquad \text{... (14.8)}$$

where d = diameter of stone in m

v = mean design velocity in m/s.

The minimum weight of stone used for apron is 0.4 kN. The width of launching apron is generally taken as 1.5 d , where d is the maximum anticipated scour depth below the bed level. The thickness of the apron is kept at 1.5 t at the inner end and at 2.25 t at the outer end, as shown in Fig. 14.22.

River training works discussed here involve heavy investment and may be cost effective only for bridges across wide rivers. They should be adopted only after careful studies on their technical feasibility and economic feasibility. After completion of the works, proper periodical inspection should be ensured to facilitate timely remedial action in case of distress.

14.18 References

1. Long, J.E., 'Bearings in structural engineering', Newnes-Buttersworth, London, 1974, 162 pp.
2. 'IRC:83-1982 Standard specifications and code of practice for road bridges, Section IX, Bearings, Part I: Metallic Bearings', Indian Roads Congress, 1982, 27 pp.
3. 'IRC:83 (Part II) - 1987 Standard specifications and code of practice for road bridges, Section IX, Bearings, Part II: Elastomeric Bearings', Indian Roads Congress, 1987, 29 pp.
4. Victor, D.J., 'The investigation, design and construction of submersible bridges', Jl. Indian Roads Congress, Vol. XXIV-1, October 1959, pp. 181-213.
5. Dillon, R.M., and Edwards, P.H.D., 'The inspection, repair and maintenance of highway bridges in London, Ontario', The Engineering Journal, Vol. 44, No. 11, Nov. 1961, pp 39-48.

6. Adke, A.S., and Chidambaran, C., Design and construction of Udyavar bridge', Jl. Indian Roads Congress, Vol. XXV-3, pp. 235-292.
7. Patil, Y.K., and Kand, C.V., 'Design and construction of prestressed concrete bridges across Kunwary river', Jl. Indian Roads Congress, Vol. XXIV-2, Nov. 1959, pp. 286-333.
8. Lee, D.J., 'The theory and practice of bearings and expansion joints for bridges', Cement and Concrete Association, London, 1971, 65 pp.
9. Nambiar, K.K., and Nambiar, P.G., 'Design and construction of a prestressed concrete bridge across river Coleroon', Jl. Indian Roads Congress, Vol. XXI-2, 1956, pp. 149-263.
10. 'IS:1076-1967, Indian standard specification on preferred numbers', Indian Standards Institution, New Delhi, 1967.
11. Ministry of Transport (Roads Wing), Govt. of India, 'Technical circulars and directives on National Highways and centrally sponsored road and bridge projects, Circular No. II-86(108)/66 dated July 7,1969', Published by Indian Roads Congress, New Delhi, Vol. II, pp. 1620/1-3.
12. 'IRC:SP:69-2005 Guidelines and specifications for expansion joints', Indian Roads Congress, New Delhi, 2005, 35 pp.
13. Watson, S.C., 'A concept of pre-engineered, prefabricated, prestressed modular and multi-modular sealing systems for our modern bridges and structures', Lecture presented to AASHO Regional Meetings, 1969, Acme Highway Products Corpn., Buffalo, N.Y. (unpublished).
14. Watson, S.C., 'Joint sealing practice for longer spans', Lecture presented to AASHO Regional Meetings, 1967, Acme Highway Products Corporation, Buffalo, N.Y. (unpublished).
15. Ministry of Road Transport and Highways, Government of India, 'Empanelment of suppliers for expansion joints', Feb. 2003.
16. WBA Corp, 'Wabo modular joint systems – STM and D Series', Watson Bowman Acme Corp., Amherst, N.Y.
17. Ministry of Transport (Roads Wing), Government of India, ' Standard plans for road bridges', Published by Indian Roads Congress, New Delhi, Vol. I, 1992, and Vol. II, 1993.
18. 'IRC:89-1985 Guidelines for design and construction of river training and control works for road bridges', Indian Roads Congress, 1997, 90 pp.

Chapter 15

Construction and Maintenance

15.1 Construction Engineering

Construction engineering in the context of bridge construction comprises design, construction operations and project management. The construction phase of a large bridge is more vulnerable to accidents and failure than the service phase, due to inadequacies in implementation of construction engineering. It is therefore worthwhile to allocate adequate efforts and resources to plan construction operations properly in advance of the work.

Construction project management is an integral part of construction engineering. It includes project design, planning, scheduling and controlling. The project design team produces drawings, specifications and special provisions for the bridge project. Planning involves analysis of the scope of work, selection of construction techniques, listing the equipment required, and determining the categories and the number of the labour force to be employed. Scheduling specifies the sequence of operations and allocation of manpower and equipment at the different stages of the project. Controlling comprises supervision, procedural instructions, maintenance of records, quality control and monitoring of costs.

The construction manager needs to apply leadership and management skills along with engineering knowledge. He should aim to so arrange his activities as to : (i) achieve even progress according to stipulated target dates; (ii) produce good quality work; (iii) avoid accidents at site; and (iv) effect economy by completing the work at low cost. His efficiency in achieving the above will depend on his level of competence in men-management and in materials-management. Maintaining a contented labour force adequate to the requirements of the job and motivating the labour to get the job done are the major factors in men-management. Materials-management requires the art of procuring materials of the right quality at the right price, to the right place, at the right time, besides preventing waste and theft of materials. Since bridge construction is a hazardous activity, the construction manager should be vigilant to prevent accidents and injuries.

On major bridge construction projects, it would be advantageous to apply proven management techniques to construction operations. Bar chart method and the Critical Path Method (CPM) are the most commonly adopted techniques. The CPM is the schematic representation of a project by means of a 'network' diagram, depicting the sequence and the interdependence of the various activities and events. The diagram helps in determining the most suitable programme for the implementation of the project leading to least cost consistent with the desired project duration. For most normal bridges, the use of bar chart programming would be adequate. Only in very large works, e.g. Ganga bridge at Patna and Pamban bridge at Rameswaram, the application of network techniques such as the CPM would be worthwhile. However, the basic logic of most of these approaches will be found to be good commonsense and should be borne in mind during the construction operations. Network operations are particularly valuable in working out initially the logical sequence of operations for any undertaking.

It is usual in bridge construction that a main contractor gives out a part of the work (even up to 30 per cent) on subcontract, e.g. foundation piling, formwork and falsework, etc. The subcontractor should also be motivated to ensure the stipulated progress, quality, safety and economy.

15.2 Construction Method Affects Total Cost

The final cost of a bridge is the sum of the cost of permanent materials, the proportionate cost to the project of plant and temporary works and the cost of labour. The cost of permanent materials can be estimated reasonably correctly. With experience, a bridge contractor can deal competently with the cost of plant and temporary works. But the labour cost does not lend itself to exact analysis. Recent competitive designs have attempted to introduce innovations in construction methods with a view to effect economy in the cost on labour by reducing temporary works and by minimizing the duration of site work.

The suitable techniques of construction of bridge superstructure will vary from site to site, and will depend on the spans and length of the bridge, type of the bridge, materials used and site conditions. For instance, cast-in-situ concrete construction could be adopted for short spans up to 40 m, if the river bed is dry for a considerable portion of the year, whereas free cantilever construction with prestressed concrete decking would be appropriate for long spans in rivers with navigational requirements. The current trend is towards the avoidance of staging as much as possible and to use precast or prefabricated components to the maximum extent. Also, construction machinery such as cranes and launching girders are coming into wider use. In the case of bridges across wide rivers, considerable saving in construction cost has been achieved by innovative use of barge mounted cranes for erection of one span at a time, as in the Delhi Noida bridge, and by adopting the incremental push launching method as in Yamuna bridge for Delhi Metro. There are greater savings to be effected by paying attention to the method of construction even from the design stage than by attacking permanent materials.

15.3 Quality Assurance for Bridge Projects

The performance of a bridge is dependent on the strength and durability of its components. These in turn depend on the quality attained at various stages of development from planning, design and construction to maintenance to meet the needs of the users. Quality in bridge engineering has to be achieved through innovative planning, diligent design, intelligent direction, competent construction and timely maintenance. For major bridges, it is desirable to prepare a quality assurance manual for compliance during design and construction stages. The most effective method of reducing maintenance costs is the implementation of an efficient quality assurance procedure during the initial construction. The designer should ensure that the structure could be built with ease and reliability under the prevailing site conditions. In the case of reinforced concrete and prestressed concrete bridges, special attention should be devoted to the "4-Cs", i.e., Constituents, Compaction, Cover and Curing.

Quality Assurance (QA) includes all those planned actions necessary to provide adequate confidence that the product (in this case, the bridge) will meet the requirements, and is essentially a system of planning, organizing and controlling human skills to assure quality[1]. Quality Control (QC) deals with operational techniques of controlling quality.

Four classes of quality assurance are specified: (i) Q-1 Nominal QA; (ii) Q-2 Normal QA; (iii) Q-3 High QA; and (iv) Q-4 Extra High QA. For bridge structures, only the three classes Q-2, Q-3 and Q-4 are acceptable. Class Q-2 may be adopted for bridges in reinforced concrete up to 60 m length with individual spans not exceeding 20 m. Class Q-3 may be adopted for reinforced concrete bridges and prestressed concrete bridges having length more than 60 m with individual spans not exceeding 45 m. Q-4 class is applicable for bridges with innovative designs or materials or construction techniques. The requirements of quality control and degree of control for the different classes of quality for project preparation, design and drawings, contractual aspects, construction organization, materials and workmanship are specified in IRC:SP:47-1998[1]. The compliance of the guidelines in the above publication will be a step towards obtaining ISO: 9000 Quality Certification.

15.4 Bridge Construction Inspection

Inspection of activities during bridge construction aids better quality assurance and promotes safety. Construction inspection includes checking of materials, operations to produce various components, and the temporary structures such as shoring systems, formwork and falsework. Material inspection should cover checking of the concrete, reinforcement and structural steel for conformance to specifications. Operation inspection relates to ensuring that the structure is being built at the correct locations, alignments and elevations according to the project plans. The inspector should ensure that the ready-mixed concrete meets the specifications and it is placed in the forms with proper vibration and consolidation. Reinforcements should be placed as required in the plans in terms of grade, size and location. Precast mortar blocks of proper quality and dimensions should be used to ensure correct cover.

The inspector is responsible for the quality assurance inspection of all welding. The welding equipment, procedures and techniques should be in accordance with the relevant specifications. Component inspection involves checking of the various components during construction for dimensions and finish. Temporary structures need special attention to prevent distortions to final structures. During placement of concrete, formwork should be inspected to prevent excessive settlement and distortion of bracings. The forms should be mortar-tight and should be strong enough to prevent excessive deflection. Safe working environment and practices should be maintained at the construction site. In addition to construction inspection, the inspector should also maintain an accurate record of work performed by contractors.

15.5 Construction of Short Span Bridges

For bridges involving spans up to about 40 m, the superstructure may be built on staging supported on the ground. Alternatively, the girders may be precast for the full span length and erected using launching girders or cranes, if the bridge has many equal spans. In the latter procedure, the additional cost on erection equipment should be less than the saving in the cost of formwork and the labour cost resulting from faster construction. Precast concrete bridge construction facilitates speedy erection, elimination of obstructive formwork, design standardization and achievement of high quality. Hence it is one of the most favoured construction techniques for bridge decks of small and medium spans.

An example of efficient site organisation using precast prestressed girders and special erection procedures is the Sone bridge at Dehri comprising 93 spans of 32.9 m each[2]. Here the beams for the superstructure were precast in a casting yard at one end of the bridge.

After prestressing, each beam in proper sequence was loaded on a tractor-trailer by a travelling gantry, moved to the span by the tractor-trailer and picked up and placed in position by a launching gantry. The repetitive nature of work and the extensive use of precast components carefully incorporated in the design, and the imaginative use of special machinery, helped to cut down construction time considerably.

15.6 Steel Bridge Construction

Typical techniques of construction of steel bridges are shown schematically in Fig 15.1[3]. A truss bridge can be assembled and erected on staging as in Fig. 15.1 (a). Continuous plate girder bridges may be erected by connecting prefabricated segments, with partial support on staging as in Fig 15.1 (b). A similar scheme suitable for a two-span continuous truss bridge is indicated in Fig 15.1 (c). Cantilever construction for a truss bridge is shown in Fig 15.1(d). The sequence of construction of the rib of an arch bridge using temporary cable support is evident from Fig 15.1 (e). In the case of a long span bridge across a river, it may be convenient to use barges to move to the site prefabricated portions of the bridge for connecting on to the bridge portion already constructed. The erection procedure chosen for any given site would depend on many factors, such as the bridge type, span, height, hydraulic conditions of the river, navigational requirements, and the erection equipment available.

In steel work erection, the plant required should be selected carefully. Due to economic reasons, existing plants in a firm should be used to the maximum extent. To effect economy, conscious effort should be made to ensure that the weight of every piece to be handled should be close to the safe loads indicated on the crane for the particular radius. The location of the plant during construction may have a significant effect on the erection stresses on the bridge members. Therefore, very careful thought must be devoted to provide for additional temporary supports or strengthening to cater to erection stresses, if necessary.

The permissible tolerances must be watched closely, especially in the following four areas: (i) the accuracy of the geometry of the bridge in the horizontal plane; (ii) the above accuracy in the vertical plane; (iii) the accuracy of matching of components at a splice; and (iv) the control of local deformation at a splice, e.g. a welded detail. The centre line and the pier positions should be established very carefully. Bearing locations should normally be placed within ± 12 mm longitudinally and within ± 3 mm transversely, and within ± 5 mm of the centre line of diaphragms and stiffeners. Levels at supports should be maintained closely, especially for continuous spans.

Very often, the pattern of stresses caused in the members of a structure during erection may be reversed with reference to the service load stresses. For example, when an arch rib is supported by temporary cables from above during erection, special stress conditions may develop. Hence the designer must check the strength and stability of the bridge members at every critical stage of erection.

The steelwork in bridges needs to be painted in order to afford protection against corrosion. The paints and the painting procedures used will depend on the nature of exposure, type of steel used, form of steel surface preparation and whether the coating is prime coat or coating over prime coat. The surface preparation may be by hand brushing, mechanical brushing, flame descaling or sand blasting. The first coat is intended to render the steel surface passive towards oxidation. The primers generally used are red lead, white lead, lead chromate and zinc chromate. The intermediate paints should be chosen on their qualities

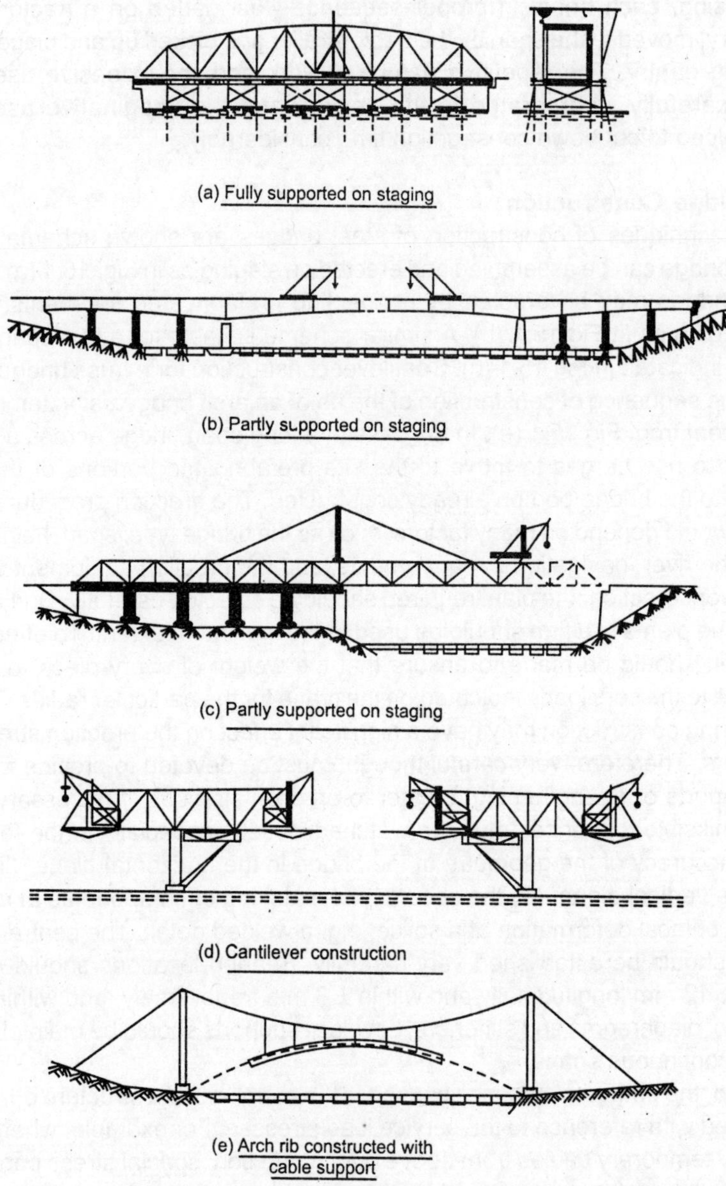

Figure 15.1 Typical Steel Bridge Construction Techniques.

of strength, durability, impermeability and tolerance to surface conditions. These paints include white lead, red iron oxide, red lead graphite and silica graphite. Final coat should be by paints which have additional qualities of water repellence, abrasion resistance and decorative finish. Metallic coatings such as aluminium, zinc (galvanizing) and lead are also used extensively.

15.7 Construction of Continuous Concrete Bridges

15.7.1 Methods of Construction

Long span concrete bridges are usually of post-tensioned prestressed concrete and are constructed either as continuous beam types or as free cantilever structures. Many methods have been developed for continuous deck construction, including: (a) cast-in-place construction on staging; (b) segmental cantilever construction using cast-in-place or precast segments; (c) span-by-span method; (d) incremental launching method; or (e) precast girders made continuous by cast-in-place slab and diaphragm. If the clearance between the ground and the bottom of deck is small and the soil is firm, the superstructure can be built on staging. This method is becoming obsolete. Currently, free-cantilever and movable scaffold systems are increasingly used to save time and to improve safety.

The movable scaffold system employs movable forms stiffened by steel frames. These forms extend one span length and are supported by steel girders which rest on a pier at one end and can be moved from span to span on a second set of auxiliary steel girders. The construction joint is placed at the point of inflection for dead load moment which is about 0.2 L from the pier. A similar but simpler arrangement using steel lattice work auxiliary supports is shown in Fig. 15.2[4]. After concreting, curing and stressing, the form for deck slab is first

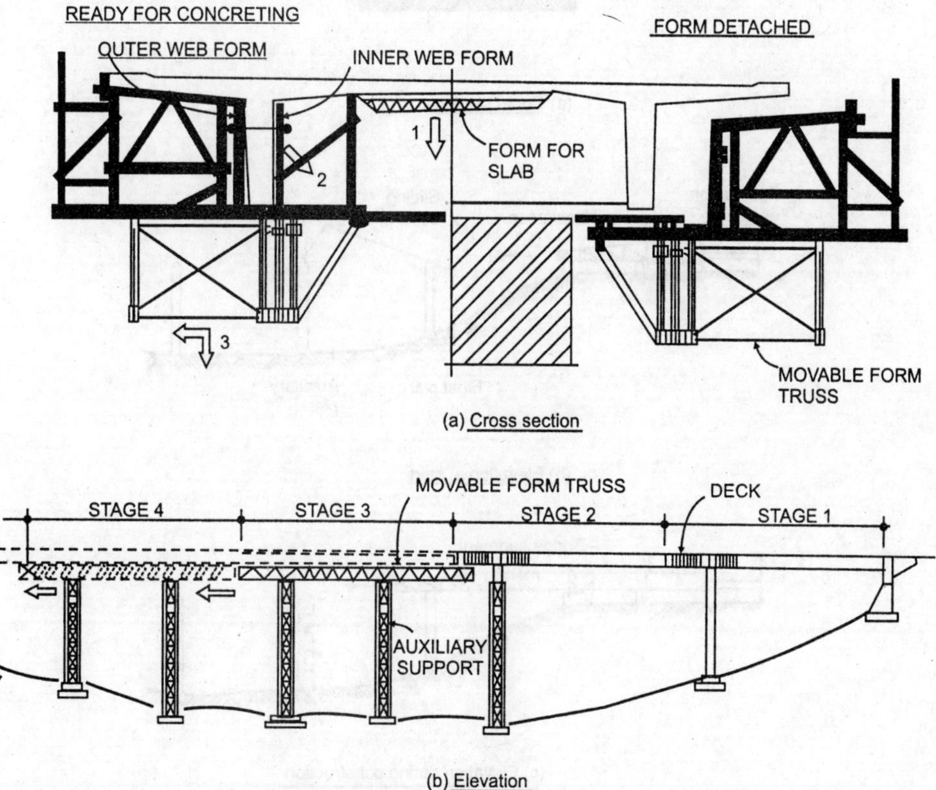

Figure 15.2 Continuous Deck Construction Using Movable Forms.

lowered. The inner web form is next loosened and rotated to free it from the web. Lastly, the outer web form and the form truss are lowered and shifted outward. The form truss is then moved lengthwise to the next stage as shown in the elevation to prepare for the next set of operations.

15.7.2 Incremental Push Launching Method

An economical construction technique known as 'incremental push-launching method' developed by Baur-Leonhardt[5] team is shown schematically in Fig.15.3. It was first used in

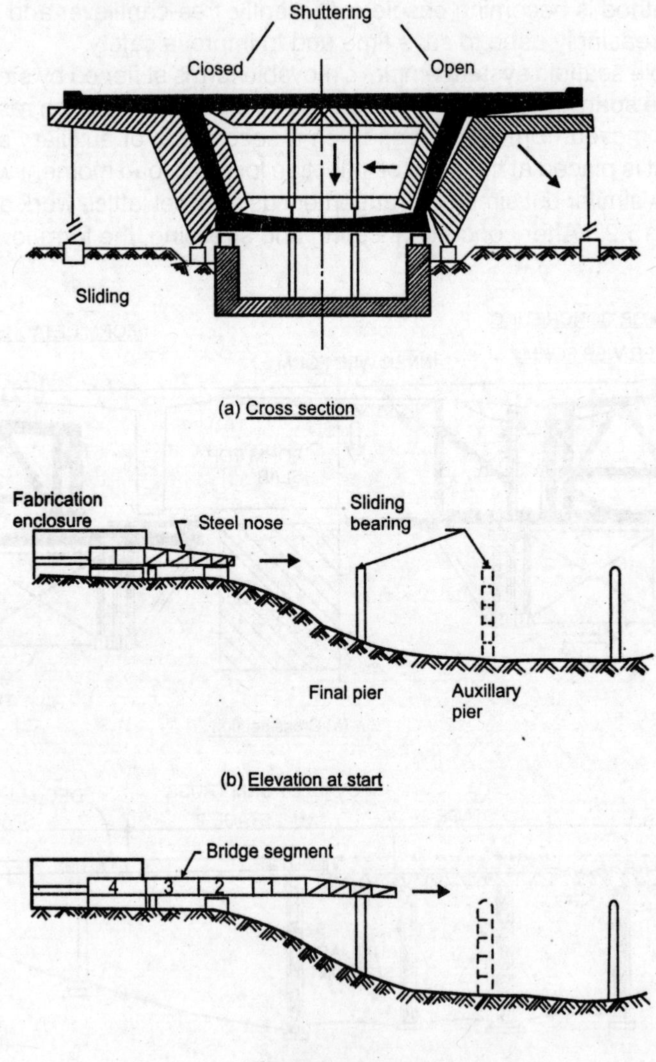

Figure 15.3 Incremental Push Launching Method.

1961 for the Rio Caroni bridge in Venezuela and has since become very popular. This highly mechanized erection method is particularly suitable in situations where construction of falsework is not feasible, such as crossing over busy traffic corridors, or spanning deep gorges or where the pier heights are large.

The total continuous deck is subdivided longitudinally into segments of 10 to 30 m length depending on the length of spans and the time available for construction. Each of these segments is concreted immediately behind the abutment of the bridge in steel framed forms, which remain in the same place for concreting all segments. The forms are so designed as to be capable of being moved transversely or rotated on hinges to facilitate easy stripping after sufficient hardening of the concrete. At the head of the first segment, a steel nose consisting of a light truss or stiffened plate girder braced assembly is attached to facilitate reaching of the first and subsequent piers without inducing a very large cantilever moment during construction. The optimum length of the launching nose is about 60% of the main span. The second and the following segments are concreted directly on the face of the hardened portion and the longitudinal reinforcement can continue across the construction joint. The pushing is achieved by hydraulic jacks which act against the abutment. Temporary sliding bearings are used on each pier and in the casting area to launch the deck. These bearings consist of a laminated elastomeric pad covered with a stainless steel plate, over which a teflon pad is fed to contact with the deck, providing a low-friction sliding surface. Since the coefficient of friction of teflon sliding bearings is only about 2 per cent, relatively low capacity hydraulic jacks would suffice to move the bridge even over long lengths of several hundred metres.

The incremental launching method can be used for straight and continuously curved bridges for spans up to 60 m and with temporary (auxiliary) piers up to a span of about 100 m. The girders must be of constant depth. When the spans are more than 15 times the girder depth, auxiliary piers are employed with a view to reduce the bending moments in the superstructure during launching. For a launched bridge, the optimal cross section is the single-cell box girder. The deck slab thickness should be carefully determined in order to keep the amount of prestressing at an optimal level. The web should be thickened at the junction with the top slab to accommodate tendon couplers and at the junction with the bottom slab to carry safely the reactions of the launching bearings. The bridge geometry should be controlled carefully during casting and launching. Longitudinal sloping grades can be accommodated, the launching being in the downward direction.

The incremental launching method facilitates the combination of the advantages of cast-in-place and precast concrete construction. The place of fabrication behind the abutment can be covered with a light-weight roof to permit continuous working independent of the weather. The fabrication area accommodates the formwork, concrete mixing plant, a rail-mounted tower crane, storage area for reinforcing and prestressing steel and the jacking equipment. The transport distances are very short. The concentration of equipment facilitates achievement of close factory conditions leading to high quality of construction. The simple and repetitive steps involved help to achieve quality work with reduced amount of labour and favourable speed of construction. A variant of the above method is to use pulling of the segments instead of pushing. With efficient site operations and control, modern constructions try to achieve a "Seven Day Cycle" for casting and launching a segment[6].

During launching, each section is subjected to continually alternating bending moments. The section will have a sagging moment while at midspan or be subject to hogging moment

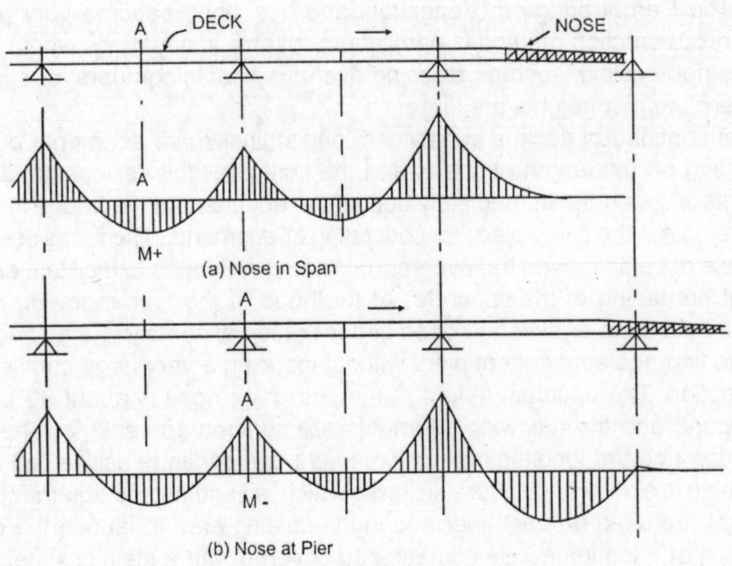

Figure 15.4 Variations in Bending Moments During Launching.

when passing over a pier as indicated for Section A in Fig. 15.4. The design of prestress should cater to the full range of moments. It is normal to provide the first stage prestress (called central prestressing) to induce a uniform compressive stress of about 5 MPa, by straight tendons placed at the top and bottom slabs, anchored at the construction joints between segments and extended to new segments by couplers[7]. On completion of launching, a second stage prestressing (called continuity prestressing) by draped cables is performed to cater to the final moments. The slight increase in the uniform prestress as required for launching helps in enhancing durability during the service life[6].

Special care should be devoted to the location of the temporary bearings on top of the piers, and their effect on the box section as the segments move during launching. The bearings should invariably be positioned below the junction of the webs and the soffit slab. The inadvertent placement of these bearings between the junctions to bear on the soffit slab has been the major cause of failure of the Injaka bridge in South Africa [8]. The design of the pier should provide for additional horizontal load on top of the pier due to friction in the temporary bearings during launching. The abutment should be designed to take the large horizontal forces towards the end of the launching operation.

The first application of the incremental push launching method in India was in the Panvel Nadi viaduct built in 1994 as part of the Konkan Railway project[9]. The launching technique was chosen here because of the deep valley site, which involved pier heights ranging between 20 m and 64 m. The bridge superstructure consists of a single-cell continuous prestressed concrete box girder with nine intermediate spans of 40 m each and two end spans of 30 m each. The bridge was cast in 20 m segments at one end in a casting yard. A steel launching nose of 30 m length was attached to the first segment to reduce the stress in the box girder during launching.

Other applications of this method in India include the new bridge on river Markanda at km 182 of NH-1 in Haryana[10] and the Yamuna bridge on Delhi Metro Railway. The superstructure of Markanda bridge is of single cell continuously prestressed box girder, consisting of 14 segments of 27.2 m each, making up 8 spans (1 x 27.2 + 6 x 54.4 + 1 x 27.2 m) for a total length of 380.8 m between centers of abutment bearings. The Yamuna bridge also has a single cell continuous box girder for a length of 550 m with main spans of 46.2 m.

15.7.3 Segmental Cantilever Construction

The segmental cantilever system, particularly suited for construction of prestressed concrete superstructure of long span bridges without the use of staging from below, has already been discussed in Section 8.10. The segments may be either cast-in-place or precast. The cast-in-place method uses a cantilever form traveller (CFT). When precast segments are used, they are erected with the use of launching trusses and cranes. For a detailed treatment of the design and construction aspects beyond the discussion in Section 8.10, the reader is referred to the Reference 11.

An innovative concept of economic construction of wide precast concrete deck by the cantilever method is to split the deck to be constructed as a spine of box section and two cantilevers[12]. The method permits handling of precast segments of manageable weights assembled using longitudinal and transverse prestressing. The spine is prestressed to take longitudinal moments and the transverse action is catered predominantly by membrane action.

15.7.4 Continuity Through Deck Slab

There are applications particularly in urban flyovers where bridges built with simply supported precast prestressed girders are made continuous through the cast-in-situ deck slab[13,14]. Two generic types of connections are suggested in IRC:SP:66-2005. In Type 1, the deck slab is continued monolithically over the piers, without continuing the girders. The deck slab is separated from the girders for a length of about 1.0 m on either girder by introducing neoprene pads of about 12 mm thickness. This provides flexibility to accommodate the rotation of the girders. In Type 2, the deck slab is hinged over the pier using partly debonded dowelling. The purpose of introducing continuity in such situations is to reduce the number of expansion joints and to enhance riding quality. This arrangement permits rotation but does not provide moment continuity.

15.8 Formwork and Falsework for Concrete Bridges

15.8.1 Formwork is the mould required to cast concrete to a desired shape. It is used to retain plastic concrete until the concrete has hardened. It is designed to resist the fluid pressure of plastic concrete and additional pressure caused by vibrators. It can be removed when the concrete hardens. Formwork may be constructed out of timber or steel plates. The total temporary structure system including formwork required to support concrete construction for the permanent structure until it becomes self-supporting is known as falsework. Scaffolding denotes any temporary elevated platform and its supporting structure used for supporting workmen and materials during construction. Scaffolding may be constructed of sawn timber, wooden poles (e.g. casuarina) or steel or aluminium tubes. Guidelines for the design and erection of false work are contained in IRC: 87 and additional particulars are available in standard texts[15-19].

15.8.2 The economical design and construction of formwork is essential for the efficient execution of concrete bridges. The appearance of the finished structure is mainly dependent on the quality of formwork. While great care is bestowed on the design calculations for the permanent structure, relatively scant attention is being paid to the design and erection of formwork and falsework. The author would recommend the adoption of German practice, which requires detailed calculations for all temporary works to be submitted by the designer/contractor prior to commencement of work for approval by the engineer-in-charge of the work.

15.8.3 The formwork for a concrete structure involves three distinct operations: manufacture, erection and stripping. Each of these operations should be given careful consideration to achieve satisfactory results. Proper coordination and communication among the design, construction and supervision agencies should be ensured in respect of all aspects of formwork. Timber forms are usually manufactured in the carpentry shop at the bridge site. Steel forms are best fabricated in a factory to close tolerances to ensure interchangeability. Special care should be given to the sequence of erection. The forms are erected first approximately to line and level; then they are correctly aligned, levelled with wedges and kept firmly intact by tightening the braces and ties. The forms should be leak proof, as leaky forms result in honeycombed concrete. Stripping should be done carefully to avoid injury to concrete or damage to the forms. Soon after stripping, the forms should be cleaned, oiled and stored for re-erection. The form bolts, nuts and washers should be cleaned and kept in buckets for reuse. Wedges should be cleaned and stacked. Proper organisation of stripping operations would lead to reduction of wastage.

15.8.4 Inspection of formwork just prior to and during concreting is very important. Prior to concreting, the formwork should be inspected by a competent engineer capable of quickly assessing the adequacy of the forms, and trained to detect the inadvertent omissions of tie wires, bolts and brace tightening. The lines must be true; and bulges and sags must be prevented. During concreting, it is good practice to detail an experienced carpenter to watch the forms, tighten the braces and be on the alert for any emergency.

15.8.5 The time interval between casting and removal of forms depends on many conditions, including the ambient temperature of the air, the setting time for the cement used, nature of stresses induced in the member and span. For normal guidance, the following may be taken as the minimum time for removal of shuttering with ordinary portland cement concrete and normal curing under normal weather[15].

	Days
Beam sides, walls, columns (unloaded)	2
Slabs (props left under)	3
Removal of props to slabs	14
Beam soffits (props left under)	7
Removal of props to beams	21

In case the atmospheric temperature reduces below 15° C, the time shown above should be increased.

15.8.6 The following factors should be given careful consideration while planning and designing formwork.

(a) *Strength*: The formwork should be capable of carrying the pressure of concrete and the weight of labour and plant engaged in its placement and compaction. The pressure due to concrete will vary depending on whether the formwork is horizontal or vertical. The factors affecting the pressure are: density of concrete, workability of the mix, rate of placing, method of concrete discharge into the forms, temperature of the concrete, extent of vibration, height of lift, dimensions of the section cast, reinforcement details and stiffness of the formwork structure. In the absence of detailed calculations, the pressure can be calculated as that due to a liquid weighing 26 kN/m^3 for horizontal surfaces and for vertical surfaces up to depth of 1.8 m. For vertical surfaces deeper than 1.8 m, the stress may be increased at 4 kN/m^2 for every additional metre. Construction loads may be taken approximately at 3.6 kN/m^2 acting vertically.

(b) *Stiffness* : The forms should be rigid enough to ensure that deflection of the completed work should not exceed 0.003 of the span and that the deflection of the form itself in any one span should not be more than 3 mm.

(c) *Repetition* : The forms should be designed in such a manner that components are easy to handle and will be reused a number of times in the same work. Since formwork cost is a considerable part of the total cost of concrete work, this aspect requires careful analysis.

(d) *Durability* : In order to ensure maximum economy, it is essential to provide for repetitive uses of the formwork. Hence careful consideration should be given to the use of durable materials so that the formwork can be used and handled without undue wear. With proper handling and care, timber form-panels should give about ten repetitive uses without major repair. Many more reuses should be possible with steel forms.

(e) *Strippability* : The ease of stripping without damage to the concrete or the forms is a requirement deserving special attention. Wedges and special insertions of smaller closing pieces are arranged to facilitate removal of forms from enclosed spaces. Unless easy stripping is ensured, the gains due to repetitive use of forms may be lost in costly repairs.

(f) *Cost* : The final cost of forming an area of concrete is the sum of the cost of materials for the forms, the cost of labour in erecting, stripping, cleaning and carrying forward to next use, and the cost of expendable material such as form ties. The aim in design is to keep the total cost to a minimum. It is usually difficult to estimate the cost of formwork reliably. With the universal rise in prices and wages, it would be well for a contractor to study the labour content of the formwork cost.

15.8.7 Innovative designs for formwork for bridge components lead to economical and efficient construction. For tall piers, slip form technique is advantageous, e.g. the tapering hollow octagonal piers involving heights up to 64 m for Panvel Nadi viaduct and the 80 m tall pier of Jammu-Udhampur Railway bridge.

15.8.8 Falsework may be defined as the temporary support work necessary to support a portion of a permanent structure during erection until it is capable of supporting itself. By normal usage, falsework refers to the part of the temporary structure on which the formwork rests. Bridge falsework may be of either conventional systems or proprietory systems. Conventional systems use falsework constructed of sawn or round timber, steel joists, scaffold tubing and include also temporary piles. Special provision should be made to facilitate easy dismantling after the completion of the permanent structures. To this end, wedges and jacks are incorporated in the falsework design at a number of suitable points. When heavy loads are expected as in falsework for arch bridges, resort may be made to sand boxes, which are metal cylinders with plungers holding the falsework and with a plugged hole through which sand can be allowed to flow out when the falsework is to be lowered. The falsework should be designed to carry its own weight, construction dead and live loads, the dead load due to the permanent structure, wind load and water pressure, if applicable. Permissible stresses for falsework may be higher than those for permanent structures, if the design is carried out in full detail and proper site supervision can be ensured; conversely, if the design is sketchy and the supervision not very efficient, it is desirable to use stresses lower than those for permanent construction. Props supporting formwork should be checked for verticality by plumbing, and no prop should be erected more than 150 mm (1 in 40) out of plumb. No runner shall be placed with its centre line more than 25 mm eccentric to centre of the prop head. Proprietory systems involve metal components assembled into modular units.

15.8.9 Falsework is especially important in the construction of arch bridges, where it is often referred to as centering. The centering for an arch is a load carrying structure and the success of an arch bridge construction depends on the design, construction and stripping of the centering. The types of centering normally used include trestles, bents and trusses, as shown diagrammatically in Fig. 15.5. Trestle centering, Fig. 15.5 (a to d), means supporting the arch by posts from the ground. This method is the least expensive, involves least deformation of the posts, gives the greatest salvage value of material and is the simplest to build. The posts

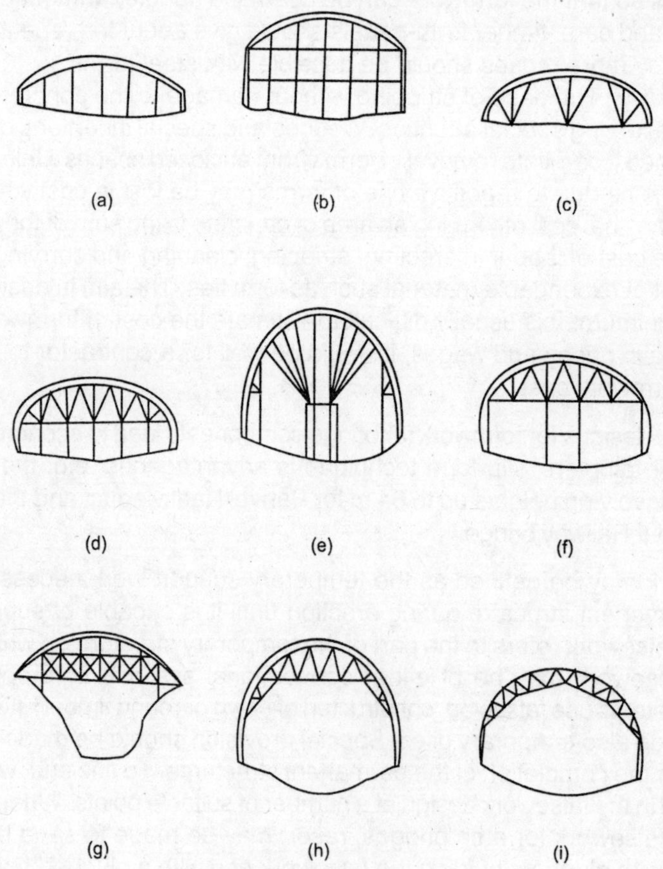

Figure 15.5 Types of Centering for Arches.

may be vertical as at (a) and (b), inclined as at (c) or combined as at (d). In high arches, the posts may be fan shaped with support at a height to avoid long inclined members, as at (e). This method also provides more clearance for passage of traffic. Trussed centres are used when simpler methods are not convenient. They are expensive to frame, low in salvage value and vulnerable for deformation. Hence, these have to be designed and executed under competent supervision of an experienced engineer. The foundations for the falsework should be checked for adequacy; and temporary piling may be used, if necessary. Striking of centering should normally be done only after three to four weeks.

15. 8.10 In the case of bowstring girder bridges, the deck forms are built first, supported on falsework. After concreting the deck, the vertical suspension members (hangers) are formed similar to columns. The arch members are formed as beams, using backforms near the ends. The arch centering is usually laid on the ground and the forms built accurately in sections and numbered to save time during erection. On completion of the whole bridge, the falsework may be removed.

15. 8.11 Tubular steel scaffolding, usually of 38 mm (1.5 in.) nominal bore, is being increasingly used for bridge centering. Its advantages include: adaptability, easy availability, simplicity in number of types of fittings, interchangeability in use for formwork or falsework, ease of erection, resistance to fire, reliable strength and good salvage value. However, it has a number of disadvantages as below: heavy first cost, costly replacement of damaged couplers and fittings, difficulty in connecting with timber formwork, possible adverse effects of human error in overlooking tightening of joints, need for additional skilled trade at the site and painting. Under normal circumstances, tubular scaffolding will be cheaper than timber centering for heights above 8 m. Each standard should rest on a sole plate. Screw jacks for use as sole plates under standards for formwork are available. For slab type bridges over river beds which are dry during summer, movable staging of tubular scaffolds on platforms with castors can be used.

15.9 Numbering of bridges

Cross-drainage works are numbered separately in each kilometre of road or railway. Each bridge or culvert is designated by a fraction in which the kilometre in which the structure is located is the numerator and the serial number of the structure within the kilometre is the denominator[20]. Thus, the third culvert in the forty-eighth kilometre (between kilometre stones 47 and 48) will be designated as 48/3 . This designation will consist of numerals of 100 mm height and will be inscribed in an area of 300 x 300 mm located near the top of the left hand side parapet wall as viewed in end elevation so as to face the traffic from each direction. If the structure has railings without end parapets, separate pillars are to be put up for inscribing the structure number. If any new culverts are to be introduced between any two existing ones, say between the 2nd and 3rd culverts in km 320, then the new culverts will be designated as 320/2A, 320/2B, etc.

15.10 Bridge Management System

15.10.1 Need for Bridge Management System

Bridges constitute key links in any road or railway system, spanning across gaps, natural and man-made. Their construction represents a very substantial national investment.

Their use involves public safety. Since bridges cannot resist indefinitely all the natural forces and hazards including time-related degradation of materials, these structures have limited service life. As the nation's bridges continue to age and as the imposed loads due to vehicles continue to increase, it is essential to develop better methods to evaluate the capacity of the existing structures and to evolve efficient and cost-effective retrofitting schemes. Bridge management aims to ensure that existing bridges remain in serviceable condition throughout their design life with the least risk of failure and at a minimum life cycle cost. The bridges have to be maintained in order to prevent premature failure and to extend the service life. The bridge design should provide for safe and easy access to the critical parts of the structure to facilitate inspection and maintenance. Since the life expectancy of an elastomeric bearing is much less than that of the bridge it supports, provision of a pedestal for the bearing to ensure good drainage and marking of adequately strengthened locations for jacking are elementary measures currently incorporated in modern designs.

A bridge may be considered unsafe under two conditions: (i) when serious damage or deterioration is detected in the main structural members (e.g. spalling of concrete with heavy corrosion of main reinforcement in longitudinal girders); and (ii) when an assessment based on calculation indicates inadequate load carrying capacity. In the first case, immediate remedial measures are warranted; in the second case, safety is essentially conceptual. Absolute safety is unattainable in structures, and inevitably there are risks of collapse and loss of life associated with any bridge. The aim in design and maintenance is to ensure the minimum acceptable level of safety for bridges during their whole life. Priority for use of resources should be devoted to the maintenance, repair and strengthening/replacement of those highway bridges with high volume of traffic and whose collapse would cause serious disruption to traffic network and to those bridges which on failure are likely to collapse suddenly without warning. The very large number of bridges existing in various states of aging due to environmental and vehicular traffic conditions requires competent management, besides modifications and rehabilitation. Successful rehabilitation of bridges calls for good engineering judgment, knowledge of the behaviour of the existing structure and experience in construction practices.

Bridge Management System (BMS) is an integral and essential part of the overall maintenance of the roads and railways. The components of BMS include: (a) Preparation of an inventory of all the bridges in the network; (b) Inspection and testing of the bridge components; (c) Structural and functional assessment; (d) Implementation of necessary repair and rehabilitation; and (e) Monitoring the bridges to ensure performance and safety. The above activities involve field inspection and testing, analytical studies and designs and computerized data management and systems analysis. BMS should serve as a strategic planning tool for policy level management and as a decision making tool for technical implementation.

The service life expectancy of a bridge may generally be about 70 years for the superstructure and about 100 years for the substructure. Bridge inspections reveal an alarming rate of deterioration of the nation's stock of bridges. The reasons for such a situation are many, including defective initial design, accidental damage, ageing, corrosive influences and overload due to increase in axle loads and volume of traffic. By adopting timely preventive and curative measures evolved under a well planned BMS, the whole life cost of new bridges can be reduced and the useful life of existing bridges can be extended.

Though the importance of proper maintenance cannot be over-emphasized, the practice of maintenance of bridges leaves much to be desired in most cases. An elementary precaution, sometimes ignored in practice, is to maintain the inside of the box section of a major prestressed concrete bridge clean without accumulation of debris and water. The obligation of assuring bridge safety and the conservation of the national investment on bridges rests with the public agencies that operate and maintain these bridges. In order to meet this responsibility, it is essential to evolve and implement suitable inspection and evaluation procedures [21-25]. The preparation of a maintenance manual for the bridge by the consultants/designer prior to opening of the bridge for use should be made mandatory for every long span bridge.

Each State Highway Department or Railway Zonal Administration should preferably have a separate cell for bridge inspection under the control of an experienced bridge engineer, capable of inspecting and pinpointing any trouble areas that could cause the abandonment, replacement or repair of a bridge. The cell should develop and maintain a proper data base and inventory record of all bridges in its jurisdiction, incorporating information on identification, administrative data, details on construction of foundation, substructure, superstructure and preventive works with detailed 'as constructed' drawings, besides data on approaches, ancillary and utility services. Details of bridge inspections, maintenance, repair and rehabilitation should also form part of the data base. The structural and functional assessment demands considerable engineering judgment and analytical expertise, especially when reliable 'as constructed' drawings are not available.

15.10.2 Inspection

The objective of bridge inspection comprises the monitoring and the evaluation of the performance of each bridge structure throughout its service life so that any deficiency in performance could be detected and corrected early. For performing this function, the design for any new bridge should be checked to incorporate adequate arrangements for easy accessibility of the bridge for inspection.

The inspections could be classified as (i) routine inspection; (ii) in-depth inspection; and (iii) special inspection. The routine inspection is particularly applicable to short span bridges. It usually involves a general examination of the structure, conducted on a regular basis, to look for obvious outward physical evidence of distress that might require repair or maintenance attention. An in-depth inspection requires a detailed visual examination of all superstructure and substructure elements. This is particularly necessary in the case of old bridges and structures of major proportions where structural failure could result in catastrophic consequence. The in-depth inspection may be scheduled once in three to five years. The special inspection is undertaken after special events such as earthquake, cyclone or passage of unusually heavy loads. For bridges subject to the constant action of the forces of nature, such as the daily ravages of a river, frequent expert inspection is essential. An inspection should be thorough and it must be conducted by knowledgeable investigators for the evaluation to be reliable.

The following equipment are required to be carried by the field staff for performing inspection of bridges: (a) Pocket tape; (b) Steel tape, 30 m; (c) Plumb bob; (d) Carpenter's level; (e) Chipping hammer; (f) Feeler gauge; (g) Crack width comparator; (h) Magnifying glass; (i) Wire brush; (j) Mirror; (k) Calipers; (l) Torch light; (m) Tool box; (n) Large screw driver; (o) Sounding line to measure water depth; (p) Nylon cord; (q) Steel scale; (r) Paint and paint brush; (s) Thermometer; (t) Shovel; (u) Binoculars; and (v) Single lens reflex camera. For major bridges requiring close inspection, additional equipment required are: (i) Ladders; (ii) Scaffolding; (iii) Truck mounted under-deck inspection platform, snooper type; (iv) Boats or barges; and (v) Diving accessories for checking scour.

Defects observed by visual inspection are only symptoms of more serious distress in the structure. In reinforced concrete bridge components, the usual defects noticed are cracking of various types, spalling of concrete and stainings. Cracks may be due to structural deficiency such as flexural cracks near the midspan or diagonal tension cracks due to shear near the supports. Corrosion of reinforcement may lead to cracking, spalling and staining of the surface. Crack widths can be measured by 'crack width comparator'. Based on crack width, cracks may be classified as fine (up to 0.3 mm), medium (between 0.3 mm and 0.5 mm) and wide (over 0.5 mm). While fine cracks are not serious, wide cracks need immediate repair. Medium cracks are to be monitored for widening and progression. A rust stain on the concrete member indicates corrosion of the reinforcing steel. The inspection report should include descriptions of the length, direction, location and extent of the cracks and rust stains.

For prestressed concrete decks, the occurrence of cracking, spalling and staining indicate serious problems. Cracks near the midspan may be due to loss of prestress, which may also result in excessive deflection. Inefficient bearings may result in cracking and spalling at the supports. Cracks at anchorages indicate inadequate detailing of reinforcement to cater to bursting forces. Special attention should be given to evidence of adverse effects of corrosion of prestressing tendons. Typically, corrosion starts at the anchorages and travels along the tendon.

The common locations of deficiencies are related to foundations, bearings, floor systems, connections and truss members. In general, the trouble spots to be checked are: (a) deterioration and cracks in concrete; (b) evidence of foundation settlement and movement; (c) metal work cracks especially at welds in tension members; (d) loose connections; (e) damaged members; (f) poorly framed structural details; (g) indiscriminate past repairs; (h) excessive vibrations; (i) distress near expansion joints; (j) inoperative expansion bearings; and (k) areas which have shown problems on other similar structures.

One of the usual defects in road bridge maintenance is the periodical addition of surface dressing resulting in dead loads much in excess of original design. Other areas of neglect include the bearings and expansion joints, which are often inoperative due to defective maintenance. Painting of steel bridges should be attended to properly.

In case of every major bridge, the inspecting engineer should perform a Structural Integrity Examination at least once in five years to evaluate the performance and adequacy of the entire foundation and the structural system. In addition, the inspecting engineer should personally inspect the bridge on the following schedule: (a) Once a year for a complete inspection from foundation up; (b) After each major flood, in case of river bridge, to examine effects of scour and changes in stream bed, banks and abutment fill; (c) During one high temperature period and one low temperature period each year to check the bearings for proper movement and joints for performance; and (d) After each accident on the bridge, to check for damages with a view to initiate immediate repairs. For major bridges across wide rivers, underwater inspection may be necessary to examine the channel bottom, to probe the extent of scour and silting, and by diving to visually inspect and measure bridge elements below water. The underwater inspector requires specialised training.

After inspecting a bridge, each major element of the bridge may be given a descriptive rating as Good, Fair, Poor or Critical [22]. The criterion for rating is to assess how well the element is fulfilling its intended function. The rating could also be given on a numerical scale ranging from 0 to 9. A rating of 'Good' corresponds to a numerical rating of 9, 8 and 7, and signifies that the component needs no repair. A component which may require minor or major rehabilitation in the near future is given a 'Fair' rating corresponding to 6 and 5. A rating of 'Poor' corresponding to a rating of 4 and 3 would indicate that the component needs immediate repair. A 'Critical' rating corresponds to a descriptive rating of 2, 1 and 0, and shows that the component is not performing its intended function. It is desirable to take colour photographs of the components under inspection so as to have a permanent record of the condition at the time of inspection. The inspector should recommend the corrective action required, listing the critical repairs first, followed by repairs to components in poor condition and attention to components in fair condition.

In order to serve the intended purpose of protecting public safety, the inspection programme should ensure that; (a) the evidence on any distress is noted intelligently; (b) the information is communicated to the appropriate authority for prompt action; and (c) efforts must be made to analyse the information in the light of the original design criteria and current standards. By close observation and careful study, the bridge inspector should develop the ability to discern between critical deterioration and cosmetic deficiencies, and to recommend the appropriate remedial measures.

15.10.3 Maintenance

Maintenance is the work necessary to keep a bridge in operating condition and to prevent potential deterioration in the future. Systematic maintenance is essential to ensure long term conservation of bridge structures. To be effective, maintenance should be based on adequate inspection, besides monitoring and assessment procedures. Maintenance solutions are normally evolved based on experience. There are four types of bridge maintenance: (a) Routine; (b) Preventive; (c) Repairs; and (d) Strengthening/Replacement. All the above options are important. The appropriate measure to be adopted at a given time for a particular bridge would depend on the circumstances. Thus preventive maintenance to protect against corrosion would only apply to a bridge where corrosion has not yet started. A combination of the methods would be needed in a case where the bridge shows corroding reinforcement: the actions to include repair the damage, arrest the corrosion process and to use a preventive measure to guard against further corrosion. The most effective way to reduce bridge maintenance costs is to include life cycle maintenance planning from the initial conceptual design through all phases of design, construction and operation, and to devote special attention to quality construction.

For steel bridges, the major maintenance effort is towards prevention of corrosion, caused primarily due to exposure of the steel to atmospheric conditions. Corrosion protection is generally achieved by providing a barrier coating of paint. While earlier practice has used multiple thin coats of paint containing lead and chromium, current environmental protection regulations do not permit such painting. Removal of lead-based paint from existing structures involves high costs to comply with requirements to contain all the hazardous waste and debris. The recent trend is to use protective coating over the existing coating with improved and environmentally safe coating systems.

15.10.4 Bearings

All bearings should be examined to ensure that they function as intended. Metallic bearings will cease to function properly unless they are maintained satisfactorily. The bearings have to be kept clean, free of dust and debris. The rockers, pins, rollers and sliding plates should be free of corrosion and should be lubricated periodically. Rattling of bearings under live load will indicate loose bearings. Tilt of rockers and horizontal travel of rollers will have to be monitored to check proper functioning.

Elastomeric bearings should be inspected to detect splitting or tearing either vertically or horizontally, which condition may be due to pads of inferior quality. The bearings should be kept free of dirt and debris. Also other conditions to look for are excessive bulging due to compression, ozone cracking, flattening and any abnormal deformation. If the bearings are deficient, they should be replaced by jacking the superstructure.

15.10.5 Expansion Joints

Expansion joints are to be checked for freedom of movement, proper clearance and correct vertical alignment. Substructure movements can cause closed or widely opened joints. Cracks are often found in the wearing coat in the neighbourhood of expansion joints. If such cracks occur, the reasons should be investigated and remedial measures taken up early.

Joint seals should be watertight to prevent water seeping to the bridge seat causing corrosion of bearings. Damage to joint seals may be caused by vehicle impact and accumulation of dirt and debris.

Finger type expansion joints and sliding plate joints should be checked for locking of joints, corrosion in sliding plate, loose anchorages due to breaking of welds and jamming of the joints due to debris or resurfacing material.

15.10.6 Maintenance of Submersible Bridges

Submersible bridges should be devoted greater maintenance attention than comparable high level bridges. The duration of submergence and number of traffic obstruction during every flood should be noted. The scour around foundations during and immediately after submergence should be recorded. The bearings should be inspected before and after submergence. When the deck is of box section, water holes are provided in the deck to allow for passage of water during the submergence. These holes should be kept clean before floods. After the submergence, the entire superstructure should be cleaned of the silt deposit and the water holes should be cleaned free of debris. The superstructure should be carefully examined for any signs of distress or cracks and any remedial measures necessary should be implemented early. The expansion joints should be inspected and maintained. During flood, collapsible railing, if provided, should be lowered in time.

15.10.7 Repairs

Repair refers to action taken to correct damages or deterioration of structural elements to restore the structures to an acceptable condition with respect to strength and serviceability.

The main defects in bridges requiring repairs are normally of the following types:

(a) *Defects in substructure and foundation, including scour and cracks in masonry:*

Scoured portions may be filled with boulders and, if necessary, aprons may be provided. Cracks in masonry may be repaired by grouting under pressure with cement grout or with epoxy grout. When the substructure shows spalling of concrete, the substructure may be strengthened by guniting with cement sand mixture under air pressure or by jacketing with additional thickness of concrete and provision of dowel bars of 20 ϕ at about 450 mm spacing both horizontally and vertically to bond with the existing structure.

(b) *Defects in concrete decks:*

An understanding of the cause of deterioration is essential for effective repair of concrete decks. The most frequent cause is the corrosion of reinforcement, which is influenced by the cover, the quality of the concrete and the environment. The repair interventions may comprise mortar repair, crack sealing and structural crack repair. The aim in repairing cracks is to prevent water or chemicals intruding into concrete and to restore the appearance of the structure. Cracks in R.C. decks can be repaired by injecting epoxy grout under pressure for cracks below 0.25 mm wide, and by filling with cement grout for cracks wider than 0.25 mm.

(c) *Excessive vibrations and deflections in prestressed concrete decks:*

This condition generally indicates loss of prestress, and can be repaired by introducing external prestressing.

(d) *Cracks and corrosion in steel work in superstructure:*

If the cracks are located in isolated places, cover plates may be added by riveting or welding. If similar cracks are noted in identical locations in all the spans in a multi-span bridge, the design should be checked. Patch repair may be attempted when corrosion is local in a girder. The defective member may be replaced by a new member when convenient.

(e) *Deterioration of kerbs and railings:*

Kerbs and railings suffer damage due to vehicle collision, cracking and spalling, corrosion and poor original design details. Collision related damage should be attended to immediately after the event. For deterioration due to cracking and corrosion, repairs as for similar defects in the superstructure are to be adopted.

15.10.8 Retrofitting

Retrofitting of a highway bridge involves procedures to improve the function of an existing bridge for a better performance or to increase the load carrying capacity. For example, many existing suspension bridges were strengthened with additional stays to reduce vibrations after the Tacoma collapse. Fairings were attached to the stiffening plate girders in the Deer Isle suspension bridge in USA and these were found to be very effective in enhancing the aerodynamic behaviour of the structure[26]. Some cable stayed bridges were provided with dampers on the stays to reduce vibrations in the cables. Addition of external prestressing to existing prestressed concrete bridges to reduce deflections and vibrations is another example of retrofit, as accomplished in the strengthening and rehabilitation of the Zuari river bridge on NH-17 in Goa[27]

Based on observations of seismic vulnerability of bridges in a number of earthquakes such as the 1971 San Fernando quake and the 1995 Kobe quake, many existing bridges have been retrofitted to improve their performance. The seismic retrofit projects included joint restrainers, confinement jackets to columns to ensure ductile behaviour, and modifications to column footings and pile caps. Bearing retrofits often involve replacement of steel rocker-type bearings with LRB elastomeric bearings along with provision of guides and seat extenders to prevent unseating of the spans.

15.10. 9 Reconstruction

Reconstruction of an existing bridge may become necessary when it fails to satisfy its functional requirements. The main causes of reconstruction include: (a) inadequate carriageway for the volume of traffic; (b) structural inadequacy due to deterioration or increase in design loadings; (c) insufficient waterway for river bridges; and (d) inadequate clearances for road underbridges. Of these causes, the predominant reason for replacement of an existing bridge is the inadequate width to carry the planned volume of traffic.

15.10.10 Design to Aid BMS

The application of BMS can be immensely aided if the initial design for the bridge takes into account the needs of maintenance, inspection and rehabilitation. The components that require most attention in maintenance are those subjected to direct impact loads, service loads, corrosive action and hydraulic forces. Also the parts that deteriorate include the roadway decks, railings, deck joints and bearings. At the design stage, care should be devoted to select the appropriate materials to ensure durability, construction procedures to assure quality and to provide for clearances to permit access for inspection and repair. For long span bridges with decks at significant heights, underdeck inspection traveler may be provided to assure access for inspection and maintenance operations.

The design should provide for the movements of the deck due to change in temperature, creep, shrinkage, loss of prestress or due to traffic or earthquake effects. While part of the movements can be absorbed by elastic action of the component, adequate and effective provision should be made for movement at expansion joints. For long span prestressed concrete bridges, it would be desirable to provide for later addition of prestress to the extent of about 20 per cent of the initial prestress. The tendon anchorages are located inside the box girder at the fillet at the web-flange junction and made accessible. The box sections should have sufficient depth and openings in the diaphragms and soffit to permit access for inspection and maintenance.

15.11 References

1. 'IRC:SP:47-1998, Guidelines on quality systems for road bridges', Indian Roads Congress, New Delhi, 1998, 88 pp.
2. 'Jawahar Setu', Pamphlet by Gammon India Ltd., Bombay, Dec. 1965, 24 pp.
3. 'Stahlerne strassenbrucken', Beratungsstelle fur stahlverwendung, Dusseldorf, Merkblatt 380, 1965, 50 pp.
4. 'Spannbetonstrassenbrucken', Technischer Bericht, Philipp Holzmann, Frankfurt/Main, Sept. 1973, 41 pp.
5. Leonhardt, F., 'General Report on Long Span Structures - Bridges', Proceedings of Seminar on Prestressed Concrete Structures, Indian National Group of International Association of Bridge and Structural Engineering, Bombay, Jan. 1975, Part II, pp. 1-28.
6. Göhler, B., and Pearson, B., 'Incrementally Launched Bridges', Ernst & Sohn, Berlin, 2000, 194 pp.
7. Hewson, N.R., 'Design of prestressed concrete bridges', In Ryall, M.J., Parke, G.A.R., and Harding, J.E., (Eds.), 'The Manual of Bridge Engineering', Thomas Telford, London, 2000, pp. 241-314.

8. Kloppenborg, L.L., 'Injaka Bridge Collapse of 6 July 1998', Investigation Report, 2002. www. makrosafe.co.za/docs/Inspectors_Incident_Report.doc.pdf, viewed on 14 Aug 2004.
9. Chakraborty, S.S., ' Innovations in design concepts and materials for bridges', State-of-art paper, Conference on 'Advances and Innovations in Bridge Engineering', Chennai, 1999, 12 pp.
10. Agarwal. M.K. and Lal, C., 'Tackling some problems of bridge distress', International Seminar on Highway Rehabilitation and Maintenance, New Delhi, Nov. 1999, Technical Papers, Part II, pp. 61-75.
11. Mathivat, J., 'The cantilever construction of prestressed concrete bridges', John Wiley, 1983, 341 pp.
12. Srinivasan, S., 'Elegance and innovation in concrete bridge design', In Dhir, R.K., et al, (Eds.), 'Role of concrete in bridges in sustainable development', Thomas Telford, London, 2003, pp. 239-247.
13. 'IRC:SP:66-2005 Guidelines for design of continuous bridges', Indian Roads Congress, 2005. 9 pp.
14. Pritchard, B., 'Bridge design for economy and durability', Thomas Telford, London, 1993.
15. 'IRC: 87-1984 Guidelines for the design and erection of falsework for road bridges', Indian Roads Congress, 1984, 36 pp.
16. Wynn, A.E., 'Design and construction of Formwork for Concrete Structures', Concrete Publications, London, Fourth Edition, 1956, 313 pp.
17. Morsch, E., Bay, H., and Deininger, K., 'Brucken aus Stahlbeton und Spannbeton, Vol. II", Verlag Konrad Wittwer, Stuttgart, 1968, 296 pp.
18. 'Report of the Joint Committee on Falsework', The Concrete Society and the Institution of Structural Engineers, July 1971.
19. Antill, J.M., and Ryan, P.W.S., 'Civil Engineering Construction', Angus and Robertson, Sydney, Third Edition, 1967, 631 pp.
20. 'Recommended practice for numbering bridges and culverts', J1. Indian Roads Congress, Vol. 24-1, Oct. 1959, 215-220.
21. 'Guidelines for inspection and maintenance of bridges', Special Publication 25, Indian Roads Congress, 1990, 54 pp.
22. 'Recording and coding guide for structure inventory and appraisal of the nation's bridges', U.S. Department of Transportation, Washington, D.C., 1988.
23. 'DIN 1076, Strassen und Wegbrucken - Richtlinien fur die Uberwachung und Prufung', Deutsche Normen, Dec. 1959.
24. Raina, V.K., ' Concrete bridges - Inspection, repair, strengthening, testing and load capacity evaluation', Tata McGraw-Hill, New Delhi, 1994, 493 pp.
25. Xanthakos, P.P., 'Bridge strengthening and rehabilitation', Prentice Hall, New Jersey, 1996, 966 pp.
26. Bosch, H. R., and Guterres, R.M., 'Effectiveness of fairings on a suspension bridge', In Mahmoud, K.M., (Ed.), 'Recent developments in Bridge Engineering', A.A. Bałkema Publishers, Lisse, The Netherlands, 2003, pp. 33-43.
27. Manjure, P.V., and Banerjee, A.K., 'Rehabilitation/Strengthening of Zuari bridge on NH-17 in Goa', Jl. Indian Roads Congress, Vol. 63-3, Dec 2002, pp. 463-504.

Chapter 16

Lessons from Bridge Failures

16.1 Major Causes

It is well known among engineers that absolute safety is unattainable, and inevitably there are risks of collapse associated with any bridge. However, the bridge engineer should take every possible precaution to avoid failures, as serious failures of bridges will often result in loss of lives, interruption to vital traffic and costly repairs. Every bridge engineer would do well to study the circumstances leading to any bridge failure that he may come across, so as to learn lessons from such failures. As bridge spans grow longer, and complex designs aim to result in lighter structures, the bridges tend to become more vulnerable to failures. Complex designs necessitate sophisticated checks to ensure careful layout and detailing of the various members by the designers and correct compliance by the construction team. Lack of communication among the various key personnel involved in the design and construction, and lapse in respect for natural forces have often proved disastrous. The failure may be total or partial. Total failure refers to the collapse of the bridge. Partial failure, on the other hand, involves deficiencies in meeting the intended requirements, necessitating reduced load limit, decreased speed, and implementation of substantial repair and rehabilitation. Total failures generally attract attention. But partial failures also merit careful study to avoid recurrence of the defects. The lessons learned from every major bridge failure would normally result in revisions to the standard specifications governing bridge design.

Based on a study of 143 bridge failures that occurred throughout the world between 1847 and 1975, Smith[1] has categorized the causes of failures as in Table 16.1. Though outdated, the qualitative inferences from the above report are still valid. About sixty per cent of the bridge failures listed were due to natural phenomena, i.e. due to flood, earthquake and wind. Other major causes of failure include: Defective design; Erection errors; Accidents (barge impact); Fatigue; and Corrosion. The failure of a bridge is normally due to a combination of several defects and errors. In this discussion, failures are grouped according to the major cause triggering the collapse.

16.2 Flood and Scour Failures

Almost half of the failures listed in Table 16.1 were due to floods. Scouring of bridge foundations is the most common cause of damage to bridges during floods. The construction of the bridge abutments and piers modifies the flow conditions in the river, resulting in new patterns of erosion and deposition. Every bridge across a river should be assessed as to its vulnerability to scour, so that prudent measures may be initiated to prevent damage to the structure. The precaution against this type of failure is to devote special attention at the stage of investigation to the correct determination of the design flood discharge, the maximum flood

Table 16.1 Categories of Bridge Failures.

Cause of failure	No. of failures During Construction	During Service	Total	Remarks
Flood and foundation movement	1	69	70	2 earth slip 1 floating debris 66 scour 1 foundation movement
Unsuitable or defective permanent material or workmanship	3	19	22	19 brittle fracture of plates or anchor bars
Overload or accident	1	13	14	10 ship or barge impact
Inadequate or unsuitable temporary works or erection procedure	12	—	12	
Earthquake	—	11	11	
Inadequate design in permanent material	5	—	5	
Wind	1	3	4	
Fatigue	—	4	4	
Corrosion	—	1	1	
Total	23	120	143	

level, the subsoil profile and the scour level. The foundation should be taken to a depth well below the scour level to ensure adequate anchorage. Scour protection measures may include provision of riprap filling at piers and abutments. The collapse of the Schoharie Creek bridge[2] in New York State in 1987 after three decades of service was due to extensive scour under pier three which rested on a shallow reinforced concrete spread footing. This failure highlighted the need for taking the foundation below the scour level, and the importance of adequate erosion protection around piers and abutments susceptible to scour.

The linear waterway should not be reduced considerably from the existing width of the river at the bridge site, especially when the bridge is located close to the mouth of the river where tidal effects should also be considered. We should realize that our calculations, however sophisticated they may appear to be, are at best approximate. We should learn to respect the river profile evolved by nature over many years. There have been cases of distress to approach embankments when abutments were brought into the river portion in an attempt to reduce the initial cost of the bridge.

Scour of piers during floods leading to bridge collapse has also occurred as a consequence of quarrying of sand from the river bed near the bridge, as reported in the case of Ponte de Ferro[3] in Portugal (2001). Bridge inspection and maintenance should include safeguarding the river environment near the bridge.

16.3 Brittle Fracture

Brittle fracture occurs as sudden failure of a steel member instead of ductile behaviour with plastic deformation. This may happen due to various causes including defective welding and sudden drop in temperature. Brittle failures in steel bridges were frequent between 1936 and 1962, when welding as a method of joining steel components was replacing riveting. With the availability of weldable notch ductile steels and better knowledge of welding techniques, this problem is now overcome.

16.4 Barge/Ship Impact

Damages to bridges across navigable rivers caused by barges or ships are on the increase. It has been reported that, during the period 1960 to 1998, there were 30 major bridge collapses worldwide due to ship or barge collision with a total loss of 321 lives[4]. A familiar example is the 1980 collapse of the Sunshine Skyway bridge in Florida, USA causing the loss of 35 lives as a result of collision by a bulk carrier. Another example is the damage to Tasman bridge in Australia in 1975. Barge tows often hit the piers across waterways, e.g. the 2002 collapse of the 1-40 bridge over the Arkansas river at Webber Falls, Oklahoma, USA. The vessels may be adrift or may hit the piers under power. The damage to the bridge can be minimized by providing properly designed protective fendering.

When potential damage due to barge impact exists, it is prudent not to use pile foundation with exposed piling above the river bed. In such cases, sturdy well and heavy caisson foundation with protective fendering will be desirable, as adopted for the new Sunshine Skyway bridge in Florida, USA and also the Zuari bridge in Goa. Bridge design for barge collision is not based on the worst-case scenario due to economic and structural constraints. A certain amount of risk is considered acceptable. The risk acceptance criteria are specified in codes with consideration of the probability of occurrence of a vessel collision and the consequences of the collision.

It is advisable to incorporate protection against vessel collision in the initial design. The horizontal and vertical clearances of the navigation span have to be determined based on a study of the anticipated vessel movements. For Storebelt East suspension bridge with a main span of 1624 m and 18000 vessel passages per year, vessel impact has been the governing criterion for the design of piers and probability based criteria were derived from comprehensive vessel simulations and collision analysis. There is scope for research study to improve our understanding of the vessel collision mechanics and the development of cost-effective protection measures.

16.5 Falsework Failures

Failures of falsework can result in loss, injury, death and interruption to traffic as much as bridge collapse. Falsework failure can cause excessive settlement and deflection, besides the catastrophic collapse of the superstructure. While flood, storm winds and earthquakes may contribute to failure, most falsework failures are attributable to human error. The problem of avoiding falsework failures is not easy to solve because of many economic and administrative factors. Falsework is a temporary structure designed and erected to last long enough to support the final structure during construction. Traditionally, this has been left to the contractor and as an economic necessity, the formwork construction needs to use secondhand materials to the extent possible, thus lacking the finesse of a finely designed structure. With increased spans of our bridges, falsework design has become more complicated. The bridge falsework design should be prepared by a competent engineer, should be checked by the government engineers and its erection should be under proper supervision. Immediately prior to and during the placing of concrete, the constructed falsework should be carefully checked for joint fits, bracing, stiffness, overturning possibilities, foundation settlement and general adequacy[5]. By improved methods of construction and constant vigilance, we can avoid falsework failures.

16.6 Erection Errors

The erection of a major bridge invariably involves special risks, which could lead to injuries and loss of lives. A major cause of erection failures in the past has been the underestimation of the construction loads and their effects on the unfinished structure. The failure during erection of the first Quebec cantilever bridge in 1907 highlighted the need for reviewing the erection stresses in the bridge members for strength and stability at each critical stage of erection. The bottom chords of the side span failed in compression at an advanced cantilever erection stage. This failure also evidenced the essentiality of the design engineer to constantly interact with the construction engineers in order to initiate timely remedial action should any defects be noticed during erection.

The failures during erection of steel box girder bridges during 1970-71 were basically due to instability of thin steel plates in compression and secondary stresses arising from minor geometric imperfections. At the West Gate bridge in Australia, the failure of the 112 m long steel box girder span resulting in the loss of 35 lives was triggered due to premature removal of 37 top flange splice bolts in an effort to facilitate connection of the two longitudinal halves of the top flange. At Koblenz, geometric imperfections caused critical secondary stresses resulting in buckling of the bottom flange. The lesson to be learnt is that the design derived from sophisticated analysis should be tempered with realistic allowances for construction methods and erection tolerances.

A recent example of errors in erection procedures is the failure in 1998 of the Injaka bridge in South Africa[6], which resulted in 14 deaths including that of the designer of the bridge. The prestressed concrete bridge of box section collapsed during erection by incremental launching method, mainly due to wrong placement of temporary bearings, which punched through the inadequately designed soffit slab. The disaster highlighted the importance of proper erection procedures.

16.7 Design Deficiencies

Bridge failures due principally to design deficiencies are relatively few. Designers attempt to achieve the required strength with the use of minimum amount of materials with a view to minimize the cost of construction. This quest for ever better strength/weight ratios, coupled with other factors, sometimes leads to catastrophic failures. According to Sibly and Walker[7], failures by weakness in design follow a 30-year cycle, as evidenced by the following examples: Dee bridge (1847), Tay bridge (1879), Quebec bridge (1907), Tacoma Narrows bridge (1940) and the steel box girder bridges including West Gate bridge (1970). The Dee bridge used an inappropriate structural system consisting of a 30 m cast iron girder trussed up with tensioned wrought iron ties. Tay bridge collapse was attributable to lack of adequate provisions for the wind effects. The Quebec railway bridge failed during construction due to buckling of the main lower chord following failure of lattice components in the web. The Tacoma Narrows bridge failed due to aerodynamic instability. In the case of the steel box girder bridges, the primary mode of failure was instability of the thin plates in compression due to minor details such as geometric imperfections. In each of these cases, a factor which was originally of secondary importance became with increasing scale of primary importance and led to failure.

16.8 Earthquake Effects

Several types of bridge failures have been noticed during earthquakes[8]. A commor failure was by span shortening. Decks slid off their supports due to violent shaking as ir Showa bridge in the 1964 Niigata earthquake (m = 7.5) and the Nishinomiya bridge in the 1995 Kobe earthquake (m = 6.9) in Japan. As abutments and piers moved together, some decks buckled, some were crushed and some collapsed. This problem is critical for bridges with simply supported spans located in soft soil. Another type was the horizontal displacement of piers due to movement of piles in liquefied soils subjected to lateral loading. A third type involved differential settlement of piers and abutments due to differences in soil characteristics due to liquefaction. Liquefaction of approach fills have resulted in settlement of fills in relation to abutments, causing accidents to motor vehicles by impact against the abutment backwall.

Columns in substructure have been found to suffer extensive damage during earthquakes. The damage is mainly attributable to inadequate detailing, which limits the ability of the column to deform inelastically. In concrete columns, inadequate ductility results from insufficient reinforcement to achieve confinement of concrete within the core. Termination of longitudinal bars at midheight led to splitting failure or shear failure near the end cutoff point in a few bridges, as noticed in the 1995 Kobe earthquake. Column failures by crushing of concrete due to extreme torsion have also been noticed. In steel columns, local buckling may cause inadequate ductility.

Several preventive measures have been suggested. Heavier and closer spaced spiral reinforcement should be provided for columns. Such reinforcement would retain the concrete in the core and prevent collapse. Restraint should be provided at expansion joints and articulations such that ordinary expansion due to temperature is permitted but larger movements under earthquake are restrained. No splices are to be allowed in columns of less than 6 m height, as lapped splices of column bars have been found to be ineffective under earthquakes. Approach slab with one end resting on abutment should be provided to permit a smooth transition in case of settlement of approaches due to liquefaction of the fill.

16.9 Failures due to Wind

Bridge failures have occurred due to wind. Major examples include the collapse of the Tay bridge in 1879, and Chester bridge over Missisipi in 1944[9]. Tay bridge consisted of eighty-five through-lattice–truss spans of malleable iron supported 26.8 m above the water level. The bridge failed due to aerostatic instability, as the design of cross bracing and its fastenings was inadequate to sustain wind forces, though the design was otherwise in conformity with contemporary practice. Sir Thomas Bough, the designer, was unjustly blamed and made the scapegoat by the Chairman of the Court of Enquiry, who summarized his report in concise and quotable form. The majority report which was supportive of Bough was voluminous and was not read by many. This example highlights the need for the bridge engineer to cultivate the art of advocacy and to learn the benefits of brevity in oral and written presentation of technical reports.

The main channel stretch of Chester bridge was a 408.8 m through truss, continuous over two equal spans. This stretch was blown off into the river during a tornado. While very little can be done to save a structure from the direct attack of a severe tornado, the damage can be minimized by providing proper anchorage of the deck with the substructure. Tacoma Narrows first bridge (1940) failed due to aerodynamic instability as discussed in Section 9.10. The recurrence of these types of failures is avoided in recent designs through streamlining the deck and adequate stiffening.

16.10 Fatigue

Fatigue may be defined as the gradual weakening of a structure due to repetitive loading and is accompanied by spreading of a crack. If the steel is corroded at the tip of the crack, progression of the crack is accelerated. An example of fatigue failure is the 1967 collapse of the Point Pleasant bridge (eye-bar suspension bridge, also known as the Silver bridge) in West Virginia, USA due to fracture at the pinhole of a single eye-bar. A minute crack formed during the casting of the concerned eye-bar grew undetected over thirty-nine years of service due to stress corrosion and corrosion fatigue, causing failure of the entire structure. This catastrophic failure, resulting in the loss of 46 lives, forcefully demonstrated the need for building-in redundancy in bridge structures. The failure led to the evolution of strict specifications for bridge inspection. In modern steel bridges, fatigue failure is avoided by paying attention to detailing of connections and ensuring ductility.

16.11 Corrosion

Corrosion of reinforcement in a reinforced concrete bridge may lead to cracking and spalling of concrete, rendering the bridge unsafe for modern traffic. Potential damage due to corrosion in a backwater area can be prevented only by careful attention to concrete cover to reinforcement, by proper placement and compaction of concrete to avoid honeycombing, and by proper curing with potable water. The risk of corrosion for concrete bridges may be reduced by adoption of a combination of preventive measures such as provision of adequate concrete cover, use of high performance concrete, application of admixtures to inhibit corrosion, use of corrosion resistant reinforcement, such as epoxy-coated bars, and exercise of good quality control.

In pretensioned structures, corrosion prevention is mainly accomplished through the use of high performance concrete and the addition of corrosion-inhibiting admixtures. In post-tensioned concrete bridges, special care should be devoted to ensuring integrity of the duct and grouting of prestressing cables soon after stressing. Delayed grouting and inadequate grouting of tendons especially near the anchorage may cause corrosion of the tendons, contributing to the failure of a prestressed concrete bridge. Corrosion of tendons may remain undetected until the loss of strength to a significant level results in major evidence of deterioration or even a sudden failure. The failure of the Mandovi first bridge at Panjim, Goa in 1986 is attributed to corrosion of prestressing steel.

Corrosion of prestressing tendons also caused the failure of the single span Ynys-y-Gwas bridge[10] in Wales, UK in 1985. This bridge built in 1953 on a minor road was a segmental post-tensioned structure of 18.3 m span with thick mortar joints and defective grouting for tendons. The corrosion took place at the transverse joints between precast segments at which chloride-containing water could penetrate. Grouting deficiencies have contributed to distress of several post-tensioned bridges in many countries[11]. While stressing operations have been devoted extensive care by construction engineers, adequate attention to subsequent grouting has been lacking. Such failures have shown the importance of durability design, besides the load and resistance based structural design. Modern segmental construction, however, uses matchcast joints sealed with epoxy.

16.12 Partial Failures

Partial failures in bridges are those necessitating major repairs to rehabilitate a deteriorating bridge. This situation is particularly prevalent in prestressed concrete bridges, where a loss of prestress due to corrosion of prestressing steel manifests in excessive deflections and vibrations, e.g. Zuari bridge on NH-17[12]. For an appreciation of the extent of distress caused by corrosion to a major prestressed concrete bridge and the extensive remedial measures undertaken to extend the service life, the reader is referred to the rehabilitation efforts for the Zuari bridge on NH-17 in Goa and the First Thana Creek bridge[12,13].

An example of successful rehabilitation of a noted deficiency in bridge performance is the case of the Millenium Footbridge in London, UK. On the opening day (10 June 2000), the bridge exhibited unacceptable sideways movement when a large number of pedestrians were crossing the bridge. This condition was attributed to 'synchronous lateral excitation'. The bridge earned the name 'The Wobbly Bridge", necessitating its closure on the third day. The situation was successfully rectified by the use of passive damping, adopting viscous dampers and tuned mass dampers located under the deck[14]. The bridge was reopened for public use in February 2002.

16.13 Summary

An unbiased engineering analysis of every bridge failure would benefit the profession in enhancing the understanding of the behaviour of bridge systems and the performance of new materials. Bridge engineers should endeavour to apply the lessons learnt from bridge disasters towards evolution of safe, cost-effective and durable bridges. Engineers have tended to place too much emphasis on sophisticated analysis of permanent members, while at times devoting inadequate attention to accuracy of design assumptions, falsework, erection methods and communication between the design consultant and the construction crew. This environment of altered priorities sometimes leads to gross errors and failure.

16.14 References

1. Smith, D.W., 'Bridge failures', Proc. Institution of Civil Engineers, Vol. 60, Pt. 1, Aug. 1976, pp. 367-382. (Also Civil Engineering -ASCE, Vol. 47, No.11, Nov.1977, pp. 58-62).
2. Storey, C., and Delatte, N. 'Lessons from the collapse of the Schoharie Creek bridge', Forensic Engineering (2003), American Society of Civil Engineers, 2003, pp. 158-167.
3. NCE Plus, 'Portuguese bridge disaster blamed on illegal dredging', NCE Plus News, 8 March 2001, http://nceplus.co.uk/news.
4. Knott, M., and Prucz, Z., 'Vehicle collision design of bridges', In Chen, W.F., and Duan, L., (Eds.), 'Bridge Engineering Handbook', CRC Press, Boca Raton, 2000, pp. (60) 1-18.
5. Elliot, A.I., 'Falsework failures: can they be prevented ?', Civil Engineering- ASCE, Vol. 43, No. 10, Oct. 1973, pp 74-76.
6. Kloppenborg, L.L., 'Injaka Bridge Collapse of 6 July 1998', Investigation Report, 2002. www.makrosafe.co.za/docs/Inspectors_Incident_Report.doc.pdf, viewed on 14 Aug 2004.
7. Sibly and Walker, 'Structural accidents and their causes', Proc. Institution of Civil Engineers, Vol 62(1), 1977.
8. Moehle, J.P., and Eberhard, M.O., 'Earthquake damage to bridges', In Chen, W.F., and Duan, L., (Eds.), 'Bridge Engineering Handbook', CRC Press, Boca Raton, 2000, pp. (34) 1-33.
9. Ross, S.S.; 'Construction disasters: design failures, causes, and prevention', McGraw-Hill, 1984, 418 pp.
10. Woodward, R.J., and Williams, F., 'Collapse of Ynys-y-Gwas bridge, West Glamorgan', Proc. Institution of Civil Engineers, London, Part I, 1988, Vol. 84, pp. 635-669.
11. Pearson-Kirk, D., 'The performance of post-tensioned bridges', In Dhir, R.K., et al, (Eds.), 'Role of concrete in bridges in sustainable development', Thomas Telford, London, 2003, pp. 129-140.

12. Manjure, P.V., and Banerjee, A.K., 'Rehabilitation/Strengthening of Zuari bridge on NH-17 in Goa', Jl. Indian Roads Congress, Vol. 63-3, Dec 2002, pp. 463-504.
13. Patil, M.V., and Sabnis, S.M., 'Rehabilitation of First Thana Creek Bridge: A success story', International Seminar on Highway Rehabilitation and Maintenance, New Delhi, 1999, Technical Papers, Part II, pp. 1-29.
14. Arup, 'The Millenium bridge', http://www.arup.com/milleniumbridge.

Chapter 17

Recent Trends in Bridge Engineering

17.1 General

Bridge engineering has registered rapid strides in recent years. Due to expansion of road and rail networks and increasing constraints on alignment and clearance requirements, the design and construction involve complex computations and special construction techniques. The basic concepts presented in this text will assist the reader in appreciating the approaches to the new techniques.

Elevated roads and rail tracks besides grade separated intersections in urban environments present special problems and provide opportunities for innovative designs. Integral bridges are increasingly adopted for short and medium span concrete bridges with a view to circumvent problems associated with expansion joints and to facilitate better riding comfort for vehicle users. Wider use of machinery has resulted in development of more efficient construction techniques for continuous construction, as already indicated for incremental launching method and segmental cantilever construction.

Recent bridges involve several adaptations to conventional configurations. For example, cable stayed bridges comprised of steel deck in early designs, whereas later applications used prestressed concrete decks. Presently, extradosed bridges are being designed adopting prestressed concrete box section deck with towers shorter than in conventional cable stayed bridges.

Advances in concrete technology aim to improve the durability and strength of structural concrete through the use of admixtures and high performance concrete. Analysis is tending to be more rational with the use of Limit States Method of design. Increasing use of digital computers facilitate voluminous calculations with relative ease and permit better precision in computations. Some of the recent trends in the area of bridge design and construction are discussed briefly in this Chapter.

17.2 Urban Flyovers and Elevated Roads

17.2.1 General

Increased congestion on urban roads at the locations of at-grade crossings has necessitated the provision of grade-separated intersections. At the locations of road-rail level crossings, increase in the frequency of train movements hampers the passage of road vehicles and warrants the provision of road over bridges. For certain stretches of roads with heavy traffic in built-up areas, the road land could not be widened, and it may become necessary to carry a part of the through traffic on an elevated road, e.g. the 2.64 km long Sirsi flyover at Bangalore[1]. Limited access expressways may have to be in the form of elevated roads for a considerable length in urban areas. The grade-separated intersections, road over bridges and elevated road stretches are often referred as flyovers. In recent years, many flyovers have been constructed in major cities, as seen in Mumbai, New Delhi and Chennai [2-4].

Flyovers are essentially bridges having a few special characteristics with respect to horizontal alignment and vertical profile. Any flyover consists of three major parts: (a) Main and obligatory span provided for the traffic crossing at the lower level; (b) Viaduct portion on either side of the main span with multiple smaller spans; and (c) Solid ramp on either side to connect the road at the abutment to the ground level. Of the many types of grade-separated intersections, urban flyovers in built-up areas are required to need minimum or no additional land acquisition, and hence are normally constrained to follow simple straight crossing at the intersection or alignment along and over the existing road in case of elevated road.

The design and construction of urban flyovers pose several challenges due to restricted working area, need to maintain uninterrupted traffic flow during construction, requirement to complete the construction within a tight time schedule, and the community expectation to have a structure that will be aesthetically pleasing. These challenges have been met successfully in most cases, adopting innovative designs and efficient construction techniques. A challenge not often faced satisfactorily is the balance of benefits to all road users, especially the pedestrians and non-motorised traffic. In this section, only the structural aspects of flyovers are discussed.

17.2.2 Design Criteria

There is a strong case to adopt modified design criteria for urban flyovers compared to the more conservative criteria prescribed in the IRC Code for bridges over rivers and creeks. Some of the modified criteria, as followed successfully for the construction of flyovers in Mumbai and Chennai [2,3], are indicated below:

(a) The flyovers need not be designed for IRC Class 70R loading, as such heavy vehicles can take the ground level road if at all they are encountered.

(b) Continuous superstructures are encouraged as foundations are generally rested on rock. Continuity results in lower depth of deck and also enhances riding comfort due to reduction in the number of expansion joints. Besides unsatisfactory performance, most bridge expansion joints leak and contribute to deck and substructure corrosion damage and hence their reduction promotes durability.

(c) When precast prestressed girders are used, continuity may be achieved by cast in situ concrete transverse beams at or near the piers[5]. The continuity connection may consist of a wide in situ integrated crosshead over the pier providing 1 m embedment of the beams. The reinforcement in the composite deck slab provides longitudinal continuity. One set of bearings is adequate. Alternatively, a narrow in situ integral crosshead may be adopted; but this will need two sets of bearings on the pier. Another type of continuity connection relies on the deck slab flexing to acommodate the rotations of the simply supported deck beams. Here the slab is separated from the support beams for about 1.5 m by a layer of compressible material. Two sets of bearings are required.

(d) High strength concretes of grade M40 and above only are used for the deck. The permissible compressive and tensile stresses are adopted as 0.33 f_{ck} and 0.033 f_{ck}, respectively, without the ceiling on the stresses. When high performance concrete is adopted, the permissible stresses are allowed with the characteristic strength scaled down by 10 MPa, as a conservative measure.

(e) For prestressed concrete superstructure, partial prestressing has been permitted with the proviso that there will be no tension under 'Dead load + 60% Live load' condition. Under full design load, tensile stress in concrete is limited to 1.0 MPa for severe exposure and 2.0 MPa for moderate exposure.

(f) As a precaution against corrosion of steel, corrosion resisting steel reinforcement or steel rods coated with anti-corrosive treatment are used. The minimum clear cover is kept at 40 mm, 50 mm and 75 mm for reinforcement in components above ground level, components below ground level and for prestressing cables, respectively.

(g) Since durability of the structure is of paramount importance, the minimum concrete grades adopted are M30, M35 and M40 for plain cement concrete, reinforced concrete and prestressed concrete, respectively, with water cement ratio limited to 0.45, 0.40 and 0.40, respectively.

(h) Approach embankment height is restricted to be less than 3.0 m. The sides of the embankment are retained with reinforced earth retaining structures avoiding massive wing walls.

(i) Pile foundations with hydraulically operated rotary drilling equipment are encouraged. Pile caps are so arranged as to have the top of the pile cap at 500 mm below the ground level.

(j) Crash barriers are essential to achieve vehicle containment. For the Mumbai flyovers, the crash barriers were designed for impact of 300 kN vehicle at 64 km/h at an angle of 20° at the top of the crash barrier, and R.C. crash barriers laid with kerb laying machinery are provided. Steel barriers are adopted in Chennai and Bangalore.

17.2.3 Innovative Construction Techniques

Emphasis on high quality and fast construction of recent flyovers necessitated the adoption of innovative construction techniques [1-4]. Typical measures are mentioned below:

(a) Pile foundations using rotary pile driving machines were made mandatory. Besides facilitating speedy construction, the method was economical. The liner plates were of 4 mm thickness instead of the usual 6 mm thickness. Also the seating of the pile on the rock was assured.

(b) Since space was not available around the flyover sites for storing and handling of materials for concrete production, use of ready mixed concrete was made compulsory. This led to many spin-off benefits in quality assurance and reduction in construction joints due to fast and continuous placement of concrete.

(c) Precast prestressed concrete girders were used extensively in flyovers in Delhi, Mumbai and Chennai. For Sirsi flyover at Bangalore, precast segments of box shape were used for the entire bridge. The adoption of precast elements for the superstructure results in considerable saving in time, as the foundations and precasting can be started simultaneously, and the superstructure beams/segments are ready for placing in position by the time the substructure is completed.

(d) To improve riding comfort, the number of expansion joints was reduced. In the case of the girder and slab system, this was achieved by providing continuity of the deck slab over the supports. For some of the flyovers, the flexibility of the deck slab in the longitudinal direction was increased by introducing neoprene pads of 12 mm thickness between the slab and the girders for a length of 1.0 to 1.5 m near the bearings. This continuity in the deck slab is not likely to reduce the total positive moment in the girders to any appreciable extent[5].

(e) Use of automatic casting machine was made compulsory for casting kerbs and crash barriers for the Mumbai flyovers. For Delhi flyovers, precast crash barriers with shear keys for interlocking were used.

17.2.4 Tyical Details of Recent Flyovers

The basic structural form for the superstructure of recent flyovers in Chennai[3] was with continuous spans of precast prestressed concrete I-girders topped by cast-in-situ reinforced concrete deck slab with single columns and pile foundations, as shown in Fig. 17.1. The span for the viaduct portion was about 20.0 m. Continuity normally extended over six spans, allowing expansion joints at about 120 m. Cast-in-situ diaphragms were provided only over the piers. The span for the obligatory opening was in the range of 30 to 40 m. Single columns were chosen to minimize the obstruction to traffic at ground level. POT-PTFE type of bearings were provided at each diaphragm location. Reinforced earth retaining system was adopted for the solid embankments, as this system facilitates fast construction, superior aesthetics and improved economy.

Typical details of the Sirsi circle flyover at Bangalore[1] are shown in Fig. 17.2. The superstructure consisted of a post-tensioned concrete continuous box girder with curved bottom spanning over eight spans of about 36.0 m each. The overall depth was limited to 2.3 m. Concrete grade of M40 was adopted for the superstructure. The deck was of segmental

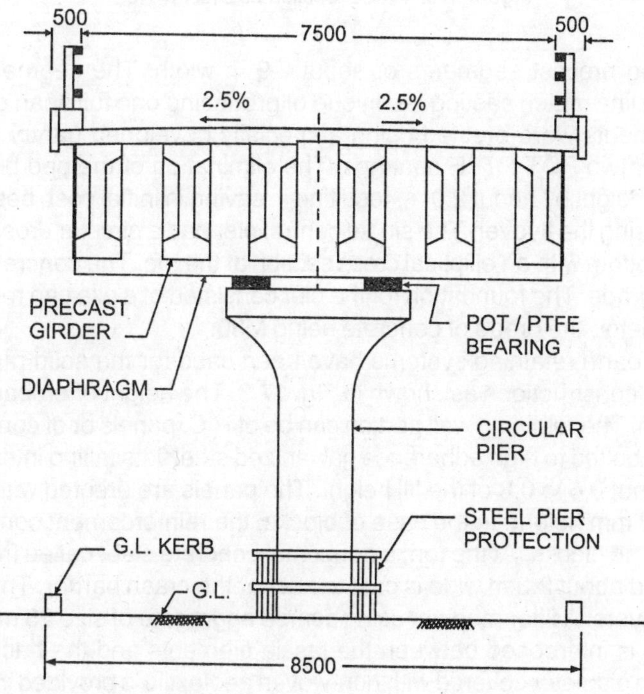

Figure 17.1 Basic Structural Scheme for Chennai Flyovers.

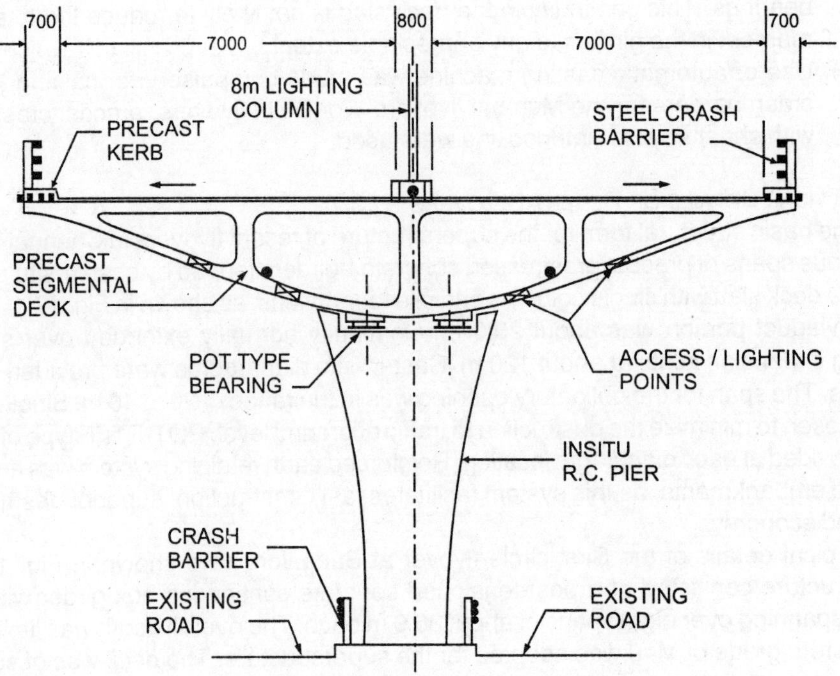

Figure 17.2 Typical Section of Sirsi Flyover.

construction using precast segments of about 2.9 m width. The segments were precast adopting the long line match casting technique of producing one full span of 12 segments at a time. The segments were erected using a specially developed gantry system. The box girder rested over two POT-PTFE bearings. The elimination of the bed block over the pier reduced the total height by about 2.0 m, resulting in saving in initial cost, besides fuel savings for the vehicles using the flyover. The single central pier had a circular cross section of 2.0 m diameter at the bottom with an elliptical cross section at the top. The concrete for the pier was normally of M40 grade. The foundation for the pier consisted of a pile cap resting on four piles of 1000 mm diameter, the grade of concrete being M30.

Reinforced earth retaining systems have been used for the solid ramp portion of the flyovers in recent constructions as shown in Fig. 17.3. The height of embankment is usually restricted to 3.0 m. The retaining wall portion can be of R.C. panels or of concrete blocks. The fascia panels are bolted to high adherence galvanized steel flats tailing into the embankment for a length of about 0.6 to 0.8 of the fill height. The panels are erected with an inward slope of 12 mm in 1500 mm height. In the case of blocks, the reinforcement consists of geogrids, which are tied to the blocks. At the top, a horizontal concrete slab, called friction slab, of 250 mm thickness and about 2.3 m wide is cast to anchor the crash barrier. The embankment is to be built up in layers. A filter medium using coarse aggregate of size 25 mm and below and of 600 mm width is interposed between the fascia elements and the backfill. A perforated PVC pipe 200 mm diameter covered with non-woven geotextile is provided in the filter medium at the ground level with a slope of 1 in 200 towards the abutment, so as to drain the collected water. The reinforced earth retaining system facilitates fast construction, enhances aesthetics,

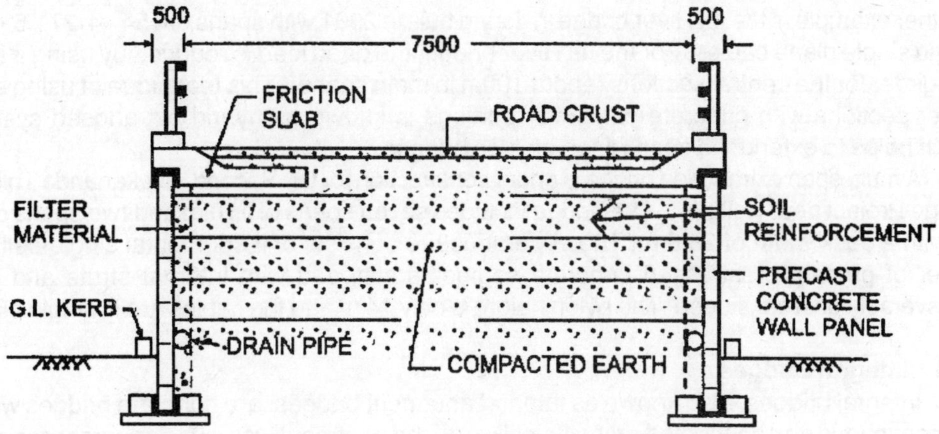

Figure 17.3 Reinforced Earth Retaining System.

avoids encroachment into the side road and results in reduced cost of construction.

17.3 Extradosed Bridges

A recent innovation combining the principles of cable stayed bridges and prestressed concrete box girders in the span range of 100 to 275 m is the evolution of extradosed bridges, a concept attributed to Mathivat in France[6]. These bridges are similar to cable stayed bridges as both use stay cables for strengthening. They differ in that the extradosed bridges generally use lower tower heights, usually about half that normally used for cable stayed bridges. In the case of cable stayed bridges, the prestressed concrete box deck structure is suspended from stay cables. Most of the load, such as the dead load and the live load, is carried through the stay cables to the top of the towers and then down to the foundation. On the other hand, in an extradosed bridge both the box deck and the stay cables share the load, with the box deck carrying the major part of the dead load and the stay cables supporting the live load and a portion of the dead load. Its structural behaviour is close to that of a girder bridge. For this reason, the relation between the centre span and the end span for an extradosed bridge should be similar to that of a normal prestressed concrete girder bridge. Also for a three-span bridge, the stay weight will be of the order of 10 to 20 kg/m^2 for an extradosed bridge compared with 40 to 50 kg/m^2 for a corresponding cable stayed bridge[6].

The stay cables may be in two planes or in a single plane. The vertical component of cable stay force is low, especially for cables in two planes. This would facilitate easy construction, reducing the need for structural diaphragms at the stay anchorages. The cable stays in extradosed bridges in Japan are stressed to 0.60 f_y, in view of the low fatigue stress range in the cables of such bridges. In the case of single plane scheme, inclined web members in the middle of the cross section may be used for transfer of stay forces to the main girder.

The Odawara Blueway bridge in Japan built in 1994 with spans of 73.3-122.3-73.3 m and two-plane fan type stay cables is an example of elegant extradosed bridges. The superstructure consists of two-cell concrete box. It is 13 m wide with depth of 2.2 m at midspan and 3.5 m at the piers. It was constructed using the segmental cantilever construction method. The height of the towers is 10.7 m above the deck and 37.2 m below the deck.

Another example is the Ibi River bridge in Japan built in 2001 with spans of 154 - 4*271.5 - 154 m and single plane cables. For the Ibi River bridge, the dead load is reduced by using a steel box girder for the central sections (about 100m in main spans). This technique of using steel girder sections with concrete cantilever portions is known as hybrid extradosed system, which helps to extend the span range of such bridges.

A multi-span extradosed bridge is under construction for the Second Vivekananda Tollway Bridge Project near Kolkata[7]. The bridge includes seven spans of 110 m and two spans of 50 m, with a deck width of 28.6 m. The superstructure consists of a segmental single-cell box girder of precast prestressed concrete segments stiffened with internal struts and also transversely post-tensioned. The pylon height is only 14 m and the cables are in single plane.

17.4 Integral Bridges

Integral bridges, also known as integral abutment bridges, are concrete bridges which are continuous and connected monolithically with the abutment wall with a moment-resisting connection[8-10]. Bearings are eliminated and expansion joints are either eliminated or provided at long distances for long bridges. The abutment wall is generally founded on a single row of vertical piles. The minimum thickness of the abutment wall is 1.0 m to provide adequate width to encase the piles and the superstructure girders. Piers for integral bridges may be constructed integrally with the superstructure adopting flexible capped-pile piers. Alternatively, the piers may rest on independent semi-rigid piers with movable bearings. The flexibility of the piling and the rigid connection between the superstructure and the substructure facilitate the transfer of the cyclical thermal movements and live load rotational displacement in the superstructure to the substructure.

The total length of the bridge, up to which an integral bridge can be built, depends on the range of temperature variation, material of the structure, geometry of the structure, pier heights, the sub-soil conditions and the allowable movement of abutment at each end. Currently, the maximum length is generally recommended as 60 m. However, integral bridges have been built up to 150 m with R.C. voided slab superstructure for two flyovers in Delhi and up to 358 m with precast prestressed bulb T-beams with composite concrete deck for the Happy Hollow Creek bridge in USA[9,10].

An approach slab for a minimum length of 3.0 m is provided at either end of the bridge. The approach slab is anchored to the abutment backwall by reinforcement. The farther end of the approach slab is normally supported on a keyed sleeper slab, where an expansion joint may be provided. This joint is located away from the main structure. The width of the joint is kept just adequate to prevent the joint from closing during hot weather, with an allowance for the minimum width of the compressed joint seal. The approach slabs move with the superstructure.

There are several advantages in using integral bridges. In a conventional bridge, joints and bearings prove to be expensive to install, maintain and repair. Leaking expansion joints and seals often lead to corrosion of bearings and deterioration of girder ends and supporting reinforced concrete substructure. Bridge deck joints suffer damage due to continual wear and heavy impact from repeated live loads with simultaneous thermal movements. Due to lower design life, bearings and joints require replacement during the service life of the bridge, besides periodical inspection. The elimination of the roadway expansion joints, seals and associated expansion bearings in an integral bridge on the other hand facilitate better riding quality for the traffic, lower demand for inspection, reduced maintenance costs and increased structural life of the bridges. Integral bridges provide additional redundancy and enable the structure to accommodate large superstructure movements during an earthquake. The integral abutments mobilize passive earth pressure which serves to dissipate significant amounts of energy. Hence integral bridges are preferable for short span bridges in highly seismic regions. Since the superstructure and the substructure are composed of one homogeneous material and are integrally connected, the resulting structure has enhanced aesthetics.

The integral abutment bridge is a significant recent development in road bridge technology. The elimination of the expansion joint in the bridge has shifted thermal movements from the joint to the end of the approach slab, necessitating additional design effort to cater to geotechnical considerations relating to soil-structure interaction at the interface of the abutment and the embankment. There is a need for developing national standards for design, detailing and construction of integral bridges.

17.5 Durability Considerations

Durability of concrete may be defined as its ability to resist deterioration from weathering action, chemical attack, abrasion and other degradation processes. Concrete should be durable in order to provide the desired performance in the conditions of exposure during its service life. It should maintain its integrity and should afford protection from corrosion to the embedded reinforcement. Concrete durability can be enhanced by careful selection of materials to control and optimize their properties; reducing variability in the mixing, transport, placement and curing of concrete; and creating and using more performance-based specifications to evaluate in-place concrete. A predominant characteristic influencing the durability of concrete is its impermeability to the ingress of water, carbon-dioxide, oxygen, chlorides, sulphates and other potentially deleterious substances. Low permeability is achieved by using proper cement content, low water-cement ratio and dense concrete obtained by thorough compaction and efficient curing.

The importance of durability of concrete bridge structures has gained better recognition in recent years, based on the experience of deterioration of the existing bridges mainly due to corrosion of prestressing steel and untensioned reinforcement. Presently it is realized that high quality concrete with the required strength and good durability can be produced with a little extra care in mix design and construction. This awareness has resulted in many precautionary measures in design and construction, including the following:

(a) Dimensions of structural elements are increased to avoid congestion of reinforcement;

(b) Concrete of low permeability is being adopted, with the use of pozzolonic materials;

(c) High performance concrete is increasingly specified, with water/cement ratio less than 0.4.

(d) The minimum concrete cover for all structural elements is increased, typically to 50 mm in superstructure and up to 100 mm for foundation in marine environment.

(e) Enhanced quality control is exercised in the various construction processes such as batching, mixing, transporting, placing, consolidation, finishing and curing.

While the current codes and standards attempt to achieve durability through a set of prescriptive specifications, the future trend is to include additional considerations such as design life, life cycle costs and specifications for planned maintenance and repair.

17.6 High Performance Concrete

High Performance Concrete (HPC) is increasingly adopted in bridge construction as a high quality concrete choice for high strength, durability and optimum life-cycle costs[11]. HPC can be designed to achieve specific requirements such as: enhanced durability; high strength greater than M60; high early gain in strength for precast bridge components; high impermeability to withstand marine exposure; air entrainment to improve resistance to freezing and thawing; and special applications like spraying, pumping and placement under water. Limited quantities (up to about seven per cent of cement) of microsilica (which is much finer than cement) can be added to produce HPC. Microsilica (also known as silica fume) and cement together constitute the binding material. The water-binder ratio of HPC is in the range of 0.3 to 0.4. The production and use of HPC demands stringent quality control to ensure a high degree of uniformity between batches and efficient curing.

The constituents of HPC are the same as in normal concrete, but proportioned and mixed (with appropriate materials such as super-plasticizer, flyash, blast furnace slag and silica fume) so as to yield a stronger and more durable product. The cement content of concrete inclusive of any mineral admixtures should not be less than 380 kg/m^3, and the cement content excluding any mineral admixtures should not exceed 450 kg/m^3. HPC structures are likely to last longer and suffer less damage from traffic and climatic conditions, resulting in reduced cost on repairs in the long run. Thus the present goals are to achieve durability and economical long-term maintenance along with economical construction. When HPC with silica fume is used for bridge piers, curing should be done with extra care to avoid surface shrinkage cracks.

HPC has been adopted for the Confederation bridge in Canada built in 1997. The bridge was designed for a service life of one hundred years. The bridge superstructure is a single cell precast prestressed trapezoidal box girder. The substructure is of precast and cast-in-place concrete. The proposed HPC was extensively tested for durability, especially through freeze-thaw cycles, sulphate resistivity and chloride diffusivity testing, checking of alkali/aggregate reactivity and evaluation of curing regimes for large components. Careful attention was devoted to production and construction practices. Precasting was chosen for improved quality and reduced construction time. In India, the first major bridge project to specify HPC was the JJ Flyover in Mumbai where HPC of M75 was used.

17.7 High Performance Steel

Research efforts initiated by Federal Highway Administration, USA are in progress to develop high performance steel (HPS) which will lead to cost effective bridge structures that will be safe and durable throughout their service lives[12, 13]. The HPS is aimed to have an optimized balance among the desired qualities such as high strength, superior weldability, toughness, ductility, corrosion resistance and formability. High strengths of the order of 480 to 590 MPa are now available. Weldability, corrosion resistance and toughness are areas being improved. Additional developmental efforts are needed to establish reliable designs of welded HPS bridges.

17.8 Limit States Method of Design

The working stress design (WSD) method is currently in use for the design of bridges, because of its simplicity and satisfactory serviceability performance. In WSD, the factor of safety is related to stress and does not give a realistic assessment of the load carrying capacity of the structure. The ultimate load method of design (ULM) enables computation of the load carrying capacity and the method generally leads to slender sections. However, ULM does not guarantee satisfactory serviceability performance in relation to deflection and crack width. This led to the evolution of the limit states method of design (LSM).

Limit states are conditions under which a structural member can no longer perform its intended function fully. There are two broad categories of limit states: (a) ultimate limit states; and (b) serviceability limit states. Ultimate limit states are limit states pertaining to structural safety. They involve the total or partial collapse of the structure. For well designed structures, the probability of the occurrence of an ultimate limit state is low. Serviceability limit states are limiting conditions affecting the function of the structure under the expected service conditions. They include conditions which restrict the intended use of the structure. Examples of serviceability limit states are excessive deflections, cracking and vibrations of superstructure, and excessive differential settlement and 'local' damage in foundations. Serviceability limit states have a higher probability of occurrence, but a lower level of danger to lives than ultimate limit states. The basic design objective in LSM is to keep the probability of a limit state being exceeded below a certain acceptable value.

The LSM aims at arriving at designs which assure adequate safety at ultimate loads and acceptable serviceability at working loads, by prescribing appropriate safety factors at the various stages considered. The design philosophy attempts to ensure safety, serviceability, durability and overall economy with appropriate degree of reliability, and thus represents an advancement over the traditional design approaches based on WSD and ULM. A distinct advantage of the LSM is the ability to give proper weight to the accuracy obtainable in the determination of the loads and resistance. The Indian Roads Congress has initiated steps towards the application of LSM for the design of road bridges in India in the near future.

17.9 Advanced Construction Methods

There is an increased awareness that substantial savings in overall costs for a bridge can be realized by applying innovative construction methods even from the design stage. A recent example of innovative adaptation of the push launching technique is the launching of the Jhajjar Khad railway bridge on the Jammu-Baramullah line[14]. This 308 m long (spans of 154 m each) open web steel bridge is the highest railway bridge in India, supported on 88.6 m high pier. A customized push launching procedure was adopted for erection of the truss superstructure as the site was in a difficult terrain with limited access. The erection scheme entailed intricate enabling works and precautionary shore tests for the maximum cantilever condition during launching. Though the push launching technique has been used extensively for concrete and steel box girder decks, it was applied for the first time for truss type steel girders at the Jhajjar bridge, testifying to the resourcefulness of Indian engineers.

17.10 References

1. Naidu, M.P., and Sankaralingam, C., ' Precast segmental construction of Sirsi flyover – Bangalore', Souvenir, IRC Diamond Jubilee Session, Highway Department, Tamil Nadu, Jan. 2000, pp. 97-108.
2. Tambe, S.R., Bongirwar, P.L., and Kanhere, D.K., 'Critical appraisal of Mumbai flyover project', Journal of Indian Roads Congress, Vol. 60-2, Oct. 1999, pp. 235-263.
3. Raghavan, N., and Reddy, R.B., 'Planning and designing urban flyovers – Chennai experience', Proceedings, Conference on Road Infrastructure Management,Bangalore, 2000, pp. 330-338.
4. Banwait, S.P., Tandon, M., Jain, N.M.D., and Goel, R., "Precast concrete segmental flyovers of

Delhi – A sea change in flyover construction technology brought about",Indian Roads Congress Journal, 2002, Vol. 63-1, pp. 147-191.

5. Pritchard, B., 'Bridge design for economy and durability', Thomas Telford, London, 1993.
6. Kasuga, A., 'Extradosed bridges in Japan', In Dhir, R.K., Newlands, M.D., and McCarthy, M.J., (Eds.), 'Role of concrete bridges in sustainable development',Thomas Telford, London, 2003, pp. 17-30.
7. 'Extradosed bridge distinguishes tollway project in India', IEI News, Institution of Engineers (India), 55N(2), May 2005, p. 5.
8. Bhowmick, A., 'Design and detailing of integral bridges: Suggested guidelines', Indian Concrete Journal, Vol 79(9), Sept. 2005, pp. 43-50.
9. Wasserman, E.P., 'Integral abutment practices in the United States', Indian Concrete Journal, Vol 79(9), Sept. 2005, pp. 11-16.
10. Panday, A., and Tandon, M., 'Integral flyovers in Delhi', Indian Concrete Journal, Vol 79(9), Sept. 2005, pp. 26-30.
11. IRC:SP:70-2005, 'Guidelines for the use of high performance concrete in bridges', Indian Roads Congress, 2005, 7 pp.
12. Wright, W., 'High performance steel: Research to practice', Public Roads Federal Highway Administration, 60(4), Spring 1997.
13. Ghosh, P.K., 'Prospect of high performance welded steel bridge[1], Advances in bridge engineering, IIT Roorkee, March 2006, pp 413-419.
14. Larsen and Toubro Limited, 'Launching the highest railway bridge completed in Jammu and Kashmir', Spotlight, Larsen and Toubro Limited – Engineering Construction and Contracts Division, No. sl 108, Feb 2, 2007.

Appendix A

Effect of Concentrated Loads on Deck Slabs

A.1 General

The effect of concentrated loads on deck slabs spanning in one or two directions or on cantilever slabs may be computed using any rational method. The value of Poisson's ratio for concrete may be assumed as 0.15.

In the case of deck slabs spanning in one direction or cantilever slabs, the bending moment per unit width of slab caused by concentrated loads can be calculated by estimating the width of slab that may be taken as effective in resisting the bending moment due to the concentrated loads. The method of assessment of effective width given here is as recommended in clause 305.16.2 of IRC Bridge Code. For precast slabs, the actual width of each precast unit should be taken as the width of slab. Slabs designed on this basis need not be checked for shear.

For slabs spanning in two directions, either Pigeaud's method or Westergaard's method can be used. Pigeaud's method is described in this Appendix.

A.2 Slab Spanning in one Direction

The maximum bending moment caused by a wheel load may be assumed to be resisted by an effective width of slab measured parallel to the supporting edges. This effective width of a single concentrated load is computed from Equation (A.1).

$$b_e = kx\left(1-\frac{x}{L}\right)+b_w \qquad \text{... (A.1)}$$

where b_e = the effective width of slab on which the load acts

L = the effective span in the case of simply supported slabs and equal to the clear span in the case of continuous slabs

x = the distance of the centre of gravity of the concentrated load from the near support

b_w = the breadth of the concentration area of the load, i.e., the dimension of the tyre or track contact area over the road surface of the slab in a direction at right angles to the span plus twice the thickness of the wearing course or surface finish above the structural slab

k = a constant having values as shown in Table A.1 depending on the ratio L'/L where L' is the width of the slab

Obviously the effective width should not exceed the actual width of the slab. Further, when a concentrated load is close to the unsupported edge of a slab, the effective width shall not exceed the above value nor half the above value plus the distance of the load from the unsupported edge.

Table A.1 Values of *k* in Equation (A.1)

$\frac{L'}{L}$	*k* for simply supported slab	*k* for continuous slab	$\frac{L'}{L}$	*k* for simply supported slab	*k* for continuous slab
0.1	0.40	0.40	1.1	2.60	2.28
0.2	0.80	0.80	1.2	2.64	2.36
0.3	1.16	1.16	1.3	2.72	2.40
0.4	1.48	1.44	1.4	2.80	2.48
0.5	1.72	1.68	1.5	2.84	2.48
0.6	1.96	1.84	1.6	2.88	2.52
0.7	2.12	1.96	1.7	2.92	2.52
0.8	2.24	2.08	1.8	2.96	2.60
0.9	2.36	2.16	1.9	3.00	2.60
1.0	2.48	2.24	2.0 and above	3.00	2.60

For two or more concentrated loads in a line in the direction of the span, the bending moment per metre width of slab shall be calculated separately for each load according to its appropriate effective width of slab from Equation (A.1).

For two or more concentrated loads in a direction perpendicular to the direction of the span, it may sometimes happen that the computed effective widths for two adjacent loads overlap. In such cases, the resultant effective width will be equal to the sum of individual widths minus the overlap.

A.3 Cantilever Slab

The effective width of dispersion, measured parallel to the supported edge, for concentrated loads on a cantilever solid slab is to be obtained from Equation (A.2)

$$b_e = 1.2\,x + b_w \qquad \text{... (A.2)}$$

where b_e = effective width

x = distance of the centre of gravity of the concentrated load from the face of the cantilever support

b_w = the breath of the concentration area of the load, i.e., the dimension of the tyre or track contact area over the road surface of the slab in a direction parallel to the suporting edge of the cantilever plus twice the thickness of the wearing course over the structural slab.

The effective width should be limited to one-third the length of the cantilever slab measured parallel to the support. Further, when the concentrated load is placed near one of the two extreme ends of the length of the cantilever slab in the direction parallel to the support, the effective width should not exceed the above value, nor should it exceed half the above value plus the distance of the concentrated load from the nearer extreme end, measured in the direction parallel to the fixed edge.

When two or more loads act on the slab, and when the effective width for one load overlaps the effective width of the adjacent load, the resultant effective width should be taken as the sum of the respective effective widths for each load minus the width of overlap.

A.4 Dispersion of Loads Along the Span

The effective length of slab on which a wheel load or track load acts shall be taken as equal to the dimension of the tyre contact area over the wearing surface of slab in the direction of the span plus twice the overall depth of the slab inclusive of the thickness of the wearing course.

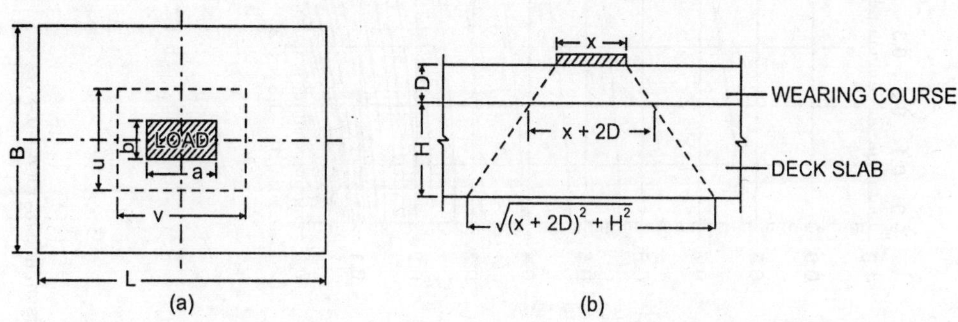

Figure A.1 Dispersion of Live Load Through Deck Slab.

A.5 Distribution Reinforcement

The distribution reinforcement in slabs spanning in one direction shall be provided at right angles to the main reinforcement. This reinforcement is provided to produce a resisting moment equal to 0.3 times the live load moment and 0.2 times the dead load moment. In cantilever slabs the distribution steel is computed to resist a moment equal to 0.3 times the live load moment and 0.2 times the dead load moment, and the steel is provided half at the top and half at the bottom of the slab.

A.6 Slabs Spanning in two Directions

For slabs spanning in two directions, the moments in the two directions can be obtained by any rational method. The use of curves given by M. Pigeaud is recommended.

Pigeaud's method is applicable to rectangular slabs supported freely on all four sides and subjected to a symmetrically placed load (Fig. A.1).

Let L and B be the span lengths in the long and short span directions.

a and b be the dimensions of the tyre contact area in the long and short span directions.

u and v be the dimensions of the load spread after allowing for dispersion through the deck slab.

K be the ratio of short span to long span.

M_1 and M_2 be the moments along the short and long spans.

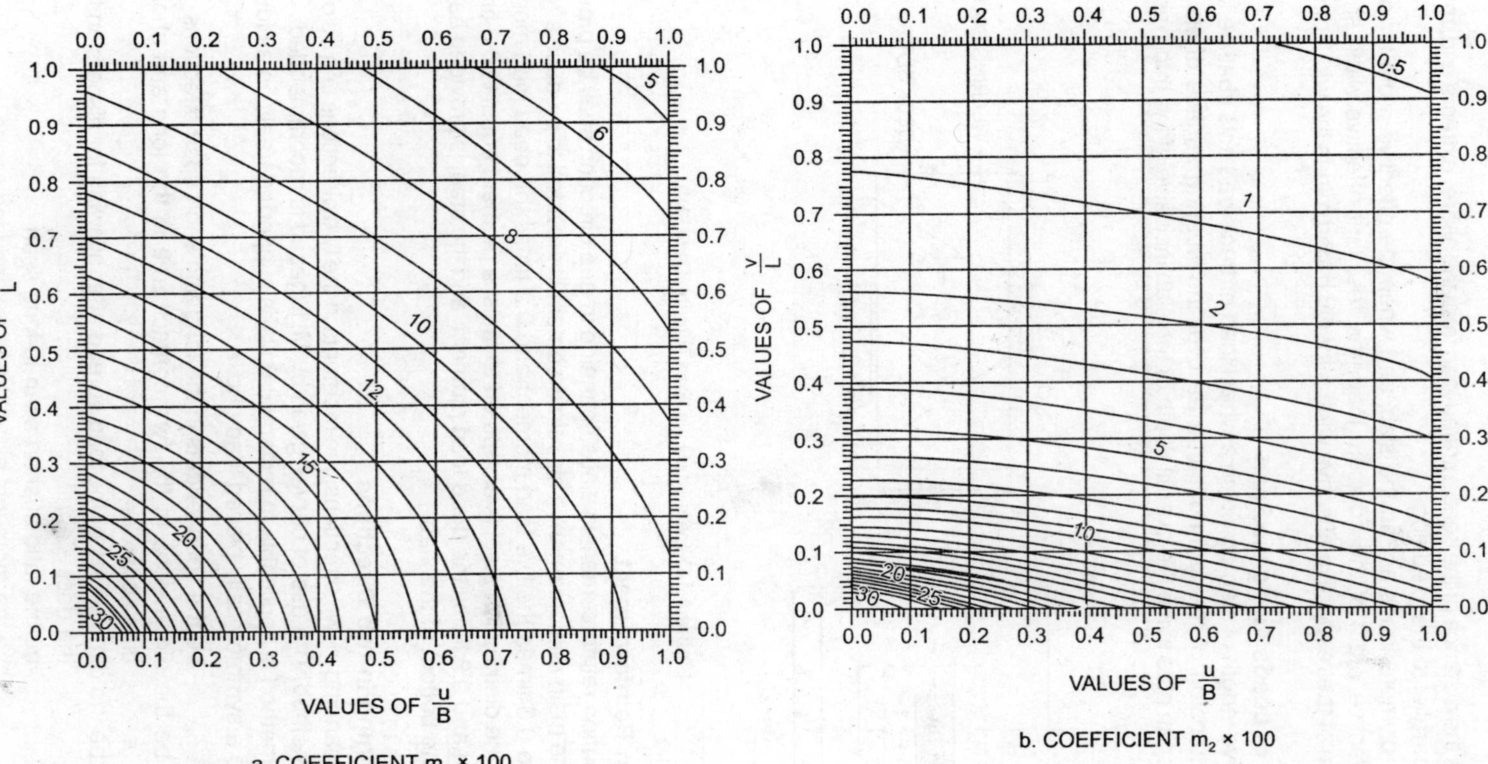

Figure A.2 Moment Coefficients m_1 and m_2 for $K = 0.4$.

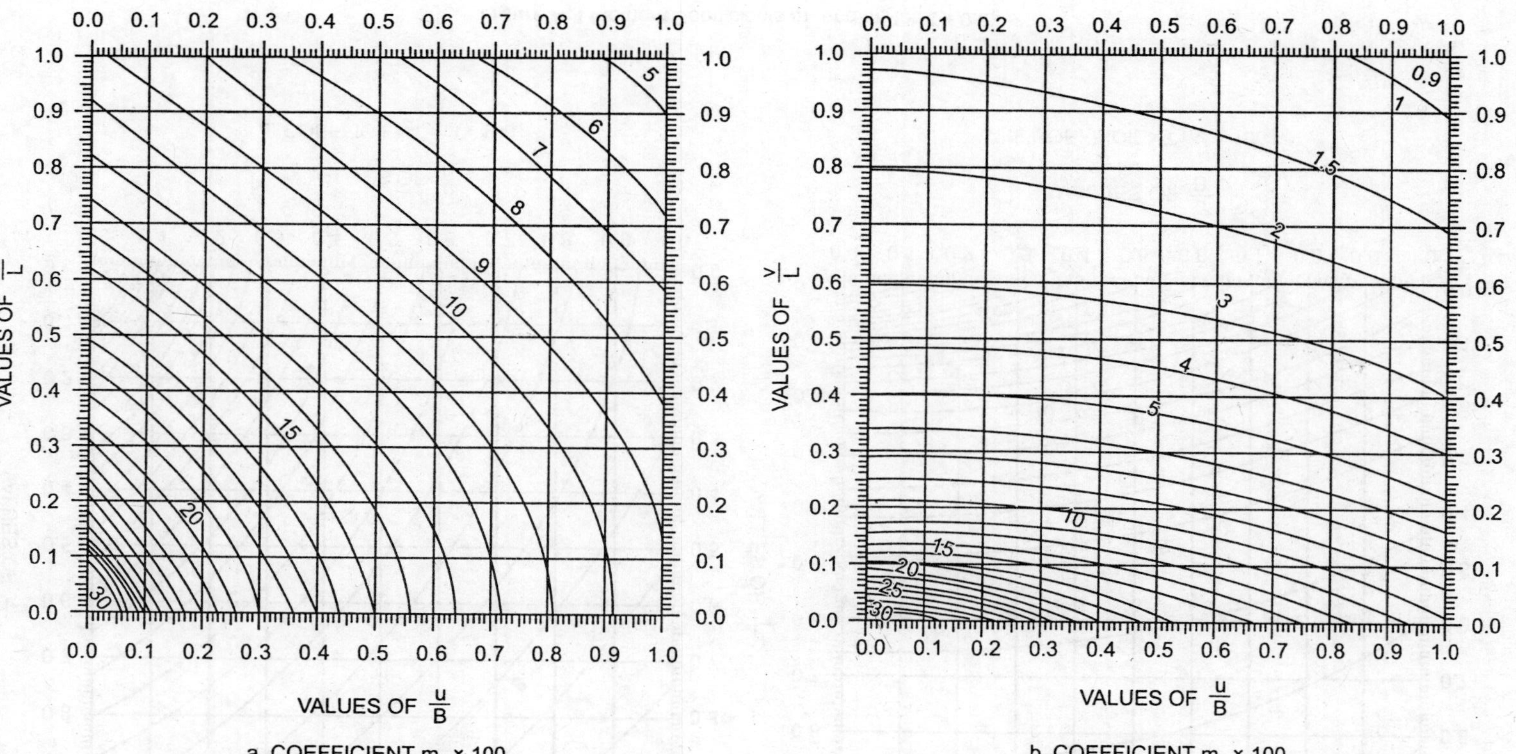

Figure A.3 Moment Coefficients m_1 and m_2 for $K = 0.5$.

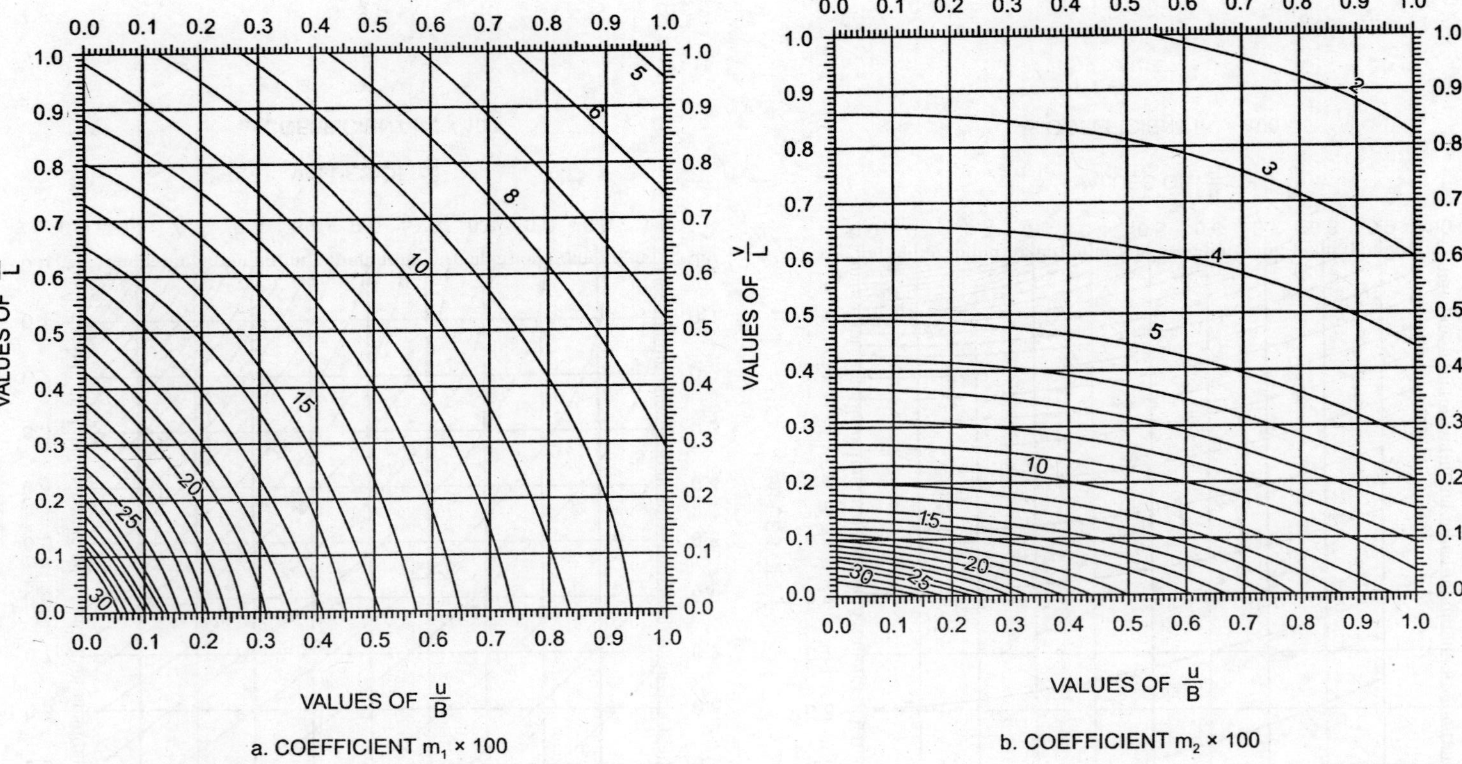

Figure A.4 Moment Coefficients m_1 and m_2 for $K = 0.6$.

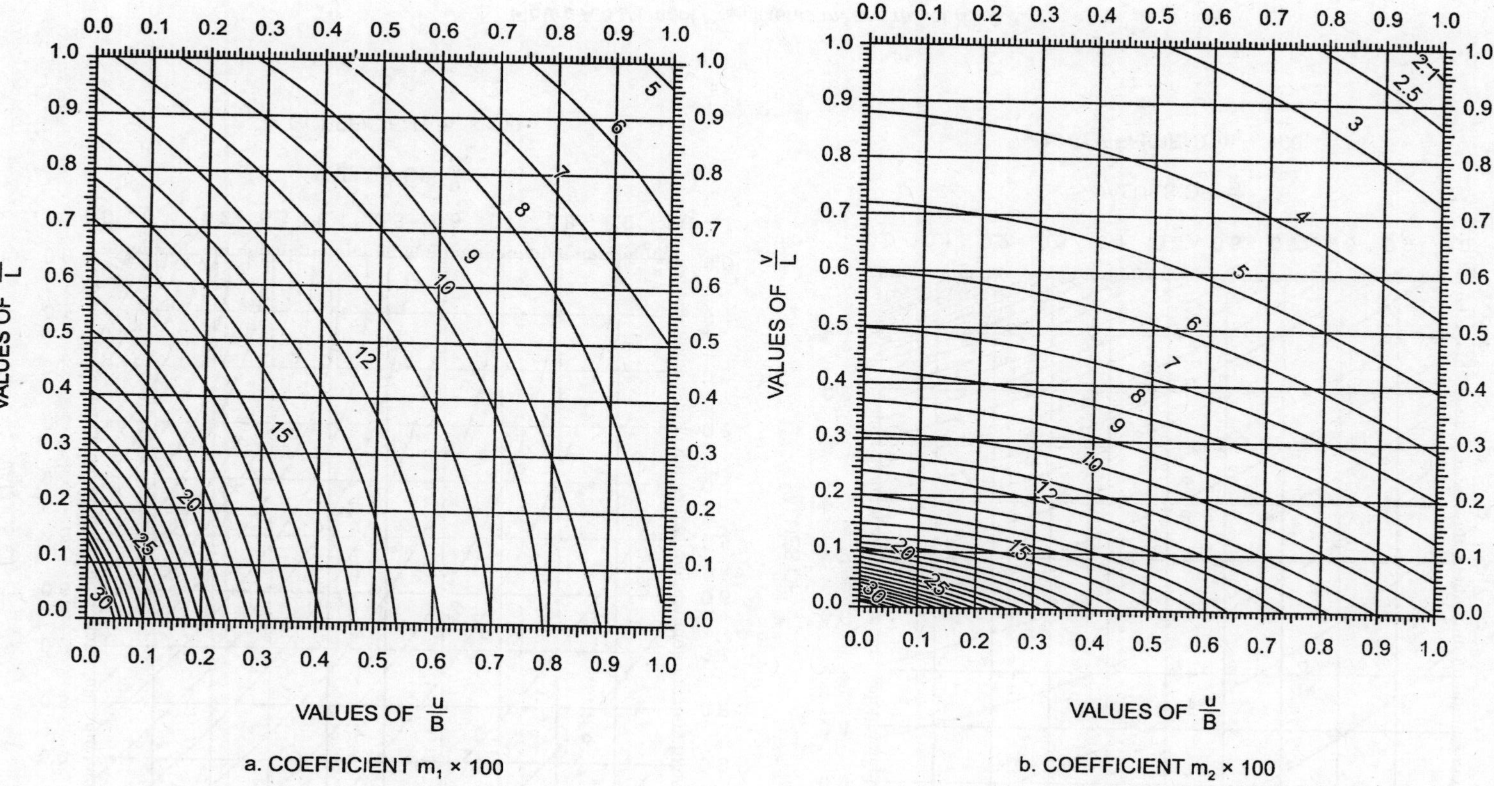

Figure A.5 Moment Coefficients m_1 and m_2 for $K = 0.7$.

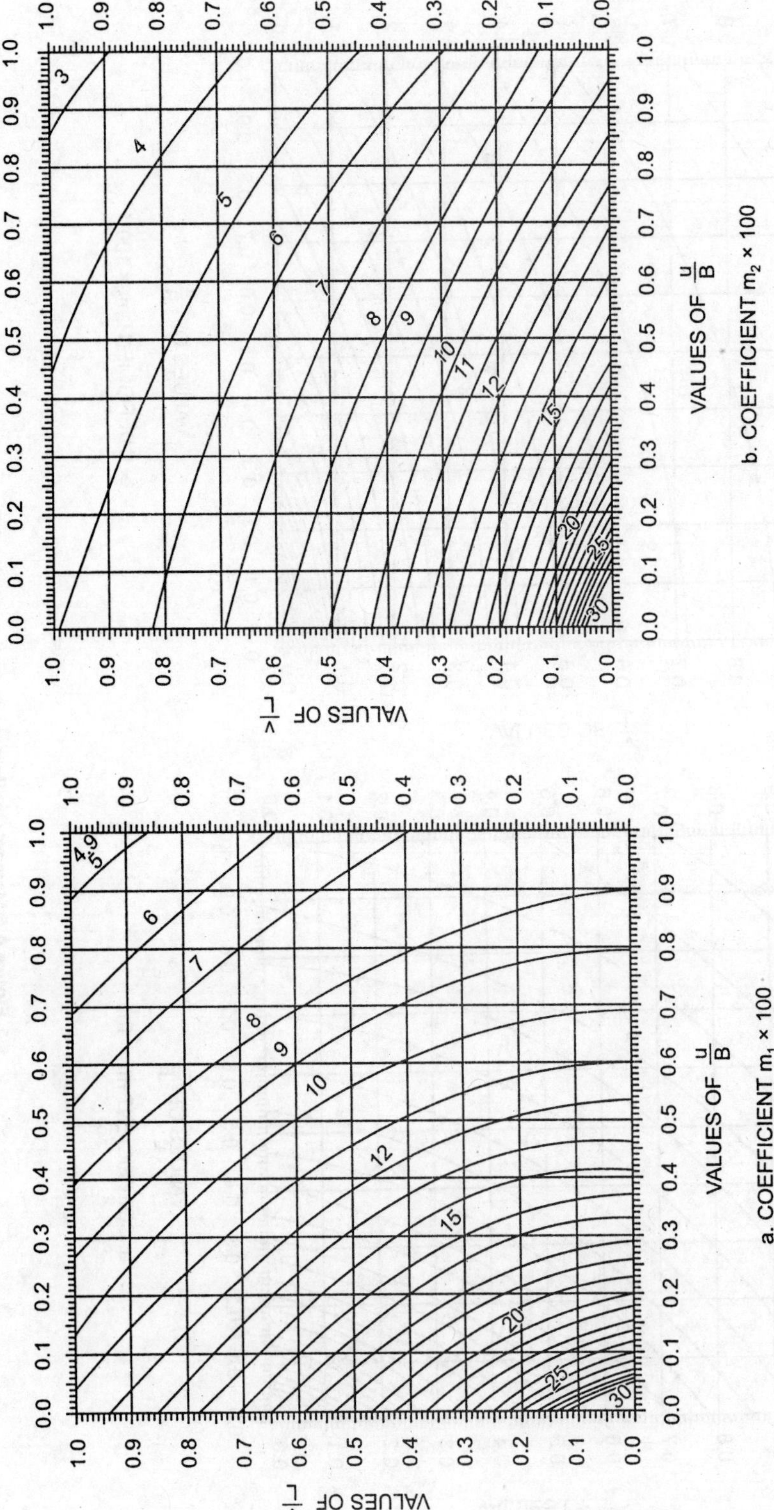

Figure A.6 Moment Coefficients m_1 and m_2 for $K = 0.8$.

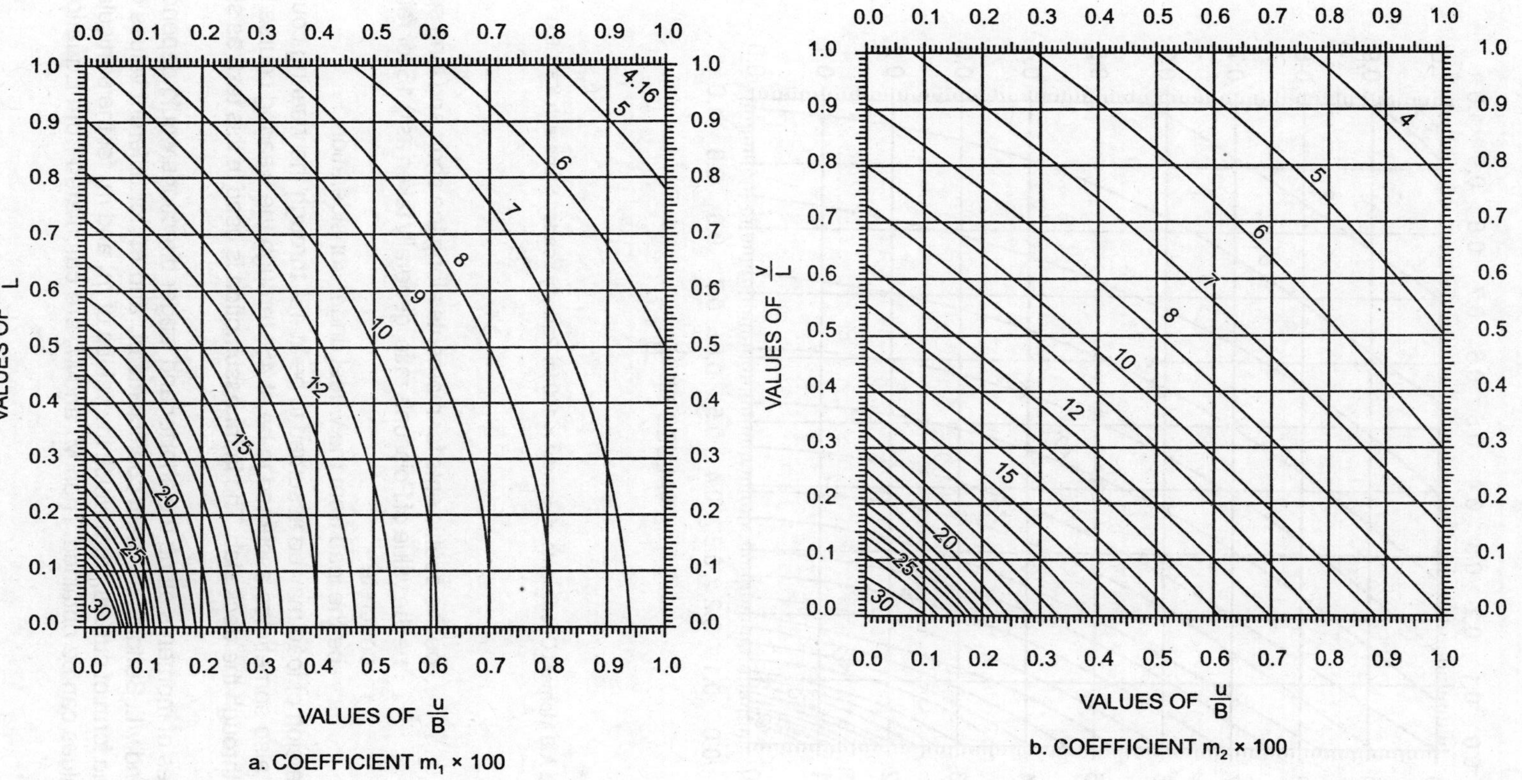

Figure A.7 Moment Coefficients m_1 and m_2 for $K = 0.9$.

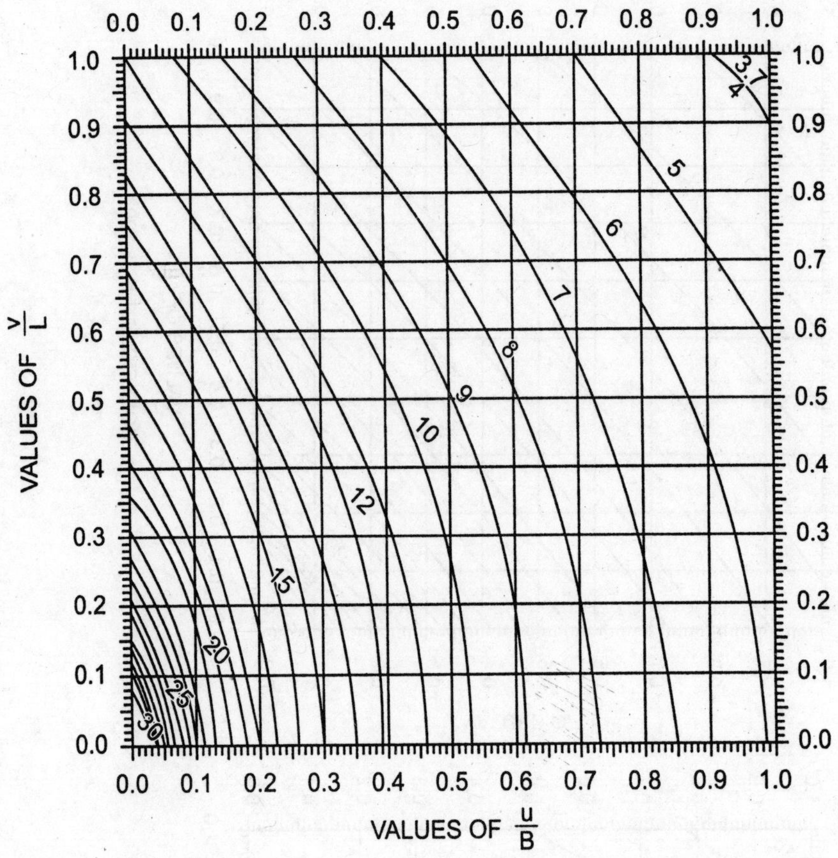

Figure A.8 Moment Coefficients m_1 (or m_2) × 100 for Partially Loaded Slabs with K = 1.0.

m_1 and m_2	be the coefficients for moments along the short and long spans.
μ	be the value of Poisson's ratio, generally taken as 0.15 for reinforced concrete.
P	be the load from the wheel under consideration.

The dispersion of load may be assumed to be at 45° through the wearing course and deck slab. It is also sometimes assumed to be at 45° through the wearing course but at a steeper angle through the deck slab. The latter assumption is used in this text, as shown in Fig. A.1(b).

The values of moments along the shorter and longer directions would depend on the ratios K, u/B and v/L. Bending moment coefficients m_1 and m_2 for different values of K are presented in the form of curves, from which the values of m_1 and m_2 for the particular set of u/B and v/L values can be obtained. Then M_1 and M_2 are calculated from Equations (A.3 and A.4).

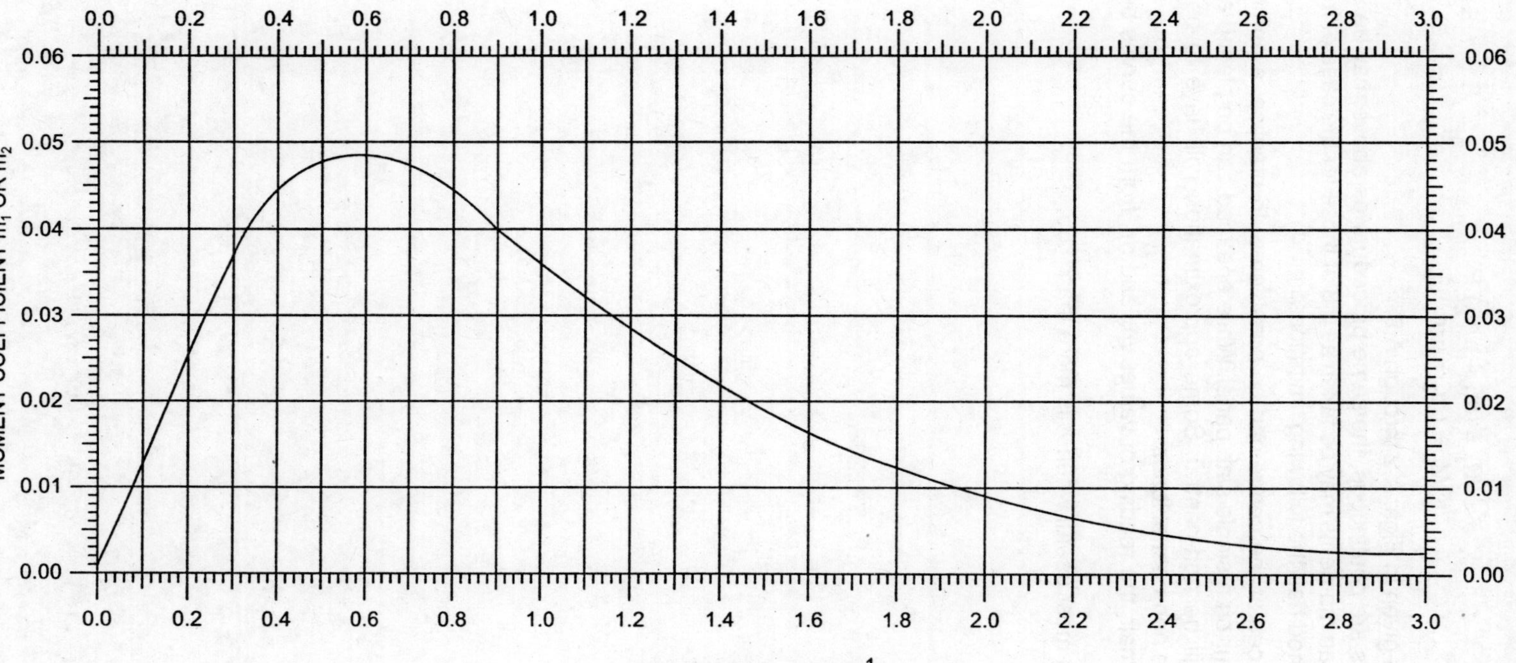

Figure A.9 Moment Coefficients for Slab Completely Loaded with Uniformly Distributed Load, Coefficient is m_1 for K and m_2 for $1/K$.

$$M_1 = (m_1 + \mu m_2)\, P \qquad \text{... (A.3)}$$

$$M_2 = (m_2 + \mu m_1)\, P \qquad \text{... (A.4)}$$

The curves are given in Figs. A.2 through A.9.

When the slabs are continuous, the same procedure as above may be adopted and the maximum midspan moments may be taken as 0.8 of those for the case of free supports.

Pigeaud's method has the following limitations:

(i) Only loads placed at the centre can be considered. In practice, a number of wheel loads may occur on a single slab panel. While one load can be placed at the centre, other loads will be non-central. Some approximation will have to be used while considering the non-central loads.

(ii) When v/L is small, the reading of values m_1 and m_2 from the curves becomes less accurate.

This method is most useful when K is more than 0.55.

Appendix B

Load Distribution Methods for Concrete Bridges

B.1 General

In order to compute the bending moment due to live load in a girder and slab bridge, the distribution of the live loads among the longitudinal girders has to be determined. When there are only two longitudinal girders, the reactions on the longitudinal can be found by assuming the supports of the deck slab as unyielding. With three or more longitudinal girders, the load distribution is estimated using any one of the rational methods. Three of these are as below:

(a) Courbon's method
(b) Hendry-Jaegar method[1]
(c) Morice and Little version of Guyon and Massonnet method[2, 3]. The important features of these three methods are described here.

B.2 Courbon's Method

According to Courbon's method, the reaction R_i of the cross beam on any girder i of a typical bridge consisting of multiple parallel beams is computed assuming a linear variation of deflection in the transverse direction. The deflection will be maximum on the exterior girder on the side of the eccentric load (or c.g. of loads if there is a system of concentrated loads) and minimum on the other exterior girder.

The reaction R_i is then given by

$$R_i = \frac{PI_i}{\Sigma I_i} + \left[\frac{PI_i}{\Sigma I_i} \cdot \frac{ed_i \Sigma I_i}{\Sigma I_i d_i^2} \right]$$

or

$$R_i = \frac{PI_i}{\Sigma I_i} \left[1 + \frac{\Sigma I_i}{\Sigma I_i d_i^2} \cdot ed_i \right] \qquad \text{... (B.1)}$$

where P = total live load
I_i = moment of inertia of longitudinal girder i
e = eccentricity of the live load (or c.g. of loads in case of multiple loads)
d_i = distance of girder i from the axis of the bridge.

When the intermediate and the end longitudinal girders have the same moment of inertia, the quantity I_i in the second term within brackets of Equation (B.1) gets cancelled and the term outside the bracket now reduces to P/n, where n is the number of longitudinal girders. This reduces the amount of computation considerably. In view of the simplicity in calculations, this method is very popular.

Courbon's method is applicable only when the following conditions are satisfied:

(a) The ratio of span to width is greater than 2 but less than 4.
(b) The longitudinal beams are interconnected by symmetrically spaced cross girders of adequate stiffness.
(c) The cross girders extend to a depth of atleast 0.75 of the depth of the longitudinal girder.

These conditions are usually satisfied for majority of modern T-beam bridges.

B.3 Hendry-Jaegar Method

Hendry and Jaegar assume that the cross beams can be replaced in the analysis by a uniform continuous transverse medium of equivalent stiffness. According to this method, the distribution of loading in an interconnected bridge deck system depends on the following three dimensionless parameters.

$$A = \frac{12}{\pi^4}\left(\frac{L}{h}\right)^3 \frac{nEI_T}{EI} \quad \text{... (B.2)}$$

$$F = \frac{\pi^2}{2n}\left(\frac{h}{L}\right)\frac{CJ}{EI_T} \quad \text{... (B.3)}$$

$$C = \frac{EI_1}{EI_2} \quad \text{... (B.4)}$$

where L = the span of the bridge
h = spacing of longitudinal girders
n = number of cross beams
EI, CJ = flexural and torsional rigidities, respectively, of one longitudinal girder
EI_1, EI_2 = flexural rigidities of the outer and inner longitudinal girders, where these are different
EI_T = flexural rigidity of one cross beam.

In case of beam and slab bridge without cross beams, nEI_T in Equations (B.2) and (B.3) is to be replaced by $L.\ EI_T$, where the latter gives the total flexural rigidity of the slab deck. Normally, for reinforced concrete T-beam bridges, the flexural rigidities of the outer and inner longitudinal girders will be nearly equal.

The parameter A is the most important of the above three parameters. It is a function of the ratio of span to the spacing of longitudinals and the ratio of transverse to longitudinal flexural rigidity. The second parameter F is a measure of the relative torsional rigidity of the longitudinals, and is difficult to determine accurately, due to uncertainties surrounding the CJ values for practical girder sections. For T-beam bridges having three or four longitudinals with a number of cross beams, it is usually permissible to employ the distribution coefficients for $F = \infty$. The torsional rigidity of the transverse system is neglected in the analysis.

Graphs giving the values of the distribution coefficients (m) for different conditions of number of longitudinals (two to six) and two extreme values of F, i.e. zero and infinity, are available in Reference 1. Coefficients for intermediate values of F may be obtained by interpolation from Equation (B.5).

$$m_F = m_0 + (m_\infty - m_0)\sqrt{\frac{F\sqrt{A}}{3 + F\sqrt{A}}} \qquad \text{... (B.5)}$$

where m_F is the required distribution coefficient and m_0 and m_∞ are respectively the coefficients for $F = 0$ and $F = \infty$.

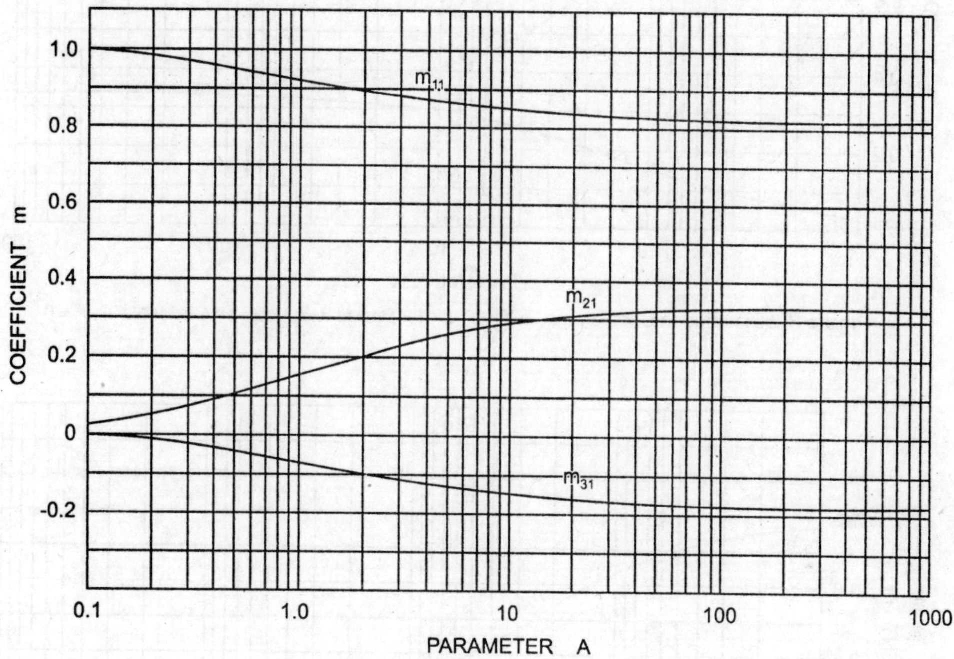

Figure B.1 Distribution Coefficients for Three-girder Bridge with Load on Girder No. 1, $F = 0$

Typical graphs for distribution coefficients for a three-girder system for $F = 0$ and $F = \infty$ are given in Figs. B1 to B4. For more detailed information the reader should refer to Reference 1.

B.4 Morice-Little Method

Morice and Little have applied the orthotropic plate theory to concrete bridge systems, using the approach first suggested by Guyon neglecting torsion and extended by Massonnet to include torsion. This approach has the merit that a single set of distribution coefficients for two extreme cases of no torsion grillage and a full torsion slab enable the distribution behaviour of any type of bridge to be found.

Complete details of the method, including derivation of the governing equations and

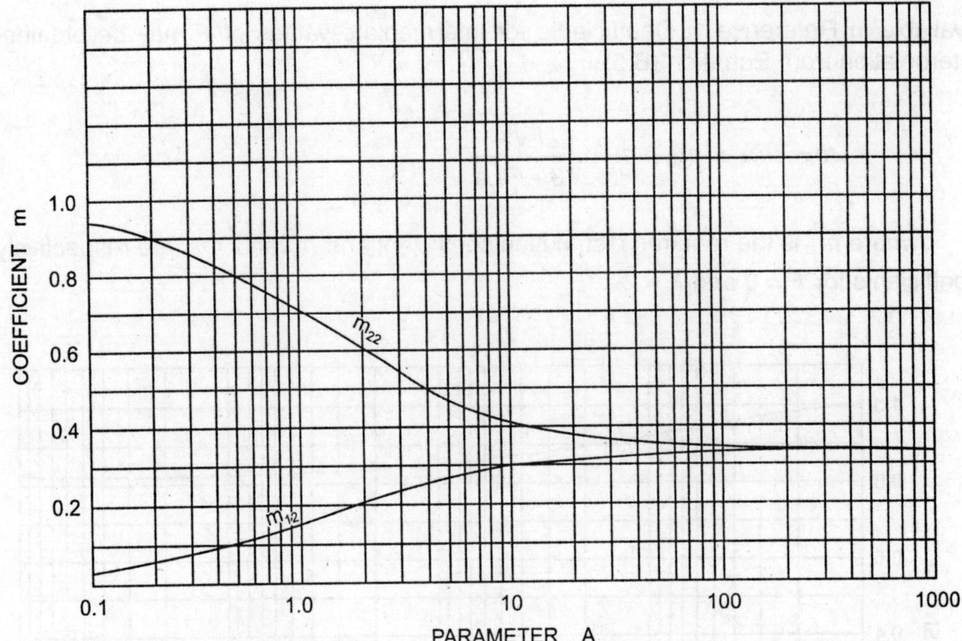

Figure B.2 Distribution Coefficients for Three-girder Bridge with Load on Girder No. 2, $F = 0$

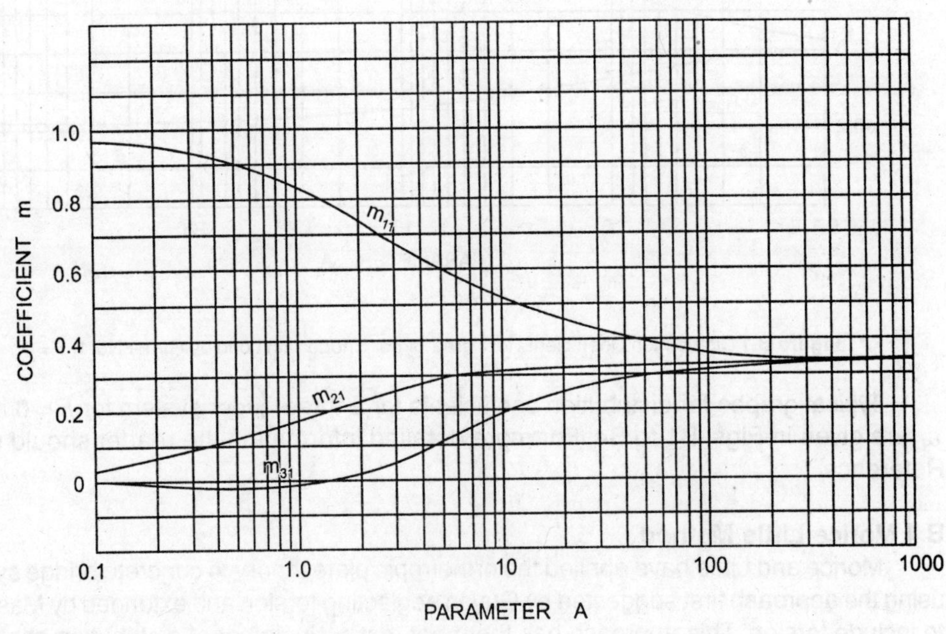

Figure B.3 Distribution Coefficients for Three-girder Bridge with Load on Girder No. 1, $F = \infty$

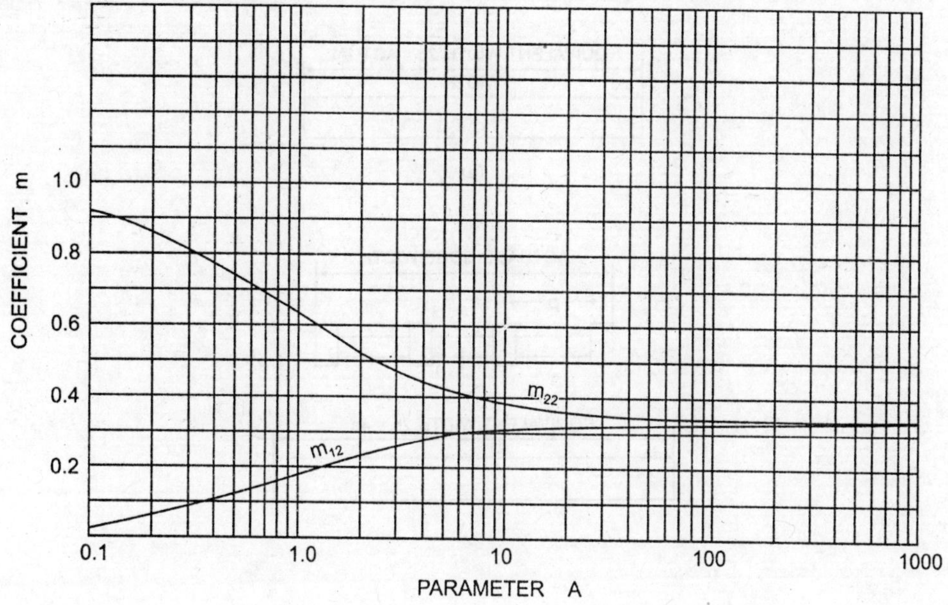

Figure B.4 Distribution Coefficients for Three-girder Bridge with Load on Girder No. 2, $F = \infty$

the full set of graphs useful in the application of this method, are available in Reference 2. Some of the important graphs are presented here, along with a brief description of the parameters involved. The structural properties of the bridge deck under analysis are defined in terms of some special parameters. The effective width of the bridge deck is defined as 2b and is taken as shown in Fig. B.5. For a slab bridge, the effective width is equal to the actual width as in Fig. B.5(a). If the bridge deck consists of a number of longitudinal girders at a spacing p, then the effective width 2b is defined as mp, where m is the number of girders. In some cases, this may lead to an effective width greater than the actual width as in Fig. B.5(b). The actual span of the bridge is taken as 2a.

The flexural properties of the bridge deck as a whole may be expressed by the parameter θ, given by Equation (B.6)

$$\theta = \frac{b}{2a}\left(\frac{i}{j}\right)^{0.25} \qquad \text{... (B.6)}$$

where i = longitudinal second moment of area of the equivalent deck per unit width

$= \dfrac{I}{p}$

I = second moment of area of each longitudinal girder

p = spacing of longitudinal girders

j = transverse second moment of area of the equivalent deck per unit length

J = the second moment of area of each transverse diaphragm or cross beam

q = spacing of diaphragms or cross beams.

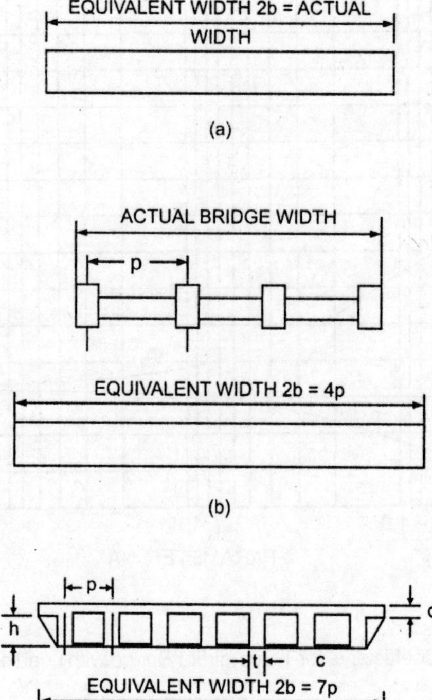

Figure B.5 Effective Width for Bridge Decks.

The torsional properties of the bridge deck as a whole may be expressed by the single parameter α, given by Equation (B.7).

$$\alpha = \frac{G(i_0 + j_0)}{2E\sqrt{ij}} \quad \text{... (B.7)}$$

where E = Young's modulus of the material of deck

G = modulus of rigidity of the material of the deck

i_0 = longitudinal torsional stiffness per unit length

$= \dfrac{I_0}{q}$

I_0 = torsional stiffness constant of a longitudinal girder

j_0 = transverse torsional stiffness per unit length

$= \dfrac{J_0}{q}$

J_0 = torsional stiffness constant of each diaphragm or cross beam.

The flexural parameter θ has a fundamental effect on the load distributing characteristics of the bridge deck structure. The value of α will be zero for a no-torsion grillage and unity for a solid slab which has the maximum torsional stiffness. For all other forms of bridge deck, the parameter α will have *a* value intermediate between 0 and 1.

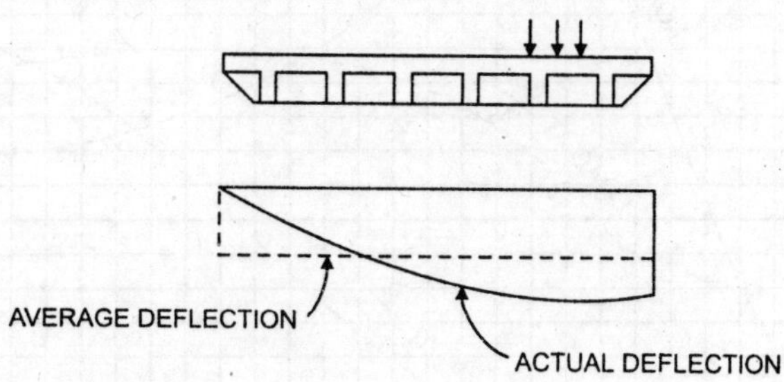

Figure B.6 Transverse Deflection of Bridge Deck.

When a concentrated load is placed on a bridge deck, the transverse section of the deck will tend to be distorted due to varying deflection across its width as shown in Fig. B.6. However, the average deflection of the deck will be the same as if the load were to be distributed uniformly across the full effective width of the deck. For a given structure, the deflection of a point under a given load will depend upon the transverse as well as longitudinal position of the load and also on the structural properties of the deck, i.e. on the values of θ and α.

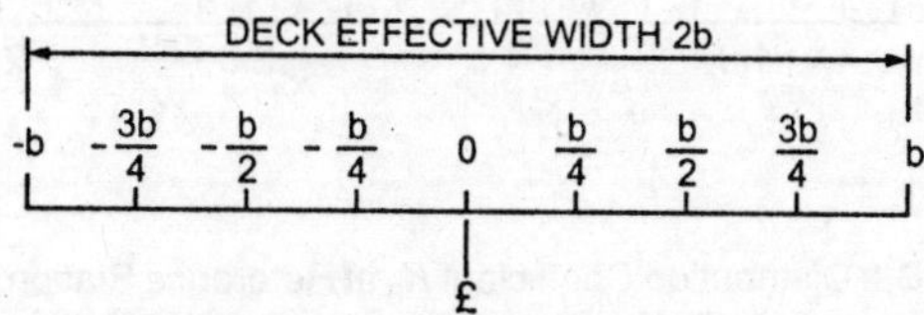

Figure B.7 The Nine Standard Positions.

In the analysis, the effective width is divided into eight equal segments, the nine boundaries of which are known as the standard positions as shown in Fig. B.7. The loadings and deflections at these nine standard positions are considered and all deflections are related to the average deflection. The actual deflection at each of these nine standard positions will be given by an arithmetical coefficient, called distribution coefficient and denoted by the symbol K, multiplied by the average deflection produced by the load distributed uniformly across the entire effective width. Figs. B.8 to B.13 give the distribution coefficient at each standard reference point for a load applied at any of the nine standard positions for a bridge deck with zero torsional stiffness. The value of K for α equal to zero is denoted by K_0. Similar curves for α equal to 1 are given in Figs. B.14 through B.18.

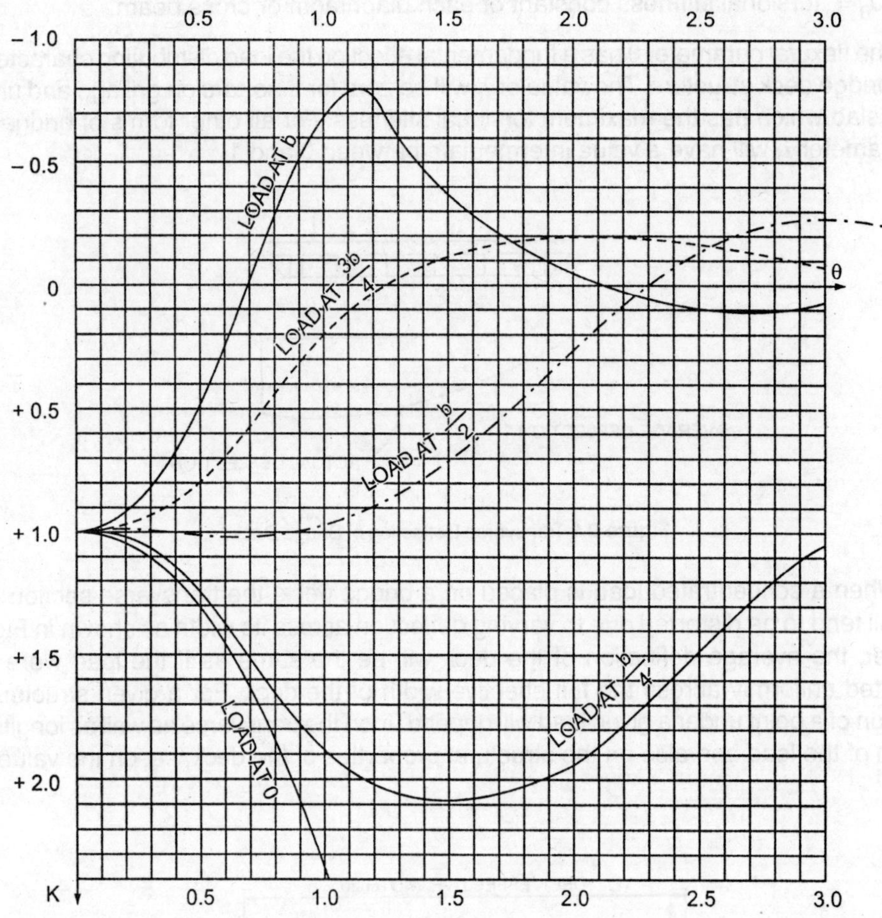

Figure B.8 Distribution Coefficient K_0 at Reference Station 0.

For any intermediate value of α, the distribution coefficient can be obtained from the interpolation relationship given in Equation (B.8).

$$K_\alpha = K_0 + (K_1 - K_0)\sqrt{\alpha} \qquad \text{... (B.8)}$$

where K_α = the distribution coefficient for the actual value of α

K_0 = distribution coefficient for α equal to zero

K_1 = distribution coefiicient for α equal to 1.

The above distribution coefficienfs have been derived for deflections. However these may also be applied for longitudinal bending moments and, therefore, to longitudinal bending stresses. Since the mathematical analysis used for the preparation of the design curves used only the first term of the harmonic series, the bending moment and stress under a

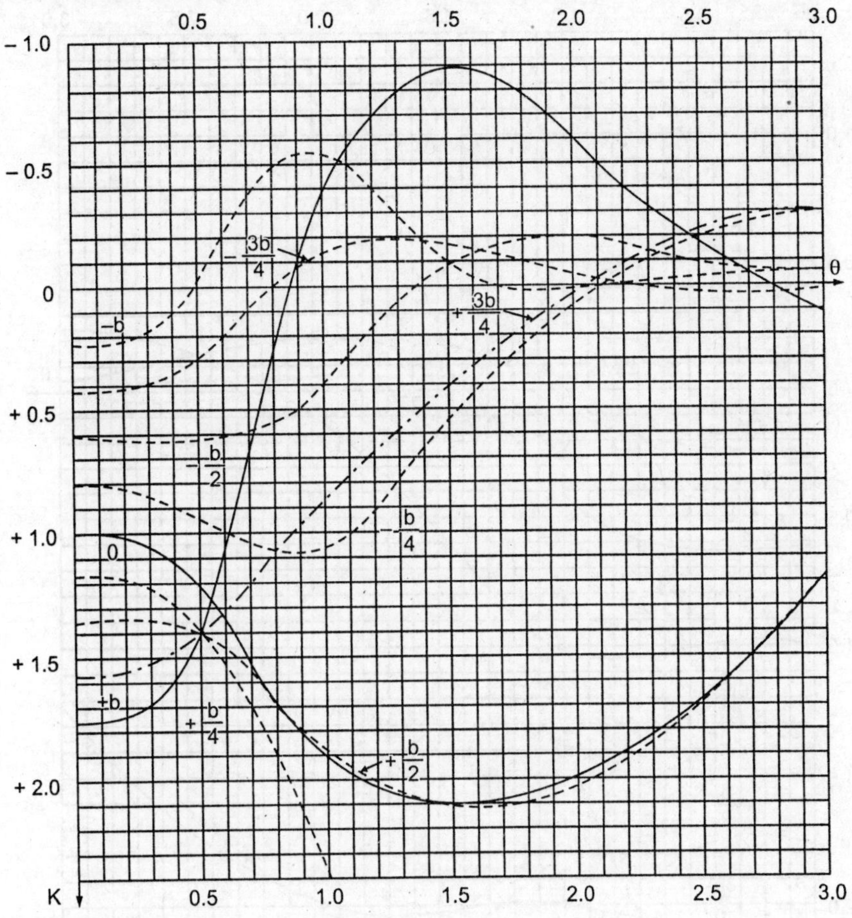

Figure B.9 Distribution Coefficient K_0 at Reference Station b/4.

concentrated load are to be increased by 10 per cent for design.

Transverse bending moments are caused by the unequal deflections across a transverse section due to the application of a concentrated load. The transverse bending moment, M_y is given by Equation (B.9).

$$M_y = \sum_{n=1}^{\infty} \mu_{n\theta} r_n b \sin \frac{n\pi x}{2a} \qquad \text{... (B.9)}$$

where $\mu_{n\theta}$ = distribution coefficient analogous to K for deflection

r_n = the nth coefficient of the Fourier series representing the longitudinal disposition of the load.

In practice, it is normally adequate to consider only five terms of the series in Equation

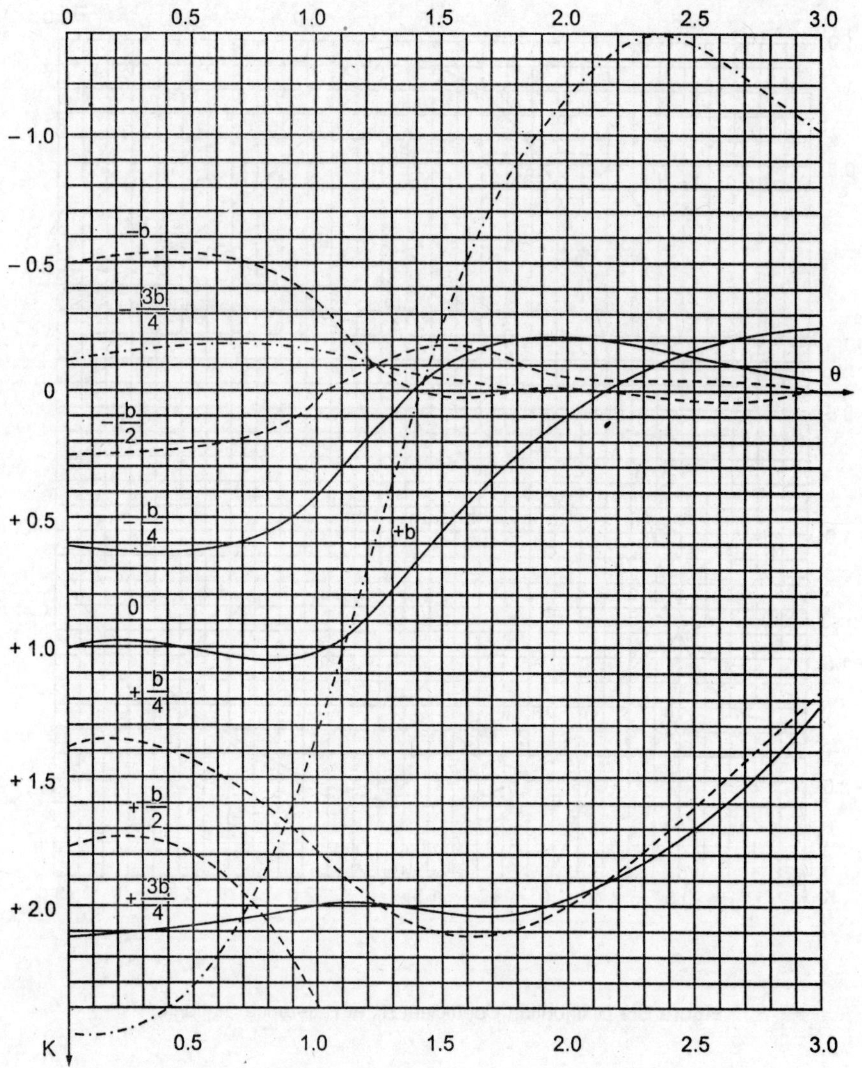

Figure B.10 Distribution Coefficient K_0 at Reference Station b/2.

(B.9), which would then reduce to Equation (B.10).

$$M_y = (\mu_\theta r_1 - \mu_{3\theta} r_3 + \mu_{5\theta} r_5) \quad \text{... (B.10)}$$

For the case of a concentrated load W at a distance u from the left support, r_n is given by Equation (B.11).

$$r_n = \frac{W}{a} \sin \frac{n \pi u}{2a} \quad \text{... (B.11)}$$

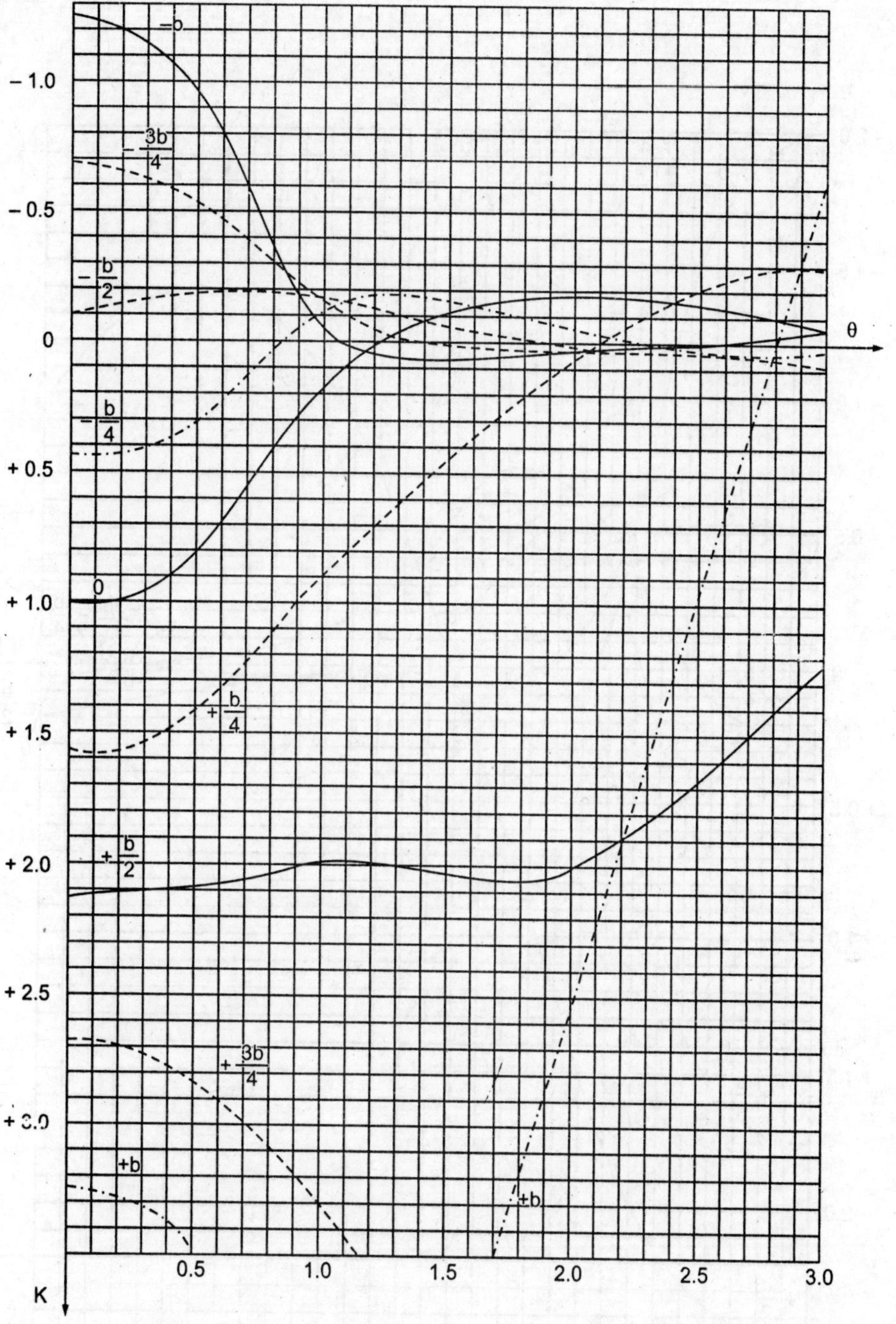

Figure B.11 Distribution Coefficient K_0 at Reference Station 3b/4.

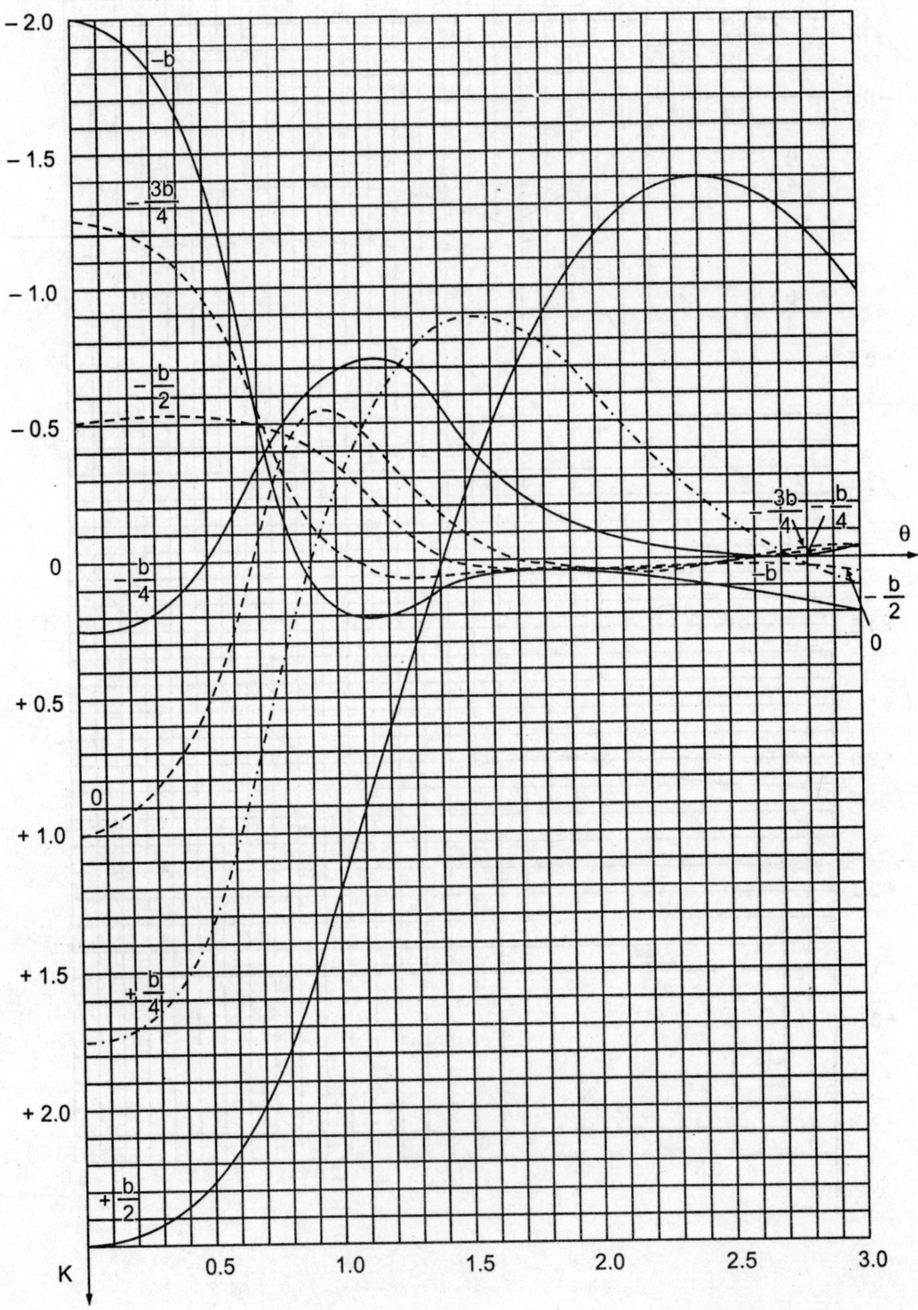

Figure B.12 Distribution Coefficient K_0 at Reference Station *b*.

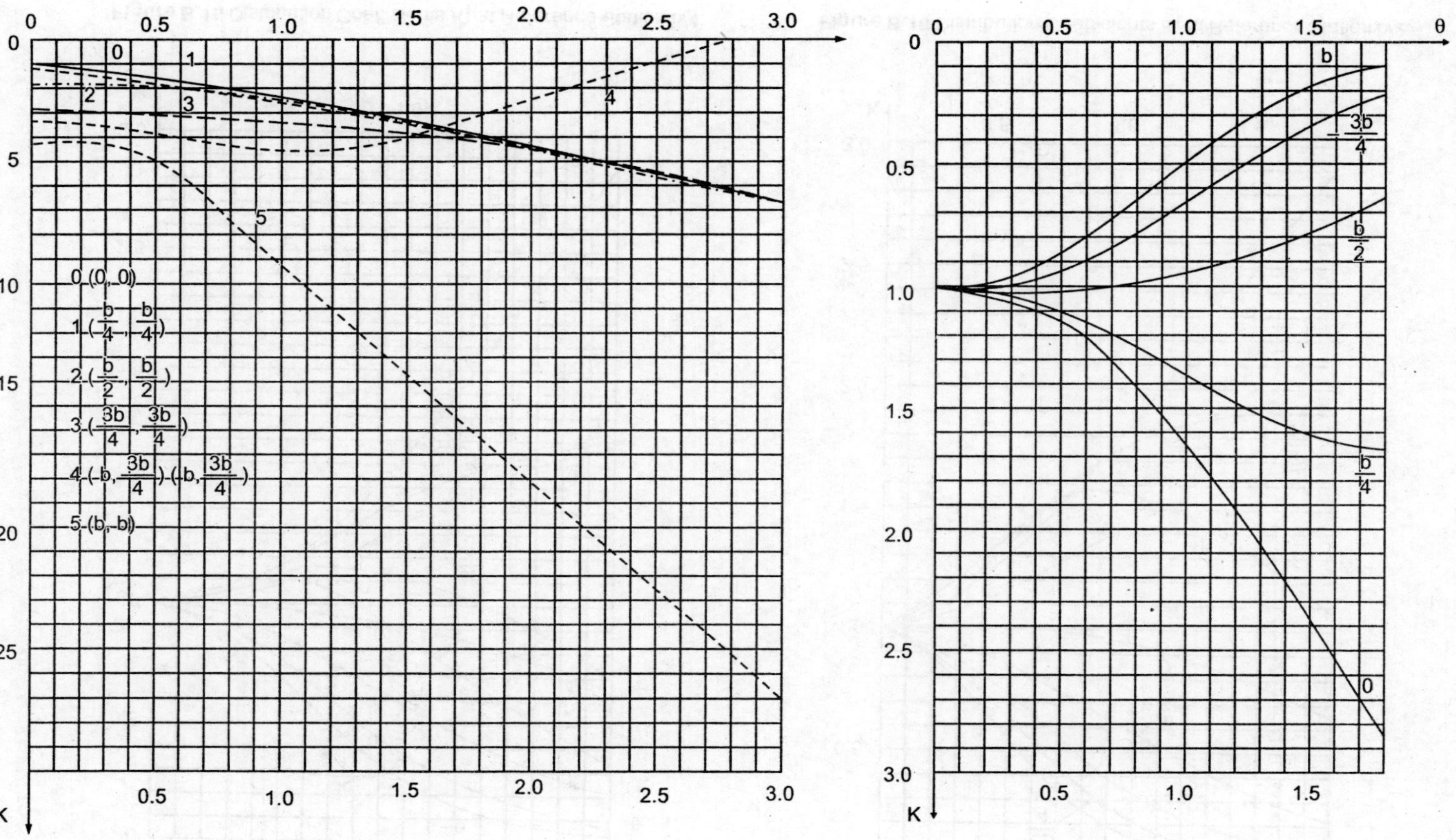

Figure B.13 Large Range Distribution Coefficients K_0.

Figure B.14 Distribution Coefficients K_1 at Reference Station 0.

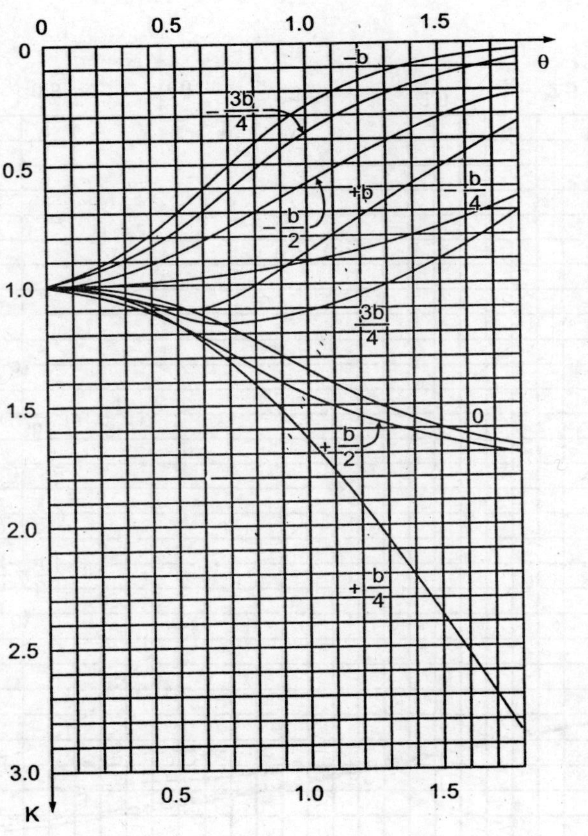

Figure B.15 Distribution Coefficients K_1 at Reference Station b/4.

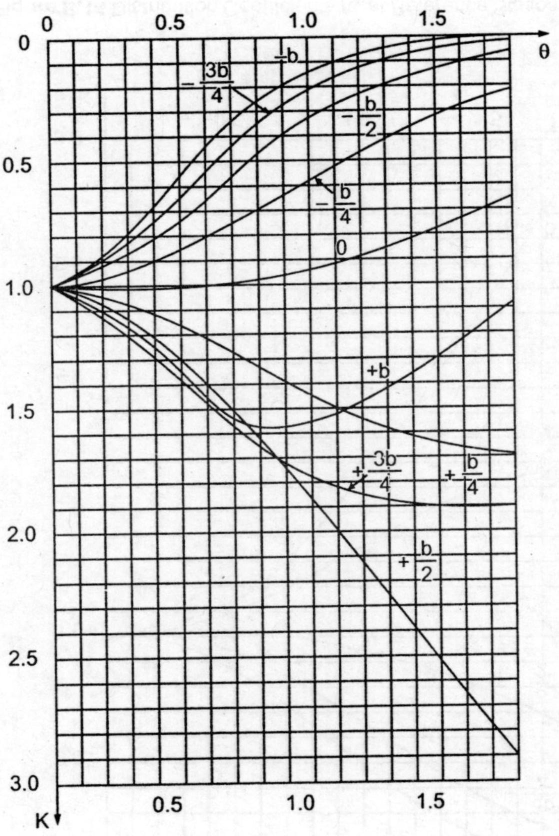

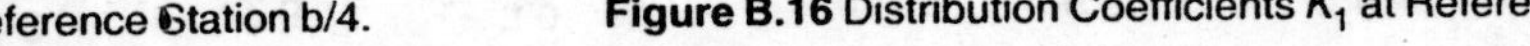

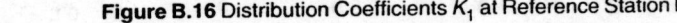

Figure B.16 Distribution Coefficients K_1 at Reference Station b/2.

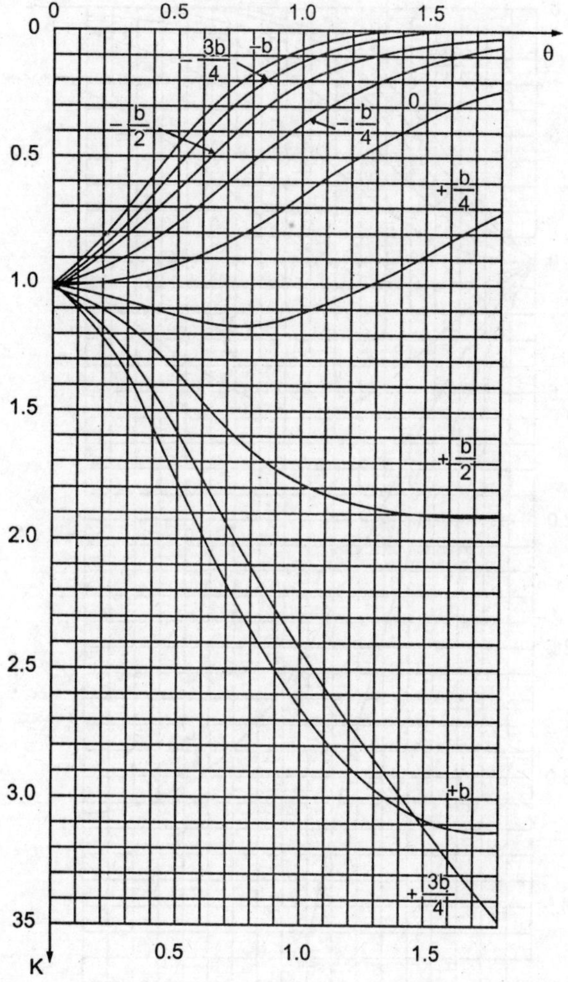

Figure B.17 Distribution Coefficient K_1 at reference station 3b/4

Thus Equation (B.10) gets modified to Equation (B.12)

$$M_y = \frac{Wb}{a}\left(\mu_\theta \sin\frac{\pi u}{2a} - \mu_{3\theta} \sin\frac{3\pi u}{2a} + \mu_{5\theta} \sin\frac{5\pi u}{2a}\right) \qquad \text{... (B.12)}$$

The greatest transverse moments occur with a concentrated load, when it is nearest to the longitudinal axis. In bridges with no central island, it is generally adequate to use the curves for the central standard position only. The curves for μ_0 and μ_1 for the standard position 0 for α equal to zero and unity, respectively, are given in Figs. B.19 and B. 20. The value of μ corresponding to any intermediate value of α can be evaluated using the interpolation relation in Equation (B.13).

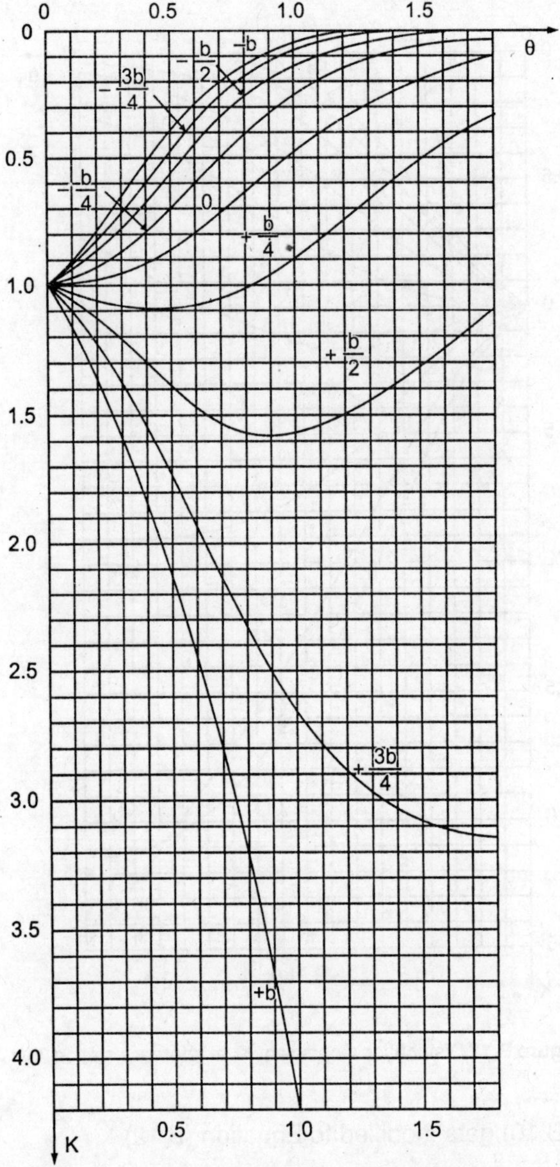

Figure B.18 Distribution Coefficient K_1 at Reference Station *b*.

$$\mu_\alpha = \mu_0 + (\mu_1 - \mu_0)\sqrt{\alpha} \qquad \text{... (B.13)}$$

B.5 Use of Morice-Little Method for T-beam Bridge Design

(i) *General*

The use of Morice-Little method for the design of a T-beam bridge is outlined here by

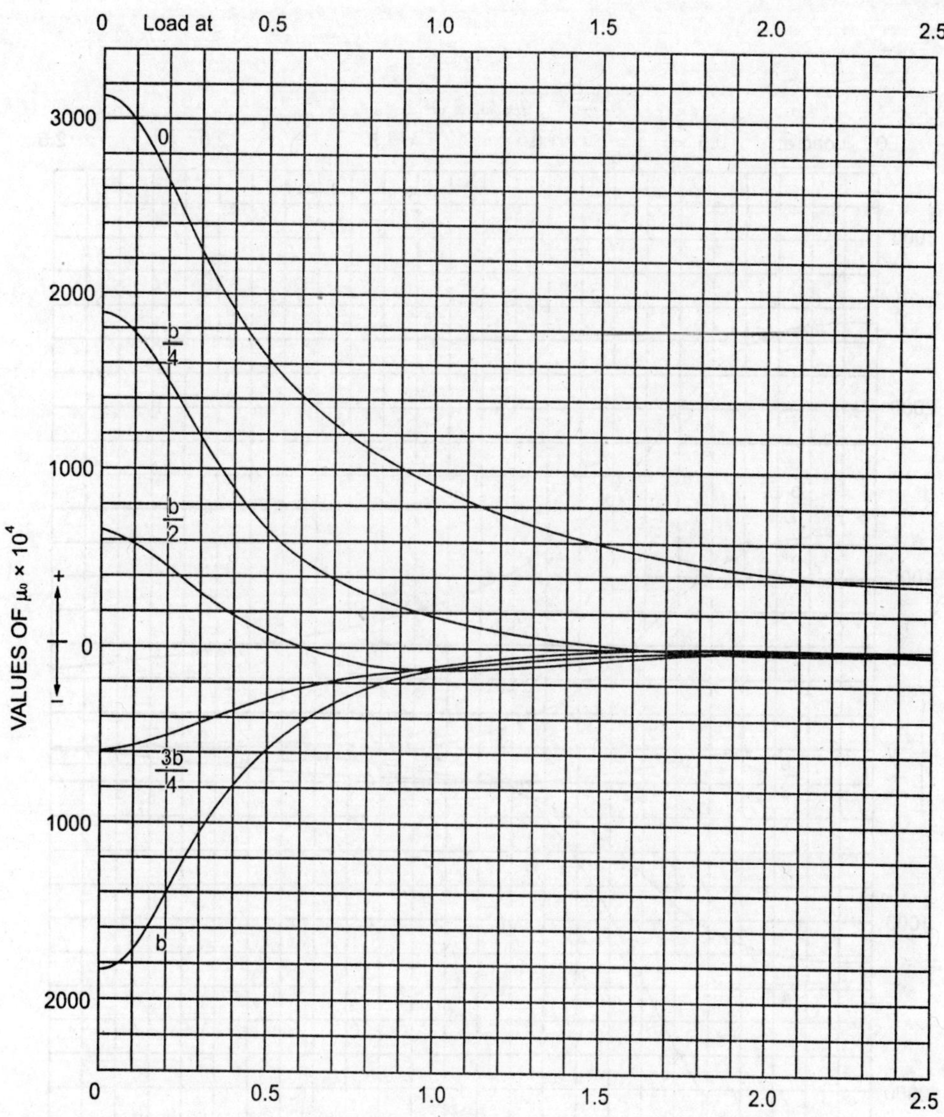

Figure B.19 Transverse Moment Coefficients μ_0 at Reference Station 0.

applying to the example of Section 7.3.

(ii) *Properties of structure*

Span 2a	=	14.5 m
No. of longitudinal girders n	=	3
Girder spacing p	=	2.5 m
Effective width 2b = np	=	7.5 m
No. of cross beams m	=	5

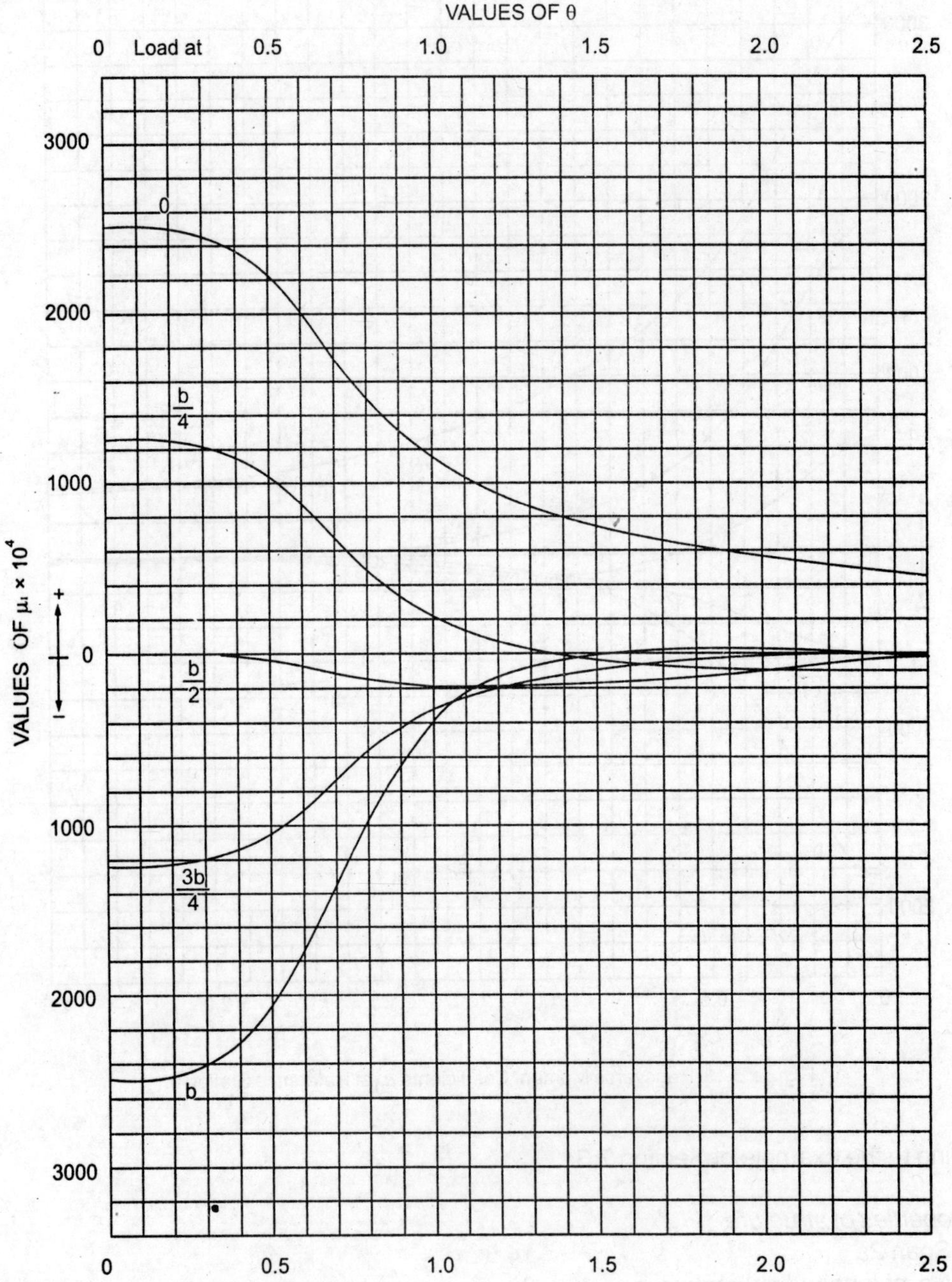

Figure B. 20 Transverse Moment Coefficients μ_1 at Reference Station 0.

Spacing of cross beams q = 3.625 m
Moment of inertia of girder I = 0.4347 m^4
Distributed longitudinal stiffness $i = I/p$ = 0.1739 m^4/m
Moment of inertia of cross beam j = 0.1229 m^4
Distributed transverse stiffness $j = J/q$ = 0.0339 m^4/m

Bending stiffness parameter $\theta = \dfrac{b}{2a} \cdot (i/j)^{0.25} = 0.39$

Torsional stiffness of longitudinal girder is computed by using Bach's approximation of considering the total stiffness as the sum of stiffnesses of component rectangles.

Torsional stiffness of girder $I_0 = 0.026\ m^4$

Distributed torsional stiffness of girder $i_0 = \dfrac{I_0}{q} = 0.0104\ m^4/m$

Torsional stiffness of cross beam $J_0 = 0.0103\ m^4$

Distributed torsional stiffness of cross beam $j_0 = \dfrac{J_0}{q} = 0.0028\ m^4/m$

Torsional parameter $\alpha = \dfrac{G}{2E} \cdot \dfrac{(i_0 + j_0)}{\sqrt{i.j}}$

where G is the shear modulus
E is Young's modulus

$$G = \frac{E}{2(1+\mu)}$$

where μ is Poisson's ratio, taken as 0.15.

$$\text{Hence } \frac{G}{2E} = \frac{1}{4(1+\mu)} = \frac{1}{4.60}$$

$$\alpha = \frac{1}{4.6} \times \frac{0.0132}{\sqrt{0.1739 \times 0.0339}}$$
$$= 0.037$$

The two main parameters θ and α defining the properties of the structure are thus evaluated.

(ii) *Unit load distribution coefficients*

The transverse width 2b is divided into 8 equal parts and the nine boundaries thus obtained are called reference stations $-b, -\frac{3}{4}b, -\frac{b}{2}, -\frac{b}{4}, 0, +\frac{b}{4}, +\frac{b}{2}, +\frac{3}{4}b$, and b, starting from the left end.

For the two extreme values of α equal to 0 for a no-torsion grillage and 1 for a full-torsion grillage, the curves given in Figs. B. 8 to B.18 for distribution coefficients K_0 or K_1, as the case may be, for the nine reference stations for loading at each of these reference stations are used. Normally, it is adequate and conservative to use the coefficients for α equal to zero. However, for a more exact calculation, the value of each coefficient corresponding to any intermediate value of α may be computed using the interpolation formula given in Equation (B.8). Here α is taken as zero.

The final unit load distribution coefficients K corresponding to θ equal to 0.39 and α equal to 0 are listed in Table B.1. The last column in this table gives the area enclosed between the influence ordinate curve and the base line, computed by simpson's rule. Each of these values should be as near to 8.0 as possible within the accuracy possible with graphical readings. Also, by Maxwell's reciprocal theorem, the distribution coefficient for station A for a unit load at *B* should be equal to the distribution coefficient for station *B* for a unit load at *A*. These two conditions facilitate checks being applied on the arithmetical work up to this stage of computations.

(iv) *Equivalent loads λP at the nine standard positions*

The actual positions of live loads will usually be different from the nine standard positions. In order to apply the coefficients of Table B.1, equivalent loads, denoted by λP, are computed assuming the portion between any two standard positions as simply supported.

Table B.1 Unit Load Distribution Coefficients K_0

Load Position	Reference Station									Row integral
	$-b$	$-\frac{3}{4}b$	$-\frac{b}{2}$	$-\frac{b}{4}$	0	$+\frac{b}{4}$	$+\frac{b}{2}$	$+\frac{3}{4}b$	$+b$	
$-b$	3.50	3.30	2.40	1.55	0.72	0.11	−0.55	−1.10	−1.65	8.28
$-\frac{3}{4}b$	3.30	2.75	2.08	1.47	0.87	0.35	−0.20	−0.60	−1.12	8.90
$-\frac{b}{2}$	2.40	2.08	1.74	1.36	0.99	0.62	0.22	−0.20	−0.55	8.66
$-\frac{b}{4}$	1.55	1.47	1.36	1.28	1.10	0.90	0.62	0.35	0.11	8.74
0	0.72	0.87	0.99	1.10	1.17	1.10	0.99	0.87	0.72	8.53
$+\frac{b}{4}$	0.11	0.35	0.62	0.90	1.10	1.28	1.36	1.47	1.55	8.74
$+\frac{b}{2}$	−0.55	−0.20	0.22	0.62	0.99	1.36	1.74	2.08	2.40	8.66
$+\frac{3}{4}b$	−1.12	−0.60	−0.20	0.35	0.87	1.47	2.08	2.75	3.30	8.90
$+b$	−1.65	−1.10	−0.55	0.11	0.72	1.55	2.40	3.30	3.50	8.28

Many alternative combinations of load positions should normally be tried, and the position giving the severest bending moment should finally be adopted for design.

In this case, only the load disposition for Class A double lane loading as shown in Fig. 7.8 will be used for the illustrative computation. The applied and equivalent loads are shown in Fig. B.21.

(v) *Distribution Coefficients K′ for Class A Loading*

Multiplying the unit load distribution coefficients K_α by the equivalent load coefficients

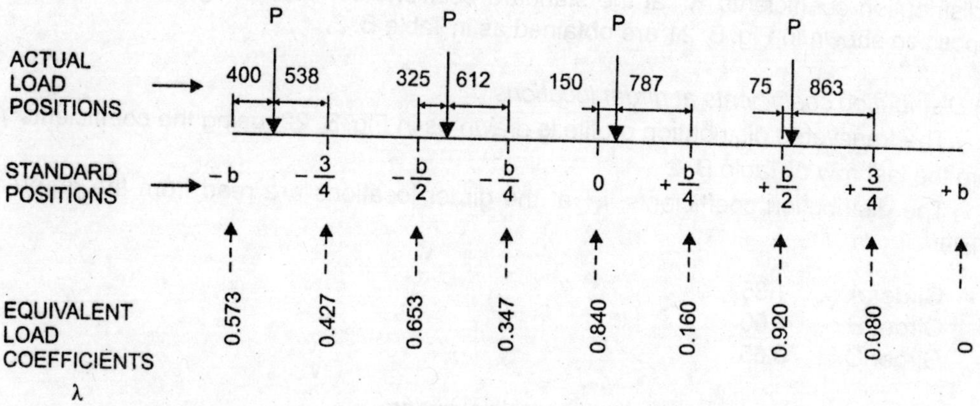

Figure B.21 Equivalent Loads λP at Standard Positions for Class A Loading.

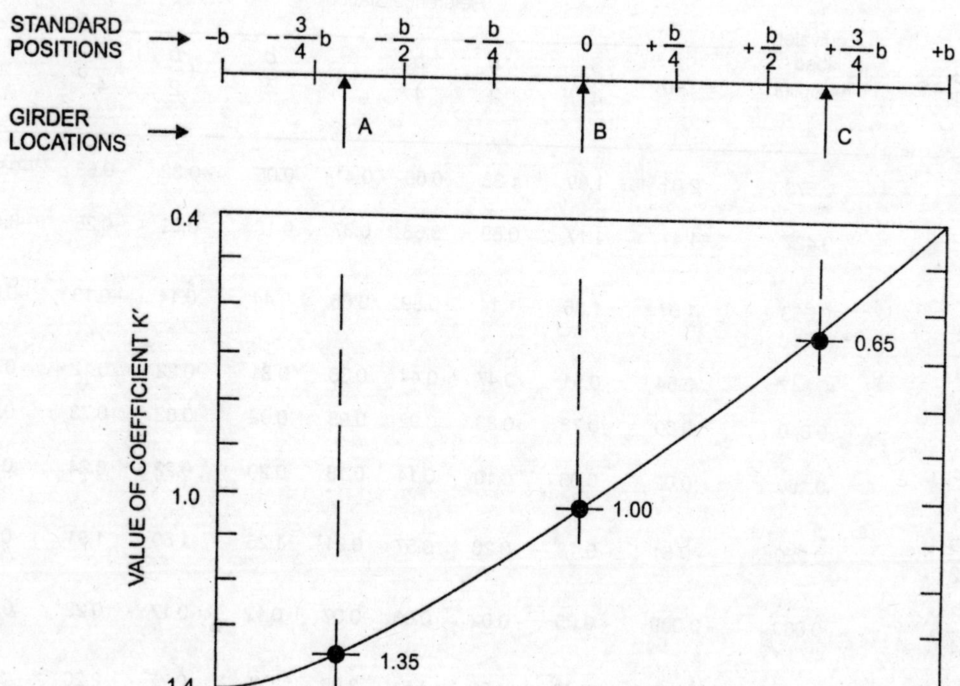

Figure B.22 Transverse Distribution Profile for Class A Loading.

λ, distribution coefficients K' at the standard positions corresponding to the actual load disposition shown in Fig. B. 21 are obtained as in Table B. 2.

(vi) *Distribution coefficients at girder locations*

The transverse distribution profile is drawn as in Fig. B. 22, using the coefficients K' from the last row of Table B. 2.

The distribution coefficients K' at the girder locations are read from the graph as below:

Girder A . . . 1.35
Girder B . . . 1.00
Girder C . . . 0.65

(vii) *Maximum bending moments on longitudinal girders*

The maximum bending moment for the structure as a whole is computed and the total moment will be divided among the three girders in the proportion shown above.

Using the load disposition shown in Fig. 7.9 and considering two lanes of Class A loading with impact, the maximum bending moment can be computed as 2257 kN.m.

Table B.2 Distribution Coefficient *K'* for Class *A* Loading.

Load Position	Equivalent load multiplier λ	Reference Station								
		$-b$	$+\frac{3}{4}b$	$-\frac{b}{2}$	$-\frac{b}{4}$	0	$+\frac{b}{4}$	$+\frac{b}{2}$	$+\frac{3}{4}b$	$+b$
$-b$	0.573	2.01	1.89	1.38	0.89	0.41	0.06	–0.32	–0.63	–0.95
$-\frac{3}{4}b$	0.427	1.41	1.17	0.89	0.63	0.37	0.15	–0.09	–0.26	–0.48
$-\frac{b}{2}$	0.653	1.57	1.36	1.14	0.89	0.65	0.44	0.14	–0.13	–0.35
$-\frac{b}{4}$	0.347	0.54	0.51	0.47	0.44	0.38	0.31	0.22	0.12	0.04
0	0.840	0.60	0.73	0.83	0.92	0.98	0.92	0.83	0.73	0.60
$+\frac{b}{4}$	0.160	0.02	0.06	0.10	0.14	0.18	0.20	0.22	0.24	0.25.
$+\frac{b}{2}$	0.920	−0.51	−0.18	0.20	0.57	0.91	1.25	1.60	1.91	0.20
$+\frac{3}{4}b$	0.080	−0.009	−0.05	−0.02	0.03	0.07	0.12	0.17	0.22	0.26
$+b$	0									
K_α		5.55	5.47	4.99	4.51	3.95	3.45	2.77	2.20	1.56
$K' = \frac{\lambda K_\alpha}{4}$		1.39	1.37	1.25	1.13	0.99	0.86	0.69	0.55	0.39

Table B.3 Comparison of Results of Bending Moments

Method	End girder kN.m	Intermediate girder kN.m
Courbon	1066	750
Hendry-Jaegar	790	753
Morice-Little	1117	828

As indicated in Section B.4, the maximum bending moment for the girders will be increased by 10%.

$$\text{Bending moment for girder } A = \frac{1.10 \times 1.35}{3.0} \times 2257 = 1117 \text{ kN.m}$$

$$\text{Bending moment for girder } B = \frac{1.10 \times 1.00}{3.0} \times 2257 = 828 \text{ kN.m}$$

B.6 Comparison of Results by Three Methods

The bending moments computed by the three methods are tabulated in Table B.3. Each of the three methods of analysis indicates that the exterior girder is subjected to severer loading than the intermediate girder. It is seen that the bending moment for the intermediate girder is nearly the same by the first two methods. The third method over-estimates the moment at the intermediate girder by about 10% as compared with the values computed from the other two methods. However, Hendry-Jaegar method under-estimates the moment at the end girder by about 27% relative to the value by the other two methods.

B.7 References

1. Hendry, A.W., and Jaegar, L.G., "The analysis of grid frameworks and related structures", Chatto and Windus, London, 1958, 308 p.
2. Rowe, R.E., "Concrete bridge design", C.R. Books Ltd., London, 1962, First Edition, 336 pp.
3. Morice, P.B., and Little, G., "Analysis of right bridge decks subjected to abnormal loading", Cement and Concrete Association, London, July 1956, 43 pp.

Appendix C

Statistics of Record Bridges

Table C.1 Suspension Bridges

Year	Bridge	Location	Main Span Metres
1998	Akashi Kaikyo	Kobe-Naruto, Japan	1991
1998	East Bridge	Great Belt, Denmark	1624
1981	Humber	Humber, England	1410
1998	Jianjyin	Jiangsu, China	1385
1997	Tsing Ma	Hong Kong, China	1377
1964	Verrazano Narrows	New York city	1298
1937	Golden Gate	San Francisco, Calif.	1280
1997	Hoga Kusten	Veda, Sweden	1210
1957	Mackinac Straits	Mackinaw city, Michigan	1158
1988	Minami Bisan Seto	Japan	1100
1988	Fatih Sultan Mehmet	Istanbul, Turkey	1090
1973	Bosphorus	Istanbul, Turkey	1074
1931	George Washington	New York city	1067
1999	Kurushima-3	Japan	1030
1999	Kurushima-2	Japan	1020
1966	Salazar	Lisbon, Portugal	1013
1964	Forth Road	Edinburgh, UK	1006
1988	Kita Bisan Seto	Japan	990
1966	Severn	Bristol, UK	988
1983	Ohnarato	Japan	876
1950	Tacoma Narrows II	Puget Sound, Washington	853
1982	Innoshima	Japan	770
1967	Angostura	Ciudad Bolivar, Venezuela	712
1973	Kanmon	Japan	712
1936	Transbay (2 spans)	San Francisco, California	704
1939	Bronx-Whitestone	New York city	701
1970	Quebec Road	Quebec, Canada	668
1951	Delaware Memorial I	Wilmington, Delaware	655
1968	Delaware Memorial II	Wilmington, Delaware	655
1957	Walt Whitman	Philadelphia, Pennsylvania	610
1965	Emmerich	Emmerich, W.Germany	500
1883	Brooklyn	New York city	486

Table C.2 Cable Stayed Bridges

Year	Bridge	Location	Main Span Metres	Deck Material
1999	Tatara	Kamiura, Japan	890	Steel
1994	Normandie	Over Seine, France	856	Steel
2001	Nanjing-2	Nanjing, China	628	Steel
1993	Yangpu	Shanghai, China	602	Composite
1997	Maiko Chuo	Nagoya, Japan	590	Steel
1999	Oresund	Sweden	490	Steel
1992	Vidyasagar Setu	Kolkata	457	Composite
1996	Second Severn	Bristol, UK	456	Composite
1987	Rama IX	Bangkok, Thailand	450	Steel
1983	Luna	Spain	440	Concrete
1975	St. Nazaire	Brittany, France	404	Steel
1978	Stretto di Rande	Vigo, Spain	400	Steel
1982	Luling	Mississippi, USA	372	Steel
1978	Dusseldorf Flehe	Dusseldorf, Germany	367	Steel
1987	Sunshine Skyway	Florida, USA	366	Concrete
1970	Duisburg-Neuenkamp	Duisburg, Germany	350	Steel
1990	Tempozan	Japan	350	Steel
1990	Glebe Island	Australia	345	Concrete
2004	Millau Viaduct	Millau, France	342	Steel
1974	West Gate	Melbourne, Australia	336	Steel
1978	Zarate-Brazo	Perana River, Argentina	330	Steel
1993	Karnali	Nepal	325	Composite
1972	Kohlbrand High Level	Hamburg, Germany	325	Steel
1969	Kniebrucke	Dusseldorf, Germany	320	Steel
1977	Brotonne	Caudebec, France	320	Concrete
1971	Erskine	Clyde River, Scotland	305	Steel
1959	Severins	Cologne, Germany	302	Steel
1987	Dongying	Yellow River, China	288	Steel
1976	Wadi Kuf	Beida, Libya	282	Concrete
1974	Tiel	Waal River, Netherlands	267	Concrete
1987	Yonghe	China	260	Concrete
1962	Maracaibo	Lake Maracaibo, Venezuela	235	Concrete
1978	Pasco	Pasco, Wash., USA	229	Concrete
1967	Polcevera	Genoa, Italy	210	Concrete
1956	Stromsund	Sweden	183	Steel
1972	Hoechst	Germany	148	Concrete

Table C.3 Steel Arch Bridges

Year	Bridge	Location	Main Span Metres
2004	Lupu	Shanghai, China	550
1977	New River Gorge	West Virginia	519
1931	Bayonne	New York city	504
1932	Sydney Harbour	Sydney, Australia	503
1972	Fremont	Portland, Oregon	383
1964	Port Mann	Vancouver, Canada	366
1962	Thatcher	Balboa, Panama	344
1967	Laviolette	Trois-Rivieres, Canada	335
1967	Zdakov	Vltava river, Czech	330
1961	Runcorn-Widnes	Mersey river, UK	330
1935	Birchenough	Sabi river, Zimbabwe	329
1990	Roosevelt Lake	Arizona	329
1955	Nagasaki	Japan	317
1959	Glen Canyon	Colorado river, Arizona	313
1962	Lawiston-Queenston	Niagara river	305
1974	Snake river	Twin Falls, Idaho	303
1917	Hell Gate	New York city	298

Table C.4 Concrete Arch Bridges

Year	Bridge	Location	Main Span Metres
1996	Wanxiang	Yangzi river, China	420
1980	KRK Island (East)	Yugoslavia	390
1964	Gladesville	Sydney, Australia	305
1964	Foz du Iguassu	Parana river, Brazil	290
1963	Arrabida	Porto, Portugal	270
1943	Sando	Angerman river, Sweden	264
1967	Shibenik	Krka River, Yugoslavia	246
1980	KRK Island (West)	Yugoslavia	244
1961	Fiumarella	Catanzaro, Italy	231
1961	Novi Sad	Danube river, Yugoslavia	210
1967	Lingenau	Bregenz, Austria	210
1970	Van Stadens Gorge	Port Elizabeth, S.Africa	198
1942	Martin Gil	Andavias, Spain	192

Table C.5 Cantilever Bridges

Year	Bridge	Location	Main Span Metres
1917	Quebec	Quebec, Canada	549
1889	Firth of Forth	Edinburgh, UK	521
1974	Osaka Port	Osaka, Japan	510
1974	J.J. Barry	Delaware river	501
1958	New Orleans	New Orleans, LA.	480
1943	Howrah	Kolkata, India	457
1936	Transbay	San Francisco, Calif.	427
1968	Baton Rouge	Baton Rouge, LA.	376
1955	Tappan Zee	Tarrytown, N.Y.	369
1930	Longview	Columbia river	366
1909	Queensboro	New York city	360
1927	Carquinez strait	California	335
1930	Harbour	Montreal, Canada.	334
1956	Richmond	San Francisco, Calif.	326
1929	Cooper river	Charleston, S.C.	320

Table C.6 Continuous Steel Truss Bridges

Year	Bridge	Location	Main Span Metres
1966	Astoria	Columbia river, Oregon	376
1977	Patapsco	Baltimore, Md.	366
1966	Tenmon	Kumamoto, Japan	300
1943	Dubuque	Mississippi river, Iowa	258
1964	Somerset	Taunton river, Mass.	256
1963	Brent Spence	Ohio river, Ohio	253
1956	Erle S. Clements	Shawneeton, Illinois	251
1953	Mathews	Jacksonville, Fla.	247
1944	St. Louis	Mississippi river, Mo.	245
1957	Kingston-Rhinecliff	Hudson river, N.Y.	244
1961	Sherman Minton	New Albany, Ind.	244

Table C.7 Simple Truss Spans

Year	Bridge	Location	Main Span Metres
1973	J.J. Barry	Delaware river	251
1973	Chester	Ohio river	228
1917	Metropolis	Ohio river	219
1929	Paducah	Ohio river	218
1922	Tanana river	Alaska	213
1911	Douglas Mac Arthur	St. Louis, Missouri	203
1933	Henderson	Ohio river	202
1919	Louisville	Ohio river	196

Table C.8 Continuous Plate and Box Girder Bridges

Year	Bridge	Location	Main Span Metres
1974	Niteroi	Rio de Janeiro, Brazil	300
1978	Neckertalbrucke	Weitingen, Germany	263
1956	Sava I	Belgrade, Yugoslavia	261
1989	Vitoria	Espirito Santo, Brazil	260
1966	Zoobrucke	Cologne, Germany	259
1970	Sava II	Belgrade, Yugoslavia	250
1991	Kaita	Hiroshima, Japan	250
1994	Shirinashi-Gawa	Osaka, Japan	250
1969	Auckland Harbour	Auckland, New Zealand	244
1996	Trans Toyo Bay	Tokyo, Japan	240
1975	Koblenz Sudbrucke	Koblenz, Germany	236
1989	Shorenji-Gawa	Osaka, Japan	235
1971	Bonn-Sud	Bonn, Germany	230
1967	San Mateo-Hayward	San Francisco, Calif.	229
1951	Dusseldorf-Neuss	Germany	206

Table C.9 Prestressed Concrete Girder Bridges

Year	Bridge	Location	Main Span Metres
1998	Stolmasundet	Austevoll, Norway	301
1998	Raftundet	Lefeton, Norway	298
1998	Humen	Pearl River, China	279
1994	Varrod	Kristiansand, Norway	260
1986	Gateway	Brisbane, Australia	260
1995	Skye	Skye Island, UK	250
1989	Schottwein	Semmering, Austria	260
1991	S. Joao	Porto, Portugal	250
1976	Hamana	Imagiri-Guchi, Japan	240
1972	Urato	Kochi, Japan	230

Table C.10 Movable Bridges

Year	Bridge	Location	Main Span Metres
Vertical Lift			
1959	Arthur Kill	Elizabeth, N.J.	170
1935	Cape Cod Canal	Massachusetts	166
1960	Delair	Delair, N.J.	165
1937	Marine Parkway	New York city	165
1930	Burlington, N.J.	Delaware river	163
Swing Span			
2001	Suez Canal Rly.	El Ferdan, Egypt	340
1927	Ft. Madison, Ia.	Mississippi river	160
1908	Portland, Ore	Willamette river	159
1893	East Omaha	Missouri river	159
1903	East Omaha	Missouri river	158
Bascule			
1914	Sault Ste. Marie	Canada	102
1940	Erie Avenue	Lorain, O.	101
1917	Chattanooga	Tennessee river	94

APPENDIX D

Chronology of Selected Developments of Bridges

Year	Event
BC	
480	Pontoon bridge with 600 ships built across the Hellespont by Xerxes
109	Ponte Milvio across the Tiber in Rome, stone arch bridge, still standing
55	Julius Caesar builds timber trestle bridge across the Rhine
AD	
14	Pont du Gard, Nimes, France – Roman aqueduct, three-tier semi-circular arches
98	Alcantara bridge, Toledo, Spain built for Roman emperor Trojan
605	Zhaozhou (Anji) bridge built in China with segmental arch of span 37.6 m
1188	Pont d'Avignon over the Rhone built with 20 arches of about 34 m span
1209	Peter Colechurch builds the Old London bridge with 19 pointed arches
1345	Ponte Vecchio segmental arch bridge at Florence, Italy
1591	Rialto bridge, Venice, built with single segmental arch of 27 m span
1667	Khaju bridge, Isfahan, Iran with 18 pointed arches carrying 26 m carriageway
1779	Coalbrookdale Ironbridge, first cast iron arch of single span 30 m
1791	Pont de la Concorda, Paris built by Perronet
1812	Lewis Wernwag builds 'Colossus', timber arch/truss bridge of 104 m span over the Schuylkill at Fairmont, Pennsylvania
1826	Telford builds the Menai Straits iron-chain suspension bridge with 177 m span
1831	Construction of the New London Bridge designed by John Rennie, after demolishing the Old London bridge
1849	Charles Ellet builds the Wheeling bridge over the Ohio, the first bridge over 1000 ft. span
1850	Robert Stephenson builds the Brittania tubular bridge, with spans 70-140-140-70 m
1855	Niagara suspension bridge (for road and rail) built by John Roebling
1864	Completion of the Clifton suspension bridge, Bristol with main span of 183 m and wrought iron chains, designed by I.K. Brunel in 1850
1867	Heinrich Gerber builds the first modern cantilever bridge across the Main at Hassfurt, Germany with main span of 129 m
1871	Adair builds the first reinforced concrete bridge of 15 m span at Hamersfield, UK
1874	The Eads bridge across the Mississipi at St. Louis, built by James B. Eads, the first major steel bridge
1876	Collapse of the Ashtabula bridge, Ohio, termed the USA's worst disaster killing 65 persons
1879	Tay bridge disaster in UK, 75 persons killed
1883	Completion of construction of Brooklyn suspension bridge (span 486 m) by the Roeblings
1884	Garabit Viaduct constructed by Gustav Eiffel with main span of 165 m

Year	Event
1889	Forth Rail cantilever bridge (2 spans of 521 m) completed by John Fowler and Benjamin Baker
1899	Upper Sone Railway bridge with 93 spans of 30.5 m each - 3.1 km long steel truss bridge in India
1907	Quebec bridge collapsed during construction
1917	Quebec bridge completed, world's longest cantilever span (549 m)
1926	Dum Dum bridge, Kolkata, first major R.C. arch bridge in India, with two spans of 24 m each
1930	Salginatobel bridge, Switzerland, built by Robert Maillart
1931	George Washington suspension bridge with main span of 1067 m and the Bayonne steel arch bridge with span of 511 m were completed by Othmar Amman
1932	Sydney Harbour bridge, Australia, completed
1932	Bixby Creek bridge in California, an elegant reinforced concrete arch bridge with a main span of 109.7 m
1937	Golden Gate bridge, San Francisco, built by Joseph Strauss with span of 1280 m
1939	Eugene Freyssinet completes the prestressed concrete Marne bridge at Esbly, France
1940	Tacoma Narrows I suspension bridge with span of 893 m collapsed due to aerodynamic instability
1941	Coronation bridge across Teesta, R.C. arch main span of 81.7 m
1943	Howrah cantilever bridge of main span 457 m completed at Kolkata
1948	Assam Rail Link bridges built, the first application of prestressed concrete to bridges in India
1949	Walnut Lane bridge, Philadelphia, the first prestressed concrete bridge in USA with main span of 48.8 m
1950	Reconstruction of Tacoma Narrows bridge with different design for stiffening girders
1952	Mosquito Creek bridge, Vancouver, BC, the first prestressed concrete bridge in Canada
1954	Palar bridge, Chingleput, the first prestressed concrete highway bridge in India, 23 spans of 27 m each
1956	Lake Pontchartrain bridge, Louisiana, USA constructed – the 38.6 m long consisting of many smaller span bridges
1957	Stromsund bridge, Sweden, the first modern cable stayed bridge with main span of 183 m, designed by F. Dischinger, Germany
1961	Barak bridge with clear span of 122 m constructed using segmental cantilever method with cast-in-place segments
1963	Maracaibo bridge, Venezuela, designed by Ricardo Morandi – Cable stayed bridge
1964	Verrazano Narrows bridge, New York suspension bridge built by Othmar Amman
1964	Gladesville bridge, Sydney, Australia, the first 1000 ft. span concrete bridge
1965	Bendorf bridge over the Rhine in Germany with 208 m main span - a breakthrough in free cantilever construction and use of short prestressing bar tendons
1965	Choisy bridge, France designed by Jean Muller was the first prestressed concrete bridge with precast segments and epoxy joints
1966	Severn bridge, Bristol, UK – the first suspension bridge using aerodynamic deck composed of all-welded steel stiffened box girder
1970	Collapse of steel box girder bridges at Milford Haven, UK and Melbourne, Australia
1971	Collapse of steel box girder bridge at Koblenz, Germany
1974	Osaka Port bridge, Japan, steel cantilever bridge with main span of 510 m
1976	Rio-Niteroi bridge, Brazil built with steel box girder deck of spans 200-300-200 m
1977	New River Gorge bridge, W. Virginia, then the world's longest span steel arch bridge with a span of 519 m
1980	KRK Island bridge, Yugoslavia, world's longest concrete arch span of 390 m
1981	Humber bridge, UK, suspension bridge of main span 1410 m
1985	Collapse of Ynys-y-Gwas bridge, Wales, UK due to corrosion of prestressing tendons, single span segmental post-tensioned structure of 18.3 m span

Year	Event
1987	Sunshine Skyway bridge, Florida, designed by E. Figg and J. Muller, with main span of 360 m, prestressed concrete deck and single plane stay cables
1987	Dames Point bridge, Florida – longest cable stayed bridge in USA with a span of 390 m
1988	Akkar bridge in Sikkim with two spans of 76.2 m each – India's first cable stayed bridge
1992	Vidyasagar Setu (Second Hooghly bridge), Kolkata, a notable cable stayed bridge in India with spans of 182.9-457.2-182.9 m
1993	Yangpu bridge, Shanghai, China, cable stayed bridge with main span of 602 m
1994	Normandie bridge, France, cable stayed bridge with main span of 856 m
1994	Panvel Nadi Viaduct, Konkan Railway Project - the first application of incremental launching technique in India, (spans 9 * 40 + 2 * 30 m)
1996	Wanxiang bridge, China, the world's longest span concrete arch bridge with span of 420 m
1996	Third Godavari Railway bridge with 28 spans of 97.5 m each – elegant concrete bow string girder bridge
1997	Confederation bridge, Canada, with 44 spans of 250 m each – used precast components and high performance concrete
1998	Akashi Kaikyo bridge, Japan, currently the world's longest span suspension bridge with span of 1991 m
1998	Great Belt East bridge, Denmark, world's second longest span (1620 m) suspension bridge
1998	Injaka bridge, S.Africa, collapsed during erection by incremental launching method due to wrong sliding path
1999	Tatara bridge, Japan, world's longest span cable stayed bridge with span of 810 m
1999	Sunniberg bridge, Switzerland completed. Multi-span curved cable stayed bridge, with main spans of 140 m.
2000	Millenium Footbridge, London, UK opened. A very shallow suspension bridge with spans 81-144-108m, and 4m wide aluminium deck. Vibrations noted on opening day were corrected using viscous dampers and tuned mass dampers. Bridge reopened in 2002.
2001	El Ferdan Swing bridge, Egypt, longest span swing bridge with span 340 m
2004	Lupu bridge, Shanghai, China, built - the world's longest span steel arch bridge with a span of 550 m
2004	Millau Viaduct, France, the world's highest bridge. Overall height of tallest pier and tower is 336.4 m. Cable stayed bridge with spans 204 + 6* 342 + 204 m

Appendix E

Miscellaneous Data

Table E.1 Perimeter of Group of Standard Bars in mm

No. of Bars	Bar Diameter in mm 6	8	10	12	16	18	20	25	28	36
1	18	25	32	38	50	56	63	78	88	113
2	37	50	62	75	100	113	125	157	175	226
3	56	75	94	113	150	169	188	235	263	339
4	75	100	125	150	201	226	251	314	351	452
5	94	125	157	188	251	282	314	392	439	565
6	113	150	188	226	301	339	376	471	527	679
8	150	201	251	301	402	452	502	628	703	904
9	169	226	282	339	452	508	565	706	791	1017
10	188	251	314	376	502	565	628	785	879	1131
12	226	301	376	452	603	678	753	942	1055	1357
15	282	376	471	565	753	848	942	1178	1319	1696
16	301	402	502	603	804	904	1005	1256	1407	1808
18	339	452	565	678	904	1017	1130	1413	1583	2036
20	376	502	628	753	1005	1130	1256	1570	1759	2260

Table E.2 Area of Group of Standard Bars in mm² and Weight per Metre

No. of Bars	Bar Diameter in mm									
	6	8	10	12	16	18	20	25	28	36
1	28	50	78	113	201	254	314	490	615	1018
2	56	100	157	226	402	508	628	981	1231	2036
3	84	150	235	339	603	763	942	1472	1847	3054
4	113	201	314	452	804	1017	1256	1963	2463	4072
5	141	251	392	565	1005	1272	1570	2454	3078	5090
6	169	301	471	678	1206	1526	1885	2945	3694	6108
8	226	402	628	904	1608	2035	2513	3927	4926	8144
9	254	452	706	1017	1809	2290	2827	4417	5541	9162
10	282	502	785	1131	2010	2544	3141	4908	6157	10180
12	339	603	942	1357	2412	3053	3769	5890	7389	12216
15	424	754	1178	1696	3015	3817	4712	7663	9236	15270
16	452	804	1256	1809	3217	4071	5026	7854	9852	16288
18	508	904	1413	2035	3619	4580	5654	8835	11083	18324
20	565	1005	1570	2262	4021	5089	6283	9817	12315	20360
Weight N/m	2.22	3.95	6.17	8.88	15.78	19.98	24.66	38.53	48.34	80.05

Table E.3 Area of Bars in Slabs in mm² per Metre

Spacing mm	Bar Diameter in mm								
	6	8	10	12	16	18	20	22	25
50	565	1005	1571	2262	4021	5089	6283	7603	9817
60	471	838	1309	1885	3351	4241	5236	6336	8181
70	404	718	1122	1616	2372	3635	4488	5430	7012
80	353	628	982	1414	2513	3181	3927	4752	6136
90	314	558	873	1257	2234	2827	3491	4224	5454
100	283	503	785	1131	2011	2545	3142	3801	4909
120	236	419	654	942	1675	2121	2618	3168	4091
140	202	359	561	808	1436	1818	2244	2715	3506
150	188	335	524	754	1340	1696	2094	2534	3272
160	177	314	491	707	1257	1590	1963	2376	3068
180	157	279	436	628	1117	1444	1745	2112	2727
200	141	251	393	565	1005	1272	1571	1901	2454
220	128	228	357	514	914	1157	1428	1728	2231
250	113	201	314	452	804	1018	1257	1520	1963
300	94	168	262	377	670	848	1047	1267	1636
400	71	126	196	283	503	636	785	950	1227

Table E.4 Conversion Factors

			Reciprocal
Length			
1 mm	=	0.0394 in.	25.4
1 cm	=	0.394 in.	2.54
1 m	=	39.37 in.	0.0254
1 m	=	3.28 ft	0.305
1 m	=	1.094 yd	0.914
1 km	=	0.621 mile	1.609
Area			
1 mm^2	=	0.00155 in^2	645
1 cm^2	=	0.155 in^2	6.452
1 m^2	=	10.76 ft^2	0.0929
1 m^2	=	1.196 yd^2	0.835
1 ha	=	2.47 acre	0.405
1 km^2	=	0.386 $mile^2$	2.590
Volume			
1 mm^3	=	0.0610 x 10^{-3} in^3	16.4 x 10^3
1 cm^3	=	0.0610 in^3	16.4
1 m^3	=	35.3 ft^3	0.0283
1 m^3	=	1.31 yd^3	0.765
1 litre	=	0.22 gal	4.55
Mass			
1 kg	=	2.20 lb.	0.454
1 kg	=	0.0197 cwt	50.8
1 t	=	0.984 ton	1.016
Density			
1 kg/cm^3	=	36.1 lb/in^3	0.028
1 kg/m^3	=	0.0624 lb/ft^3	16.0
1 t/m^3	=	0.0361 lb/in^3	27.7
Flexural Stiffness			
1 N mm^2	=	0.348 x 10^{-3} lb. $in.^2$	2.87 x 10^3
Second Moment of Area			
1 mm^4	=	2.40 x 10^{-6} in^4	416 x 10^3
1 cm^4	=	0.0240 in^4	41.6
1 m^4	=	2.40 x 10^6 in^4	0.416 x 10^{-6}
m^4	=	115.9 ft^4	0.00863

Velocity

1 m/s	=	3.28 ft/s	0.305
1 km/h	=	0.911 ft/s	1.097
1 km/h	=	0.621 mile/h	1.61

Force

1 N	=	0.102 kgf	9.81
1 N	=	0.225 lbf	4.45
1 kN	=	0.1004 tonf	9.96
1 kN	=	102 kgf	0.01

Stress

1 N/m²	=	0.0209 lbf/ft²	47.9
1 N/mm²	=	145 lbf/in²	6.89×10^{-3}
1 N/mm²	=	0.0647 tonf/in²	15.4
1 N/mm²	=	10.19 kgf/cm²	0.0981
1 kN/m²	=	20.9 lbf/ft²	0.0478
1 kgf/cm²	=	14.22 lbf/in²	0.0703
1 kgf/m²	=	0.205 lbf/ft²	4.882

Force/Unit Length

1 kN/m	=	0.0306 tonf/ft	32.68
1 t/m	=	0.300 tonf/ft	3.33
1 N/mm	=	5.71 lbf/in	0.175
1 N/mm	=	0.102 kgf/mm	9.81

Moment and Torque

1 N m	=	0.738 lbf ft	1.36
1 kg m	=	7.23 lbf ft	0.1383
1 kg m	=	9.81 N m	0.102
1 kg cm	=	0.868 lbf in	1.152
1 kg cm	=	0.0723 lbf ft	13.82
1 t m	=	3.23 tonf ft	0.310

Important Quantities

	SI unit	*English unit*
Acceleration of gravity	9.81 m/s²	32.2 ft/sec²
Density of water (4° C, 39.4° F)	1000 kg/m³	
Specific weight of water (15° C, 50° F)	5.81 kN/m³	62.4 lb/ft³
Standard sea-level atmosphere	101.32 kN/m², abs. 760 mm Hg 10.33 m H_2O 1013.2 millibars	14.7 psia 29.92 in. Hg 33.9 ft H_2O

APPENDIX F

Review Questions

Chapter 1—Introduction

(a) What are the basic forms of a bridge structure?
(b) What are the components of a bridge structure?
(c) Discuss the different ways in which bridges may be classified?
(d) Using a flow chart, outline the historical development of bridges.
(e) What are the special features of the following historical bridges?
- (i) Ponte Milvio over the Tiber in Rome
- (ii) Pont du Gard at Nimes, France
- (iii) Ponte Vecchio at Florence, Italy
- (iv) Colossus bridge of Fairmont, Pennsylvania
- (v) Coalbrookdale bridge over the Severn in England
- (vi) Brittania tubular bridge over Menai Straints
- (vii) Eads Bridge at St. Louis, Missouri
- (viii) Firth of Forth railway bridge in Scotland
- (ix) Quebec bridge over St. Lawrence river
- (x) Howrah bridge at Kolkata
- (xi) Sydney harbour bridge
- (xii) Bayonne bridge over Kill van Kull at New York
- (xiii) Brooklyn bridge
- (xiv) Tacoma Narrows bridge I at Puget Sound, Washington
- (xv) Humber Estuary bridge
- (xvi) Second Hooghly bridge (Vidyasagar Setu) at Kolkata
- (xvii) Akashi Kaikyo bridge.

(e) State the name of the bridge and the length of its main span for the world's longest span bridge under each of the following types:
- (i) Cable suspension
- (ii) Cantilever
- (iii) Steel arch
- (iv) Concrete arch
- (v) Continuous truss
- (vi) Simple truss

(vii) Prestressed concrete girder
(viii) Reinforced concrete girder
(ix) Masonry arch
(x) Cable stayed bridge

(f) Write a brief note on any major bridge built in India, indicating the special features incorporated into its design or construction.

Chapter 2—Investigation for Bridges

(a) What are the characteristics of an ideal site for a major bridge across a river?
(b) List the preliminary data to be collected by an engineer conducting investigation for a major bridge.
(c) State the details to be incorporated in: (i) Site Plan; and (ii) Cross section at a proposed site.
(d) What are the factors influencing flood discharge in a river?
(e) Discuss briefly the salient features of the rational method for the determination of the flood discharge of a river at a bridge site.
(f) Critically review the methods normally used for the estimation of the design discharge at any bridge site.
(g) Define 'unit hydrograph'.
(h) Explain briefly how the peak runoff can be estimated by the use of unit hydrograph method.
(i) State the limitations of unit hydrograph method.
(j) Determine the design discharge after computing the maximum discharge by (i) Empirical Method; (ii) Rational Method: and (iii) Area-Velocity Method, for a bridge site having the following characteristics:

Catchment area = 160 km^2 (f = 0.67)
Distance of site from coast = 12 km (C = 6.5)
Distance of critical point of watershed from bridge site = 16 km
Difference in elevation between the critical point and the bridge site = 96 m
Peak intensity of rainfall = 55 mm/hour
The catchment is in loam soil, largely cultivated. (p = 0.30)

At a typical natural cross section, the stream has a flow area of 115 m^2 below HFL with a wetted perimeter of 81 m. The banks are clean and straight and the stream is in fair condition (n = 0.30). The slope of stream based on LWL at different cross sections is determined as 1/500.

(k) How is afflux caused and how is its value estimated?
(l) How does the magnitude of afflux influence the design of a bridge?
(m) What are the factors to be considered in the determination of linear waterway and span length?
(n) What is meant by 'economical span'? Derive the condition for an economical span, stating clearly the assumptions made in the derivation.
(o) Why should a vertical clearance above HFL be provided? State typical values of clearance for girder bridges and arch bridges.
(p) Distinguish between 'vertical clearance' and 'freeboard'.
(q) What is the importance of subsoil exploration in the design of a major bridge? List the data to be obtained from such an exploration.

(r) Describe how you would obtain the subsoil profile at a proposed bridge site.
(s) Define 'normal depth of scour'.
(t) How would you estimate the maximum scour depth for any bridge pier?
(u) State how you would determine the depth of foundation required for a pier of a bridge located near the mouth of a river.
(v) Discuss the factors influencing the choice of the type of bridge and its basic features.
(w) Write a brief note on the importance of proper investigation for a major bridge.

Chapter 3—Standard Specifications for Road Bridges

(a) List the IRC Codes to be used while designing road bridges on a National Highway.
(b) What is the width of carriageway for roadway of: (i) Single lane; and (ii) Multilane.
(c) Sketch the requirements of minimum clearances when constructing a subway to carry a two-lane roadway below a set of railway tracks.
(d) List the loads and stresses to be considered while designing highway bridges. State the conditions under which increased stresses may be permitted over the normal permissible working stresses.
(e) Describe the IRC standard loadings and indicate the conditions under which each should be used.
(f) Describe how you would compute the maximum moment caused by a concentrated load acting on a bridge slab: (i) Supported on two opposite sides, and (ii) Cantilevered beyond an adjacent supporting beam?
(g) Explain the procedure of Pigeaud's method for the determination of maximum moments on a bridge slab due to a concentrated load. What are the limitations of the method? How would you take into account a non-symmetrical load?
(h) What is the significance of the impact factor and how is it estimated for (i) Design of superstructure, and (ii) Design of substructure?
(i) Give a critical review of IRC loadings for bridges.
(j) What are the considerations in determining the effect of wind loads?
(k) What are the causes for longitudinal forces on bridges?
(l) How would you compute the centrifugal force due to movement of vehicles on a bridge?
(m) How would you compute the pressure on a pier due to water currents?
(n) Discuss the procedure for computing earth pressure on an abutment.
(o) What are seismic zones? In which zone is your city located?
(p) Discuss how you would compute the seismic forces on a bridge.

Chapter 4—Standards for Railway Bridges

(a) List the Indian Railway standards to be followed in the design of railway bridges.
(b) State the gauges used on Indian Railways.
(c) List the loads to be considered in the design of a railway bridge.
(d) Write a short note on the wheel loads and equivalent uniformly distributed loadings for railway bridges.
(e) State how the dynamic-effect is considered in railway bridge design.
(f) What is meant by 'racking force'?

Chapter 5—General Design Considerations

(a) What is meant by: (i) Concrete grade: (ii) Steel grade?

(b) What are the requirements of bar sizes and spacing to ensure crack control in concrete?
(c) Describe how you would design the mix for M20 grade concrete for a T-beam deck for a bridge.
(d) Discuss anchorage length.
(e) How would you determine the effective width of flange for a T-beam and an L-beam?
(f) What harm would result if the main bars are curtailed in the tension zone at the theoretical cut-off points?
(g) List the three types of stressed concrete members according to IS: 1343.
(h) List the special features of prestressed concrete as against normal reinforced concrete.
(i) Draw a diagram showing typical stress-strain relations for tensioned steel and untensioned steel used in structural concrete construction.
(j) What are the types of tendons used for prestressing?
(k) Discuss how you would estimate the losses in prestress in a post-tensioned precast prestressed concrete girder for a bridge.
(l) Using the suggested notation, state what is meant by: (i) B-12ϕ 200; and (ii) 12-28ϕ.
(m) Write a brief note on the permissible stresses in steel construction for bridges. How would the permissible stresses be altered under combined loading conditions?
(n) State the values of effective lengths of struts under different end conditions.
(o) State how you would determine the number of lanes of roadway and cycle-track required for a city road?
(p) Discuss the relative merits of underpasses and overhead bridges for use of pedestrians.
(q) What are the minimum sight distances required for design speeds of 50 km/h and 100 km/h?
(r) What precautions would you take while deciding on the location of lighting posts on a prestressed concrete cantilever bridge with hinged connections at midspan?
(s) Discuss a few basic criteria regarding aesthetic design of bridges.
(t) Discuss the cost components of superstructure, substructure and foundation in relation to the overall cost of a prestressed concrete bridge in the span range 20 m to 50 m.

Chapter 6—Culverts

(a) Define culvert.
(b) List a few types of culverts.
(c) List the components of: (i) R.C. slab culvert; (ii) Pipe culvert; and (iii) Box culvert.
(d) Sketch typical sections of (i) Abutment, and (ii) Wing wall for a culvert.
(e) Describe the procedure for the design of the deck slab of a culvert on a State Highway.
(f) Sketch the general arrangement for a R.C. slab culvert of 5.0 m clear span without footpath.
(g) Sketch typical reinforcement details for the deck slab of a reinforced concrete culvert with clear span of 5 m.
(h) What are the advantages in the use of the author's charts for the design of deck slabs for slab bridges.
(i) How would you design the deck slab of a skew culvert if the angle of skew is (i) 10 degrees, and (ii) 20 degrees.
(j) Sketch the typical layout of reinforcement for a skew slab bridge with skew angle of (i) 10 degrees, and (ii) 20 degrees.

(k) Sketch typical details of a pipe culvert (i) for shallow embankment, and (ii) for high embankment.
(l) How would you determine the pipe and the bedding required for a pipe culvert (i) across a road, and (ii) across a railway track.
(m) Discuss the different loading cases for the design of a single vent reinforced concrete box culvert.
(n) Define submersible bridge.
(o) State the conditions when a submersible bridge would be desirable.
(p) Describe how you would determine the road level at the location of a proposed submersible bridge.

Chapter 7—Reinforced Concrete Bridges

(a) State the usual types of reinforced concrete bridges and indicate the span range in which each type would be applicable.
(b) Discuss the relative merits of the following three arrangements of a three girder T-beam bridge: (i) Without any cross beam or diaphragm, (ii) With three cross beams, and (iii) With three diaphragms.
(c) How would you decide on the number of longitudinal girders and transverse beams to be used in a highway bridge of T-beam type?
(d) Sketch the cross section of a T-beam superstructure and indicate the different components. Describe briefly how you would design each component.
(e) List three methods for load distribution among the longitudinal girders of a T-beam type concrete bridge deck.
(f) Describe Courbon's method for load distribution and indicate the limitations.
(g) What are the advantages of the hollow girder superstructure?
(h) Describe briefly the use and advantages of a balanced cantilever type of superstructure.
(i) What are the advantages and disadvantages of reinforced concrete continuous girder bridges over simply supported girder bridges?
(j) Indicate the situations in which rigid frame bridges can be used advantageously.
(k) State the considerations in the analysis of reinforced concrete arches.
(l) Sketch a bowstring girder bridge and indicate how the different components are designed.

Chapter 8—Prestressed Concrete Bridges

(a) Discuss the construction procedures used in pre-tensioning and post-tensioning as applied to bridge girders.
(b) Sketch typical cross sections of bridges with (i) Pre-tensioned members, and (ii) Post-tensioned members.
(c) Describe the application of pre-tensioning and post-tensioning to bridge girders, citing examples of actual construction in India.
(d) Explain briefly the principles of design of: (i) Midspan section of girder, (ii) End block of girder, and (iii) Deck slab in 'gap slab' type of deck.
(e) Sketch a typical cross section of a composite prestressed concrete bridge deck of two-lane carriageway for a span of 35 m.
(f) Sketch typical details of the end block for a precast prestressed concrete girder of a composite bridge deck.

(g) Discuss the methods of erection of precast bridge girders.
(h) Describe the sequence of operations for the erection of precast prestressed girders for a bridge using a launching truss.
(i) Sketch the typical forms of prestressed concrete superstructure used in continuous bridges and elevated expressways.
(j) Discuss any two methods of construction of a long span continuous bridge.
(k) What are the advantages and special features of the free cantilever method of construction of prestressed concrete bridges?
(l) Describe the salient features of a stressed ribbon bridge.
(m) Why is it necessary to grout the ducts for prestressing tendons in girders of prestressed concrete bridges?
(n) State the precautions to be observed for grouting of ducts in prestressed concrete girders.
(o) Discuss the precautions to be observed by a bridge engineer in the design and construction of a prestressed concrete girder bridge.
(p) Discuss the situations when external prestressing can be adopted.

Chapter 9—Steel Bridges

(a) List the different types of steel bridges, and indicate the span range applicable to each type.
(b) Sketch the cross section of a single track B.G.M.L. railway culvert of span 5 m using rolled steel joists?
(c) Describe how you would design a plate girder railway bridge to carry a B.G.M.L. standard single track.
(d) Sketch typical cross sections of steel box girder highway bridges. What are the advantages of box girder construction?
(e) Discuss the different types of trusses used in bridge construction.
(f) Draw a schematic diagram of a steel truss bridge and indicate the various components.
(g) Describe the procedure of design for a railway steel truss bridge.
(h) Write brief notes on: (i) Steel arch bridges and (ii) Cantilever bridges.
(i) Describe the main features of a cable stayed bridge.
(j) Sketch the main types of cable stayed bridges with one main span in the middle and one anchor span on either side.
(k) Sketch the different types of towers for cable stayed bridges.
(l) What are the main requirements of stay cables?
(m) Sketch a suspension bridge and indicate its components. Briefly describe the function and design of each of the components.
(n) Write a brief note on the salient features of Tacoma Narrows first bridge, Verrazano Narrows bridge and Severn bridge.
(o) State why cold drawn wire is specified for suspension cable and not heat treated wire of the same strength.
(p) Comment on the design of the Tacoma Narrows first bridge which failed.

Chapter 10—Masonry and Composite Bridges

(a) What are the different types of masonry arches?
(b) Describe the general principles of design of a masonry arch bridge.
(c) Draw a sketch of a typical single span masonry segmental arch bridge and indicate the various components of the bridge.

(d) How do you determine the thickness of the arch ring of a masonry arch?
(e) Discuss the merits and the principles of design of composite construction for bridges.
(f) What is the function of shear connectors in composite construction?
(g) Illustrate with sketches the different types of shear connectors used in bridge construction.

Chapter 11—Temporary and Movable Bridges

(a) List a few situations in which temporary bridges will have to be built.
(b) Why are timber bridges designated as temporary bridges?
(c) Describe the points to be considered in timber construction for bridges.
(d) Sketch an elevation and a cross section of a typical timber truss bridge for a span of 12 m.
(e) Sketch a sectional isometric view showing the details of a temporary causeway on an emergency road in a forward combat area.
(f) List the classifications of military bridges.
(g) Briefly discuss the salient features of a Bailey bridge.
(h) Mention two examples of application of Bailey bridge under difficult site conditions by Indian army engineers.
(i) State the appropriate correlation between the IRC and military loading classifications.
(j) What are the advantages of a military floating bridge over a fixed bridge?
(k) List the general requirements of a military floating bridge.
(l) State four methods of erection of floating bridges.
(m) Describe with neat sketches the main features of: (i) Swing bridge, (ii) Bascule bridge, (iii) Vertical lift bridge, and (iv) Transporter bridge.

Chapter 12—Substructure

(a) Sketch the reinforcement details for the pier cap over a masonry pier with a top width of 1.5 m.
(b) Describe the different types of bridge pier.
(c) Discuss the special features of: (i) Hammer head pier and (ii) Framed pier.
(d) What are the loads and forces to be considered in the design of a bridge pier?
(e) Sketch typical shapes of reinforced concrete abutments.
(f) Sketch a typical cross section of a masonry abutment and indicate the forces acting on the abutment.
(g) State how the backfill behind an abutment is to be constructed.
(h) What is the function of an approach slab?

Chapter 13—Foundations

(a) What are the different types of foundations?
(b) How do you determine the grip length required for the well foundation for a bridge across a river?
(c) Under what conditions can we use shallow foundations?
(d) How do you compute the ultimate load bearing capacity of a pile?
(e) How would you estimate the safe load per pile in a group of piles under a bridge pier?
(f) Discuss the relative merits of a precast concrete pile and a cast-in-place pile.
(g) What are the considerations in handling a precast concrete pile?

(h) Sketch the reinforcement details of an R.C. precast pile of about 9 m length.
(i) State the principles of design of a pile foundation for a bridge pier and sketch the details of a typical foundation.
(j) Describe a simple procedure for splicing precast concrete piles at a bridge site.
(k) Why is it necessary to use M20 grade concrete for the bottom plug in a well for a bridge foundation?
(l) Draw a typical section through a well foundation and indicate the different components. Describe briefly the design of each component.
(m) Draw a typical section through a pneumatic caisson for a bridge pier and explain the operation of sinking such a caisson.
(n) Write a brief note on sinking of wells for a bridge across a shallow tidal river.

Chapter 14—Bearings, Joints and Appurtenances

(a) What is the function of bearings in bridges?
(b) Sketch the details of bearings for a submersible slab bridge.
(c) What are the different types of expansion bearings for girder bridges? State the circumstances under which each would be appropriate.
(d) Describe the various types of fixed bearings.
(e) What are the considerations for the design of an elastomeric bearing for a girder bridge?
(f) List the procedure for the design of an elastomeric pad bearing for a girder bridge.
(g) Sketch the basic deformations of elastomeric pad bearing under load.
(h) State what you understand by 'an elastomeric pot bearing' and describe its working and applications.
(i) What is the function of an expansion joint? Describe the details of joints suitable for movements of (i) 20 mm, (ii) 50 mm, and (iii) 150 mm.
(j) State the performance criteria for an effective joint sealing system.
(k) Describe the modular compression sealing system and state its advantages.
(l) Sketch typical handrails for a reinforced concrete T-beam bridge.
(m) How would you provide kerbs for a submersible bridge?
(n) Sketch an arrangement of railing for a bridge approach which is on an embankment.
(o) Sketch typical details of a footpath: (i) on a concrete slab bridge, and (ii) on a T-beam bridge.
(p) Sketch the details of a typical drainage spout on a concrete bridge.
(q) State the typical details of the wearing course over a reinforced concrete bridge deck.
(r) Sketch typical details of an approach slab for a T-beam bridge without footpath.
(s) Describe the provision of guide bunds for a major river at a bridge site.

Chapter 15—Construction and Maintenance

(a) Discuss how the construction method affects the total cost of a bridge.
(b) State what is meant by quality assurance for bridge construction.
(c) List the four classes of quality assurance.
(d) Describe an example of efficient site organisation for a prestressed concrete bridge of many spans of 30 m across a major river whose bed is dry for a major part of the year.
(e) Describe with sketches typical methods of construction of steel bridges.
(f) What are the considerations in the selection of erection plant in steel bridge erection?

(g) Write a note on the permissible tolerances in the steel work in a bridge.
(h) What are the factors considered in selecting paints for steel bridges?
(i) Describe the movable scaffold system for long span concrete bridges.
(j) Discuss briefly the incremental push-launching method of construction and state its special merits.
(k) What are the special features of segmental cantilever system of construction? Highlight the advantages of the system.
(l) Discuss the relative merits and applicability of segmental cantilever system using cast-in-place segments and precast segments.
(m) Distinguish between formwork and falsework.
(n) Why is inspection of formwork important?
(o) State when forms may be removed for a T-beam bridge superstructure.
(p) List and discuss the factors to be considered while planning and designing formwork for a bridge superstructure.
(q) Discuss the different types of centering for arches.
(r) State the sequence of forming and casting concrete for a bowstring girder bridge.
(s) What are the advantages of tubular scaffolding over timber falsework?
(t) Write a brief note on construction management.
(u) State how you would number a bridge on a highway.
(v) Discuss the main features of Bridge Management System.
(v) List the equipment needed for bridge inspection.
(w) What is meant by life expectancy of a bridge?
(x) Why is bridge inspection important?
(y) State what is meant by 'routine inspection' and 'in-depth' inspection.
(z) List the trouble spots to be checked during bridge inspection.

Chapter 16— Lessons from Bridge Failures

(a) Discuss the major causes of bridge failures.
(b) State the precautions you would take in design and construction to protect a bridge against impact from ships passing below.
(c) Why is corrosion prevention important in maintenance of prestressed concrete bridges?

Chapter 17— Recent Trends in Bridge Engineering

(a) List the modified design criteria adopted for recent flyovers in Mumbai and Chennai.
(b) What are the special features of extradosed bridges?
(c) List some of the precautionary measures adopted to enhance durability of concrete bridges.
(d) How is high performance concrete different from ordinary concrete.?
(e) Discuss the advantages of using high performance concrete for bridges.
(f) State some situations when integral bridges will be advantageous.
(g) What are the precautions to be taken in bridge design and construction to enhance durability?
(h) Write a brief note on Limit States Method of Design.

Index